Yoshio Sone

Kinetic Theory and Fluid Dynamics

Birkhäuser
Boston • Basel • Berlin

Yoshio Sone
Professor Emeritus
Kyoto University
230-133 Iwakura-Nagatani-cho
Sakyo-ku Kyoto 606-0026
Japan

Library of Congress Cataloging-in-Publication Data

Sone, Yoshio, 1936-
Kinetic theory and fluid dynamics / Yoshio Sone
p. cm. – (Modeling and simulation in science, engineering and technology)
Includes bibliographical references and index.
ISBN 0-8176-4284-6-ISBN 3-7643-4284-6
1. Kinetic theory of gases. 2. Fluid dynamics. I. Title. II Modelling and simulation in science, engineering & technology

OC175.S663 2002
533'7-dc21
2002071059
CIP

AMS Subject Classifications: Primary: 76P05, 76A02, 82C40, 82C70, 35Q99, 35Q35, 35Q30; Secondary: 76-01, 76-02, 76M99, 76M45, 76N15, 76N20, 76D06, 76T99, 45K 05,45L05, 41A60, 65M99

ISBN 0-8176-4284-6 ISBN 3-7643-4284-6

Birkhauser Boston• Basel• Berlin
A member of BertelsmannSpringer Science + Business Media GmbH

Sone: Kinetic Theory and Fluid Dynamics

First Indian Reprint 2011

ISBN 978-81-8489-842-2

This edition is published by Springer (India) Private Limited, (a part of Springer Science+Business Media), Registered Office: 3rd Floor, Gandharva Mahavidyalaya, 212, Deen Dayal Upadhyaya Marg, New Delhi – 110 002, India

Printed in India by Rashtriya Printers, Delhi.

Contents

Preface

This monograph is intended to provide a comprehensive description of the relation between kinetic theory and fluid dynamics for a time-independent behavior of a gas in a general domain. A gas in a steady (or time-independent) state in a general domain is considered, and its asymptotic behavior for small Knudsen numbers is studied on the basis of kinetic theory. Fluid-dynamic-type equations and their associated boundary conditions, together with their Knudsen-layer corrections, describing the asymptotic behavior of the gas for small Knudsen numbers are presented. In addition, various interesting physical phenomena derived from the asymptotic theory are explained.

The background of the asymptotic studies is explained in Chapter 1, according to which the fluid-dynamic-type equations that describe the behavior of a gas in the continuum limit are to be studied carefully. Their detailed studies depending on physical situations are treated in the following chapters. What is striking is that the classical gas dynamic system is incomplete to describe the behavior of a gas in the continuum limit (or in the limit that the mean free path of the gas molecules vanishes). Thanks to the asymptotic theory, problems for a slightly rarefied gas can be treated with the same ease as the corresponding classical fluid-dynamic problems. In a rarefied gas, a temperature field is directly related to a gas flow, and there are various interesting phenomena which cannot be found in a gas in the continuum limit. Flows accompanying evaporation or condensation on a boundary are important problems of fluid dynamics, for which the fluid-dynamic equations and their boundary conditions on an interface of a gas and its condensed phase are prepared by the asymptotic theory. These problems are discussed as applications of the asymptotic theory.

The author owes a great deal to many people. He was influenced by fruitful discussions with the late Harold Grad, who offered the author a chance to work with him at the Courant Institute for two years. H. Cabannes's invitation to the author to Université Paris VI (Université Pierre et Marie Curie) to be a visiting professor initiated the collaboration with C. Bardos and F. Golse, which was stimulating and enjoyable. The author enjoyed many discussions with L. Arkeryd, who offered him chances of writing some of this monograph in the comfortable climate of Swedish summer. The comments by or discussion with M. N. Kogan, T.-P. Liu, B. Perthame, P. Degond, and many other people on the draft of the manuscript encouraged the author to publish the monograph in its present form. Collaboration with K. Aoki on activities in

Kyoto was helpful. T. Doi, T. Kataoka, and M. Handa, separately, read the manuscript carefully and pointed out many misprints. H. Sugimoto helped the author in preparing figures and formatting the manuscript, in addition to a local examination of the manuscript. S. Takata also helped the author with his local but careful examination of the manuscript. Many flaws were eliminated by their assistance. The editorial group of T. Grasso at Birkhäuser, with their quick responses and advice to the author's inquiries, processed the manuscript promptly. The author expresses his hearty thanks for the various courtesies and contributions.

Yoshio Sone
Kyoto, 2002

Kinetic Theory
and
Fluid Dynamics

Chapter 1

Introduction

The study of the relation of the two systems describing the behavior of a gas, the fluid-dynamic system and the Boltzmann system, has a long history (Boltzmann [1896–98], Hilbert [1900, 1912], Chapman [1916], Enskog [1917], Grad [1949, 1963a, 1963b, 1969], Chapman & Cowling [1952], Darrozes [1969], Sone [1969, 1971, 1991a], Nishida [1978], Ukai & Asano [1983], Bardos, Golse & Levermore [1993, 2000], Maslova [1993], Bellomo [1995], Sone, Aoki, Takata, Sugimoto & Bobylev [1996], Arkeryd & Nouri [2000], Sone, Bardos, Golse & Sugimoto [2000], Bouchut, Golse & Pulvirenti [2000], etc.). The first mathematical analysis is the work of Hilbert, where the solution of the Boltzmann equation is obtained by expanding the velocity distribution function and the macroscopic variables, such as density, flow velocity, and temperature, in a power series of the Knudsen number (say, Kn). In this solution (or Hilbert solution or expansion), the velocity distribution function is expressed in terms of the component functions of the expansion of the macroscopic variables: density, flow velocity, and temperature (or pressure); a series of equations for these component functions of the macroscopic variables is given. [The set of equations for (component functions of) macroscopic variables that determines the behavior of a gas will be called fluid-dynamic-type equations.] The set of equations appearing at the leading order of the expansion is the Euler set of equations for the leading terms of the expansion of the macroscopic variables. The fluid-dynamic-type equations in the following orders consist of two parts: (i) the terms that are derived from the Euler set of equations by substituting the expansions of the macroscopic variables in the Euler set of equations and arranging the same-order terms of the Knudsen number and (ii) the inhomogeneous terms that enter as the contribution of stress and heat flow expressed by derivatives of lower-order quantities.

In this system the Navier–Stokes set of equations never appears. Probably for this reason, Chapman and Enskog, independently, developed a skillful expansion, where the velocity distribution function is expanded, but the macroscopic variables are not, and obtained a series of fluid-dynamic-type equations. The leading set of equations is the Euler set of equations, the second is the Navier–Stokes set of equations, the third is the Burnett set of equations, and so on. Owing to the derivation of the Navier–Stokes set of equations, Chapman–Enskog expansion is mainly referred to when the relation between fluid dynamics and kinetic theory is mentioned. However, disadvantages of this expansion are also mentioned. For example, a non-well-posed set of equations appears after the Navier–Stokes set (e.g., Sone [1984b]); it shows some unfavorable behavior of the solution when a boundary-value problem is considered (Cercignani [1988]); the order of the differential of the equations increases as the approximation advances, which introduces problems for the construction of its boundary condition. Furthermore, it is not so well noticed that the Navier–Stokes set is derived as the next-order set of equations (or just a perturbation) of the Euler set. Do we need the Navier–Stokes set, the set derived at the second step, to obtain the leading-order information as $\mathrm{Kn} \to 0$ (or the information in the continuum limit)? What is the status of the series of the macroscopic variables obtained from the successively derived sets of equations by Chapman–Enskog expansion? What step of the approximation is required or is sufficient to obtain a solution that is correct up to order Kn^n? These kinds of questions can be made clear by Hilbert expansion, without taking a step via the technical Chapman–Enskog expansion, if one specifies the situation that one wants to discuss, and if the correct ordering of the sizes of the variables is made beforehand (see, for example, Sone & Aoki [1994], Sone [1998]). According to this analysis, a serious result is obtained. That is, there is an important class of problems in the continuum limit ($\mathrm{Kn} \to 0$) for which neither the Euler set nor Navier–Stokes set gives a correct answer, which will be explained a little more concretely in the paragraph after next. This result makes the status of the Chapman–Enskog expansion more unclear. Incidentally, the structure of Chapman–Enskog expansion, for which the original articles are not clear-cut and are lengthy, is clearly and briefly explained in Grad [1958].

In these two expansions themselves, an initial- or boundary-value problem is not taken into account. A general theory for initial-value problems was developed by Grad [1963a], where the initial layer is first introduced and the initial slip condition is analyzed systematically. The boundary-value problem for steady (or time-independent) behavior of a gas for small Knudsen numbers in a general domain is also discussed (e.g., Darrozes [1969], Sone [1969, 1971, 1991a], Sone & Aoki [1987], Sone et al. [1996]). In Sone [1969, 1971, 1991a], the case where the state of gas is close to a uniform state at rest is considered, and its asymptotic behavior for small Knudsen numbers is analyzed on the basis of the linearized Boltzmann equation. The overall behavior of the gas is described by the Stokes set of equations at any order of approximation. The boundary condition for the Stokes set and the Knudsen-layer correction in the neighborhood of the boundary are obtained up to the second order of the Knudsen number. At the second order of the Knudsen number, a thin layer with thickness of the order of the mean free path squared divided by the radius of the curvature of the boundary appears at the bottom of the Knudsen layer over a convex boundary (Sone [1973]). This is due to the discontinuity of the velocity distribution function of the gas molecules over such a boundary (Sone & Takata [1992]). The slip boundary conditions up to this order as well as the fluid-dynamic-type equations are not affected by the existence of this layer.

Sone [1971] extended the linearized theory, which corresponds to the case where the Reynolds number of the system is very small (Re $\ll$ 1), to the case where the Reynolds number is of the order of unity, for which the linearized Boltzmann equation is no longer applicable, and the expansion is carried out by taking into account the von Karman relation among three important parameters that Mach number Ma is of the same order as the product of Reynolds number Re and the Knudsen number Kn (Ma $\propto$ ReKn). The leading fluid-dynamic-type equations are the Navier–Stokes set of equations for incompressible fluids.[1] The next-order equations are the second set of equations of the Mach number expansion of the Navier–Stokes set for compressible fluids with an additional thermal stress term in the momentum equation owing to a rarefaction effect of the gas. The boundary conditions for these sets are the nonslip condition for "the incompressible Navier–Stokes set" and, for the next-order set, the slip condition consisting of tangential velocity slip due to the shear of flow and the temperature gradient along the boundary and temperature jump due to the temperature gradient normal to the boundary. It should be noted that the slip condition alone for the Navier–Stokes set is not sufficient to obtain the correct first-order effect of gas rarefaction. The restriction imposed in this work that the variation of the temperature of the boundary be small (or the same order as the Knudsen number) is eliminated in Sone & Wakabayashi [1988] and Sone, Aoki, Takata, Sugimoto & Bobylev [1996], where an important result, which shows incompleteness of continuum gas dynamics (classical gas dynamics), is derived. That is, it is discovered that something that does not exist in the limit Kn $\to$ 0, but exists for a finite Kn, gives a finite effect on the behavior of a gas in this limit (ghost effect).

The above-mentioned works on the boundary-value problem are limited to small flow velocity (or a small Mach number). Darrozes [1969] considered the case of a finite Mach number. He claimed that the behavior of the gas is conveniently described by splitting the domain into three regions: overall domain described by Euler-type equations, viscous boundary layer in the neighborhood of a boundary, with thickness of the order of the square root of the mean free path, and the Knudsen layer at the bottom of the viscous boundary layer with thickness of the order of the mean free path. One of his main conclusions is that the boundary-layer equations describing the leading effect of gas rarefaction contain terms that are not contained in the Navier–Stokes set for compressible fluids. However, recent analysis by Sone, Bardos, Golse & Sugimoto [2000] shows that the non-Navier–Stokes terms do not exist in the boundary layer equations expressing the leading effect of gas rarefaction (or those determining the quantities of the order of the square root of the Knudsen number), in contrast to the statement of Darrozes. Incidentally, when a finite Mach number is mentioned, one may think that this is a general case that covers a small Mach number case. This is not so in the case of small Knudsen numbers, because a finite Mach

[1] See Footnote 6 in Section 4.3 for the strict statement. In a time-dependent case, as will be shown in Sections 4.9 and 4.10.3, there is an important difference in the energy equation from the Navier–Stokes set for an incompressible fluid.

number means a very large Reynolds number owing to the von Karman relation among the three parameters. Thus, a finite Mach number is, rather, a special case to be treated separately.

The asymptotic analysis of the boundary-value problem of the Boltzmann equation, briefly summarized above, is extended to the case where the boundary is made of condensed phase of a gas, on which evaporation or condensation takes place, and the boundary condition for fluid-dynamic-type equations on an interface of a gas and its condensed phase is derived (Sone & Onishi [1978], Onishi & Sone [1979], Sone & Aoki [1987], Aoki & Sone [1991]). Similar classification to the above discussion applies to the present case, but there are also some important differences. For example, the Knudsen layer does not vanish in the continuum limit though its thickness vanishes, and its analysis is required to obtain the boundary condition for the fluid-dynamic-type equations in this limit; thus, there is a temperature jump in the limit of $\mathrm{Kn} \to 0$, which introduces the negative (or inverted) temperature-gradient phenomenon (Pao [1971]); when the speed of evaporation or condensation is of the order of the speed of sound, the Knudsen layer is directly connected with the Euler region without intervention of a viscous boundary layer, and this Knudsen layer is governed by a nonlinear equation, which introduces interesting features of the solution; another example of the ghost effect is found (Sone, Takata & Sugimoto [1996]).

In the present monograph, the comprehensive description of the asymptotic theory of the solution of the time-independent boundary-value problem of the Boltzmann equation briefly described in this introduction is presented together with its physically interesting applications. Chapter 2 is preparation for discussions in the following chapters. The notation used globally is explained there. We do not hesitate to repeat the explanation of notation used semi-globally. Those who are interested in more mathematical results, most of which are for time-dependent problems without boundaries, can find a clear description of the subject by Golse in Bouchut, Golse & Pulvirenti [2000] and another in Maslova [1993].

Chapter 2

Boltzmann Equation

For the convenience of discussions in this monograph, the definitions, formulas, and fundamental properties related to the Boltzmann equation are summarized without their detailed explanations and derivations. They are found, for example, in introductory parts of Grad [1958], Kogan [1969], Cercignani [1988, 2000], or Sone & Aoki [1994], but not all the materials in one book. The basic notations used in this monograph also are explained here for easier reference.

2.1 Velocity distribution function and macroscopic variables

Let X_i (or $\boldsymbol{X}$) be the Cartesian coordinates of our physical space, and ξ_i (or $\boldsymbol{\xi}$) the molecular velocity.[1] Let the number $\mathrm{d}N$ of molecules in the six-dimensional volume element $\mathrm{d}X_1\mathrm{d}X_2\mathrm{d}X_3\mathrm{d}\xi_1\mathrm{d}\xi_2\mathrm{d}\xi_3$ ($\mathbf{d}\boldsymbol{X}\mathbf{d}\boldsymbol{\xi}$ for short) be expressed as

$$\mathrm{d}N = \frac{1}{m}f(\boldsymbol{X},\boldsymbol{\xi},t)\mathbf{d}\boldsymbol{X}\mathbf{d}\boldsymbol{\xi}, \tag{2.1}$$

where m is the mass of a molecule and t is the time. Then, f or f/m, which is a function of the seven variables $\boldsymbol{X}$, $\boldsymbol{\xi}$, and t, is called the velocity distribution function of the gas molecules.

The macroscopic variables: the density of the gas ρ, the flow velocity v_i, the temperature T, the pressure p, the specific internal energy e, the stress tensor p_{ij}, and the heat-flow vector q_i, at a point $\boldsymbol{X}$ and at time t, are defined by the

[1] In this monograph, the Cartesian-tensor notation (Jeffreys [1965]) is mainly used. However, the vector notation is sometimes convenient, for example, to express the arguments of a function. The subscripts of these variables are cumbersome and confusing with subscripts used in real operations. Thus, we introduce the vector notation without notice. What should be noted in the Cartesian-tensor notation is the summation convention: that is, double indices mean the summation without $\sum$ sign (e.g., $a_ib_i = a_1b_1 + a_2b_2 + a_3b_3$, $a_{ii} = a_{11} + a_{22} + a_{33}$).

following moments of f:

$$\rho = \int f(\boldsymbol{X}, \boldsymbol{\xi}, t) \mathrm{d}\boldsymbol{\xi}, \tag{2.2a}$$

$$v_i = \frac{1}{\rho} \int \xi_i f(\boldsymbol{X}, \boldsymbol{\xi}, t) \mathrm{d}\boldsymbol{\xi}, \tag{2.2b}$$

$$3RT = \frac{1}{\rho} \int (\xi_i - v_i)^2 f(\boldsymbol{X}, \boldsymbol{\xi}, t) \mathrm{d}\boldsymbol{\xi}, \tag{2.2c}$$

$$p = \frac{1}{3} \int (\xi_i - v_i)^2 f(\boldsymbol{X}, \boldsymbol{\xi}, t) \mathrm{d}\boldsymbol{\xi} = R\rho T, \tag{2.2d}$$

$$e = \frac{1}{\rho} \int \frac{1}{2} (\xi_i - v_i)^2 f(\boldsymbol{X}, \boldsymbol{\xi}, t) \mathrm{d}\boldsymbol{\xi} = \frac{3}{2} RT, \tag{2.2e}$$

$$p_{ij} = \int (\xi_i - v_i)(\xi_j - v_j) f(\boldsymbol{X}, \boldsymbol{\xi}, t) \mathrm{d}\boldsymbol{\xi}, \tag{2.2f}$$

$$q_i = \int \frac{1}{2} (\xi_i - v_i)(\xi_j - v_j)^2 f(\boldsymbol{X}, \boldsymbol{\xi}, t) \mathrm{d}\boldsymbol{\xi}, \tag{2.2g}$$

where R is the specific gas constant [the Boltzmann constant k_B $(= 1.3806503 \times 10^{-23} \mathrm{J{\cdot}K^{-1}})$ divided by m] and the three dimensional integration with respect to $\boldsymbol{\xi}$ is, hereafter, carried out over the whole space of $\boldsymbol{\xi}$ unless otherwise stated. These definitions are compatible with those in the continuum gas dynamics.

The mass M, momentum P_i, and energy EF transferred from the gas to its (real or imaginary) boundary, at a point $\boldsymbol{X}$, per its unit area and per unit time are given by

$$\begin{aligned} M &= -\int (\xi_j - v_{wj}) n_j f(\boldsymbol{X}, \boldsymbol{\xi}, t) \mathrm{d}\boldsymbol{\xi} \\ &= -n_j \rho (v_j - v_{wj}), \end{aligned} \tag{2.3a}$$

$$\begin{aligned} P_i &= -\int \xi_i (\xi_j - v_{wj}) n_j f(\boldsymbol{X}, \boldsymbol{\xi}, t) \mathrm{d}\boldsymbol{\xi} \\ &= -n_j [p_{ij} + \rho v_i (v_j - v_{wj})], \end{aligned} \tag{2.3b}$$

$$\begin{aligned} EF &= -\int \frac{1}{2} \xi_i^2 (\xi_j - v_{wj}) n_j f(\boldsymbol{X}, \boldsymbol{\xi}, t) \mathrm{d}\boldsymbol{\xi} \\ &= -n_j \left[q_j + p_{ij} v_i + \rho \left(e + \frac{1}{2} v_i^2 \right) (v_j - v_{wj}) \right], \end{aligned} \tag{2.3c}$$

where v_{wi} is the velocity of the boundary and n_i is the unit normal vector to the boundary, pointed to the gas. When there is no mass flux ($M = 0$) through a boundary,

$$P_i = -n_j p_{ij}, \qquad EF = -n_j (q_j + p_{ij} v_i). \tag{2.4}$$

2.2 Boltzmann equation

The behavior of the velocity distribution function f is determined by the following Boltzmann equation:

$$\frac{\partial f}{\partial t}+\xi_i\frac{\partial f}{\partial X_i}+\frac{\partial F_i f}{\partial \xi_i}=J(f,f), \tag{2.5}$$

$$J(f,f)=\frac{1}{m}\int_{\text{all }\alpha_i,\text{ all }\xi_{i*}}(f'f'_*-ff_*)B(|\alpha_j V_j|/V,\ V)\mathrm{d}\Omega(\boldsymbol{\alpha})\mathbf{d}\boldsymbol{\xi}_*, \tag{2.6}$$

where

$$\left.\begin{aligned} &f=f(X_i,\xi_i,t), && f_*=f(X_i,\xi_{i*},t),\\ &f'=f(X_i,\xi'_i,t), && f'_*=f(X_i,\xi'_{i*},t),\\ &\xi'_i=\xi_i+\alpha_i\alpha_jV_j, && \xi'_{i*}=\xi_{i*}-\alpha_i\alpha_jV_j,\\ &V_i=\xi_{i*}-\xi_i, && V=(V_i^2)^{1/2}=|V_i|, \end{aligned}\right\} \tag{2.7}$$

and mF_i is the external force on a molecule, α_i (or $\boldsymbol{\alpha}$) is a unit vector, expressing the variation of the direction of the molecular velocity owing to an intermolecular collision, $\mathrm{d}\Omega(\boldsymbol{\alpha})$ is the solid-angle element in the direction of α_i, and $B(|\alpha_iV_i|/V,\ V)$ is a nonnegative function of $|\alpha_iV_i|/V$ and V, whose functional form is determined by the intermolecular force (e.g., for a gas consisting of hard-sphere molecules with diameter d_m, $B=d_m^2|\alpha_iV_i|/2$). The integrations with respect to ξ_{i*} and α_i are carried out over the whole space of ξ_{i*} and over the whole direction of α_i (the whole spherical surface) respectively. The integral $J(f,f)$ is called the collision integral or the collision term of the Boltzmann equation (2.5).[2] A formal derivation of the Boltzmann equation from the Liouville equation is given in Appendix A.1 on the basis of Grad [1958].

The generalized form $J(f,g)$ of the collision integral $J(f,f)$,

$$J(f,g)=\frac{1}{2m}\int_{\text{all }\alpha_i,\text{ all }\xi_{i*}}(f'g'_*+f'_*g'-fg_*-f_*g)B(|\alpha_jV_j|/V,\ V)\mathrm{d}\Omega(\boldsymbol{\alpha})\mathbf{d}\boldsymbol{\xi}_*, \tag{2.8}$$

is often used in the following chapters. The following properties of the integral $J(f,g)$ are also frequently used in the analysis of the Boltzmann equation. The moment $\int\varphi(\boldsymbol{\xi})J(f,g)\mathbf{d}\boldsymbol{\xi}$, where φ is an arbitrary function of $\boldsymbol{\xi}$, satisfies the

[2]The first part $m^{-1}\int_{\text{all }\alpha_i,\text{ all }\xi_{i*}}f'f'_*B\mathrm{d}\Omega\mathbf{d}\boldsymbol{\xi}_*$ of the collision integral (2.6) is called the gain term and the second $m^{-1}\int_{\text{all }\alpha_i,\text{ all }\xi_{i*}}ff_*B\mathrm{d}\Omega\mathbf{d}\boldsymbol{\xi}_*$ the loss term. Each of them is infinite when the intermolecular potential extends up to infinity, though the integral $J(f,f)$ as a whole is finite for sufficiently rapidly decaying potential. The derivation of the Boltzmann equation is carried out for an intermolecular potential with a finite range, and the gain term and the loss term are discussed separately (Appendix A.1). The function B is obtained by analyzing the binary collision process [Eq. (A.18)]. The formula for B can easily be extended to an intermolecular potential of infinite range. The Boltzmann equation with this extended B is taken as the Boltzmann equation for the corresponding infinite-range intermolecular potential. A modified $B(|\alpha_iV_i|/V,V)$ for an infinite-range intermolecular potential, where $B(|\alpha_iV_i|/V,V)$ in some neighborhood of $|\alpha_iV_i|/V=0$ is set to be zero, is often considered (Grad [1963b]). This is what is called a cutoff potential.

symmetry relation

$$\int \varphi(\boldsymbol{\xi})J(f,g)\mathrm{d}\boldsymbol{\xi}$$
$$= \frac{1}{8m}\int(\varphi+\varphi_*-\varphi'-\varphi'_*)(f'g'_*+f'_*g'-fg_*-f_*g)B\mathrm{d}\Omega(\boldsymbol{\alpha})\mathrm{d}\boldsymbol{\xi}_*\mathrm{d}\boldsymbol{\xi}, \quad (2.9)$$

where the integration is carried out over the whole spaces of $\boldsymbol{\xi}$, $\boldsymbol{\xi}_*$, and $\boldsymbol{\alpha}$. By choosing 1, ξ_i, or ξ_i^2 for φ, it is easily seen that

$$\int \begin{pmatrix} 1 \\ \xi_i \\ \xi_i^2 \end{pmatrix} J(f,g)\mathrm{d}\boldsymbol{\xi} = 0, \quad (2.10)$$

since $\xi_i + \xi_{i*} = \xi'_i + \xi'_{i*}$ and $\xi_i^2 + \xi_{i*}^2 = \xi_i'^2 + \xi_{i*}'^2$.

2.3 Conservation equations

Multiplying the Boltzmann equation (2.5) by 1, ξ_i, or ξ_i^2, and integrating the result over the whole space of $\boldsymbol{\xi}$, we obtain the following conservation equations:[3]

$$\frac{\partial \rho}{\partial t} + \frac{\partial}{\partial X_i}(\rho v_i) = 0, \quad (2.11)$$

$$\frac{\partial}{\partial t}(\rho v_i) + \frac{\partial}{\partial X_j}(\rho v_i v_j + p_{ij}) = \rho F_i, \quad (2.12)$$

$$\frac{\partial}{\partial t}\left[\rho\left(e+\frac{1}{2}v_i^2\right)\right] + \frac{\partial}{\partial X_j}\left[\rho v_j\left(e+\frac{1}{2}v_i^2\right) + v_i p_{ji} + q_j\right] = \rho v_j F_j, \quad (2.13)$$

where the force F_i is assumed to be independent of the molecular velocity $\boldsymbol{\xi}$. The collision term vanishes on integration [Eq. (2.10)]. These are the conservation equations of mass, momentum, and energy in the classical gas dynamics, where p_{ij} and q_i are assumed to be in appropriate forms to close the system (2.11)–(2.13). For example,

$$p_{ij} = p\delta_{ij}, \quad q_i = 0, \quad (2.14)$$

or

$$p_{ij} = p\delta_{ij} - \mu\left(\frac{\partial v_i}{\partial X_j} + \frac{\partial v_j}{\partial X_i} - \frac{2}{3}\frac{\partial v_k}{\partial X_k}\delta_{ij}\right) - \mu_B\frac{\partial v_k}{\partial X_k}\delta_{ij}, \quad q_i = -\lambda\frac{\partial T}{\partial X_i}, \quad (2.15)$$

where δ_{ij} is Kronecker's delta (i.e., $\delta_{ij} = 1$ for $i = j$, $\delta_{ij} = 0$ for $i \neq j$), and μ, μ_B, and λ, called the viscosity, bulk viscosity, and thermal conductivity of the gas respectively, are functions of temperature. The set of equations with the former stress and heat flow is called the Euler set, and the set with the latter the Navier–Stokes set.

[3] The velocity distribution function is generally discontinuous, as will be explained in Section 3.7. The integration should be performed a little bit carefully.

2.4 Maxwell distribution (Equilibrium distribution)

The solution f_e of the Boltzmann equation (2.5) with $F_i = 0$ for an equilibrium state ($\partial f/\partial t = \partial f/\partial X_i = 0$) is given by the following Maxwell distribution (or Maxwellian) with five constant parameters ρ, v_i, and T:

$$f_e = \frac{\rho}{(2\pi RT)^{3/2}} \exp\left(-\frac{(\xi_i - v_i)^2}{2RT}\right). \tag{2.16}$$

If ρ, v_i, and T depend on X_i or t, the distribution does not satisfy Eq. (2.5) with $F_i = 0$ except for some special cases (see Grad [1949]). The distribution (with the variable parameters ρ, v_i, and T) is called local Maxwellian.

2.5 Mean free path

The collision frequency $\nu_c(\boldsymbol{\xi})$ of a molecule with velocity ξ_i per unit time is[4]

$$\nu_c(\boldsymbol{\xi}) = \frac{1}{m}\int f(\boldsymbol{\xi}_*)B\mathrm{d}\Omega(\boldsymbol{\alpha})\mathbf{d}\boldsymbol{\xi}_*. \tag{2.17}$$

Its average over all the molecules (the mean collision frequency $\bar{\nu}_c$) is

$$\bar{\nu}_c = \frac{1}{\rho m}\int f(\boldsymbol{\xi})f(\boldsymbol{\xi}_*)B\mathrm{d}\Omega(\boldsymbol{\alpha})\mathbf{d}\boldsymbol{\xi}_*\mathbf{d}\boldsymbol{\xi}. \tag{2.18}$$

The inverses of $\nu_c(\boldsymbol{\xi})$ and $\bar{\nu}_c$ are called, respectively, the free time $\tau_c(\boldsymbol{\xi})$ [$= 1/\nu_c(\boldsymbol{\xi})$] of the molecule $\boldsymbol{\xi}$ and the mean free time $\bar{\tau}_c$ of the gas molecules. The mean free time multiplied by the average speed $\bar{c}$ [$= \rho^{-1}\int(\xi_i^2)^{1/2} f\mathbf{d}\boldsymbol{\xi}$] of the gas molecules is called the mean free path ℓ of the gas molecules:

$$\ell = \bar{c}\bar{\tau}_c. \tag{2.19}$$

If the velocity distribution function is a Maxwellian with $v_i = 0$, then $\bar{\nu}_c$, $\bar{c}$, and ℓ are given by

$$\bar{\nu}_c = 4\sqrt{\pi}d_m^2(\rho/m)(RT)^{1/2}, \tag{2.20a}$$

$$\bar{c} = (8RT/\pi)^{1/2}, \tag{2.20b}$$

$$\ell = \frac{1}{\sqrt{2}\pi d_m^2(\rho/m)}, \tag{2.20c}$$

where d_m is the radius of the influence range of the intermolecular force. [For a hard-sphere gas, the diameter of a molecule corresponds to d_m.]

---

[4] We have not shown the argument(s) of a function when there is no fear of confusion or misunderstanding. This rule is applied to a special argument or arguments. That is, $\nu_c(\boldsymbol{\xi})$ and $f(\boldsymbol{\xi})$ instead of $\nu_c(\boldsymbol{X}, \boldsymbol{\xi}, t)$ and $f(\boldsymbol{X}, \boldsymbol{\xi}, t)$.

The above definitions of the collision frequency, the mean collision frequency, the mean free time, and the mean free path are not useful when the intermolecular potential extends to infinity, for the first and second ones are infinite for any such potential, however fast it decays. In such a case its effective value, as a quantitative symbol of the collision effect, is defined (see, e.g., Kogan [1969]). For example, the range of integral in Eq. (2.18) is limited in a domain where a range $|\alpha_i(\xi_{i*}-\xi_i)|/|\xi_{j*}-\xi_j| \ll 1$ is excluded (or the intermolecular interactions that induce only small velocity changes are not counted as collision); the mean free time $\bar{\tau}_c$ and mean free path ℓ are defined on the basis of this $\bar{\nu}_c$ by the relations shown above.

2.6 Boundary condition

2.6.1 Simple boundary

On a boundary or wall where there is no mass flux across it [or $M = 0$; see Eq. (2.3a)], which will be called a simple boundary, the following condition (the Maxwell-type condition) is widely used:

$$f(\boldsymbol{X},\boldsymbol{\xi},t) = (1-\alpha)f(X_i,\xi_i - 2(\xi_j - v_{wj})n_j n_i, t) + \frac{\alpha\sigma_w}{(2\pi RT_w)^{3/2}}\exp\left(-\frac{(\xi_j - v_{wj})^2}{2RT_w}\right) \quad [(\xi_j - v_{wj})n_j > 0], \tag{2.21a}$$

$$\sigma_w = -\left(\frac{2\pi}{RT_w}\right)^{1/2}\int_{(\xi_i - v_{wi})n_i<0}(\xi_j - v_{wj})n_j f(\boldsymbol{X},\boldsymbol{\xi},t)\mathbf{d}\boldsymbol{\xi}, \tag{2.21b}$$

where T_w and v_{wi} are, respectively, the temperature and velocity of the boundary, n_i is the unit normal vector to the boundary, pointed to the gas, and α $(0 \le \alpha \le 1)$ is the accommodation coefficient.[5] These quantities depend on the position of the boundary. The case $\alpha = 1$ is called the **diffuse reflection**, and $\alpha = 0$ the **specular reflection**.

More generally, the boundary condition is expressed in terms of a scattering kernel $K_B(\boldsymbol{\xi},\boldsymbol{\xi}_*,\boldsymbol{X},t)$ as

$$f(\boldsymbol{X},\boldsymbol{\xi},t) = \int_{(\xi_{i*}-v_{wi})n_i<0} K_B(\boldsymbol{\xi},\boldsymbol{\xi}_*,\boldsymbol{X},t)f(\boldsymbol{X},\boldsymbol{\xi}_*,t)\mathbf{d}\boldsymbol{\xi}_* \quad [(\xi_i - v_{wi})n_i > 0]. \tag{2.22}$$

[5] When the condition of the boundary is specified, it is generally rather loosely mentioned that the temperature or velocity of the body or the condensed phase under consideration is so and so. This means that the surface temperature or velocity of the body or the condensed phase is so and so; then the temperature or velocity inside of it is not important. Hereafter we do not repeat this type of note.

The kernel $K_B(\boldsymbol{\xi}, \boldsymbol{\xi}_*, \boldsymbol{X}, t)$ is required to satisfy the following conditions:

$$\text{(i)}\ K_B(\boldsymbol{\xi}, \boldsymbol{\xi}_*) \geq 0 \quad [(\xi_i - v_{wi})n_i > 0,\ (\xi_{i*} - v_{wi})n_i < 0]. \tag{2.23a}$$

$$\text{(ii)}\ -\int_{(\xi_i - v_{wi})n_i > 0} \frac{(\xi_k - v_{wk})n_k}{(\xi_{j*} - v_{wj})n_j} K_B(\boldsymbol{\xi}, \boldsymbol{\xi}_*)\mathrm{d}\boldsymbol{\xi} = 1 \quad [(\xi_{i*} - v_{wi})n_i < 0], \tag{2.23b}$$

which corresponds to the condition of a simple boundary (or $M = 0$). When the kernel K_B is determined by the local condition of the boundary,[6]

$$\text{(iii)}\ f_B(\boldsymbol{\xi}) = \int_{(\xi_{i*} - v_{wi})n_i < 0} K_B(\boldsymbol{\xi}, \boldsymbol{\xi}_*) f_B(\boldsymbol{\xi}_*)\mathrm{d}\boldsymbol{\xi}_* \quad [(\xi_i - v_{wi})n_i > 0], \tag{2.23c}$$

where

$$f_B(\xi_i) = \frac{\rho}{(2\pi RT_w)^{3/2}} \exp\left(-\frac{(\xi_i - v_{wi})^2}{2RT_w}\right),$$

with ρ being arbitrary, and the other Maxwellians do not satisfy the relation (2.23c). (The last uniqueness condition excludes the specular condition.) The condition (2.23c) is the result of the local property of the kernel K_B and the natural requirement that the equilibrium state at temperature $\bar{T}_w$ and velocity $\bar{v}_{wi}$ is established in a box with a uniform temperature $\bar{T}_w$ and moving with a uniform velocity $\bar{v}_{wi}$.

For the Maxwell-type condition (2.21a) with (2.21b), the scattering kernel K_B is given by

$$\begin{aligned} K_B(\boldsymbol{\xi}, \boldsymbol{\xi}_*) &= K_{BM}(\boldsymbol{\xi}, \boldsymbol{\xi}_*) \\ &= \frac{-\alpha}{2\pi (RT_w)^2}(\xi_{j*} - v_{wj})n_j \exp\left(-\frac{(\xi_k - v_{wk})^2}{2RT_w}\right) \\ &\quad + (1-\alpha)\delta(\xi_{i*} - [\xi_i - 2(\xi_j - v_{wj})n_j n_i]), \end{aligned}$$

where $\delta(\xi_i)$ is the Dirac delta function.

2.6.2 Interface

On the interface of a gas and its condensed phase, the following condition is often used:

$$\begin{aligned} f(\boldsymbol{X}, \boldsymbol{\xi}, t) = {}& \frac{\alpha_c \rho_w}{(2\pi RT_w)^{3/2}} \exp\left(-\frac{(\xi_j - v_{wj})^2}{2RT_w}\right) \\ & + (1-\alpha_c)\Bigg[(1-\alpha) f(X_i, \xi_i - 2(\xi_j - v_{wj})n_j n_i, t) \\ & \qquad + \frac{\alpha \sigma_w}{(2\pi RT_w)^{3/2}} \exp\left(-\frac{(\xi_j - v_{wj})^2}{2RT_w}\right)\Bigg] \quad [(\xi_j - v_{wj})n_j > 0], \end{aligned} \tag{2.24a}$$

[6]The kernel is determined by the velocity v_{wi}, temperature T_w, and the other properties of the boundary at the position $\boldsymbol{X}$ and time t under consideration and is independent of their derivatives.

$$\sigma_w = -\left(\frac{2\pi}{RT_w}\right)^{1/2} \int_{(\xi_i - v_{wi})n_i<0} (\xi_j - v_{wj})n_j f(\boldsymbol{X},\boldsymbol{\xi},t)\mathrm{d}\boldsymbol{\xi}, \tag{2.24b}$$

where ρ_w is the saturated gas density at temperature T_w, α_c $(0 \le \alpha_c \le 1)$ is the condensation coefficient, and α $(0 \le \alpha \le 1)$ is the accommodation coefficient. Especially the case $\alpha_c = 1$ (the **complete condensation** condition) is widely used. The experimental discussion is found in Takens, Mischke, Korving & Beenakker [1984]. The saturated gas pressure p_w $(= R\rho_w T_w)$ at temperature T_w is more often used instead of ρ_w.

More generally, the condition is expressed in terms of a scattering kernel $K_I(\boldsymbol{\xi},\boldsymbol{\xi}_*,\boldsymbol{X},t)$ as

$$f(\boldsymbol{X},\boldsymbol{\xi},t) = g_B(\boldsymbol{X},\boldsymbol{\xi},t) + \int_{(\xi_{i*} - v_{wi})n_i<0} K_I(\boldsymbol{\xi},\boldsymbol{\xi}_*,\boldsymbol{X},t) f(\boldsymbol{X},\boldsymbol{\xi}_*,t)\mathrm{d}\boldsymbol{\xi}_* \quad [(\xi_i - v_{wi})n_i > 0], \tag{2.25}$$

where g_B, independent of f, corresponds to the term including ρ_w in Eq. (2.24a). The g_B and $K_I(\boldsymbol{\xi},\boldsymbol{\xi}_*,\boldsymbol{X},t)$ are required to satisfy the following conditions:

$$\text{(i)}\ g_B(\boldsymbol{X},\boldsymbol{\xi},t) \ge 0 \quad [(\xi_i - v_{wi})n_i > 0]. \tag{2.26a}$$

$$\text{(ii)}\ K_I(\boldsymbol{\xi},\boldsymbol{\xi}_*) \ge 0 \quad [(\xi_i - v_{wi})n_i > 0,\ (\xi_{i*} - v_{wi})n_i < 0]. \tag{2.26b}$$

When the kernel K_I is determined by the local condition of the boundary (see Footnote 6),

$$\text{(iii)}\ f_w(\boldsymbol{\xi}) = g_B(\boldsymbol{X},\boldsymbol{\xi},t) + \int_{(\xi_{i*} - v_{wi})n_i<0} K_I(\boldsymbol{\xi},\boldsymbol{\xi}_*) f_w(\boldsymbol{\xi}_*)\mathrm{d}\boldsymbol{\xi}_* \quad [(\xi_i - v_{wi})n_i > 0], \tag{2.26c}$$

where

$$f_w = \frac{\rho_w}{(2\pi RT_w)^{3/2}} \exp\left(-\frac{(\xi_i - v_{wi})^2}{2RT_w}\right),$$

and the other Maxwellians do not satisfy the relation (2.26c). The condition (2.26c) is the result of the local property of the kernel K_I and the natural requirement that the equilibrium state at temperature $\bar{T}_w$, density $\bar{\rho}_w$ (the saturated gas density at temperature $\bar{T}_w$), and velocity $\bar{v}_{wi}$ is established in a box, with a uniform temperature $\bar{T}_w$ and moving with a uniform velocity $\bar{v}_{wi}$, made of the condensed phase of the gas.

For the condition (2.24a) with (2.24b), g_B and K_I are given by

$$g_B(\boldsymbol{\xi}) = \frac{\alpha_c \rho_w}{(2\pi RT_w)^{3/2}} \exp\left(-\frac{(\xi_j - v_{wj})^2}{2RT_w}\right),$$

$$K_I(\boldsymbol{\xi},\boldsymbol{\xi}_*) = (1-\alpha_c) K_{BM}(\boldsymbol{\xi},\boldsymbol{\xi}_*).$$

2.7 H theorem

Consider the following functional, called H function, of the velocity distribution function f:

$$H(X_i,t) = \int f \ln(f/c_0)\mathbf{d}\boldsymbol{\xi}, \tag{2.27}$$

where c_0 is a constant with the dimension of f. The time variation of H is given by

$$\frac{\partial H}{\partial t} + \frac{\partial H_i}{\partial X_i} = G, \tag{2.28}$$

where

$$H_i = \int \xi_i f \ln(f/c_0)\mathbf{d}\boldsymbol{\xi}, \tag{2.29a}$$

$$G = -\frac{1}{4m}\int (f'f'_* - ff_*)\ln\left(\frac{f'f'_*}{ff_*}\right) B\mathrm{d}\Omega\mathbf{d}\boldsymbol{\xi}_*\mathbf{d}\boldsymbol{\xi} \leq 0. \tag{2.29b}$$

The equality in the last relation holds when and only when f is a (local) Maxwellian. The time variation of the integral $\overline{H}$ of H over a domain D bounded by ∂D, which may be a moving boundary, is given by

$$\frac{\mathrm{d}\overline{H}}{\mathrm{d}t} - \int_{\partial D}(H_i - Hv_{wi})n_i\mathrm{d}S = \int_D G\mathbf{d}\boldsymbol{X} \leq 0, \tag{2.30}$$

where

$$\overline{H} = \int_D H\mathbf{d}\boldsymbol{X},$$

and v_{wi} is the velocity of the boundary and n_i is the unit normal vector to the boundary, pointed to the gas.

From these equations, we have

(i) If the state, or f, is spatially uniform, then H never increases.
(ii) If $(H_i - Hv_{wi})n_i = 0$ on the boundary ∂D, then $\overline{H}$ never increases.
In both cases, H or $\overline{H}$ remains constant only when f is a (local) Maxwellian.

These statements comprise the Boltzmann H theorem, which shows that the time evolution of a solution of the Boltzmann equation has a direction.

If the boundary is a simple boundary, the second statement is made more explicit by Darrozes & Guiraud [1966] (or in Cercignani [1988]) by the estimate of the boundary contribution:

$$\int_{\partial D}(H_i - Hv_{wi})n_i\mathrm{d}S \leq -\int_{\partial D}\frac{n_i[q_i + p_{ij}(v_j - v_{wj})]}{RT_w}\mathrm{d}S. \tag{2.31}$$

Then,

$$\frac{\mathrm{d}\overline{H}}{\mathrm{d}t} \leq -\int_{\partial D}\frac{n_i[q_i + p_{ij}(v_j - v_{wj})]}{RT_w}\mathrm{d}S, \tag{2.32}$$

where the equality holds only when f is the Maxwellian that satisfies the Boltzmann equation and the boundary condition.[7] In this case the right-hand side vanishes. From this relation, we have

(ii$'$) If there is no heat flow that flows into the boundary on each point of a simple boundary,[8] then $\overline{H}$ never increases.

2.8 Model equation

The following model equation [Boltzmann–Krook–Welander, BKW, or BGK equation (Bhatnagar, Gross & Krook [1954], Welander [1954], Kogan [1958], Sone & Aoki [1994])], where the collision term in Eq. (2.5) is simplified, is widely used in analyses of rarefied gas flows:

$$\frac{\partial f}{\partial t} + \xi_i \frac{\partial f}{\partial X_i} + \frac{\partial F_i f}{\partial \xi_i} = A_c \rho (f_e - f), \tag{2.33}$$

where A_c is a constant and f_e is the local Maxwellian with ρ, v_i, and T derived from f [Eqs. (2.2a)–(2.2c)] as its five parameters.[9] The loss term $A_c \rho f$ of the BKW equation is equivalent to that of the cutoff Maxwell molecule, where the intermolecular potential is inversely proportional to the fourth power of the intermolecular distance, but the gain term $A_c \rho f_e$ is a crude assumption. Identification of the $A_c \rho f$ term in Eq. (2.33) as the loss term gives the physical meaning of A_c, i.e., $A_c \rho$ is the collision frequency, which is independent of $\boldsymbol{\xi}$ in accordance with the cutoff Maxwell molecule. Thus, the mean free path ℓ of the equilibrium state at rest with density ρ and temperature T is related to A_c as

$$\ell = \frac{(8RT/\pi)^{1/2}}{A_c \rho}. \tag{2.34}$$

In the above simplification of the collision integral, the velocity distribution function f of the gas molecules is not assumed to be close to the local Maxwellian f_e, but the difference $|f_{\mathrm{af}} - f_e|$ between the velocity distribution function f_{af} of the molecules just after collision during the mean free time $(A_c \rho)^{-1}$ and the local Maxwellian f_e is assumed to be small compared with the difference $|f - f_e|$ of the velocity distribution function f and the local Maxwellian f_e (thus, f may differ considerably from f_e).

For the BKW equation, the same conservation equations as in Section 2.3, where the collision term has dropped off by integration (this is an important property of the collision term), are derived, and the H theorem (Section 2.7), an important property of the Boltzmann equation, holds.

---

[7] This Maxwellian does not generally exist.

[8] On a simple boundary, the energy transferred to the boundary from the gas is $-n_i(q_i + p_{ij} v_j)$ [Eq. (2.4)], and the work done on the boundary by the gas is $-n_i p_{ij} v_{wj}$. Thus, their difference $-n_i[q_i + p_{ij}(v_j - v_{wj})]$ is the heat flow into the boundary.

[9] Surprisingly, many people misunderstand the BKW equation to be linear. Besides the factor ρ before the parentheses, f_e is highly nonlinear.

2.9 Nondimensional expressions I

The nondimensional variables and equations, which are used in the following chapters, are listed here.

Let L, p_0, T_0, and t_0 be, respectively, the reference length, pressure, temperature, and time, and put

$$\rho_0 = p_0(RT_0)^{-1}. \tag{2.35}$$

Then, the nondimensional variables are defined as follows:

$$\left.\begin{array}{lll}
x_i = X_i/L, & \hat{t} = t/t_0, & \zeta_i = \xi_i/(2RT_0)^{1/2}, \\
\hat{f} = f/[\rho_0(2RT_0)^{-3/2}], & \hat{F}_i = F_i/(2RT_0/L), & \\
\hat{\rho} = \rho/\rho_0, & \hat{v}_i = v_i/(2RT_0)^{1/2}, & \hat{T} = T/T_0, \\
\hat{p} = p/p_0, & \hat{p}_{ij} = p_{ij}/p_0, & \hat{q}_i = q_i/p_0(2RT_0)^{1/2}, \\
\hat{v}_{wi} = v_{wi}/(2RT_0)^{1/2}, & \hat{T}_w = T_w/T_0, & \hat{\rho}_w = \rho_w/\rho_0, \\
\hat{p}_w = p_w/p_0, & (\hat{p}_w = \hat{\rho}_w\hat{T}_w), &
\end{array}\right\} \tag{2.36}$$

where $(2RT_0)^{1/2}$ and $p_0(2RT_0)^{1/2}$ are chosen as the reference magnitude of velocity and heat flow vector. With the notation $E(\zeta)$:

$$E(\zeta) = \frac{1}{\pi^{3/2}}\exp(-\zeta^2), \quad \zeta = (\zeta_i^2)^{1/2} = |\zeta_i|, \tag{2.37}$$

the Maxwellian distribution f_0 with $v_i = 0$, $p = p_0$, and $T = T_0$, i.e.,

$$f_0 = \frac{\rho_0}{(2\pi RT_0)^{3/2}}\exp\left(-\frac{\xi_i^2}{2RT_0}\right), \tag{2.38}$$

is expressed in the form

$$f_0 = \frac{\rho_0}{(2RT_0)^{3/2}}E(\zeta). \tag{2.39}$$

The nondimensional form of the Boltzmann equation for $\hat{f}$ is

$$\mathrm{Sh}\frac{\partial \hat{f}}{\partial \hat{t}} + \zeta_i\frac{\partial \hat{f}}{\partial x_i} + \frac{\partial \hat{F}_i\hat{f}}{\partial \zeta_i} = \frac{1}{k}\hat{J}(\hat{f}, \hat{f}), \tag{2.40a}$$

$$\hat{J}(\hat{f}, \hat{g}) = \frac{1}{2}\int(\hat{f}'\hat{g}'_* + \hat{f}'_*\hat{g}' - \hat{f}\hat{g}_* - \hat{f}_*\hat{g})\widehat{B}\,\mathrm{d}\Omega(\alpha)\,\mathrm{d}\boldsymbol{\zeta}_*, \tag{2.40b}$$

where

$$\mathrm{Sh} = \frac{L}{t_0(2RT_0)^{1/2}}, \tag{2.41a}$$

$$k = \frac{(2RT_0)^{1/2}}{(\rho_0/m)B_0L} = \frac{\sqrt{\pi}\ell_0}{2L} = \frac{\sqrt{\pi}}{2}\mathrm{Kn}, \tag{2.41b}$$

$$\widehat{B} = \widehat{B}\left(|\boldsymbol{\alpha}\cdot(\boldsymbol{\zeta}_* - \boldsymbol{\zeta})|/|\boldsymbol{\zeta}_* - \boldsymbol{\zeta}|, |\boldsymbol{\zeta}_* - \boldsymbol{\zeta}|\right) = \frac{B(|\boldsymbol{\alpha}\cdot(\boldsymbol{\xi}_* - \boldsymbol{\xi})|/|\boldsymbol{\xi}_* - \boldsymbol{\xi}|, |\boldsymbol{\xi}_* - \boldsymbol{\xi}|)}{B_0}, \tag{2.41c}$$

$$\begin{aligned} B_0 &= \frac{1}{\rho_0^2}\int f_0 f_{0*} B(|\boldsymbol{\alpha}\cdot(\boldsymbol{\xi}_* - \boldsymbol{\xi})|/|\boldsymbol{\xi}_* - \boldsymbol{\xi}|, |\boldsymbol{\xi}_* - \boldsymbol{\xi}|)\mathrm{d}\Omega(\boldsymbol{\alpha})\,\mathbf{d}\boldsymbol{\xi}\mathbf{d}\boldsymbol{\xi}_* \\ &= \int EE_* B(|\boldsymbol{\alpha}\cdot(\boldsymbol{\zeta}_* - \boldsymbol{\zeta})|/|\boldsymbol{\zeta}_* - \boldsymbol{\zeta}|, (2RT_0)^{1/2}|\boldsymbol{\zeta}_* - \boldsymbol{\zeta}|)\mathrm{d}\Omega(\boldsymbol{\alpha})\,\mathbf{d}\boldsymbol{\zeta}\mathbf{d}\boldsymbol{\zeta}_* \\ &= 4\sqrt{\pi}d_m^2(RT_0)^{1/2}, \end{aligned} \tag{2.41d}$$

$$\mathbf{d}\boldsymbol{\zeta} = \mathrm{d}\zeta_1\mathrm{d}\zeta_2\mathrm{d}\zeta_3, \quad \mathbf{d}\boldsymbol{\zeta}_* = \mathrm{d}\zeta_{1*}\mathrm{d}\zeta_{2*}\mathrm{d}\zeta_{3*},$$

$$\left.\begin{aligned} &\hat{f} = \hat{f}(\zeta_i), \quad \hat{f}_* = \hat{f}(\zeta_{i*}), \quad \hat{f}' = \hat{f}(\zeta_i'), \quad \hat{f}_*' = \hat{f}(\zeta_{i*}'), \\ &\zeta_i' = \zeta_i + \alpha_i\alpha_j(\zeta_{j*} - \zeta_j), \qquad \zeta_{i*}' = \zeta_{i*} - \alpha_i\alpha_j(\zeta_{j*} - \zeta_j), \end{aligned}\right\} \tag{2.41e}$$

and a similar rule to Eq. (2.41e) is applied to other functions of ζ_i, the dot $\cdot$ between bold letters indicates their scalar products, and d_m is the radius of the influence range of the intermolecular force.[10] The Sh is the Strouhal number, $\rho_0 B_0 m^{-1}$ is the mean collision frequency at the equilibrium state at rest with pressure p_0 and temperature T_0 [or the state given by Eq. (2.38)], and ℓ_0 is the mean free path at the state. The Knudsen number Kn is an important parameter characterizing the rate of rarefaction of the gas. It should be noted that the nondimensionalized form $\widehat{B}$ of B, a function of $|\alpha_i(\zeta_{i*} - \zeta_i)|/|\zeta_{k*} - \zeta_k|$ and $|\zeta_{i*} - \zeta_i|$, generally depends on T_0 as well as the intermolecular potential.[11] For a hard-sphere gas,

$$\widehat{B} = \frac{|\alpha_i(\zeta_{i*} - \zeta_i)|}{4(2\pi)^{1/2}}. \tag{2.42}$$

The generalized collision integral $\hat{J}(\hat{f}, \hat{g})$ satisfies the symmetry relation: For

---

[10] When the intermolecular potential extends up to infinity ($d_m \to \infty$), B_0 is infinite and therefore the above-defined nondimensionalization is useless. In such a case, the range of integral of the definition of B_0 is limited, for example, in a domain where the range $|\alpha_i(\zeta_{i*} - \zeta_i)|/|\zeta_{k*} - \zeta_k| \ll 1$ is excluded. The other quantities, i.e., ℓ_0, k, Kn, and $\widehat{B}$, are defined on the basis of this B_0 by the relations shown above (see also Section 2.5).

[11] When the intermolecular potential of the finite range d_m is given, for example, by $U_0 U(r/d_m)$, where r is the intermolecular distance, the function $\widehat{B}$ is a function of $|\alpha_i(\zeta_{i*} - \zeta_i)|/|\zeta_{k*} - \zeta_k|$, $|\zeta_{i*} - \zeta_i|$, and $U_0/k_B T_0$, i.e., $\widehat{B}(|\alpha_i(\zeta_{i*} - \zeta_i)|/|\zeta_{k*} - \zeta_k|, |\zeta_{i*} - \zeta_i|, U_0/k_B T_0)$ and its functional form depends on U (see Sone & Aoki [1994]). Thus, for a given intermolecular potential, the nondimensional Boltzmann equation contains the reference temperature T_0 in addition to the Strouhal and Knudsen numbers Sh and Kn (and external force). The Boltzmann equation for a hard-sphere gas and the BKW equation are exceptional cases, where this quantity is absent in their nondimensional forms [see Eqs. (2.42) and (2.54a)]. This fact is not widely mentioned, but it should be noted that the solution of the nondimensional Boltzmann equation depends on this parameter as well as Sh and Kn. If $B(|\alpha_i V_i|/V, V)$ in Eq. (2.6) is in the form $B(|\alpha_i V_i|/V, aV) = a^m B(|\alpha_i V_i|/V, V)$, then $\widehat{B}(|\alpha_i(\zeta_{i*} - \zeta_i)|/|\zeta_{k*} - \zeta_k|, |\zeta_{i*} - \zeta_i|)$ is independent of the reference temperature T_0.

any $\varphi(\zeta)$, $\hat{f}(\zeta)$, and $\hat{g}(\zeta)$,

$$\int \varphi(\zeta)\hat{J}(\hat{f},\hat{g})\mathbf{d}\zeta = \frac{1}{8}\int (\varphi+\varphi_*-\varphi'-\varphi'_*)(\hat{f}'\hat{g}'_*+\hat{f}'_*\hat{g}'-\hat{f}\hat{g}_*-\hat{f}_*\hat{g})\widehat{B}\mathrm{d}\Omega\mathbf{d}\zeta_*\mathbf{d}\zeta, \tag{2.43}$$

from which

$$\int \begin{pmatrix} 1 \\ \zeta_i \\ \zeta_i^2 \end{pmatrix} \hat{J}(\hat{f},\hat{g})\mathbf{d}\zeta = 0. \tag{2.44}$$

The collision-integral operator $\hat{J}(*,*)$ has the following isotropic property (see Sone & Aoki [1994]): Put

$$\hat{F}(\zeta_i) = \hat{J}(\hat{f}(\zeta_i), \hat{g}(\zeta_i))$$

and

$$\hat{f}_R(\zeta_i) = \hat{f}(l_{ij}\zeta_j), \quad \hat{g}_R(\zeta_i) = \hat{g}(l_{ij}\zeta_j),$$

where

$$l_{ik}l_{jk} = \delta_{ij}. \tag{2.45}$$

Then, it is easily shown that

$$\hat{F}(l_{ij}\zeta_j) = \hat{J}(\hat{f}_R(\zeta_i), \hat{g}_R(\zeta_i)). \tag{2.46}$$

This is the definition of the isotropic property of an operator.

Let

$$\hat{F}_{i_1,\ldots,i_m,i_n,\ldots,i_s}(\zeta_i) = \hat{J}(\zeta_{i_1}\cdots\zeta_{i_m}\hat{f}(\zeta), \zeta_{i_n}\cdots\zeta_{i_s}\hat{g}(\zeta)),$$

where $\hat{F}_{i_1,\ldots,i_m,i_n,\ldots,i_s}(\zeta_i)$ is symmetric with respect to the subscripts $i_1,\ldots,i_m$ and to $i_n,\ldots,i_s$. Then, from Eq. (2.46) and the bilinearity of $\hat{J}(\hat{f}(\zeta_i),\hat{g}(\zeta_i))$ with respect to $\hat{f}(\zeta_i)$ and $\hat{g}(\zeta_i)$,

$$\hat{F}_{i_1,\ldots,i_m,i_n,\ldots,i_s}(l_{ij}\zeta_j) = l_{i_1j_1}\cdots l_{i_mj_m}l_{i_nj_n}\cdots l_{i_sj_s}\hat{F}_{j_1,\ldots,j_m,j_n,\ldots,j_s}(\zeta_i). \tag{2.47}$$

Generally, a tensor field $\hat{F}_{j_1,\ldots,j_m,j_n,\ldots,j_s}(\zeta_i)$ that satisfies the relation (2.47) is called spherically symmetric. The tensor field $\hat{J}(\zeta_{i_1}\cdots\zeta_{i_m}\hat{f}(\zeta), \zeta_{i_n}\cdots\zeta_{i_s}\hat{g}(\zeta))$ is spherically symmetric. Incidentally, a tensor field $\hat{F}_{j_1,\ldots,j_m,j_n,\ldots,j_s}(\zeta_i)$ is called axially symmetric with respect to the axis a_i, when the relation (2.47) holds for the limited set of l_{ij} that satisfies the relation

$$a_i = l_{ij}a_j,$$

for a fixed a_i in addition to Eq. (2.45).

The relations between the nondimensional macroscopic variables $\hat{\rho}, \hat{v}_i, \hat{T}$, etc. and the nondimensional velocity distribution function $\hat{f}$ are

$$\hat{\rho} = \int \hat{f} d\boldsymbol{\zeta}, \tag{2.48a}$$

$$\hat{\rho}\hat{v}_i = \int \zeta_i \hat{f} d\boldsymbol{\zeta}, \tag{2.48b}$$

$$\frac{3}{2}\hat{\rho}\hat{T} = \int (\zeta_i - \hat{v}_i)^2 \hat{f} d\boldsymbol{\zeta}, \tag{2.48c}$$

$$\hat{p} = \hat{\rho}\hat{T}, \tag{2.48d}$$

$$\hat{p}_{ij} = 2\int (\zeta_i - \hat{v}_i)(\zeta_j - \hat{v}_j)\hat{f} d\boldsymbol{\zeta}, \tag{2.48e}$$

$$\hat{q}_i = \int (\zeta_i - \hat{v}_i)(\zeta_j - \hat{v}_j)^2 \hat{f} d\boldsymbol{\zeta}. \tag{2.48f}$$

The local Maxwell distribution function $\hat{f}_e$ in the present nondimensional expression is given by

$$\hat{f}_e = \frac{\hat{\rho}}{(\pi\hat{T})^{3/2}} \exp\left(-\frac{(\zeta_i - \hat{v}_i)^2}{\hat{T}}\right), \tag{2.49}$$

for which

$$\hat{J}(\hat{f}_e, \hat{f}_e) = 0. \tag{2.50}$$

The conservation equations (2.11), (2.12), and (2.13) are

$$\mathrm{Sh}\frac{\partial \hat{\rho}}{\partial \hat{t}} + \frac{\partial \hat{\rho}\hat{v}_i}{\partial x_i} = 0, \tag{2.51}$$

$$\mathrm{Sh}\frac{\partial \hat{\rho}\hat{v}_i}{\partial \hat{t}} + \frac{\partial}{\partial x_j}\left(\hat{\rho}\hat{v}_i\hat{v}_j + \frac{1}{2}\hat{p}_{ij}\right) = \hat{\rho}\hat{F}_i, \tag{2.52}$$

$$\mathrm{Sh}\frac{\partial}{\partial \hat{t}}\left[\hat{\rho}\left(\hat{v}_i^2 + \frac{3}{2}\hat{T}\right)\right] + \frac{\partial}{\partial x_j}\left[\hat{\rho}\hat{v}_j\left(\hat{v}_i^2 + \frac{3}{2}\hat{T}\right) + \hat{p}_{ij}\hat{v}_i + \hat{q}_j\right] = 2\hat{\rho}\hat{v}_j\hat{F}_j. \tag{2.53}$$

For the BKW equation the collision integral $\hat{J}(\hat{f}, \hat{f})$ and k are given by[12]

$$\hat{J}(\hat{f}, \hat{f}) = \hat{\rho}(\hat{f}_e - \hat{f}), \tag{2.54a}$$

$$k = \frac{\sqrt{\pi}\ell_0}{2L} = \frac{(2RT_0)^{1/2}}{A_c\rho_0 L}. \tag{2.54b}$$

[12]For the BKW equation, the expression $\hat{J}(\hat{f}, \hat{g})$ is not defined unless $\hat{g} = \hat{f}$, and $\hat{J}(\hat{f}, \hat{f})$ is taken as the collision integral as a whole. Various expressions obtained by the use of the bilinear character of $\hat{J}(\hat{f}, \hat{g})$ in the following chapters are not applied to the BKW equation, including the case where the same arguments happen to appear. Some $\hat{J}(\hat{f}, \hat{g})$ expressions in equations correspond to linearized collision integrals linearized around some Maxwellian distributions (see Footnote 26 in Appendix A.7) and play an important role there. Then it is noted at key places in each chapter.

The Maxwell-type boundary condition is expressed as

$$\hat{f}(x_i,\zeta_i,\hat{t}) = (1-\alpha)\hat{f}(x_i,\zeta_i - 2(\zeta_j-\hat{v}_{wj})n_jn_i,\hat{t}) + \frac{\alpha\hat{\sigma}_w}{(\pi\hat{T}_w)^{3/2}}\exp\left(-\frac{(\zeta_i-\hat{v}_{wi})^2}{\hat{T}_w}\right) \quad [(\zeta_j-\hat{v}_{wj})n_j>0], \tag{2.55a}$$

$$\hat{\sigma}_w = -2\left(\frac{\pi}{\hat{T}_w}\right)^{1/2}\int_{(\zeta_j-\hat{v}_{wj})n_j<0}(\zeta_j-\hat{v}_{wj})n_j\hat{f}(x_i,\zeta_i,\hat{t})\mathbf{d}\boldsymbol{\zeta}. \tag{2.55b}$$

The condition (2.24a) with (2.24b) on the interface is reduced to the following form:

$$\hat{f}(x_i,\zeta_i,\hat{t}) = \frac{\alpha_c\hat{\rho}_w}{(\pi\hat{T}_w)^{3/2}}\exp\left(-\frac{(\zeta_i-\hat{v}_{wi})^2}{\hat{T}_w}\right) + (1-\alpha_c)\left[(1-\alpha)\hat{f}(x_i,\zeta_i-2(\zeta_j-\hat{v}_{wj})n_jn_i,\hat{t}) + \frac{\alpha\hat{\sigma}_w}{(\pi\hat{T}_w)^{3/2}}\exp\left(-\frac{(\zeta_i-\hat{v}_{wi})^2}{\hat{T}_w}\right)\right] \quad [(\zeta_j-\hat{v}_{wj})n_j>0], \tag{2.56a}$$

$$\hat{\sigma}_w = -2\left(\frac{\pi}{\hat{T}_w}\right)^{1/2}\int_{(\zeta_j-\hat{v}_{wj})n_j<0}(\zeta_j-\hat{v}_{wj})n_j\hat{f}(x_i,\zeta_i,\hat{t})\mathbf{d}\boldsymbol{\zeta}. \tag{2.56b}$$

2.10 Nondimensional expressions II

When we consider the case where the state of a gas is not much different from an equilibrium state at rest in a system with $F_i = 0$, it is convenient to choose the variables expressing the perturbation from this state. The Maxwellian distribution f_0 given by Eq. (2.38) is taken as the reference state. The nondimensional perturbed variables are chosen as follows:

$$\left.\begin{aligned}
&\phi = f/f_0 - 1 && \omega = \rho/\rho_0 - 1 && u_i = v_i/(2RT_0)^{1/2}\\
&\quad = \hat{f}/E - 1, && \quad = \hat{\rho} - 1, && \quad = \hat{v}_i,\\
&\tau = T/T_0 - 1 && P = p/p_0 - 1 && P_{ij} = p_{ij}/p_0 - \delta_{ij}\\
&\quad = \hat{T} - 1, && \quad = \hat{p} - 1, && \quad = \hat{p}_{ij} - \delta_{ij},\\
&Q_i = q_i/p_0(2RT_0)^{1/2} && u_{wi} = v_{wi}/(2RT_0)^{1/2} && \tau_w = T_w/T_0 - 1\\
&\quad = \hat{q}_i, && \quad = \hat{v}_{wi}, && \quad = \hat{T}_w - 1,\\
&\omega_w = \rho_w/\rho_0 - 1 && P_w = p_w/p_0 - 1 && (P_w = \omega_w + \tau_w + \omega_w\tau_w).\\
&\quad = \hat{\rho}_w - 1, && \quad = \hat{p}_w - 1, &&
\end{aligned}\right\} \tag{2.57}$$

These nondimensional variables are used in Chapters 3, 4, and 8.

The nondimensional form of the Boltzmann equation with $F_i = 0$ is given as

$$\mathrm{Sh}\frac{\partial \phi}{\partial \hat{t}} + \zeta_i \frac{\partial \phi}{\partial x_i} = \frac{1}{k}[\mathcal{L}(\phi) + \mathcal{J}(\phi, \phi)], \tag{2.58a}$$

$$\mathcal{L}(\phi) = \int E_*(\phi' + \phi'_* - \phi - \phi_*)\widehat{B}\, \mathrm{d}\Omega(\boldsymbol{\alpha})\, \mathrm{d}\boldsymbol{\zeta}_*, \tag{2.58b}$$

$$\mathcal{J}(\phi, \psi) = \frac{1}{2}\int E_*(\phi'\psi'_* + \phi'_*\psi' - \phi\psi_* - \phi_*\psi)\widehat{B}\, \mathrm{d}\Omega(\boldsymbol{\alpha})\, \mathrm{d}\boldsymbol{\zeta}_*, \tag{2.58c}$$

where

$$\phi = \phi(\zeta_i), \quad \phi_* = \phi(\zeta_{i*}), \quad \phi' = \phi(\zeta'_i), \quad \phi'_* = \phi(\zeta'_{i*}),$$

$$\zeta'_i = \zeta_i + \alpha_i\alpha_j(\zeta_{j*} - \zeta_j), \qquad \zeta'_{i*} = \zeta_{i*} - \alpha_i\alpha_j(\zeta_{j*} - \zeta_j).$$

The operator $\mathcal{J}$ is related to $\mathcal{L}$ and $\hat{J}$ in the following way:

$$2\mathcal{J}(1, \phi) = \mathcal{L}(\phi), \tag{2.59a}$$

$$E\mathcal{J}(\phi, \psi) = \hat{J}(E\phi, E\psi). \tag{2.59b}$$

The relations between the nondimensional macroscopic variables and the nondimensional velocity distribution function ϕ are

$$\omega = \int \phi E \mathrm{d}\boldsymbol{\zeta}, \tag{2.60a}$$

$$(1+\omega)u_i = \int \zeta_i \phi E \mathrm{d}\boldsymbol{\zeta}, \tag{2.60b}$$

$$\frac{3}{2}(1+\omega)\tau = \int \left(\zeta_i^2 - \frac{3}{2}\right)\phi E \mathrm{d}\boldsymbol{\zeta} \ - (1+\omega)u_i^2, \tag{2.60c}$$

$$P = \omega + \tau + \omega\tau, \tag{2.60d}$$

$$P_{ij} = 2\int \zeta_i\zeta_j \phi E \mathrm{d}\boldsymbol{\zeta} - 2(1+\omega)u_i u_j, \tag{2.60e}$$

$$Q_i = \int \zeta_i\zeta_j^2 \phi E \mathrm{d}\boldsymbol{\zeta} - \frac{5}{2}u_i - u_j P_{ij} - \frac{3}{2}Pu_i - (1+\omega)u_i u_j^2. \tag{2.60f}$$

The linear part $\mathcal{L}(\phi)$ of the collision integral, called linearized collision integral, satisfies the following relations, which are the linearized versions of Eqs. (2.43), (2.44), and (2.50):

$$\int E\psi(\boldsymbol{\zeta})\mathcal{L}(\phi)\mathrm{d}\boldsymbol{\zeta} = \frac{1}{4}\int EE_*(\psi + \psi_* - \psi' - \psi'_*)(\phi' + \phi'_* - \phi - \phi_*)\widehat{B}\mathrm{d}\Omega\mathrm{d}\boldsymbol{\zeta}_*\mathrm{d}\boldsymbol{\zeta}$$

$$\text{for any } \phi \text{ and } \psi, \tag{2.61a}$$

$$\int g(\boldsymbol{\zeta})\mathcal{L}(\phi)E\mathrm{d}\boldsymbol{\zeta} = 0 \ \text{ for any } \phi, \tag{2.61b}$$

$$\mathcal{L}(g(\boldsymbol{\zeta})) = 0, \tag{2.61c}$$

where $g(\boldsymbol{\zeta})$ is 1, ζ_i, or ζ_i^2. From Eq. (2.61a), the operator is found to be self-adjoint:

$$\int \psi\mathcal{L}(\phi)E\mathbf{d}\boldsymbol{\zeta} = \int \phi\mathcal{L}(\psi)E\mathbf{d}\boldsymbol{\zeta} \quad \text{for any } \phi \text{ and } \psi. \tag{2.62}$$

From Eqs. (2.44) and (2.59b),

$$\int g(\boldsymbol{\zeta})\mathcal{J}(\phi,\psi)E\mathbf{d}\boldsymbol{\zeta} = 0 \quad \text{for any } \phi \text{ and } \psi. \tag{2.63}$$

From Eq. (2.61a),

$$\int \phi\mathcal{L}(\phi)E\mathbf{d}\boldsymbol{\zeta} \leq 0, \tag{2.64}$$

where the equality sign holds for $\phi = 1$, ζ_i, or ζ_i^2.

The linearized collision-integral operator $\mathcal{L}(*)$ has the following isotropic property (see Sone & Aoki [1994]): Put

$$F(\zeta_i) = \mathcal{L}(\phi(\zeta_i)), \tag{2.65}$$

and

$$\phi_R(\zeta_i) = \phi(l_{ij}\zeta_j),$$

where

$$l_{ik}l_{jk} = \delta_{ij}.$$

Then, it is easily shown that

$$F(l_{ij}\zeta_j) = \mathcal{L}(\phi_R(\zeta_i)). \tag{2.66}$$

This is the definition of the isotropic property of an operator.

Let

$$F_{i_1,\dots,i_m}(\zeta_i) = \mathcal{L}(\zeta_{i_1}\cdots\zeta_{i_m}\phi(\zeta)), \tag{2.67}$$

where $F_{i_1,\dots,i_m}(\zeta_i)$ is symmetric with respect to the subscripts $i_1,\dots,i_m$. Then, from Eq. (2.66) and the linearity of $\mathcal{L}(\phi(\zeta_i))$ with respect to $\phi(\zeta_i)$,

$$F_{i_1,\dots,i_m}(l_{ij}\zeta_j) = l_{i_1j_1}\cdots l_{i_mj_m}F_{j_1,\dots,j_m}(\zeta_i). \tag{2.68}$$

Thus the tensor field $\mathcal{L}(\zeta_{i_1}\cdots\zeta_{i_m}\phi(\zeta))$ is spherically symmetric.

For a hard-sphere gas, the linearized collision integral $\mathcal{L}(\phi)$ is expressed in terms of the collision kernel in Grad [1963b] as follows:

$$\mathcal{L}(\phi) = \mathcal{L}_1(\phi) - \mathcal{L}_2(\phi) - \nu(\zeta)\phi, \tag{2.69}$$

where

$$\mathcal{L}_1(\phi) = \frac{1}{\sqrt{2}\pi}\int \frac{1}{|\zeta_j - \zeta_{j*}|}\exp\left(-\zeta_*^2 + \frac{(\epsilon_{klm}\zeta_l\zeta_{m*})^2}{(\zeta_j-\zeta_{j*})^2}\right)\phi(x_i,\zeta_{i*},\hat{t})\mathbf{d}\boldsymbol{\zeta}_*, \tag{2.70a}$$

$$\mathcal{L}_2(\phi) = \frac{1}{2^{3/2}\pi}\int |\zeta_j - \zeta_{j*}|\exp(-\zeta_*^2)\phi(x_i,\zeta_{i*},\hat{t})\mathbf{d}\boldsymbol{\zeta}_*, \tag{2.70b}$$

$$\nu(\zeta) = \frac{1}{2^{3/2}}\left[\exp(-\zeta^2) + \left(2\zeta + \frac{1}{\zeta}\right)\int_0^\zeta \exp(-\zeta_*^2)\mathrm{d}\zeta_*\right], \tag{2.70c}$$

where ϵ_{ijk} is Eddington's epsilon[13] and $\zeta = (\zeta_i^2)^{1/2}$.

The (local) Maxwellian distribution ϕ_e in the present nondimensional perturbed expression is

$$E\phi_e = \frac{1+\omega}{\pi^{3/2}(1+\tau)^{3/2}} \exp\left(-\frac{(\zeta_i - u_i)^2}{1+\tau}\right) - E. \tag{2.71}$$

Sometimes, its parameters ω, u_i, and τ are explicitly shown as $\phi_e(\omega,\ u_i,\ \tau)$. Corresponding to Eq. (2.50),

$$\mathcal{L}(\phi_e) + \mathcal{J}(\phi_e, \phi_e) = 0. \tag{2.72}$$

The BKW equation is

$$\mathrm{Sh}\frac{\partial \phi}{\partial \hat{t}} + \zeta_i \frac{\partial \phi}{\partial x_i} = \frac{1}{k}(1+\omega)(\phi_e - \phi), \tag{2.73}$$

where

$$k = (2RT_0)^{1/2}(\rho_0 A_c L)^{-1}[= (\sqrt{\pi}/2)\ell_0 L^{-1}]. \tag{2.74}$$

The conservation equations are expressed as

$$\mathrm{Sh}\frac{\partial \omega}{\partial \hat{t}} + \frac{\partial (1+\omega)u_i}{\partial x_i} = 0, \tag{2.75}$$

$$\mathrm{Sh}\frac{\partial (1+\omega)u_i}{\partial \hat{t}} + \frac{\partial}{\partial x_j}\left[(1+\omega)u_i u_j + \frac{1}{2}P_{ij}\right] = 0, \tag{2.76}$$

$$\mathrm{Sh}\frac{\partial}{\partial \hat{t}}\left[(1+\omega)u_i^2 + \frac{3}{2}P\right] + \frac{\partial}{\partial x_j}\left[\frac{5}{2}u_j + u_i P_{ij} + \frac{3}{2}P u_j + (1+\omega)u_j u_i^2 + Q_j\right] = 0. \tag{2.77}$$

The Maxwell-type boundary condition in a general case is just a transcription of the formula (2.55a) with (2.55b) by Eq. (2.57) and not a simple form, since the Maxwellian $E(\zeta)$ with ζ_i replaced by $\zeta_i - 2(\zeta_j - \hat{v}_{wj})n_j n_i$ [or $\zeta_i - 2(\zeta_j - u_{wj})n_j n_i$]

---

[13]Eddington's epsilon ϵ_{ijk} is defined by the rules

$$\epsilon_{123} = 1 \text{ and } \epsilon_{ijk} = -\epsilon_{jik} = -\epsilon_{ikj}.$$

More plainly, it is specified by the rules: (i) if any two of the i, j, k are equal, $\epsilon_{ijk} = 0$; (ii) if they are all different and occur in succession in the order 123123..., $\epsilon_{ijk} = 1$; (iii) if they are all different and occur in succession in the order 213213..., $\epsilon_{ijk} = -1$. That is,

$$\epsilon_{123} = \epsilon_{231} = \epsilon_{312} = 1, \quad \epsilon_{213} = \epsilon_{132} = \epsilon_{321} = -1.$$

is not a simple form unless $\hat{v}_{wi}n_i = 0$ (or $u_{wi}n_i = 0$), that is,

$$E(\zeta)[1+\phi(x_i,\zeta_i,\hat{t})] = (1-\alpha)E(\check{\zeta})[1+\phi(x_i,\check{\zeta}_i,\hat{t})] + \alpha E(\zeta)[1+\phi_e(\check{\sigma}_w,u_{wi},\tau_w)]$$
$$[(\zeta_j - u_{wj})n_j > 0], \tag{2.78a}$$

$$\check{\sigma}_w = \hat{\sigma}_w - 1$$
$$= -2\left(\frac{\pi}{1+\tau_w}\right)^{1/2}\int_{(\zeta_j-u_{wj})n_j<0}(\zeta_j-u_{wj})n_jE(\zeta)[1+\phi(x_i,\zeta_i,\hat{t})]\mathbf{d}\boldsymbol{\zeta} - 1, \tag{2.78b}$$

$$\check{\zeta}_i = \zeta_i - 2(\zeta_j - u_{wj})n_jn_i, \quad \check{\zeta} = (\check{\zeta}_i^2)^{1/2}. \tag{2.78c}$$

When $u_{wi}n_i = 0$, the variable $\check{\zeta}$ is reduced to ζ, and the formula is a little simplified as

$$\phi(x_i,\zeta_i,\hat{t}) = (1-\alpha)\phi(x_i,\check{\zeta}_i,\hat{t}) + \alpha\phi_e(\check{\sigma}_w,u_{wi},\tau_w) \quad (\zeta_jn_j > 0), \tag{2.79a}$$

$$\check{\sigma}_w = -2\left(\frac{\pi}{1+\tau_w}\right)^{1/2}\int_{\zeta_jn_j<0}\zeta_jn_jE(\zeta)\phi(x_i,\zeta_i,\hat{t})\mathbf{d}\boldsymbol{\zeta} + \left(\frac{1}{1+\tau_w}\right)^{1/2} - 1, \tag{2.79b}$$

$$\check{\zeta}_i = \zeta_i - 2\zeta_jn_jn_i. \tag{2.79c}$$

The nondimensional perturbed formula corresponding to Eq. (2.56a) on the interface of a gas and its condensed phase is

$$E(\zeta)[1+\phi(x_i,\zeta_i,\hat{t})] = \alpha_cE(\zeta)[1+\phi_e(\omega_w,u_{wi},\tau_w)]$$
$$+ (1-\alpha_c)\{(1-\alpha)E(\check{\zeta})[1+\phi(x_i,\check{\zeta}_i,\hat{t})]$$
$$+\alpha E(\zeta)[1+\phi_e(\check{\sigma}_w,u_{wi},\tau_w)]\}$$
$$[(\zeta_j-u_{wj})n_j>0], \tag{2.80}$$

where $\check{\sigma}_w$ and $\check{\zeta}_i$ are given by the same expressions as Eqs. (2.78b) and (2.78c). When $u_{wi}n_i = 0$, similarly to Eq. (2.79a), this formula is reduced to

$$\phi(x_i,\zeta_i,\hat{t}) = \alpha_c\phi_e(\omega_w,u_{wi},\tau_w)$$
$$+ (1-\alpha_c)[(1-\alpha)\phi(x_i,\check{\zeta}_i,\hat{t}) + \alpha\phi_e(\check{\sigma}_w,u_{wi},\tau_w)] \quad (\zeta_jn_j>0), \tag{2.81}$$

where $\check{\sigma}_w$ and $\check{\zeta}_i$ are given by the same simplified expressions as Eqs. (2.79b) and (2.79c).

Here some notes may be in order.

(i) The formulas (equations, boundary conditions, etc.) in this section are expressed in the perturbed quantities defined by Eq. (2.57), and are conveniently used in analyzing problems where the state of the gas is close to an equilibrium state at rest. The expressions are the simple transformations of the corresponding formulas in the dimensional variables in Sections 2.1–2.8 or in the nondimensional variables in Section 2.9, and no approximation is introduced.

(ii) In a problem consisting of a gas and its condensed phase, the density of the gas at an equilibrium state at rest is given by the saturated gas density at its temperature. The choice of the basic equilibrium state [or f_0 in Eqs. (2.57) and (2.38)] in perturbation analyses is not unique. The variables ρ_0 and T_0 may be chosen at any values such that the perturbed velocity distribution function ϕ is small, and ρ_0 may not necessarily be the saturated gas density at temperature T_0.

2.11 Linearized Boltzmann equation

In analyzing the behavior of a gas which deviates only slightly from an equilibrium state at rest of a system with $F_i = 0$, the linearized version of the Boltzmann equation, which is obtained by neglecting the nonlinear terms of ϕ in Eq. (2.58a), is widely used. It is given as

$$\mathrm{Sh}\frac{\partial \phi}{\partial \hat{t}} + \zeta_i \frac{\partial \phi}{\partial x_i} = \frac{1}{k}\mathcal{L}(\phi). \tag{2.82}$$

The linearized version of Eqs. (2.60a)–(2.60f) is

$$\omega = \int \phi E \mathbf{d}\boldsymbol{\zeta}, \tag{2.83a}$$

$$u_i = \int \zeta_i \phi E \mathbf{d}\boldsymbol{\zeta}, \tag{2.83b}$$

$$\frac{3}{2}\tau = \int \left(\zeta_i^2 - \frac{3}{2}\right) \phi E \mathbf{d}\boldsymbol{\zeta}, \tag{2.83c}$$

$$P = \omega + \tau, \tag{2.83d}$$

$$P_{ij} = 2\int \zeta_i \zeta_j \phi E \mathbf{d}\boldsymbol{\zeta}, \tag{2.83e}$$

$$Q_i = \int \zeta_i \zeta_j^2 \phi E \mathbf{d}\boldsymbol{\zeta} - \frac{5}{2}u_i. \tag{2.83f}$$

The linearized expression for the perturbed (local) Maxwellian ϕ_e is

$$\phi_e = \omega + 2\zeta_i u_i + \left(\zeta_i^2 - \frac{3}{2}\right)\tau. \tag{2.84}$$

The linearized forms of the conservation equations are

$$\mathrm{Sh}\frac{\partial \omega}{\partial \hat{t}} + \frac{\partial u_i}{\partial x_i} = 0, \tag{2.85}$$

$$2\mathrm{Sh}\frac{\partial u_i}{\partial \hat{t}} + \frac{\partial P_{ij}}{\partial x_j} = 0, \tag{2.86}$$

$$\frac{3}{2}\mathrm{Sh}\frac{\partial P}{\partial \hat{t}} + \frac{\partial}{\partial x_i}\left(\frac{5}{2}u_i + Q_i\right) = 0. \tag{2.87}$$

The linearized BKW equation is

$$\mathrm{Sh}\frac{\partial \phi}{\partial \hat{t}} + \zeta_i \frac{\partial \phi}{\partial x_i} = \frac{1}{k}\left[-\phi + \omega + 2\zeta_i u_i + \left(\zeta_i^2 - \frac{3}{2}\right)\tau\right]. \tag{2.88}$$

With the same notation $\mathcal{L}(\phi)$ as the standard Boltzmann equation for the expression in the brackets,

$$\mathcal{L}(\phi) = \int \left[1 + 2\zeta_i\zeta_{i*} + \frac{2}{3}\left(\zeta_i^2 - \frac{3}{2}\right)\left(\zeta_{j*}^2 - \frac{3}{2}\right)\right]\phi(\zeta_*)E(\zeta_*)\mathbf{d}\boldsymbol{\zeta}_* - \phi. \tag{2.89}$$

The linearized form of the boundary condition (2.78a) with (2.78b) on a simple boundary is

$$\begin{aligned}\phi(x_i, \zeta_i, \hat{t}) = (1-\alpha)&\left[\phi(x_i, \zeta_i - 2(\zeta_j - u_{wj})n_j n_i, \hat{t}) + 4\zeta_j u_{wk} n_j n_k\right] \\ &+\alpha\left[\check{\sigma}_w + 2\zeta_j u_{wj} + \left(\zeta_j^2 - \frac{3}{2}\right)\tau_w\right] \quad [(\zeta_j - u_{wj})n_j > 0],\end{aligned} \tag{2.90a}$$

$$\check{\sigma}_w = \sqrt{\pi}u_{wj}n_j - \frac{1}{2}\tau_w - 2\sqrt{\pi}\int_{(\zeta_k - u_{wk})n_k < 0} (\zeta_j - u_{wj})n_j \phi E \mathbf{d}\boldsymbol{\zeta}. \tag{2.90b}$$

The linearized form of the boundary condition (2.80) on an interface of a gas and its condensed phase is

$$\begin{aligned}\phi(x_i, \zeta_i, \hat{t}) &= \alpha_c\left[\omega_w + 2\zeta_j u_{wj} + \left(\zeta_j^2 - \frac{3}{2}\right)\tau_w\right] \\ &+ (1-\alpha_c)\Big\{(1-\alpha)\left[\phi(x_i, \zeta_i - 2(\zeta_j - u_{wj})n_j n_i, \hat{t}) + 4\zeta_j u_{wk} n_j n_k\right] \\ &+ \alpha\left[\check{\sigma}_w + 2\zeta_j u_{wj} + \left(\zeta_j^2 - \frac{3}{2}\right)\tau_w\right]\Big\} \quad [(\zeta_j - u_{wj})n_j > 0],\end{aligned} \tag{2.91}$$

where $\check{\sigma}_w$ is given by the same expression as Eq. (2.90b).

Here some notes may be in order.

(i) The formulas given in this section are the simplified version of those in Section 2.10 obtained by neglecting the second and higher terms of the perturbed quantities introduced in Eq. (2.57).

(ii) The choice of the basic equilibrium state [f_0 in Eqs. (2.57) and (2.38)] in perturbation analyses is not unique, and it can be chosen freely as far as the perturbed velocity distribution function ϕ is so small that its nonlinear terms can be neglected [see also note (ii) in the last paragraph in Section 2.10].

2.12 Boltzmann equation in the cylindrical and spherical coordinate systems

It is convenient to list the formulas of the transport derivative term

$$\frac{\partial f}{\partial t}+\xi_i\frac{\partial f}{\partial X_i}+\frac{\partial F_i f}{\partial \xi_i}, \tag{2.92}$$

of the Boltzmann equation in the cylindrical and spherical coordinate systems. The formula in a general coordinate system is given in the textbook by Kogan [1969] using Lamé coefficients (see Cartan [1946]).

In the cylindrical coordinate system (r,θ,z), where the relation to the rectangular coordinate X_i is given, for example, as

$$X_1=r\cos\theta,\ X_2=r\sin\theta,\ X_3=z,$$

the following relations hold:

$$\frac{\partial f}{\partial t}+\xi_i\frac{\partial f}{\partial X_i}=\frac{\partial f}{\partial t}+\xi_r\frac{\partial f}{\partial r}+\frac{\xi_\theta}{r}\frac{\partial f}{\partial \theta}+\xi_z\frac{\partial f}{\partial z}+\frac{\xi_\theta^2}{r}\frac{\partial f}{\partial \xi_r}-\frac{\xi_r\xi_\theta}{r}\frac{\partial f}{\partial \xi_\theta}, \tag{2.93a}$$

$$\frac{\partial F_i f}{\partial \xi_i}=\frac{\partial F_r f}{\partial \xi_r}+\frac{\partial F_\theta f}{\partial \xi_\theta}+\frac{\partial F_z f}{\partial \xi_z}, \tag{2.93b}$$

where ξ_r, ξ_θ, and ξ_z are, respectively, the r, θ, and z components of ξ_i (or $\boldsymbol{\xi}$), and F_r, F_θ, and F_z are, respectively, the corresponding components of F_i (or $\boldsymbol{F}$).

In the spherical coordinate system (r,θ,φ), where the relation to the rectangular coordinate X_i is given, for example, as

$$X_1=r\sin\theta\cos\varphi,\ X_2=r\sin\theta\sin\varphi,\ X_3=r\cos\theta,$$

the following relations hold:

$$\begin{aligned}\frac{\partial f}{\partial t}+\xi_i\frac{\partial f}{\partial X_i}=&\frac{\partial f}{\partial t}+\xi_r\frac{\partial f}{\partial r}+\frac{\xi_\theta}{r}\frac{\partial f}{\partial \theta}+\frac{\xi_\varphi}{r\sin\theta}\frac{\partial f}{\partial \varphi}\\&+\frac{\xi_\theta^2+\xi_\varphi^2}{r}\frac{\partial f}{\partial \xi_r}+\left(\frac{\xi_\varphi^2}{r}\cot\theta-\frac{\xi_r\xi_\theta}{r}\right)\frac{\partial f}{\partial \xi_\theta}\\&-\left(\frac{\xi_\theta\xi_\varphi}{r}\cot\theta+\frac{\xi_r\xi_\varphi}{r}\right)\frac{\partial f}{\partial \xi_\varphi},\end{aligned} \tag{2.94a}$$

$$\frac{\partial F_i f}{\partial \xi_i}=\frac{\partial F_r f}{\partial \xi_r}+\frac{\partial F_\theta f}{\partial \xi_\theta}+\frac{\partial F_\varphi f}{\partial \xi_\varphi}, \tag{2.94b}$$

where ξ_r, ξ_θ, and ξ_φ are, respectively, the r, θ, and φ components of ξ_i (or $\boldsymbol{\xi}$), and F_r, F_θ, and F_φ are, respectively, the corresponding components of F_i (or $\boldsymbol{F}$).

Chapter 3

Linear Theory – Small Reynolds Numbers

From this chapter, we start the discussion of the main topics of this monograph, i.e., the relation between kinetic theory and fluid dynamics. When a system, subject to no external force, deviates slightly from an equilibrium state at rest, the linearized Boltzmann equation (2.82) is often used to analyze the system. In this chapter, we discuss the asymptotic behavior of the solution for small Knudsen numbers of a time-independent boundary-value problem in a general domain of the linearized Boltzmann equation. Both kinds of boundaries, a simple boundary and a boundary made of the condensed phase of the gas, are considered. First, we derive the solution expressing the overall behavior of the gas in the form of a power series of the Knudsen number under the assumption that the length scale of variation of the velocity distribution function is the geometric characteristic length of the system. Its solution is reduced to solving a series of fluid-dynamic-type sets of equations (Stokes sets of equations in the linearized problem) for the component functions of the expansion of the macroscopic variables: density (or pressure), velocity, and temperature. The solution, however, cannot generally be made to satisfy the kinetic boundary condition. Thus, its correction in the neighborhood of the boundary (Knudsen-layer correction) is introduced. From analysis of the Knudsen layer, the boundary conditions for the fluid-dynamic-type equations as well as the Knudsen-layer correction are obtained. In addition, general features of the behavior of the gas derived from this asymptotic analysis and its application are explained. The basic notation and relations used in this chapter are introduced in Sections 2.10 and 2.11.

3.1 Problem

We consider a rarefied gas in a steady state in an arbitrary domain except that its boundary shape is smooth. No external force is acting on the gas [or F_i in Eq. (2.5) is zero]. The behavior of the gas that slightly deviates from an equilibrium state at rest with a uniform temperature T_0 and density ρ_0 [the state expressed by Eq. (2.38)], that is, the case $\phi \ll 1$ [see Eq. (2.57)] is considered here. Thus, the behavior of the gas is well described by the linearized Boltzmann equation (2.82).[1] Then, the mathematical problem is the following time-independent boundary-value problem of the linearized Boltzmann equation (2.82): The basic equation is

$$\zeta_i \frac{\partial \phi}{\partial x_i} = \frac{1}{k}\mathcal{L}(\phi). \tag{3.1}$$

The boundary condition is, symbolically,

$$\phi = \phi_w \quad (\zeta_i n_i > 0), \tag{3.2}$$

where n_i is the unit normal vector to the boundary, pointed to the gas region, and the condition $u_{wi} n_i = 0$, which is required in a time-independent problem, is used. For the diffuse reflection [Eq. (2.90a) with (2.90b) and $\alpha = 1$],

$$\phi(x_i, \zeta_i) = \check{\sigma}_w + 2\zeta_j u_{wj} + \left(\zeta_j^2 - \frac{3}{2}\right)\tau_w \quad (\zeta_i n_i > 0), \tag{3.3a}$$

$$\check{\sigma}_w = -\frac{1}{2}\tau_w - 2\sqrt{\pi}\int_{\zeta_k n_k < 0} \zeta_j n_j \phi E d\boldsymbol{\zeta}, \tag{3.3b}$$

and for the complete condensation condition [Eq. (2.91) with $\alpha_c = 1$],

$$\phi(x_i, \zeta_i) = \omega_w + 2\zeta_j u_{wj} + \left(\zeta_j^2 - \frac{3}{2}\right)\tau_w \quad (\zeta_j n_j > 0). \tag{3.4}$$

In this chapter we investigate the asymptotic behavior for small Knudsen numbers[2] (or small k) of the solution ϕ of Eq. (3.1) with Eq. (3.2). The explicit form of the boundary condition (3.2) is not specified for the moment; the boundary may be a simple boundary or a boundary made of the condensed phase of the gas. The method of analysis is due to Sone [1969, 1971].

3.2 Grad–Hilbert solution and fluid-dynamic-type equations

Putting aside the boundary condition, we look for a moderately varying solution of Eq. (3.1), whose length scale of variation is of the order of the characteristic length L of the system [or $\partial\phi/\partial x_i = O(\phi)$], in a power series of k:

$$\phi_G = \phi_{G0} + \phi_{G1}k + \phi_{G2}k^2 + \cdots, \tag{3.5}$$

---

[1] There are some cases where this assumption is not true (see Sections 4.10.1 and 4.10.2 and Sone, Kataoka, Ohwada, Sugimoto & Aoki [1994]) even when the state is close to an equilibrium state at rest.

[2] There may be various Knudsen numbers of different order corresponding to various geometrical reference lengths, e.g., sizes of bodies in the gas, distances between boundaries, and radii of curvatures of boundaries. We consider the case where the maximum Knudsen number of them is small.

where the subscript G is attached for discrimination of the moderately varying solution of the linearized equation. The solution or expansion will be called the Grad–Hilbert solution or expansion. Corresponding to this expansion, the macroscopic variables ω, u_i, τ, etc. [Eqs. (2.83a)–(2.83f)] are also expanded in k:

$$h_G = h_{G0} + h_{G1}k + h_{G2}k^2 + \cdots, \tag{3.6}$$

where $h = \omega, u_i, \ldots$. The h_{Gm} is related to ϕ_{Gm} by the expansions of Eqs. (2.83a)-(2.83f) with $\phi = \phi_G$ as follows:

$$\omega_{Gm} = \int \phi_{Gm} E \mathbf{d}\boldsymbol{\zeta}, \tag{3.7a}$$

$$u_{iGm} = \int \zeta_i \phi_{Gm} E \mathbf{d}\boldsymbol{\zeta}, \tag{3.7b}$$

$$\frac{3}{2}\tau_{Gm} = \int \left(\zeta_i^2 - \frac{3}{2}\right) \phi_{Gm} E \mathbf{d}\boldsymbol{\zeta}, \tag{3.7c}$$

$$P_{Gm} = \omega_{Gm} + \tau_{Gm}, \tag{3.7d}$$

$$P_{ijGm} = 2\int \zeta_i\zeta_j \phi_{Gm} E \mathbf{d}\boldsymbol{\zeta}, \tag{3.7e}$$

$$Q_{iGm} = \int \zeta_i\zeta_j^2 \phi_{Gm} E \mathbf{d}\boldsymbol{\zeta} - \frac{5}{2}u_{iGm}. \tag{3.7f}$$

Substituting the series (3.5) into the linearized Boltzmann equation (3.1) and arranging the terms by the order in k, we obtain a series of integral equations for ϕ_{Gm}:

$$\mathcal{L}(\phi_{G0}) = 0, \tag{3.8}$$

$$\mathcal{L}(\phi_{Gm}) = \zeta_i \frac{\partial \phi_{Gm-1}}{\partial x_i} \quad (m = 1, 2, 3, \ldots). \tag{3.9}$$

The homogeneous equation (3.8) has five independent solutions $1, \zeta_i$, and ζ_i^2 [Eq. (2.61c)]. In view of the relations (3.7a)–(3.7c) for $m = 0$, the solution ϕ_{G0} is expressed as

$$\phi_{G0} = \omega_{G0} + 2\zeta_j u_{jG0} + \left(\zeta_j^2 - \frac{3}{2}\right)\tau_{G0}. \tag{3.10}$$

For the inhomogeneous equation (3.9) to have a solution, its inhomogeneous term should satisfy the solvability condition

$$\int g\zeta_k \frac{\partial \phi_{Gm-1}}{\partial x_k} E \mathbf{d}\boldsymbol{\zeta} = 0, \tag{3.11}$$

where

$$g = 1,\ \zeta_i,\ \text{or}\ \zeta_i^2, \tag{3.12}$$

since $\int g\mathcal{L}(\phi) E \mathbf{d}\boldsymbol{\zeta} = 0$ [Eq. (2.61b)]. Then, the solution ϕ_{Gm} is given in the form

$$\phi_{Gm} = \omega_{Gm} + 2\zeta_j u_{jGm} + \left(\zeta_j^2 - \frac{3}{2}\right)\tau_{Gm} + \hat{\phi}_{Gm}, \tag{3.13}$$

where the first part $\omega_{Gm} + 2\zeta_j u_{jGm} + (\zeta_j^2 - 3/2)\tau_{Gm}$ is the component function ϕ_{eGm} of the expansion in k of the linearized version of the Maxwellian (2.84) with $(\omega, u_j, \tau) = (\omega_G, u_{jG}, \tau_G)$ (say, ϕ_{eG}), and $\hat{\phi}_{Gm}$ is the particular solution of Eq. (3.9) orthogonal to g:

$$\int \hat{\phi}_{Gm} g E \mathbf{d}\boldsymbol{\zeta} = 0. \tag{3.14}$$

The inhomogeneous term of the integral equation for ϕ_{G1} [Eq. (3.9) with $m = 1$] is

$$\zeta_k \frac{\partial \phi_{G0}}{\partial x_k} = \zeta_k \frac{\partial \omega_{G0}}{\partial x_k} + 2\zeta_j \zeta_k \frac{\partial u_{jG0}}{\partial x_k} + \zeta_k \left(\zeta_j^2 - \frac{3}{2}\right) \frac{\partial \tau_{G0}}{\partial x_k}. \tag{3.15}$$

The two relations for $g = 1$ and $g = \zeta_i^2$ in the corresponding solvability condition (3.11) with $m = 1$ degenerate into the same equation

$$\frac{\partial u_{iG0}}{\partial x_i} = 0, \tag{3.16}$$

and the relation for $g = \zeta_i$ is

$$\frac{\partial P_{G0}}{\partial x_i} = 0. \tag{3.17}$$

With these conditions and Eq. (3.7d), the inhomogeneous term (3.15) is reduced to

$$\zeta_k \frac{\partial \phi_{G0}}{\partial x_k} = 2\left(\zeta_j \zeta_k - \frac{1}{3}\zeta^2 \delta_{jk}\right) \frac{\partial u_{jG0}}{\partial x_k} + \zeta_k \left(\zeta_j^2 - \frac{5}{2}\right) \frac{\partial \tau_{G0}}{\partial x_k}.$$

With this form of the inhomogeneous term, the first-order solution ϕ_{G1} is expressed by a linear combination of $\partial u_{jG0}/\partial x_k$ and $\partial \tau_{G0}/\partial x_k$, whose coefficients are functions of ζ_i. Furthermore, according to the general discussion in Appendix A.2, the coefficients of $\partial u_{jG0}/\partial x_k$ and $\partial \tau_{G0}/\partial x_k$ are, respectively, of the forms $(\zeta_j \zeta_k - \zeta^2 \delta_{jk}/3) g_1(\zeta)$ and $\zeta_k g_2(\zeta)$, where $g_1(\zeta)$ and $g_2(\zeta)$ are determined by the linearized collision operator. The analysis is continued in a similar way (i.e, the repetition of the process of construction of the inhomogeneous term from the solution, application of the solvability condition on the inhomogeneous term, and construction of the solution of the integral equation with the aid of the discussion in Appendix A.2). Then, we find that $\hat{\phi}_{Gm}$ is expressed by a linear combination of the $(m - s)$-th partial derivatives of ω_{Gs}, u_{iGs}, and τ_{Gs} with respect to x_j ($j = 1, 2$, or 3; $s = 0, 1, \ldots, m - 1$), whose coefficients are functions of ζ_i. Further, according to Appendix A.2, these coefficient functions are polynomials of ζ_i ($i = 1, 2$, and 3), whose coefficients are functions of ζ. From the solvability condition (3.11) with this form of ϕ_{Gm} (without the explicit forms of the latter coefficient functions), the following Stokes set of partial differential equations for the set of variables $(u_{iGm}, \tau_{Gm}\ P_{Gm+1})$, of the same form for any m, is derived.[3] (A detailed description of the process is given in

[3]To obtain the equations for u_{iGn}, τ_{Gn}, and P_{Gn+1}, the solvability conditions (3.11) up to $m = n + 2$ are required. This is due to the fact that τ_{Gn} is absent in Eq. (3.11) for $m \leq n + 1$, and that u_{iGn} appears only in the scalar equation with $g = 1$. This degeneracy at $m = 1$ results in: (a) the two relations of Eq. (3.11), with $g = 1$ and $g = \zeta_i^2$, degenerate into the single equation (3.16); (b) u_{iG0} does not appear in the equation with $g = \zeta_i$ [or Eq. (3.17)].

Appendix A.3.)

$$\frac{\partial P_{G0}}{\partial x_i} = 0, \tag{3.18}$$

$$\frac{\partial u_{iGm}}{\partial x_i} = 0, \tag{3.19a}$$

$$\frac{\partial P_{Gm+1}}{\partial x_i} = \gamma_1 \frac{\partial^2 u_{iGm}}{\partial x_j^2}, \tag{3.19b}$$

$$\frac{\partial^2 \tau_{Gm}}{\partial x_j^2} = 0, \tag{3.19c}$$

$$(m = 0, 1, 2, \dots),$$

where

$$P_{Gm} = \omega_{Gm} + \tau_{Gm}, \tag{3.20}$$

and γ_1 is a constant related to the collision integral and to be given in Eq. (3.33).

The solvability condition being satisfied, the solution ϕ_{Gm} is given in the following form:

$$\phi_{G0} = \phi_{eG0}, \tag{3.21}$$

$$\phi_{G1} = \phi_{eG1} - \zeta_i\zeta_j B(\zeta)\frac{\partial u_{iG0}}{\partial x_j} - \zeta_i A(\zeta)\frac{\partial \tau_{G0}}{\partial x_i}, \tag{3.22}$$

$$\begin{aligned}\phi_{G2} = \phi_{eG2} &- \zeta_i\zeta_j B(\zeta)\frac{\partial u_{iG1}}{\partial x_j} - \zeta_i A(\zeta)\frac{\partial \tau_{G1}}{\partial x_i} + \frac{1}{\gamma_1}\zeta_i D_1(\zeta)\frac{\partial P_{G1}}{\partial x_i} \\ &+ \zeta_i\zeta_j\zeta_k D_2(\zeta)\frac{\partial^2 u_{iG0}}{\partial x_j \partial x_k} - \zeta_i\zeta_j F(\zeta)\frac{\partial^2 \tau_{G0}}{\partial x_i \partial x_j},\end{aligned} \tag{3.23}$$

$$\cdots\cdots\cdots\cdots,$$

where

$$\phi_{eGm} = \omega_{Gm} + 2\zeta_i u_{iGm} + \left(\zeta^2 - \frac{3}{2}\right)\tau_{Gm}, \tag{3.24}$$

and the functions $A(\zeta)$, $B(\zeta)$, $D_1(\zeta)$, $D_2(\zeta)$, and $F(\zeta)$ are the solutions of the following integral equations:[4]

$$\left.\begin{aligned}&\mathcal{L}[\zeta_i A(\zeta)] = -\zeta_i\left(\zeta^2 - \frac{5}{2}\right), \\ &\text{subsidiary condition} \quad \int_0^\infty \zeta^4 A(\zeta)\exp(-\zeta^2)d\zeta = 0,\end{aligned}\right\} \tag{3.25}$$

$$\mathcal{L}\left[\left(\zeta_i\zeta_j - \frac{1}{3}\zeta^2\delta_{ij}\right)B(\zeta)\right] = -2\left(\zeta_i\zeta_j - \frac{1}{3}\zeta^2\delta_{ij}\right), \tag{3.26}$$

[4]These functions generally depend on the reference temperature T_0, except for a hard-sphere gas and the BKW model, according to Footnote 11 in Section 2.9. The confusion of similar notations $B(\zeta)$ and $B(|\alpha_i V_i|/V, V)$ introduced in Section 2.9 may be avoided from their contexts.

Table 3.1. Functions $A(\zeta)$, $B(\zeta)$, etc.

ζ	$A(\zeta)$	$B(\zeta)$	$D_1(\zeta)$	$D_2(\zeta)$	$F(\zeta)$
0.0	−6.137070	3.510384	−2.970357	5.094481	10.446110
0.2	−6.000493	3.488592	−2.929588	5.016259	10.125883
0.4	−5.601358	3.426008	−2.810643	4.796309	9.217954
0.6	−4.968023	3.330004	−2.622546	4.472115	7.857892
0.8	−4.138507	3.210186	−2.377495	4.089129	6.213220
1.0	−3.152155	3.076042	−2.088289	3.687574	4.437018
1.2	−2.044107	2.935479	−1.766585	3.296146	2.644323
1.4	−0.843026	2.794342	−1.422149	2.931792	0.909371
1.6	0.428951	2.656552	−1.062759	2.602411	−0.726525
1.8	1.755233	2.524509	−0.694443	2.310013	−2.244748
2.0	3.123547	2.399517	−0.321799	2.053253	−3.640744
2.2	4.524827	2.282139	0.051699	1.829137	−4.917909
2.4	5.952341	2.172469	0.423445	1.634051	−6.083685
2.6	7.401032	2.070310	0.791504	1.464336	−7.147220
2.8	8.867055	1.975305	1.154458	1.316567	−8.118052
3.0	10.347443	1.887008	1.511283	1.187672	−9.005409
3.2	11.839877	1.804946	1.861263	1.074970	−9.817868
3.4	13.342516	1.728639	2.203918	0.976148	−10.563225
3.6	14.853884	1.657625	2.538948	0.889234	−11.248471
3.8	16.372779	1.591466	2.866195	0.812552	−11.879827
4.0	17.898215	1.529757	3.185604	0.744679	−12.462804
4.2	19.429375	1.472124	3.497204	0.684411	−13.002275
4.4	20.965572	1.418224	3.801083	0.630726	−13.502544
4.6	22.506231	1.367745	4.097373	0.582757	−13.967417
4.8	24.050858	1.320404	4.386241	0.539765	−14.400261
5.0	25.599033	1.275944	4.667874	0.501122	−14.804060

$$\mathcal{L}\left[\left(\zeta_i\zeta_j - \frac{1}{3}\zeta^2\delta_{ij}\right)F(\zeta)\right] = \left(\zeta_i\zeta_j - \frac{1}{3}\zeta^2\delta_{ij}\right)A(\zeta), \tag{3.27}$$

$$\left.\begin{aligned}&\mathcal{L}[(\zeta_i\delta_{jk}+\zeta_j\delta_{ki}+\zeta_k\delta_{ij})D_1(\zeta)+\zeta_i\zeta_j\zeta_kD_2(\zeta)]\\&\qquad\qquad = \gamma_1(\zeta_i\delta_{jk}+\zeta_j\delta_{ki}+\zeta_k\delta_{ij})-\zeta_i\zeta_j\zeta_kB(\zeta),\\&\text{subsidiary condition}\quad\int_0^\infty[5\zeta^4D_1(\zeta)+\zeta^6D_2(\zeta)]\exp(-\zeta^2)\mathrm{d}\zeta=0.\end{aligned}\right\} \tag{3.28}$$

The compatibility of these special forms of solutions of the above integral equations is given in Appendix A.2. The numerical data of $A(\zeta)$, $B(\zeta)$, $D_1(\zeta)$, $D_2(\zeta)$, and $F(\zeta)$ for a hard-sphere gas (Ohwada & Sone [1992][5]) are shown in Table 3.1 (see also Pekeris & Alterman [1957], where the properties of these functions are discussed). For the BKW model,

$$A(\zeta)=\zeta^2-\tfrac{5}{2},\ B(\zeta)=2,\ D_1(\zeta)=-1,\ D_2(\zeta)=2,\ F(\zeta)=-\zeta^2+\tfrac{5}{2}. \tag{3.29}$$

3.3 Stress tensor and heat-flow vector of the Grad–Hilbert solution

From the expressions for the component function ϕ_{Gm} of the velocity distribution function, we obtain the stress tensor and heat-flow vector of the Grad–

[5]The function $\nu(\zeta)$ in Eqs. (A1)–(A5) on page 409 in Ohwada & Sone [1992] is a misprint for $2\sqrt{2}\nu(\zeta)$.

Hilbert solution as follows:

$$P_{ijG0} = P_{G0}\delta_{ij}, \tag{3.30a}$$

$$P_{ijG1} = P_{G1}\delta_{ij} + \gamma_1 S_{ijG0}, \tag{3.30b}$$

$$P_{ijG2} = P_{G2}\delta_{ij} + \gamma_1 S_{ijG1} + \gamma_3 \frac{\partial^2 \tau_{G0}}{\partial x_i \partial x_j}, \tag{3.30c}$$

$$\begin{aligned} P_{ijG3} &= P_{G3}\delta_{ij} + \gamma_1 S_{ijG2} + \gamma_3 \frac{\partial^2 \tau_{G1}}{\partial x_i \partial x_j} + \gamma_6 \frac{\partial^2 S_{ijG0}}{\partial x_k^2} \\ &= P_{G3}\delta_{ij} + \gamma_1 S_{ijG2} + \gamma_3 \frac{\partial^2 \tau_{G1}}{\partial x_i \partial x_j} - \frac{2\gamma_6}{\gamma_1} \frac{\partial^2 P_{G1}}{\partial x_i \partial x_j}, \end{aligned} \tag{3.30d}$$

$$Q_{iG0} = 0, \tag{3.31a}$$

$$Q_{iG1} = \frac{5}{4}\gamma_2 G_{iG0}, \tag{3.31b}$$

$$\begin{aligned} Q_{iG2} &= \frac{5}{4}\gamma_2 G_{iG1} - \frac{\gamma_3}{2} \frac{\partial S_{ijG0}}{\partial x_i} \\ &= \frac{5}{4}\gamma_2 G_{iG1} + \frac{\gamma_3}{2\gamma_1} \frac{\partial P_{G1}}{\partial x_i}, \end{aligned} \tag{3.31c}$$

$$\begin{aligned} Q_{iG3} &= \frac{5}{4}\gamma_2 G_{iG2} - \frac{\gamma_3}{2} \frac{\partial S_{ijG1}}{\partial x_i} \\ &= \frac{5}{4}\gamma_2 G_{iG2} + \frac{\gamma_3}{2\gamma_1} \frac{\partial P_{G2}}{\partial x_i}. \end{aligned} \tag{3.31d}$$

Here

$$S_{ijGm} = -\left(\frac{\partial u_{iGm}}{\partial x_j} + \frac{\partial u_{jGm}}{\partial x_i}\right), \qquad G_{iGm} = -\frac{\partial \tau_{Gm}}{\partial x_i}, \tag{3.32}$$

and γ_1, γ_2, γ_3, and γ_6, which may be called nondimensional transport coefficients (viscosity, thermal conductivity, etc.), are expressed by the solutions $A(\zeta)$, $B(\zeta)$, etc. of the integral equations (3.25)–(3.28) as follows:[6]

$$\left.\begin{aligned} &\gamma_1 = I_6(B), \qquad \gamma_2 = 2I_6(A), \\ &\gamma_3 = I_6(AB) = 5I_6(D_1) + I_8(D_2) = -2I_6(F), \\ &\gamma_6 = \tfrac{1}{2}I_6(BD_1) + \tfrac{3}{14}I_8(BD_2), \end{aligned}\right\} \tag{3.33}$$

where $I_n(Z)$, with $Z = A, B, \ldots$, is defined by the integral

$$I_n(Z) = \frac{8}{15\sqrt{\pi}} \int_0^\infty \zeta^n Z(\zeta) \exp(-\zeta^2) d\zeta. \tag{3.34}$$

For a hard-sphere gas,

$$\left.\begin{aligned} &\gamma_1 = 1.270042427, \qquad \gamma_2 = 1.922284066, \\ &\gamma_3 = 1.947906335, \qquad \gamma_6 = 1.419423836. \end{aligned}\right\} \tag{3.35}$$

[6] These nondimensional transport coefficients generally depend on the reference temperature T_0, except for a hard-sphere gas and the BKW model, since the functions $A(\zeta)$, $B(\zeta)$, etc. depend on T_0 [see Footnotes 11 (in Section 2.9) and 4].

For the BKW model,

$$\gamma_1 = \gamma_2 = \gamma_3 = \gamma_6 = 1. \tag{3.36}$$

It is noted that the stress $P_{ijG3} - P_{G3}\delta_{ij}$ and the heat flow Q_{iG3} at the third order are expressed only by derivatives of u_{iGm} and τ_{Gm} with $m \leq 2$. That is, the Grad–Hilbert parts of the stress and heat flow are obtained up to the third order by analysis of the sets of fluid-dynamic-type equations to the second order. The stress and heat-flow formulas are obtained without solving the third-order distribution function ϕ_{G3} explicitly with the aid of the self-adjoint property (2.62) of the linearized collision operator $\mathcal{L}(*)$ as follows:

$$\begin{aligned} P_{ijG3} - P_{G3}\delta_{ij} &= 2\int \left(\zeta_i\zeta_j - \frac{1}{3}\zeta^2\delta_{ij}\right)\phi_{G3}E\mathbf{d}\boldsymbol{\zeta} \\ &= -\int \mathcal{L}\left[\left(\zeta_i\zeta_j - \frac{1}{3}\zeta^2\delta_{ij}\right)B(\zeta)\right]\phi_{G3}E\mathbf{d}\boldsymbol{\zeta} \quad \text{[Eq. (3.26)]} \\ &= -\int \left(\zeta_i\zeta_j - \frac{1}{3}\zeta^2\delta_{ij}\right)B(\zeta)\mathcal{L}(\phi_{G3})E\mathbf{d}\boldsymbol{\zeta} \quad \text{[Eq. (2.62)]} \\ &= -\int \zeta_k\left(\zeta_i\zeta_j - \frac{1}{3}\zeta^2\delta_{ij}\right)B(\zeta)\frac{\partial\phi_{G2}}{\partial x_k}E\mathbf{d}\boldsymbol{\zeta} \quad \text{[Eq. (3.9)]}, \end{aligned} \tag{3.37}$$

and

$$\begin{aligned} Q_{iG3} &= \int \zeta_i\left(\zeta_j^2 - \frac{5}{2}\right)\phi_{G3}E\mathbf{d}\boldsymbol{\zeta} \\ &= -\int \mathcal{L}[\zeta_i A(\zeta)]\phi_{G3}E\mathbf{d}\boldsymbol{\zeta} \quad \text{[Eq. (3.25)]} \\ &= -\int \zeta_i A(\zeta)\mathcal{L}(\phi_{G3})E\mathbf{d}\boldsymbol{\zeta} \quad \text{[Eq. (2.62)]} \\ &= -\int \zeta_i\zeta_j A(\zeta)\frac{\partial\phi_{G2}}{\partial x_j}E\mathbf{d}\boldsymbol{\zeta} \quad \text{[Eq. (3.9)]}. \end{aligned} \tag{3.38}$$

Using Eq. (3.23) in the last expressions of Eqs. (3.37) and (3.38), we obtain the desired expressions. These third-order formulas are useful when discussing the force or energy transfer on a closed (bounded) body (Section 3.8). The second formulas for P_{ijG3}, Q_{iG2}, and Q_{iG3} in Eqs. (3.30d), (3.31c), and (3.31d) are derived from the first with the aid of Eqs. (3.19a) and (3.19b).

In the formulas (3.30a)–(3.30d) of the stress, the terms proportional to S_{ijGm} correspond to viscous stress in the classical gas dynamics, the higher-order terms proportional to $\partial^2\tau_{Gm}/\partial x_i\partial x_j$ are called thermal stress, and the last term of P_{ijG3} (proportional to $\partial^2 S_{ijG0}/\partial x_k^2$ or $\partial^2 P_{G1}/\partial x_i\partial x_j$) may be called nonharmonic-shear stress or pressure stress. In the formulas (3.31a)–(3.31d) of the heat-flow vector, the terms proportional to G_{iGm} correspond to the heat flow given by Fourier's law in the classical gas dynamics, and the heat flow depends on pressure gradient as well as temperature gradient at the second and higher orders. A numerical example of the heat flow proportional to pressure gradient is given in the Poiseuille flow analysis by Ohwada, Sone & Aoki [1989b].

3.4 Analysis of the Knudsen layer

The component function ϕ_{Gm} of the Grad–Hilbert expansion is of a very limited form of ζ_i. Thus, this solution does not have enough freedom to be fitted to the kinetic boundary condition (3.2). Therefore, the solution of the boundary-value problem cannot be expressed only by the Grad–Hilbert solution. More explicitly, in the case of the diffuse reflection [or Eq. (3.3a) with (3.3b)], for example, where the velocity distribution function of the outgoing molecules from the boundary is the corresponding part of a Maxwellian, ϕ_{G0} given by Eq. (3.21) will be easily seen to be fitted to the boundary distribution function (3.3a) with (3.3b) by choosing the boundary data of u_{iG0} and τ_{G0}, with the aid of which u_{iG0} and τ_{G0} are determined from the Stokes set. In the next order, however, ϕ_{G1} given by Eq. (3.22) is no longer a Maxwellian owing to the terms $\zeta_i\zeta_j B(\zeta)\partial u_{iG0}/\partial x_j$ and $\zeta_i A(\zeta)\partial\tau_{G0}/\partial x_i$. We cannot kill these terms to fit ϕ_{G1} to the boundary distribution function, since they are already determined quantities. The absence of this freedom in the Grad–Hilbert solution can also be seen by the fact that Eq. (3.1) is of a singular type, where the differential term $\zeta_i\partial\phi/\partial x_i$ is, in practice, multiplied by the small parameter k. Thus, we try to obtain the solution of the boundary-value problem by introducing a correction to the Grad–Hilbert solution in the neighborhood of the boundary. In other words, the condition of the length scale of variation imposed on the Grad–Hilbert solution is loosened in the neighborhood of the boundary. That is, we put the solution in the form

$$\phi = \phi_G + \phi_K, \tag{3.39}$$

where ϕ_K, called the Knudsen-layer correction, makes an appreciable change in the direction normal to the boundary over the distance of the order of the mean free path $[kn_i\partial\phi_K/\partial x_i = O(\phi_K)]$ and is appreciable only in a thin layer (Knudsen layer), with thickness of the order of the mean free path, adjacent to the boundary. The Grad–Hilbert solution ϕ_G is called the fluid-dynamic part.

In the linearized problem, the equation for ϕ_K is the same as Eq. (3.1). That is,

$$\zeta_i\frac{\partial\phi_K}{\partial x_i} = \frac{1}{k}\mathcal{L}(\phi_K). \tag{3.40}$$

Here the following Knudsen-layer variables (η, χ_1, χ_2) are introduced:

$$x_i = k\eta n_i(\chi_1, \chi_2) + x_{wi}(\chi_1, \chi_2), \tag{3.41}$$

where x_{wi} is the boundary surface, η is a stretched coordinate normal to the boundary, χ_1 and χ_2 are (unstretched) coordinates within a parallel surface $\eta = \text{const}$, and the normal vector n_i is a function of χ_1 and χ_2. Then, Eq. (3.40) is rewritten as

$$\zeta_i n_i\frac{\partial\phi_K}{\partial\eta} = \mathcal{L}(\phi_K) - k\zeta_i\left(\frac{\partial\chi_1}{\partial x_i}\frac{\partial\phi_K}{\partial\chi_1} + \frac{\partial\chi_2}{\partial x_i}\frac{\partial\phi_K}{\partial\chi_2}\right). \tag{3.42}$$

The Knudsen-layer correction ϕ_K is also expanded in a power series of k:

$$\phi_K = \phi_{K0} + \phi_{K1}k + \cdots . \tag{3.43}$$

Each term ϕ_{Km} of the expansion is governed by a spatially one-dimensional (homogeneous or inhomogeneous) linearized Boltzmann equation, i.e.,

$$\zeta_i n_i \frac{\partial \phi_{K0}}{\partial \eta} = \mathcal{L}(\phi_{K0}), \tag{3.44}$$

$$\zeta_i n_i \frac{\partial \phi_{K1}}{\partial \eta} = \mathcal{L}(\phi_{K1}) - \zeta_i \left[\left(\frac{\partial \chi_1}{\partial x_i} \right)_0 \frac{\partial \phi_{K0}}{\partial \chi_1} + \left(\frac{\partial \chi_2}{\partial x_i} \right)_0 \frac{\partial \phi_{K0}}{\partial \chi_2} \right], \tag{3.45}$$

$$\cdots\cdots\cdots ,$$

where the parentheses $(\ )_0$ with the subscript 0 indicate that the quantity in the parentheses ( ) is evaluated at $\eta = 0$.

The Knudsen-layer part h_K $(= h - h_G)$ of the macroscopic variables, where h represents ω, u_i, τ, etc., is also expanded in k as

$$h_K = h_{K0} + h_{K1}k + \cdots . \tag{3.46}$$

The component function h_{Km} is related to ϕ_{Km} in the same way as h_{Gm} to ϕ_{Gm}:

$$\omega_{Km} = \int \phi_{Km} E \mathbf{d}\boldsymbol{\zeta},$$

$$u_{iKm} = \int \zeta_i \phi_{Km} E \mathbf{d}\boldsymbol{\zeta},$$

$$\frac{3}{2}\tau_{Km} = \int \left(\zeta_i^2 - \frac{3}{2} \right) \phi_{Km} E \mathbf{d}\boldsymbol{\zeta},$$

$$P_{Km} = \omega_{Km} + \tau_{Km},$$

$$P_{ijKm} = 2 \int \zeta_i \zeta_j \phi_{Km} E \mathbf{d}\boldsymbol{\zeta},$$

$$Q_{iKm} = \int \zeta_i \zeta_j^2 \phi_{Km} E \mathbf{d}\boldsymbol{\zeta} - \frac{5}{2} u_{iKm}.$$

The boundary condition for ϕ_{Km} at $\eta = 0$ is

$$\phi_{Km} = \phi_{wm} - \phi_{Gm} \qquad (\zeta_i n_i > 0), \tag{3.47}$$

where ϕ_{wm} is defined by[7]

$$\phi_w = \phi_{w0} + \phi_{w1}k + \cdots . \tag{3.48}$$

[7]The boundary data may have some undetermined factor, which depends on the Knudsen number, for example, the surface temperature of a particle set free in a gas. To include the case, the boundary condition on the velocity distribution function is allowed to have Knudsen-number dependence. Correspondingly, u_{wi}, τ_w, ω_w, and P_w are expanded in power series of k.

For the diffuse reflection,

$$\begin{aligned}\phi_{Km} = \check{\sigma}_{wm} - \omega_{Gm} + 2[u_{wim} - (u_{iGm} - u_{jGm}n_jn_i)]\zeta_i \\ + (\tau_{wm} - \tau_{Gm})\left(\zeta_j^2 - \frac{3}{2}\right) - 2u_{jGm}n_jn_i\zeta_i - \hat{\phi}_{Gm} \quad (\zeta_in_i > 0),\end{aligned} \tag{3.49}$$

$$\check{\sigma}_{wm} = -\frac{1}{2}\tau_{wm} - 2\sqrt{\pi}\int_{\zeta_kn_k<0}\zeta_jn_j(\phi_{Gm} + \phi_{Km})E\mathbf{d}\boldsymbol{\zeta}, \tag{3.50}$$

and for the complete condensation condition,

$$\begin{aligned}\phi_{Km} = \omega_{wm} - \omega_{Gm} + 2[u_{wim} - (u_{iGm} - u_{jGm}n_jn_i)]\zeta_i \\ + (\tau_{wm} - \tau_{Gm})\left(\zeta_j^2 - \frac{3}{2}\right) - 2u_{jGm}n_jn_i\zeta_i - \hat{\phi}_{Gm} \qquad (\zeta_in_i > 0),\end{aligned} \tag{3.51}$$

where the boundary data u_{wi}, τ_w, and ω_w are considered to depend on k and expanded in k.

Since ϕ_K is assumed to be the correction to ϕ_G, the condition at infinity is

$$\phi_{Km} \to 0 \qquad \text{as } \eta \to \infty, \tag{3.52}$$

where the decay is assumed to be faster than any inverse power of η. This condition of rapid decay corresponds to the complete separation of the two types of solutions, the Grad–Hilbert solution and the Knudsen-layer correction, that the tail of the Knudsen-layer correction may not be counted as some-order term of the Grad–Hilbert solution. The existence of the solution of the rapid decay is confirmed by the existence and uniqueness theorem to be explained in the next paragraph.

The boundary condition (3.47) contains undetermined boundary values of the macroscopic variables u_{iGm}, τ_{Gm}, and ω_{Gm}, as well as those of the $(m-s)$-th partial derivatives of u_{iGs}, τ_{Gs}, and $\omega_{Gs}(s < m)$ at the previous stages of approximation, through ϕ_{Gm}.[8] The existence and uniqueness theorem (Bardos, Caflisch & Nicolaenko [1986], Cercignani [1986], Coron, Golse & Sulem [1988], Golse & Poupaud [1989]) of the half-space boundary-value problem of the linearized Boltzmann equation shows that the solution ϕ_{Km} exists when and only when the macroscopic variables satisfy some relations. These relations give the boundary condition for the Stokes set of partial differential equations (3.18)–(3.19c). The process of derivation of the boundary condition is first explained

[8] For simplicity, consider the case of the diffuse reflection or complete condensation condition. In the case of the specular reflection condition, the $u_{jGm}(\delta_{ij} - n_in_j)$, τ_{Gm}, and ω_{Gm} terms (or ϕ_{eGm} term except for $2\zeta_in_iu_{jGm}n_j$ term in it) contained in ϕ_{Gm} are canceled out in Eq. (3.47). In this case (or for the specular reflection condition), the procedure of the analysis and its result, which are given by Sone & Aoki [1977a], are different from the following ones. Their analysis is extended to the case for the Maxwell-type boundary condition with the accommodation coefficient α being of the order of the Knudsen number (or k) by Aoki, Inamuro & Onishi [1979].

for the cases of the diffuse reflection and complete condensation boundary conditions and then is outlined for the general boundary condition expressed in terms of a scattering kernel [see Eqs. (2.22) and (2.25)] in the following several paragraphs.

The existence and uniqueness theorem for the half-space problem of the linearized Boltzmann equation [Eq. (3.44) or (3.45) and Eqs. (3.47) and (3.52)] by Bardos, Caflisch & Nicolaenko [1986] is as follows. Let the boundary condition for the linearized Boltzmann equation (3.44) or (3.45) at $\eta = 0$ be given by

$$\phi_{Km} = c_0 + c_i(\zeta_i - \zeta_j n_j n_i) + c_4\zeta_j^2 + g(\zeta_i) \qquad (\zeta_i n_i > 0), \tag{3.53}$$

where c_0, c_i, and c_4 are undetermined constants, which are practically four because of no contribution of $c_i n_i$ to Eq. (3.53), and $g(\zeta_i)$ is a given function. Then the solution of this boundary-value problem, where exponential decay of the inhomogeneous term as $\eta \to \infty$ is assumed, exists when and only when the four undetermined constants $c_0, c_i - c_j n_j n_i$, and c_4 take special values.[9] The solution is unique including the values of $c_0, c_i - c_j n_j n_i$, and c_4. It is also shown that the speed of decay of the solution as $\eta \to \infty$ is exponential.[10]

In the case of the complete condensation condition (3.51), $\omega_{wm} - \omega_{Gm} - 3(\tau_{wm} - \tau_{Gm})/2$, $2[u_{wim} - (u_{iGm} - u_{jGm}n_j n_i)]$, $\tau_{wm} - \tau_{Gm}$, and $-2\zeta_i u_{jGm} n_i n_j - \hat{\phi}_{Gm}$ correspond, respectively, to $c_0, c_i - c_j n_j n_i, c_4$, and $g(\zeta_i)$, where the condition $u_{wim} n_i = 0$ should be noted. Thus, the solution ϕ_{Km} and the boundary data $\omega_{wm} - \omega_{Gm}$, $u_{wim} - (u_{iGm} - u_{jGm} n_j n_i)$, and $\tau_{wm} - \tau_{Gm}$ are determined by $2\zeta_i u_{jGm} n_i n_j + \hat{\phi}_{Gm}$ and ϕ_{Kr} $(r < m)$ in the inhomogeneous term of the equation for ϕ_{Km}, e.g., Eq. (3.45) for $m = 1$. In view of the form of $\hat{\phi}_{Gm}$ [see Eqs. (3.13), (3.21)–(3.23)], the way that ϕ_{Km} is determined from ϕ_{Kr} $(r < m)$, and the linear property of the problem, the undetermined constants $\omega_{wm} - \omega_{Gm}$, $u_{wim} - (u_{iGm} - u_{jGm} n_j n_i)$, and $\tau_{wm} - \tau_{Gm}$ and the solution ϕ_{Km} are expressed by the linear combinations of the boundary values of $u_{jGm} n_j$ and the $(m-r)$-th or the lower-order partial derivatives of ω_{Gr}, u_{iGr}, and τ_{Gr} $(r < m)$, where the coefficients in the expression for ϕ_{Km} are functions of η and ζ_i. The expressions for $\omega_{wm} - \omega_{Gm}$, $u_{wim} - (u_{iGm} - u_{jGm} n_j n_i)$, and $\tau_{wm} - \tau_{Gm}$ (the relations among the boundary values of ω_{Gm}, u_{iGm}, and τ_{Gm}) correspond to the boundary conditions for the Stokes set. More explicitly for $m = 0$, where $\hat{\phi}_{G0} = 0$, noting the property (2.68) for (2.67) of the linearized collision operator $\mathcal{L}$ and the result in Appendix B.4.2, we find that the solution can be obtained as the sum of two independent solutions of different symmetry with respect to $\zeta_i - \zeta_j n_j n_i$.[11]

[9] This was first conjectured by Grad [1969]. Those who are not interested in detailed mathematical discussion can find a simple description that the constants cannot be arbitrary in Sone & Yamamoto [1968] and Sone, Aoki & Onishi [1977].

[10] This guarantees the exponential decay of the inhomogeneous term in the next-order Knudsen-layer equation.

[11] From the above-mentioned property of the linearized collision operator, the operator $\zeta_i n_i \partial/\partial\eta - \mathcal{L}$ conserves the symmetry with respect to $\zeta_i - \zeta_j n_j n_i$. That is, $\zeta_i n_i \partial\varphi(\zeta, \zeta_l n_l)/\partial\eta - \mathcal{L}(\varphi(\zeta, \zeta_l n_l)) = g(\zeta, \zeta_l n_l)$ and $\zeta_k n_k \partial(\zeta_i - \zeta_j n_j n_i)\varphi_i(\zeta, \zeta_l n_l)/\partial\eta - \mathcal{L}((\zeta_i - \zeta_j n_j n_i)\varphi_i(\zeta, \zeta_l n_l)) = (\zeta_i - \zeta_j n_j n_i) g_i(\zeta, \zeta_l n_l)$. Thus, each part, $\varphi(\zeta, \zeta_l n_l)$ and $(\zeta_i - \zeta_j n_j n_i)\varphi_i(\zeta, \zeta_l n_l)$, of the solution ϕ_{K0} satisfies the original equation independently. The boundary condition (3.51) is rewritten

Thus, the relations among the boundary data and the Knudsen-layer solution are obtained in the following forms:

$$\omega_{G0} - \omega_{w0} = c_{01} u_{iG0} n_i,$$

$$\tau_{G0} - \tau_{w0} = c_{02} u_{iG0} n_i,$$

$$(u_{iG0} - u_{jG0} n_j n_i) - u_{wi0} = 0,$$

$$\phi_{K0} = u_{iG0} n_i \phi_{c1}(\eta, \chi_1, \chi_2, \zeta_i n_i, \zeta),$$

where c_{01} and c_{02} are determined constants.[12]

In the case of the diffuse reflection (3.49) with (3.50), besides the conditions on $\check{\sigma}_{wm} - \omega_{Gm}$, $u_{wim} - (u_{iGm} - u_{jGm} n_j n_i)$, and $\tau_{wm} - \tau_{Gm}$ corresponding to those on $\omega_{wm} - \omega_{Gm}$, $u_{wim} - (u_{iGm} - u_{jGm} n_j n_i)$, and $\tau_{wm} - \tau_{Gm}$ in the case of the complete condensation condition, $\check{\sigma}_{wm}$ is subject to another condition (3.50). Thus, $\check{\sigma}_{wm}$ being eliminated from these two relations on $\check{\sigma}_{wm}$, a condition on $u_{iGm} n_i$ is obtained. After all, the conditions on $u_{iGm} - u_{wim}$ and $\tau_{Gm} - \tau_{wm}$, which are the slip conditions on velocity and temperature, are obtained. For $m = 0$, where $\hat{\phi}_{G0} = 0$, the set $u_{iG0} = u_{wi0}$, $\tau_{G0} = \tau_{w0}$, and $\phi_{K0} = 0$ obviously satisfies the Knudsen-layer equation (3.44) and the boundary conditions (3.52) and (3.49) with (3.50). From Bardos's theorem, it is the unique set of the condition (nonslip condition) and Knudsen-layer correction (absence of Knudsen layer). When the boundary is a solid boundary, the following relations simplifying the results in the steps $m \geq 1$ are derived from the nonslip condition ($u_{iG0} = u_{wi0}$):[13]

$$\left(\frac{\partial u_{iG0}}{\partial x_j} + \frac{\partial u_{jG0}}{\partial x_i}\right) t_i s_j = 0, \tag{3.54}$$

and, with the aid of Eq. (3.19a),

$$\left(\frac{\partial u_{iG0}}{\partial x_j} + \frac{\partial u_{jG0}}{\partial x_i}\right) n_i n_j = 0, \tag{3.55}$$

where t_i and s_i are unit vectors tangential to the boundary (s_i may be equal to t_i). For $m = 1$, where $\hat{\phi}_{G1} = -\zeta_i \zeta_j B(\zeta) \partial u_{iG0}/\partial x_j - \zeta_i A(\zeta) \partial \tau_{G0}/\partial x_i$, the

---

in the following form:

$$\phi_{K0} = \omega_{w0} - \omega_{G0} + (\tau_{w0} - \tau_{G0})\left(\zeta^2 - \tfrac{3}{2}\right) - 2u_{jG0} n_j \zeta_i n_i$$
$$+ 2[u_{wi0} - (u_{iG0} - u_{jG0} n_j n_i)](\zeta_i - \zeta_k n_k n_i) \qquad (\zeta_i n_i > 0).$$

[12] In the formulas in the following parts of this book, $P_{wm} - P_{Gm}$, where $P_{wm} = \omega_{wm} + \tau_{wm}$ in the linear theory, is used exclusively instead of $\omega_{wm} - \omega_{Gm}$.

[13] Let the solid motion determined by u_{wi} of a solid boundary be $\tilde{u}_{wi}$. Then $\partial \tilde{u}_{wi}/\partial x_j + \partial \tilde{u}_{wj}/\partial x_i = 0$, from which

$$\left(s_j \frac{\partial u_{wi} t_i}{\partial x_j} + t_i \frac{\partial u_{wj} s_j}{\partial x_i}\right) - u_{wi}\left(s_j \frac{\partial t_i}{\partial x_j} + t_j \frac{\partial s_i}{\partial x_j}\right) = 0,$$

since the data of $\tilde{u}_{wi}$ and its tangential derivatives on the boundary can be replaced by those of u_{wi}. With the aid of the nonslip condition, the desired relation is obtained.

relations required by Bardos's theorem and ϕ_{K1} are expressed in the following forms:[14]

$$\omega_{G1} - \check{\sigma}_{w1} = c_{11} u_{iG1} n_i + c_{13} n_i \frac{\partial \tau_{G0}}{\partial x_i},$$

$$\tau_{G1} - \tau_{w1} = c_{12} u_{iG1} n_i + c_{14} n_i \frac{\partial \tau_{G0}}{\partial x_i},$$

$$(u_{iG1} - u_{jG1} n_j n_i) - u_{wi1} = c_{15}(\delta_{ik} - n_i n_k) n_j \left(\frac{\partial u_{kG0}}{\partial x_j} + \frac{\partial u_{jG0}}{\partial x_k} \right) + c_{16}(\delta_{ij} - n_i n_j) \frac{\partial \tau_{G0}}{\partial x_j},$$

$$\phi_{K1} = u_{iG1} n_i \varphi_{d1} + n_i \frac{\partial \tau_{G0}}{\partial x_i} \varphi_{d2} + (\zeta_i - \zeta_k n_k n_i) \left[n_j \left(\frac{\partial u_{iG0}}{\partial x_j} + \frac{\partial u_{jG0}}{\partial x_i} \right) \varphi_{d3} + \frac{\partial \tau_{G0}}{\partial x_i} \varphi_{d4} \right],$$

where $c_{11}, ..., c_{16}$ are constants and $\varphi_{d1}, \varphi_{d2}, \varphi_{d3}$, and φ_{d4} are functions of η, $\zeta_k n_k$, and ζ. Instead of the condition (3.50), we take an equivalent but simpler way. Integration of Eq. (3.45) multiplied by E over the whole $\boldsymbol{\zeta}$ leads to

$$\frac{\partial u_{iK1} n_i}{\partial \eta} = 0.$$

Therefore,

$$u_{iK1} n_i = 0,$$

since the Knudsen layer vanishes as $\eta \to \infty$.[15] From the condition $u_i n_i = 0$ on the boundary, which is equivalent to the condition (3.50), we have

$$u_{iG1} n_i = 0.$$

[14]First rewrite ζ_i and $\zeta_i \zeta_j$ in $\hat{\phi}_{G1}$ in the following way:

$$\zeta_i = \zeta_i - \zeta_k n_k n_i + \zeta_k n_k n_i = \zeta_k (\delta_{ki} - n_k n_i) + \zeta_k n_k n_i,$$
$$\zeta_i \zeta_j = (\zeta_i - \zeta_k n_k n_i + \zeta_k n_k n_i)(\zeta_j - \zeta_l n_l n_j + \zeta_l n_l n_j)$$
$$= \zeta_k \zeta_l (\delta_{ki} - n_k n_i)(\delta_{lj} - n_l n_j) + \zeta_k n_k [(\zeta_j - \zeta_l n_l n_j) n_i + (\zeta_i - \zeta_l n_l n_i) n_j] + (\zeta_k n_k)^2 n_i n_j.$$

Owing to the above relations and Eqs. (3.54) and (3.55), the $\hat{\phi}_{G1}$ is rewritten in the form

$$\hat{\phi}_{G1} = -(\zeta_i - \zeta_l n_l n_i) \left[\zeta_k n_k n_j B(\zeta) \left(\frac{\partial u_{iG0}}{\partial x_j} + \frac{\partial u_{jG0}}{\partial x_i} \right) + A(\zeta) \frac{\partial \tau_{G0}}{\partial x_i} \right] - \zeta_k n_k n_i A(\zeta) \frac{\partial \tau_{G0}}{\partial x_i}.$$

Then, the solution is obtained as the sum of two independent solutions of different symmetry with respect to $(\zeta_i - \zeta_l n_l n_i)$ in a similar procedure to that explained in Footnote 11.

[15]This simpler process to obtain $u_{iK1} n_i$ can be extended to obtain $u_{iKm} n_i$, $P_{ijKm} n_i$, and $Q_{iKm} n_i$. That is, the equation for ϕ_{Km} [Eqs. (3.44), (3.45), etc.] multiplied by $(1, \zeta_j, \zeta^2)E$ being integrated over the whole space of $\boldsymbol{\zeta}$, simple equations for $(u_{iKm} n_i, P_{ijKm} n_i, Q_{iKm} n_i)$ are obtained, where there is no contribution from the collision term $\mathcal{L}(\phi_{Km})$ owing to the general relation (2.61b).

Thus, we find that the boundary conditions for u_{iG1} and τ_{G1} and the Knudsen-layer correction ϕ_{K1} have the following forms:

$$(u_{iG1} - u_{jG1}n_jn_i) - u_{wi1} = c_{15}(\delta_{ik} - n_in_k)n_j\left(\frac{\partial u_{kG0}}{\partial x_j} + \frac{\partial u_{jG0}}{\partial x_k}\right) + c_{16}(\delta_{ij} - n_in_j)\frac{\partial \tau_{G0}}{\partial x_j},$$

$$u_{iG1}n_i = 0,$$

$$\tau_{G1} - \tau_{w1} = c_{14}n_i\frac{\partial \tau_{G0}}{\partial x_i},$$

$$\phi_{K1} = n_i\frac{\partial \tau_{G0}}{\partial x_i}\varphi_{d2} + (\zeta_i - \zeta_kn_kn_i)\left[n_j\left(\frac{\partial u_{iG0}}{\partial x_j} + \frac{\partial u_{jG0}}{\partial x_i}\right)\varphi_{d3} + \frac{\partial \tau_{G0}}{\partial x_i}\varphi_{d4}\right].$$

For the general boundary conditions (2.22) and (2.25), the formal analysis can be carried out in a similar way, as outlined here. First take the condition (2.25). Let $\omega_{wm} - \omega_{Gm}$, $u_{wim} - u_{iGm}$, and $\tau_{wm} - \tau_{Gm}$ be denoted, respectively, by $-\Delta\omega_{Gm}$, $-\Delta u_{iGm}$, and $-\Delta\tau_{Gm}$. In view of the condition (iii) [or Eq. (2.26c) and the uniqueness condition] for K_I in Section 2.6.2, the boundary condition (3.47) for ϕ_{Km} on the interface is expressed in the following form:[16]

$$\phi_{Km} = -\Delta\omega_{Gm} - 2\sum_{i=2}^{3}\zeta_i\Delta u_{iGm} - \left(\zeta^2 - \tfrac{3}{2}\right)\Delta\tau_{Gm} + G(\zeta) + \mathcal{K}(\phi_{Km}), \quad (3.56a)$$

---

[16]The boundary condition (2.25) on an interface can be expressed in the nondimensional form

$$E(1+\phi_w) = \hat{g}_B + \int_{\zeta_{i*}n_i<0}\hat{K}_I(\boldsymbol{\zeta},\boldsymbol{\zeta}_*)E_*(1+\phi)_*\mathbf{d}\boldsymbol{\zeta}_* \quad (\zeta_in_i>0),$$

and the corresponding expression of the condition (2.26c) is

$$E(1+\phi_{ew}) = \hat{g}_B + \int_{\zeta_{i*}n_i<0}\hat{K}_I(\boldsymbol{\zeta},\boldsymbol{\zeta}_*)E_*(1+\phi_{ew})_*\mathbf{d}\boldsymbol{\zeta}_* \quad (\zeta_in_i>0),$$

where

$$\phi_{ew} = \phi_e(\omega_w, u_{wi}, \tau_w).$$

Putting

$$\phi_w = \phi_{ew} + \phi_w - \phi_{ew}, \quad \phi = \phi_{ew} + \phi_G - \phi_{ew} + \phi_K,$$

and taking the above relation for ϕ_{ew} into account, we have ϕ_w in the form

$$E\phi_w = E\phi_{ew} + \int_{\zeta_{i*}n_i<0}\hat{K}_I(\boldsymbol{\zeta},\boldsymbol{\zeta}_*)E_*(\phi_G - \phi_{ew} + \phi_K)_*\mathbf{d}\boldsymbol{\zeta}_*.$$

Thus, its component function ϕ_{wm} is given by

$$E\phi_{wm} = E\phi_{ewm} + \int_{\zeta_{i*}n_i<0}\hat{K}_IE_*(\phi_{Gm} - \phi_{ewm} + \phi_{Km})_*\mathbf{d}\boldsymbol{\zeta}_*$$
$$= E\phi_{ewm} + \mathcal{K}(\phi_{eGm} - \phi_{ewm}) + \mathcal{K}(\hat{\phi}_{Gm}) + \mathcal{K}(\phi_{Km}),$$

where $\mathcal{K}(*)$ indicates the integral operator and the expression $\phi_{Gm} = \phi_{eGm} + \hat{\phi}_{Gm}$ is used. With this form of ϕ_{wm} and Eqs. (3.24) and (2.84) for ϕ_{eGm} and ϕ_{ewm} into Eq. (3.47), we obtain the desired form, where $a_0 = \mathcal{K}(1)$, $a_i = \mathcal{K}(2\zeta_i)$, and $a_4 = \mathcal{K}(\zeta^2 - 3/2)$.

where

$$G(\zeta) = a_0(\zeta)\Delta\omega_{Gm} + \sum_{i=2}^{3} a_i(\zeta)\Delta u_{iGm} + a_4(\zeta)\Delta\tau_{Gm} + g_G(\zeta), \tag{3.56b}$$

$$\text{with} \quad g_G(\zeta) = [-2\zeta_1 + a_1(\zeta)]u_{1Gm} - \hat{\phi}_{Gm} + \mathcal{K}(\hat{\phi}_{Gm}),$$

and the normal vector n_i is taken as $(1, 0, 0)$ to shorten the expressions. Here a_0, a_1, a_2, a_3, and a_4 are functions of ζ and $\mathcal{K}(\phi_{Km})$ is a linear functional of ϕ_{Km}; they are the new terms that enter the general boundary condition. The addition of $\mathcal{K}(\phi_{Km})$ is important, for which we need a more general theory than Bardos's theorem (e.g., Coron, Golse & Sulem [1988]). Let the same theorem as Bardos's hold for the boundary condition (3.53) with the additional term $\mathcal{K}(\phi_{Km})$, i.e.,

$$\phi_{Km} = c_0 + \sum_{i=2}^{3} c_i\zeta_i + c_4\zeta^2 + g(\zeta) + \mathcal{K}(\phi_{Km}) \quad (\zeta_1 > 0). \tag{3.57}$$

Let the Knudsen-layer equation for ϕ_{Km} with the inhomogeneous term set as zero [or Eq. (3.44) with $\phi_{K0} = \phi_{Km}$] be called the corresponding homogeneous equation for ϕ_{Km}. Let the constants c_0, c_2, c_3, and c_4 determined by the corresponding homogeneous equation for ϕ_{Km} with $g = a_r$ (r = 0, 2, 3, or 4) in the generalized boundary condition (3.57) be, respectively, c_0^r, c_2^r, c_3^r, and c_4^r, and let c_0, c_2, c_3, and c_4 determined by the Knudsen-layer equation for ϕ_{Km} (with the inhomogeneous term) with $g = g_G$ in the boundary condition (3.57) be, respectively, c_0^G, c_2^G, c_3^G, and c_4^G, each of which consists of three terms: the term proportional to u_{1Gm}, the term determined by $-\hat{\phi}_{Gm} + \mathcal{K}(\hat{\phi}_{Gm})$, and the term determined by the inhomogeneous term of the equation for ϕ_{Km}. Then, according to the generalized Bardos theorem,

$$c_r = c_r^0\Delta\omega_{Gm} + \sum_{i=2}^{3} c_r^i\Delta u_{iGm} + c_r^4\Delta\tau_{Gm} + c_r^G.$$

This relation being compared with the boundary condition (3.56a) with (3.56b), the following simultaneous linear equations for $\Delta\omega_{Gm}$, Δu_{2Gm}, Δu_{3Gm}, and $\Delta\tau_{Gm}$ are obtained:

$$(1 + c_0^0)\Delta\omega_{Gm} + \sum_{k=2}^{3} c_0^k\Delta u_{kGm} + (-\tfrac{3}{2} + c_0^4)\Delta\tau_{Gm} = -c_0^G, \tag{3.58a}$$

$$c_i^0\Delta\omega_{Gm} + \sum_{k=2}^{3}(2\delta_{ik} + c_i^k)\Delta u_{kGm} + c_i^4\Delta\tau_{Gm} = -c_i^G \quad (i = 2, 3), \tag{3.58b}$$

$$c_4^0\Delta\omega_{Gm} + \sum_{k=2}^{3} c_4^k\Delta u_{kGm} + (1 + c_4^4)\Delta\tau_{Gm} = -c_4^G. \tag{3.58c}$$

If the determinant of the coefficients of $\Delta\omega_{Gm}$, Δu_{2Gm}, Δu_{3Gm}, and $\Delta\tau_{Gm}$ does not vanish, which is conveniently called here the case with a finite complete condensation part, $\Delta\omega_{Gm}$, Δu_{2Gm}, Δu_{3Gm}, and $\Delta\tau_{Gm}$ (i.e., the boundary values of ω_{Gm}, u_{2Gm}, u_{3Gm}, and τ_{Gm}) are determined uniquely with u_{1Gm} as the parameter. That is, four relations hold among the boundary values of the five quantities ω_{Gm}, u_{1Gm}, u_{2Gm}, u_{3Gm}, and τ_{Gm}.

For the condition (2.22) on a simple boundary, the boundary condition (3.47) for ϕ_{Km} is expressed formally in the same form as Eq. (3.56a) with (3.56b), with the reservation that ω_{wm} in $\Delta\omega_{Gm}$ is left undetermined.[17] The analysis can be carried out in a similar way to that for the condition (2.25). The ambiguity of ω_{wm} is compensated with the condition $M \equiv 0$ [see Eq. (2.3a)] [or $\int \zeta_1(\phi_{Gm} + \phi_{Km})E d\boldsymbol{\zeta} = 0$ or Eq. (2.23b)] of a simple boundary. The role of this condition can be understood more simply in the same way as in the case of the diffuse reflection. From the integral of the Knudsen-layer equation for ϕ_{Km} over the whole space of $\boldsymbol{\zeta}$ and the condition at infinity, the normal velocity u_{1Km} is expressed by the solution ϕ_{Kn} $(n < m)$ of the previous stages through the inhomogeneous term of the equation for ϕ_{Km}. The boundary value of u_{1Gm} is given by $u_{1Gm} = -u_{1Km}(\eta = 0)$ from $M = 0$. The parameter u_{1Gm} being fixed, the $\Delta\omega_{Gm}$, Δu_{2Gm}, Δu_{3Gm}, and $\Delta\tau_{Gm}$ are determined. From these four, the boundary values of u_{1Gm}, u_{2Gm}, u_{3Gm}, and τ_{Gm} and the undetermined ω_{wm} are determined. With this ω_{wm}, the boundary value of ϕ_{Km} $(\zeta_1 > 0)$ is found. The case where the corresponding determinant of the equations for $\Delta\omega_{Gm}$, Δu_{2Gm}, Δu_{3Gm}, and $\Delta\tau_{Gm}$, which is independent of the choice of ω_{wm}, is not zero, is called the case with a finite diffuse reflection part.

For a locally isotropic boundary,[18] $a_0(\boldsymbol{\zeta})$, $a_1(\boldsymbol{\zeta})$, and $a_4(\boldsymbol{\zeta})$ in Eq. (3.56b) are reduced to functions of ζ_1 and ζ in place of $\boldsymbol{\zeta}$, and $a_2(\boldsymbol{\zeta})$ and $a_3(\boldsymbol{\zeta})$ are reduced, respectively, to $\zeta_2 a^c(\zeta_1, \zeta)$ and $\zeta_3 a^c(\zeta_1, \zeta)$ with a common $a^c(\zeta_1, \zeta)$. Then, the axially symmetric property with respect to the n_i direction of the corresponding homogeneous equation for ϕ_{Km} being taken into account (see Footnote 11),[19] it is found that $c_0^2 = c_0^3 = c_4^2 = c_4^3 = 0$, $c_2^0 = c_2^4 = c_3^0 = c_3^4 = 0$, $c_2^3 = c_3^2 = 0$, and therefore the equations (3.58a)–(3.58c) are separated into three sets of equations for Δu_{2Gm}, for Δu_{3Gm}, and for $\Delta\omega_{Gm}$ and $\Delta\tau_{Gm}$ as in the case of the diffuse reflection or complete condensation condition. Further, the inhomogeneous terms (or c_r^G) of Eqs. (3.58a)–(3.58c) consist of three kinds of

---

[17] Note the difference between the two conditions (iii) on K_B in Section 2.6.1 and K_I in Section 2.6.2.

[18] The "locally isotropic" means that the scattering kernel $K_B(\boldsymbol{\xi}, \boldsymbol{\xi}_*, \boldsymbol{X})$ or $K_I(\boldsymbol{\xi}, \boldsymbol{\xi}_*, \boldsymbol{X})$ and the inhomogeneous function $g_B(\boldsymbol{X}, \boldsymbol{\xi})$ in the kinetic boundary condition (2.22) or (2.25) contain the velocity $\boldsymbol{v}_w$ of the boundary in the forms $\boldsymbol{\xi} - \boldsymbol{v}_w$ and $\boldsymbol{\xi}_* - \boldsymbol{v}_w$, and that they are invariant for the following two kinds of transformations of $\boldsymbol{\xi} - \boldsymbol{v}_w$ and $\boldsymbol{\xi}_* - \boldsymbol{v}_w$: a rotation around the normal n_i of the boundary and the reflection with respect to a plane containing the normal n_i. The possible functional form for a tensor field such as $a_i(\boldsymbol{\zeta})$ defined by this type of operator is discussed in Appendix B.4.2. In the linearized problem in this chapter (or in a general perturbation problem), the contribution of $\boldsymbol{v}_w$ to a perturbed quantity, such as $a_i(\boldsymbol{\zeta})$, can be neglected (or is of a higher order).

[19] It should be noted for the general boundary condition that the symmetric property of ϕ_{Km} with respect to ζ_2 and ζ_3 is conserved in $\mathcal{K}(\phi_{Km})$ in the boundary condition (3.56a) on the locally isotropic boundary.

terms: the term proportional to u_{1Gm}, the linear functional $-\hat{\phi}_{Gm}+\mathcal{K}(\hat{\phi}_{Gm})$ of $\hat{\phi}_{Gm}$, and the term resulting from the inhomogeneous term of the Knudsen-layer equation for ϕ_{Km}; the first one appears only in the equations for $\Delta\omega_{Gm}$ and $\Delta\tau_{Gm}$, the second is determined by the Grad–Hilbert solution, and the third is determined by ϕ_{Kn} $(n < m)$. From the structure or way of determination of these terms, the inhomogeneous terms have the same character as those in the diffuse reflection or complete condensation condition. Therefore, on an interface, u_{2Gm}, u_{3Gm} and two relations among ω_{Gm}, u_{1Gm}, and τ_{Gm} are expressed by the same types of terms as those for the complete condensation condition; and on a simple boundary, u_{1Gm}, u_{2Gm}, u_{3Gm}, and τ_{Gm} are expressed by the same types of terms as those for the diffuse reflection condition.

3.5 Slip condition and Knudsen-layer correction

Here we summarize the boundary condition for the Stokes set of equations on a simple solid boundary or on an interface of a gas and its condensed phase and the Knudsen-layer correction to the macroscopic variables of the Grad–Hilbert solution, which are determined by the Stokes set of equations and its associated boundary condition. The formulas apply to a general locally isotropic boundary (Footnote 18), where the reflection law is expressed by Eq. (2.22) with a finite diffuse reflection part or by Eq. (2.25) with a finite complete condensation part. The boundary condition is called a slip or jump condition.

3.5.1 On a simple solid boundary

On a simple solid boundary, the slip boundary condition for the Stokes set of equations and the Knudsen-layer correction for the macroscopic variables are given as follows (Sone [1969,1971,1991a]):[20]

$$u_{iG0} - u_{wi0} = 0, \tag{3.59a}$$

$$\tau_{G0} - \tau_{w0} = 0, \tag{3.59b}$$

$$u_{iK0} = 0, \quad \omega_{K0} = \tau_{K0} = 0, \tag{3.59c}$$

$$\begin{bmatrix} (u_{iG1} - u_{wi1})t_i \\ u_{iK1}t_i \end{bmatrix} = S_{ijG0}n_i t_j \begin{bmatrix} k_0 \\ Y_0(\eta) \end{bmatrix} + G_{iG0}t_i \begin{bmatrix} K_1 \\ \frac{1}{2}Y_1(\eta) \end{bmatrix}, \tag{3.60a}$$

$$\begin{bmatrix} u_{iG1}n_i \\ u_{iK1}n_i \end{bmatrix} = 0, \tag{3.60b}$$

[20] a) There is a misprint in the third equation of the slip boundary condition (24) in Sone [1991a], where the term $-d_3 n_i n_j \partial G_{iG0}/\partial x_j$ is missing. b) For the specular reflection or the Maxwell-type condition with very small accommodation coefficient, the formulas take different forms (Sone & Aoki [1977a], Aoki, Inamuro & Onishi [1979]). c) In some literature, a temperature-jump term proportional to $S_{ijG0}n_i n_j$ is retained in $\tau_{G1} - \tau_{w1}$ of Eq. (3.60c). However, $S_{ijG0}n_i n_j = 0$ [Eq. (3.55)] on a solid simple boundary. More terms appear in the following formulas for the slip condition and Knudsen-layer correction unless the conditions (3.54) and (3.55) hold on the boundary.

$$\begin{bmatrix} \tau_{G1} - \tau_{w1} \\ \omega_{K1} \\ \tau_{K1} \end{bmatrix} = -G_{iG0} n_i \begin{bmatrix} d_1 \\ \Omega_1(\eta) \\ \Theta_1(\eta) \end{bmatrix}, \tag{3.60c}$$

$$\begin{aligned}\begin{bmatrix} (u_{iG2} - u_{wi2}) t_i \\ u_{iK2} t_i \end{bmatrix} &= S_{ijG1} n_i t_j \begin{bmatrix} k_0 \\ Y_0(\eta) \end{bmatrix} + \frac{\partial S_{ijG0}}{\partial x_r} n_j n_r t_i \begin{bmatrix} a_1 \\ Y_{a1}(\eta) \end{bmatrix} \\ &+ \bar{\kappa} S_{ijG0} n_i t_j \begin{bmatrix} a_2 \\ Y_{a2}(\eta) \end{bmatrix} + \kappa_{ij} S_{jrG0} n_r t_i \begin{bmatrix} a_3 \\ Y_{a3}(\eta) \end{bmatrix} \\ &+ \frac{\partial G_{iG0}}{\partial x_j} n_j t_i \begin{bmatrix} a_4 \\ Y_{a4}(\eta) \end{bmatrix} + \bar{\kappa} G_{iG0} t_i \begin{bmatrix} a_5 \\ Y_{a5}(\eta) \end{bmatrix} \\ &+ \kappa_{ij} G_{jG0} t_i \begin{bmatrix} a_6 \\ Y_{a6}(\eta) \end{bmatrix} - \frac{\partial \tau_{w1}}{\partial x_i} t_i \begin{bmatrix} K_1 \\ \frac{1}{2} Y_1(\eta) \end{bmatrix}, \end{aligned} \tag{3.61a}$$

$$\begin{aligned}\begin{bmatrix} u_{iG2} n_i \\ u_{iK2} n_i \end{bmatrix} &= \frac{\partial S_{ijG0}}{\partial x_r} n_i n_j n_r \begin{bmatrix} b_1 \\ \frac{1}{2} \int_\infty^\eta Y_0(\eta_0) d\eta_0 \end{bmatrix} \\ &+ \left(\frac{\partial G_{iG0}}{\partial x_j} n_i n_j + 2\bar{\kappa} G_{iG0} n_i \right) \begin{bmatrix} b_2 \\ \frac{1}{2} \int_\infty^\eta Y_1(\eta_0) d\eta_0 \end{bmatrix}, \end{aligned} \tag{3.61b}$$

$$\begin{aligned}\begin{bmatrix} \tau_{G2} - \tau_{w2} \\ \omega_{K2} \\ \tau_{K2} \end{bmatrix} &= -G_{iG1} n_i \begin{bmatrix} d_1 \\ \Omega_1(\eta) \\ \Theta_1(\eta) \end{bmatrix} - \frac{\partial S_{ijG0}}{\partial x_r} n_i n_j n_r \begin{bmatrix} d_4 \\ \Omega_4(\eta) \\ \Theta_4(\eta) \end{bmatrix} \\ &- \frac{\partial G_{iG0}}{\partial x_j} n_i n_j \begin{bmatrix} d_3 \\ \Omega_3(\eta) \\ \Theta_3(\eta) \end{bmatrix} - \bar{\kappa} G_{iG0} n_i \begin{bmatrix} d_5 \\ \Omega_5(\eta) \\ \Theta_5(\eta) \end{bmatrix}, \end{aligned} \tag{3.61c}$$

$$\bar{\kappa} = \frac{1}{2}(\kappa_1 + \kappa_2), \qquad \kappa_{ij} = \kappa_1 l_i l_j + \kappa_2 m_i m_j. \tag{3.62}$$

Here the following notes are in order:
(i) As already explained, u_{wim} and τ_{wm} are the component functions of the expansions of u_{wi} and τ_w, where $(2RT_0)^{1/2} u_{wi}$ ($u_{wi} n_i = 0$)[21] and $T_0(1 + \tau_w)$ are, respectively, the velocity and temperature of the boundary [see Eqs. (2.57) and Footnote 7]:

$$u_{wi} = u_{wi0} + u_{wi1} k + \cdots, \qquad \tau_w = \tau_{w0} + \tau_{w1} k + \cdots. \tag{3.63}$$

(ii) The t_i is a unit vector tangential to the boundary.
(iii) The κ_1/L and κ_2/L are the principal curvatures of the boundary, where κ_1 or κ_2 is taken negative when the corresponding center of curvature lies on

[21] As noted just after Eq. (3.2), the velocity of the boundary must satisfy the condition $u_{wi} n_i = 0$ for the state to be steady.

the side of the gas; the l_i and m_i in Eq. (3.62) are the direction cosines of the principal directions corresponding to κ_1 and κ_2 respectively; and the curvature tensor κ_{ij} is related to the variation of the normal n_i as follows:

$$\kappa_{ij} = \left(\frac{\partial \chi_1}{\partial x_i}\right)_0 \frac{\partial n_j}{\partial \chi_1} + \left(\frac{\partial \chi_2}{\partial x_i}\right)_0 \frac{\partial n_j}{\partial \chi_2}. \tag{3.64}$$

(iv) The quantities with the subscript G are evaluated on the boundary surface; and S_{ijGm} and G_{iGm} are defined by Eq. (3.32).
(v) The k_0, K_1, $a_1, \ldots, a_6$, b_1 $[= \frac{1}{2}\int_0^\infty Y_0(\eta_0)\mathrm{d}\eta_0]$, b_2 $[= \frac{1}{2}\int_0^\infty Y_1(\eta_0)\mathrm{d}\eta_0]$, d_1, d_3, d_4, and d_5 are constants called slip coefficients, determined by the molecular model (e.g., hard-sphere, BKW) and the reflection law on the boundary (e.g., diffuse reflection); $Y_0(\eta)$, $Y_1(\eta)$, $\Omega_1(\eta)$, $\Theta_1(\eta)$, etc. are functions of η, called Knudsen-layer functions, whose functional forms are determined by the molecular model and the reflection law.[22]

For a hard-sphere gas under the diffuse reflection, the first-order slip coefficients and Knudsen-layer functions and some of them of the second order are obtained in Ohwada, Sone & Aoki [1989a], Sone, Ohwada & Aoki [1989a], and Ohwada & Sone [1992]:[23]

$$\left.\begin{aligned} &k_0 = -1.2540, && K_1 = -0.6463, && d_1 = 2.4001, \\ &a_4 = 0.0330, && b_1 = 0.1068, && b_2 = 0.4776, \end{aligned}\right\} \tag{3.65}$$

and the Knudsen-layer functions $Y_0(\eta)$, $Y_1(\eta)$, $\Omega_1(\eta)$, $\Theta_1(\eta)$, and $Y_{a4}(\eta)$ are tabulated in Table 3.2.

For the BKW model under the diffuse reflection, the slip coefficients and Knudsen-layer functions are obtained up to the second order in Sone [1969, 1971].[24] The slip coefficients are

$$\left.\begin{aligned} &k_0 = -1.01619, && K_1 = -0.38316, && d_1 = 1.30272, \\ &a_1 = 0.76632, && a_2 = 0.50000, && a_3 = -0.26632, \\ &a_4 = 0.27922, && a_5 = 0.26693, && a_6 = -0.76644, \\ &b_1 = 0.11684, && b_2 = 0.26693, && d_3 = 0, \\ &d_4 = 0.11169, && d_5 = 1.82181. \end{aligned}\right\} \tag{3.66}$$

The Knudsen-layer functions are tabulated in Table 3.3. Some of the functions

[22] The slip coefficients and Knudsen-layer functions generally depend on the reference temperature T_0 except for a hard-sphere molecular gas and the BKW model (see Footnote 11 in Section 2.9).

[23] The data for the Maxwell-type condition are found in Ohwada & Sone [1992] and Wakabayashi, Ohwada & Golse [1996].

[24] The shear flow slip problem corresponding to k_0 for the BKW equation is studied analytically by Cercignani [1962], and the temperature jump problem corresponding to d_1 is studied numerically by Welander [1954]. In the latter, the BKW equation is proposed. Many people refer to that equation as the BGK equation from Bhatnagar, Gross & Krook [1954]. In the present monograph, the BKW equation (abbreviation of Boltzmann–Krook–Welander) is preferred.

Table 3.2. Knudsen-layer functions for the Boltzmann equation for hard-sphere molecules under the diffuse reflection condition or under the complete condensation condition. The functions of the same symbol, e.g., $Y_0(\eta)$, are common to the diffuse reflection and the complete condensation.

η	$Y_0(\eta)$	$Y_1(\eta)/2$	$\Omega_1(\eta)$	$-\Theta_1(\eta)$	$H_A(\eta)$	$H_B(\eta)$	$Y_{a4}(\eta)$	$\Omega_4^*(\eta)$	$\Theta_4^*(\eta)$
0.0000	0.34522	0.44508	0.51641	0.73783	0.13384	1.38948	0.29625	0.37815	0.05206
0.02503	0.30466	0.41870	0.46886	0.67376	0.12660	1.28950	0.28278	0.33624	0.04783
0.05011	0.28008	0.40046	0.43974	0.63360	0.12161	1.22172	0.27284	0.31065	0.04490
0.10108	0.24424	0.37107	0.39661	0.57323	0.11350	1.11477	0.25613	0.27306	0.04026
0.15138	0.21806	0.34744	0.36451	0.52766	0.10689	1.03068	0.24218	0.24538	0.03662
0.20226	0.19691	0.32686	0.33809	0.48977	0.10107	0.95884	0.22971	0.22286	0.03352
0.30081	0.16525	0.29328	0.29753	0.43097	0.09143	0.84434	0.20878	0.18886	0.02861
0.41305	0.13848	0.26186	0.26200	0.37886	0.08226	0.74030	0.18860	0.15979	0.02420
0.60049	0.10657	0.21995	0.21757	0.31303	0.06979	0.60615	0.16086	0.12468	0.01862
0.76791	0.08628	0.19014	0.18766	0.26839	0.06076	0.51404	0.14059	0.10205	0.01487
0.99185	0.06653	0.15806	0.15673	0.22204	0.05089	0.41800	0.11827	0.07973	0.01107
1.17977	0.05426	0.13621	0.13625	0.19136	0.04409	0.35452	0.10280	0.06570	0.00865
1.40407	0.04310	0.11471	0.11642	0.16171	0.03731	0.29358	0.08733	0.05279	0.00641
1.66967	0.03327	0.09416	0.09766	0.13382	0.03078	0.23688	0.07233	0.04130	0.00443
1.81952	0.02891	0.08444	0.08879	0.12072	0.02767	0.21058	0.06515	0.03615	0.00356
1.98156	0.02493	0.07517	0.08032	0.10828	0.02468	0.18585	0.05826	0.03142	0.00277
2.54689	0.01523	0.05063	0.05759	0.07538	0.01672	0.12204	0.03977	0.01974	0.00094
2.99541	0.01052	0.03733	0.04484	0.05739	0.01237	0.08856	0.02959	0.01397	0.00015
3.50590	0.00702	0.02658	0.03410	0.04257	0.00883	0.06213	0.02125	0.00960	−0.00035
4.08156	0.00453	0.01825	0.02528	0.03076	0.00607	0.04209	0.01471	0.00642	−0.00060
6.21632	0.00099	0.00472	0.00883	0.00993	0.00157	0.01057	0.00390	0.00167	−0.00054
7.97384	0.00030	0.00160	0.00385	0.00414	0.00053	0.00355	0.00135	0.00062	−0.00031
9.96662	0.00008	0.00048	0.00154	0.00159	0.00016	0.00106	0.00042	0.00022	−0.00015
15.00000	0.00000	0.00002	0.00016	0.00016	0.00001	0.00005	0.00003	0.00002	−0.00002
20.77429	0.00000	0.00000	0.00001	0.00001	0.00000	0.00000	0.00000	0.00000	−0.00000
25.39373	0.00000	0.00000	0.00000	0.00000	0.00000	0.00000	0.00000	0.00000	0.00000

are or can be put in the following forms:

$$\left.\begin{array}{lll} Y_{a1} = -Y_1, & Y_{a2} = -2\widetilde{Y}_0, & Y_{a3} = -Y_1 - k_0 Y_0, \\ Y_{a4} = 0, & Y_{a5} = Y_2 - \widetilde{Y}_1, & Y_{a6} = \frac{1}{2}Y_2 - (K_1 + \frac{1}{4})Y_0, \\ \Omega_3 = \Theta_3 = 0, & (\Omega_4, \Theta_4) = (-\frac{1}{4}\Omega_4^*, -\frac{1}{4}\Theta_4^*), & \end{array}\right\} \tag{3.67}$$

and the functions on the right-hand sides are shown there.

The slip conditions (3.59a) and (3.59b) show that there is neither velocity slip nor temperature jump at the leading order, and are called nonslip conditions. In the higher orders, there are velocity slips and temperature jumps consisting of various terms; the curvatures of the boundary appear in the second-order condition. The second term on the right-hand side of the slip condition (3.60a) shows that a flow is induced over a wall with a temperature gradient along it. The flow is called thermal creep flow (Section 3.11.1), whose simple experimental demonstration is given in Sone [1991b] and Sone, Sawada & Hirano [1994]. The fifth term on the right-hand side of Eq. (3.61a) shows the existence of another type of flow, called thermal-stress slip flow (Sone [1972]). These flows, not expected in the classical gas dynamics and discussed in more detail in Sections 3.11.1 and 3.11.2, are typical examples of flows induced by the effect of gas rarefaction and

Table 3.3. Knudsen-layer functions for the BKW equation under the diffuse reflection condition or under the complete condensation condition. The functions of the same symbol, e.g., $Y_0(\eta)$, are common to the diffuse reflection and the complete condensation.

η	Y_0	Y_1	Y_2	$\check{Y}_0$	$\check{Y}_1$	$\int_\eta^\infty Y_0 d\eta_0$	$\int_\eta^\infty Y_1 d\eta_0$
0.00	0.30908	0.54777	0.92525	0.3907	1.1922	0.23368	0.53385
0.05	0.25727	0.48117	0.83415	0.3569	1.1004	0.21974	0.50837
0.10	0.22827	0.43961	0.77446	0.3350	1.0391	0.20764	0.48539
0.20	0.18878	0.37876	0.68370	0.3017	0.9439	0.18693	0.44465
0.40	0.14003	0.29679	0.55502	0.2536	0.8032	0.15449	0.37770
0.60	0.10929	0.24081	0.46264	0.2181	0.6970	0.12976	0.32425
0.80	0.08773	0.19932	0.39151	0.1898	0.6115	0.11017	0.28042
1.00	0.07175	0.16720	0.33475	0.1666	0.5403	0.09430	0.24390
1.20	0.05948	0.14167	0.28842	0.1471	0.4800	0.08123	0.21310
1.40	0.04982	0.12096	0.25000	0.1305	0.4281	0.07033	0.18691
1.60	0.04209	0.10394	0.21777	0.1162	0.3832	0.06117	0.16447
1.80	0.03580	0.08978	0.19048	0.1038	0.3439	0.05340	0.14514
2.00	0.03063	0.07790	0.16720	0.0930	0.3093	0.04677	0.12841
2.50	0.02118	0.05551	0.12224	0.0712	0.2393	0.03399	0.09543
3.00	0.01498	0.04028	0.09068	0.0551	0.1870	0.02506	0.07172
4.00	0.00787	0.02206	0.05153	0.0338	0.1164	0.01407	0.04157
5.00	0.00434	0.01256	0.03023	0.0212	0.0740	0.00815	0.02474
10.00	0.00032	0.00104	0.00277	0.0025	0.0092	0.00071	0.00236
coef. of $\eta \ln\eta$*	0.39894	0.40540	0.50519	0.2027	0.5052	0	0

η	Ω_1	$-\Theta_1$	Ω_4^*	$\Theta_4^* \times 10$	$-\int_\eta^\infty(\Omega_1+\Theta_1)d\eta_0$	$\int_\eta^\infty(\Omega_4^*+\Theta_4^*)d\eta_0$
0.00	0.34771	0.44920	0.36303	0.3717	0.11609	0.23886
0.05	0.29177	0.38521	0.28722	0.3771	0.11123	0.22106
0.10	0.26121	0.34842	0.24683	0.3639	0.10672	0.20592
0.20	0.21987	0.29703	0.19399	0.3328	0.09852	0.18060
0.40	0.16877	0.23114	0.13269	0.2757	0.08466	0.14249
0.60	0.13610	0.18770	0.09669	0.2297	0.07331	0.11479
0.80	0.11276	0.15607	0.07296	0.1927	0.06385	0.09377
1.00	0.09507	0.13181	0.05634	0.1627	0.05587	0.07738
1.20	0.08118	0.11258	0.04424	0.1382	0.04908	0.06438
1.40	0.06999	0.09700	0.03519	0.1180	0.04325	0.05392
1.60	0.06080	0.08417	0.02828	0.1012	0.03822	0.04542
1.80	0.05315	0.07345	0.02293	0.0871	0.03387	0.03844
2.00	0.04671	0.06440	0.01872	0.0752	0.03007	0.03267
2.50	0.03442	0.04717	0.01156	0.0528	0.02254	0.02210
3.00	0.02588	0.03521	0.00735	0.0376	0.01707	0.01523
4.00	0.01524	0.02042	0.00315	0.0198	0.01004	0.00754
5.00	0.00932	0.01231	0.00143	0.0108	0.00606	0.00389
10.00	0.00107	0.00134	0.00004	0.0007	0.00063	0.00021
coef. of $\eta \ln\eta$*	0.46451	0.47587	0.63047	−0.5610	0	0

η	Ω_5	Θ_5	$-\Omega_6$	$-\Theta_6$	Ω_7	Θ_7
0.00	0.3516	−0.5956	0.31622	0.21690	0.5166	0.3136
0.05	0.2853	−0.4931	0.26242	0.19447	0.4441	0.2837
0.10	0.2452	−0.4292	0.23040	0.17970	0.3999	0.2640
0.20	0.1883	−0.3368	0.18563	0.15738	0.3364	0.2340
0.40	0.1153	−0.2163	0.12971	0.12619	0.2529	0.1916
0.60	0.0686	−0.1383	0.09488	0.10417	0.1976	0.1611
0.80	0.0363	−0.0838	0.07112	0.08746	0.1577	0.1375
1.00	0.0130	−0.0444	0.05413	0.07428	0.1276	0.1186
1.20	−0.0040	−0.0154	0.04161	0.06363	0.1043	0.1030
1.40	−0.0166	0.0061	0.03220	0.05487	0.0859	0.0900
1.60	−0.0260	0.0222	0.02501	0.04758	0.0712	0.0791
1.80	−0.0328	0.0341	0.01947	0.04146	0.0592	0.0697
2.00	−0.0378	0.0428	0.01515	0.03626	0.0495	0.0617
2.50	−0.0444	0.0550	0.00799	0.02632	0.0320	0.0459
3.00	−0.0460	0.0587	0.00400	0.01943	0.0209	0.0346
4.00	−0.0420	0.0545	0.00052	0.01097	0.0091	0.0203
5.00	−0.0349	0.0452	−0.00050	0.00643	0.0039	0.0122
10.00	−0.0093	0.0116	−0.00027	0.00061	−0.0001	0.0012
coef. of $\eta \ln\eta$*	0.4291	−0.6043	0.34843	0.11459	0.4646	0.1528

* The Knudsen-layer functions generally have the singularity $\eta \ln\eta$ at $\eta = 0$. Its coefficient is shown in this row.

cause thermophoresis[25], which is one of the most interesting subjects in aerosol science. The second term of the slip condition (3.61c) is a temperature jump proportional to the normal gradient of the normal stress. Thus, for example, in a uniform flow past a body with the same temperature as that of the flow, the temperature distribution is nonuniform, because of this jump, even in a slow flow governed by the linearized equation. This phenomenon is called thermal polarization (Bakanov, Vysotskij, Derjaguin & Roldughin [1983], Takata, Sone & Aoki [1993]).

The Knudsen-layer parts of the stress tensor P_{ij} and the heat-flow vector Q_i are given by

$$P_{ijK0} = 0, \tag{3.68}$$

$$P_{ijK1} = -\frac{3}{2}G_{kG0}n_k[\Omega_1(\eta) + \Theta_1(\eta)](\delta_{ij} - n_i n_j), \tag{3.69}$$

$$P_{ijK2}n_i t_j = \frac{3}{2}\left(\frac{\partial G_{iG0}}{\partial x_j}n_i t_j + G_{iG0}\kappa_{ij}t_j\right)\int_\infty^\eta [\Omega_1(\eta_0) + \Theta_1(\eta_0)]\mathrm{d}\eta_0, \tag{3.70a}$$

$$P_{ijK2}n_i n_j = -3\bar{\kappa}G_{iG0}n_i \int_\infty^\eta [\Omega_1(\eta_0) + \Theta_1(\eta_0)]\mathrm{d}\eta_0, \tag{3.70b}$$

$$Q_{iK0} = 0, \tag{3.71}$$

$$Q_{iK1}n_i = 0, \quad Q_{iK1}t_i = -S_{ijG0}t_i n_j H_A(\eta) - G_{iG0}t_i H_B(\eta), \tag{3.72}$$

$$\begin{aligned} Q_{iK2}n_i = &-\frac{1}{2}\frac{\partial S_{ijG0}}{\partial x_k}n_i n_j n_k \int_\infty^\eta H_A(\eta_0)\mathrm{d}\eta_0 \\ &-\left(\frac{\partial G_{iG0}}{\partial x_j}n_i n_j + 2\bar{\kappa}G_{iG0}n_i\right)\int_\infty^\eta H_B(\eta_0)\mathrm{d}\eta_0, \end{aligned} \tag{3.73}$$

where the quantities with the subscript G are evaluated on the boundary, and the functions $H_A(\eta)$ and $H_B(\eta)$ for a hard-sphere gas are shown in Table 3.2. For the BKW model,

$$H_A(\eta) = \frac{1}{2}Y_0(\eta), \quad H_B(\eta) = \frac{1}{4}Y_1(\eta) + \frac{1}{\sqrt{\pi}}\int_0^\infty \exp\left(-y^2 - \frac{\eta}{y}\right)\mathrm{d}y. \tag{3.74}$$

3.5.2 On an interface of a gas and its condensed phase with evaporation or condensation

On an interface of a gas and its condensed phase, the boundary condition for the Stokes set of equations and the Knudsen-layer correction for the macroscopic

[25]Thermophoresis is a phenomenon that a body in a rarefied gas (or a small particle in a gas) with a temperature gradient is subject to a force and drifts in the gas. Simple examples of application of the asymptotic theory are given in Section 3.11.5. More detailed analyses and references are found, for example, in Sone [1972], Sone & Aoki [1977a, b, 1979, 1981, 1983], Loyalka [1983], Bakanov [1991], Kogan [1992], Takata, Aoki & Sone [1994], Takata & Sone [1995], Sone [2000a].

variables are given as follows (Sone & Onishi [1978], Sone & Aoki [1987], Sone [1991a]):

$$(u_{iG0} - u_{wi0})t_i = 0, \tag{3.75a}$$

$$u_{iK0} = 0, \tag{3.75b}$$

$$\begin{bmatrix} P_{G0} - P_{w0} \\ \tau_{G0} - \tau_{w0} \\ \omega_{K0} \\ \tau_{K0} \end{bmatrix} = u_{iG0}n_i \begin{bmatrix} C_4^* \\ d_4^* \\ \Omega_4^*(\eta) \\ \Theta_4^*(\eta) \end{bmatrix}, \tag{3.75c}$$

$$\begin{bmatrix} (u_{iG1} - u_{wi1})t_i \\ u_{iK1}t_i \end{bmatrix} = S_{ijG0}n_it_j \begin{bmatrix} k_0 \\ Y_0(\eta) \end{bmatrix} + G_{iG0}t_i \begin{bmatrix} K_1 \\ \frac{1}{2}Y_1(\eta) \end{bmatrix} + t_j \frac{\partial u_{iG0}n_i}{\partial x_j} \begin{bmatrix} K_2 \\ Y_{K2}(\eta) \end{bmatrix}, \tag{3.76a}$$

$$u_{iK1}n_i = 0, \tag{3.76b}$$

$$\begin{bmatrix} P_{G1} - P_{w1} \\ \tau_{G1} - \tau_{w1} \\ \omega_{K1} \\ \tau_{K1} \end{bmatrix} = u_{iG1}n_i \begin{bmatrix} C_4^* \\ d_4^* \\ \Omega_4^*(\eta) \\ \Theta_4^*(\eta) \end{bmatrix} - G_{iG0}n_i \begin{bmatrix} C_1 \\ d_1 \\ \Omega_1(\eta) \\ \Theta_1(\eta) \end{bmatrix} - S_{ijG0}n_in_j \begin{bmatrix} C_6 \\ d_6 \\ \Omega_6(\eta) \\ \Theta_6(\eta) \end{bmatrix} - 2\bar{\kappa}u_{iG0}n_i \begin{bmatrix} C_7 \\ d_7 \\ \Omega_7(\eta) \\ \Theta_7(\eta) \end{bmatrix}. \tag{3.76c}$$

Here the following notes are in order:
(i) See note (i) after Eq. (3.62) for u_{wim} and τ_{wm}; the P_{wm} is the component function of the expansion of P_w, where $p_0(1 + P_w)$ is the saturated gas pressure at temperature $T_0(1 + \tau_w)$ [see Eqs. (2.57) and Footnote 7]:

$$P_w = P_{w0} + P_{w1}k + \cdots . \tag{3.77}$$

(ii) The t_i is a unit vector tangential to the boundary.
(iii) The $\bar{\kappa}/L$ is the mean curvature of the boundary [see Eq. (3.62) and the note (iii) after it].
(iv) The quantities with the subscript G are evaluated on the interface; and S_{ijGm} and G_{iGm} are defined by Eq. (3.32).
(v) The C_4^*, d_4^*, K_2, C_1, C_6, C_7, d_6, and d_7, as well as k_0, K_1, and d_1, are constants determined by the molecular model and the reflection law on the boundary; $\Omega_4^*(\eta)$, $\Theta_4^*(\eta)$, $Y_{K2}(\eta)$, $\Omega_6(\eta)$, $\Theta_6(\eta)$, $\Omega_7(\eta)$, and $\Theta_7(\eta)$, as well as $Y_0(\eta)$, $Y_1(\eta)$, $\Omega_1(\eta)$, and $\Theta_1(\eta)$, are functions of η, whose functional forms are determined by the molecular model and the reflection law. Among them, k_0, K_1, d_1, Y_0, Y_1, Ω_1, and Θ_1 under the complete condensation condition are the same as those in Eqs. (3.60a)–(3.60c) under the diffuse reflection condition;

that is, the slip coefficients and Knudsen-layer functions of the same symbol are common to the diffuse reflection and the complete condensation. (See Footnote 22.)

For a hard-sphere gas under the complete condensation condition, some of the slip coefficients and Knudsen-layer functions are obtained in Sone, Ohwada & Aoki [1989b].[26] That is,

$$C_4^* = -2.1412, \qquad d_4^* = -0.4557, \qquad C_1 = 1.0947, \tag{3.78}$$

and the Knudsen-layer functions $\Omega_4^*(\eta)$ and $\Theta_4^*(\eta)$ are tabulated in Table 3.2. For the BKW model under the complete condensation condition, the slip (or jump) coefficients and Knudsen-layer functions are obtained up to the first order of k in Sone & Onishi [1978]. The slip coefficients are

$$\left.\begin{array}{llll} C_1 = 0.55844, & C_4^* = -2.13204, & C_6 = 0.82085, & C_7 = -0.38057, \\ K_2 = -0.79519, & d_4^* = -0.44675, & d_6 = 0.33034, & d_7 = -0.13157. \end{array}\right\} \tag{3.79}$$

The Knudsen-layer functions, among which Y_{K2} is given by

$$Y_{K2} = 2Y_0 + \frac{1}{2}d_4^* Y_1, \tag{3.80}$$

are tabulated in Table 3.3.

Eq. (3.75c) shows that there is a temperature jump at the leading order and that Knudsen-layer corrections are required there, when evaporation or condensation is taking place.

The Knudsen-layer parts of the stress tensor P_{ij} and the heat-flow vector Q_i are given by

$$P_{ijK0} = \frac{3}{2}u_{kG0}n_k[\Omega_4^*(\eta) + \Theta_4^*(\eta)](\delta_{ij} - n_i n_j), \tag{3.81}$$

$$P_{ijK1}n_i n_j = 3\bar{\kappa}u_{iG0}n_i \int_\infty^\eta [\Omega_4^*(\eta_0) + \Theta_4^*(\eta_0)]\mathrm{d}\eta_0, \tag{3.82a}$$

$$P_{ijK1}n_i t_j = -\frac{3}{2}\frac{\partial u_{iG0}n_i}{\partial x_j}t_j \int_\infty^\eta [\Omega_4^*(\eta_0) + \Theta_4^*(\eta_0)]\mathrm{d}\eta_0, \tag{3.82b}$$

$$Q_{iK0} = 0, \tag{3.83}$$

$$Q_{iK1}n_i = 0, \tag{3.84}$$

where the quantities with the subscript G are evaluated on the boundary.

---

[26]The slip coefficients, corresponding to k_0, K_1, d_1, C_1, and d_4^*, obtained in Loyalka & Hickey [1989] and Loyalka [1989, 1991] are not literally those for a hard-sphere gas. In their analysis, the kernel of the collision integral $\mathcal{L}(\phi)$ [Eqs. (2.69)–(2.70c)] is expanded in the associated Legendre functions and truncated with two terms (and modified further) in Loyalka & Hickey [1989] and with four terms in Loyalka [1989] (not clear in Loyalka [1991]). It is not shown there how close the approximated kernels are to the exact kernel. According to Wakabayashi, Ohwada & Golse [1996] and H. Chihara (private communication), there is considerable difference between the exact kernel and its four term approximation. Thus, the solutions there are those of model equations, such as the BKW equation, different from the Boltzmann equation for hard-sphere molecules. In fact, the Knudsen-layer profile in Loyalka & Hickey [1989] differs considerably from that of Ohwada, Sone & Aoki [1989a], although their slip coefficients are close to those in Eqs. (3.65) and (3.78).

3.6 Determination of macroscopic variables

The equations and their boundary conditions for the component functions of the Grad–Hilbert expansion of the macroscopic variables are obtained in Sections 3.2 and 3.5. Here we explain how the component functions are determined successively from the lowest order. The discussion is made separately for a simple boundary and for an interface of a gas and its condensed phase.

First we start with the discussion of the case of a simple boundary. From Eq. (3.18), P_{G0} is uniform, i.e.,

$$P_{G0} = P_0.$$

In an open-(or unbounded-)domain problem, where the pressure is specified, for example, at infinity, the constant P_0 is determined by the pressure there. In a closed-(or bounded-)domain problem, where the total mass in the domain is specified,[27] P_0 cannot be determined at this stage. Now consider the set of equations (3.19a)–(3.19c) for $m = 0$, where u_{iG0}, τ_{G0}, and P_{G1} are contained. The variables u_{iG0}, τ_{G0}, and P_{G1} are determined by this set of equations under the boundary conditions (3.59a) and (3.59b), leaving an undetermined integration constant (say, P_1) in P_{G1}. From Eq. (3.20), ω_{G0} is expressed by P_{G0} and τ_{G0} as

$$\omega_{G0} = P_0 - \tau_{G0}.$$

At this stage the constant P_0, if undetermined previously, is determined, for example, from the total mass condition with the aid of ω_{G0}. In the next step, the variables u_{iG1}, τ_{G1}, ω_{G1}, and P_{G2} and the constant P_1 (if undetermined) are determined by Eqs. (3.19a)–(3.19c) with Eq. (3.20) for $m = 1$ under the boundary conditions (3.60a)–(3.60c) leaving an undetermined integration constant in P_{G2}, where P_1 and ω_{G1} are determined in a similar way to P_0 and ω_{G0}. In this way, we can determine the macroscopic variables successively.

For a boundary of an interface of a gas and its condensed phase, the process is similar but there is some difference for determination of the integration constant in P_{Gm}. From Eq. (3.18), P_{G0} is uniform, i.e.,

$$P_{G0} = P_0.$$

The constant P_0 is determined by the condition at infinity in an open problem or undetermined at this stage in a closed problem. From the set of equations (3.19a)–(3.19c) for $m = 0$ and the boundary conditions (3.75a) and (3.75c), the variables u_{iG0}, τ_{G0}, and P_{G1} are determined, leaving an undetermined integration constant (say, P_1) in P_{G1}; the result contains the (determined or undetermined) parameter P_{G0} (or P_0) through the boundary condition (3.75c). The constant P_0 cannot be arbitrary, because it is determined by the condition at

[27] In a closed domain problem bounded by a simple boundary, we can put any amount of the gas in the domain, and once it is put in, it remains in the domain irrespective of its state (steady or unsteady). Thus, the total mass is a quantity that can naturally be specified and should be specified to fix the problem.

infinity in an open-domain problem, or because the quantity $u_{iG0}n_i$ integrated over the boundary must be zero from Eq. (3.19a) in a closed-domain problem. Then, ω_{G0} is determined by Eq. (3.20). In the next step, the variables u_{iG1}, τ_{G1}, ω_{G1}, and P_{G2} and the constant P_1 (if undetermined) are determined by Eqs. (3.19a)–(3.19c) with Eq. (3.20) for $m = 1$ under the boundary conditions (3.76a) and (3.76c) leaving an undetermined integration constant in P_{G2}, where P_1 and ω_{G1} are determined in a similar way to P_0 and ω_{G0}. In this way, we can determine the macroscopic variables of the fluid-dynamic part successively.

3.7 Discontinuity of the velocity distribution function and S layer

The Boltzmann equation (3.1) determines the variation of the velocity distribution function ϕ in the ζ_i direction (or along the characteristic of the equation) in x_i space. Thus, if there is a discontinuity of ϕ at $(x_i = x_i^{(0)}, \zeta_i = \zeta_i^{(0)})$, then the discontinuity propagates from $x_i^{(0)}$ in the direction of $\zeta_i^{(0)}$.[28] This discontinuity decays owing to molecular collisions over the distance of the order of the free path of the molecules with $\zeta_i = \zeta_i^{(0)}$.[29] From this property, we find that the velocity distribution function has discontinuity in a gas around a boundary with a convex part [e.g., around a closed (or bounded) body], irrespective of the continuity of the boundary data, for the following reason. The velocity distribution of the molecules ($\zeta_i n_i > 0$) leaving the boundary, which is determined by the boundary condition (3.2), is different in nature from that of the impinging molecules ($\zeta_i n_i < 0$), which is formed by interactions with the surrounding molecules. Thus, the velocity distribution function generally has discontinuity at the velocity ($\zeta_i n_i = 0$) tangent to the boundary. This discontinuity propagates into the gas on a convex part of the boundary (Fig. 3.1). On a concave part of the boundary, on the other hand, the characteristic does not enter the gas region, and therefore the discontinuity does not propagate into the gas (Sone & Takata [1992]). The discontinuity of the velocity distribution function around a convex body is analyzed by detailed numerical computation in drag and thermal force problems around a sphere and in a strong evaporation problem from a spherical or cylindrical condensed phase (Takata, Sone & Aoki [1993], Sone, Takata & Wakabayashi [1994], Takata & Sone [1995], Sugimoto & Sone [1992], Sone & Sugimoto [1993], and Sone & Sugimoto [1995]).

Now it is clear that there is generally a discontinuity of the velocity distribution function in a gas. Thus, the proof by Arkeryd [1972] that the equality in Eq. (2.30) holds only when the velocity distribution function is Maxwellian for

[28]The discontinuity of the velocity distribution function is not particular to the linearized or steady problem. Examples of the discontinuity in time-dependent problems are given in Sone & Sugimoto [1990] and Aoki, Sone, Nishino & Sugimoto [1991].

[29]Many people seem to misunderstand that the discontinuity disappears immediately by molecular collisions. This is not correct because all the molecules do not collide at once. This misunderstanding comes from confusion with the viscous effect, which enters the highest differential term of the Navier–Stokes set of equations.

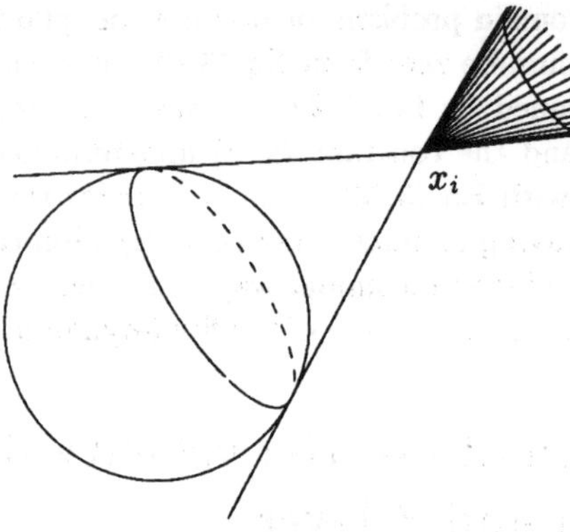

Figure 3.1. Discontinuity of the velocity distribution function in a gas around a convex boundary. At point x_i, the velocity distribution function is discontinuous on the shaded cone in ζ_i space.

a wide class of functions where discontinuous functions are included is essential to the statement of the H theorem. This is an example showing that theorems apparently only of mathematical interest are very important in physical statements.

When the mean free path ℓ of the gas molecules is comparable to or larger than the radius of curvature (say, R_w) of a convex body, the discontinuity is appreciable in the region where the distance from the boundary is of the order of the mean free path [Fig. 3.2 (a)]. This is the transition region where the gas molecules leaving the boundary are accommodated to the state of the surrounding gas by molecular collisions. That is, if ℓ/R_w (Knudsen number based on the radius of curvature) is not small, the region with the discontinuity extends in the same area as the transition region. On the other hand, if ℓ/R_w is small, the discontinuity line is almost parallel to the boundary at the distance of the order of several ℓ along the discontinuity, where the discontinuity has almost vanished. Thus, the discontinuity extends only in the region within $O(\ell^2/R_w)$ from the boundary [Fig. 3.2 (b)]. The discontinuity exists only at the bottom of the transition region (called Knudsen layer for small Knudsen numbers).

The discontinuity of the velocity distribution function requires corrections to the asymptotic analysis. That is, we have to take into account the discontinuity on the cone in the analysis of the Knudsen-layer equation (3.44), (3.45), etc. The error introduced by the discontinuity is estimated in Sone & Takata [1992]. At the order next to that where the Knudsen-layer correction first appears (at k^2 order for the diffuse reflection or at k order for the complete condensation condition), corrections for macroscopic variables are required in a thin layer with thickness of the order of ℓ^2/R_w at the bottom of the Knudsen layer. However, the correction to the slip condition and the variables $u_i n_i$, $P_{ij} n_j$, and $Q_i n_i$ are of higher order, and therefore the formulas for the slip condition and these variables given in this chapter are not subject to any correction. The correction in the

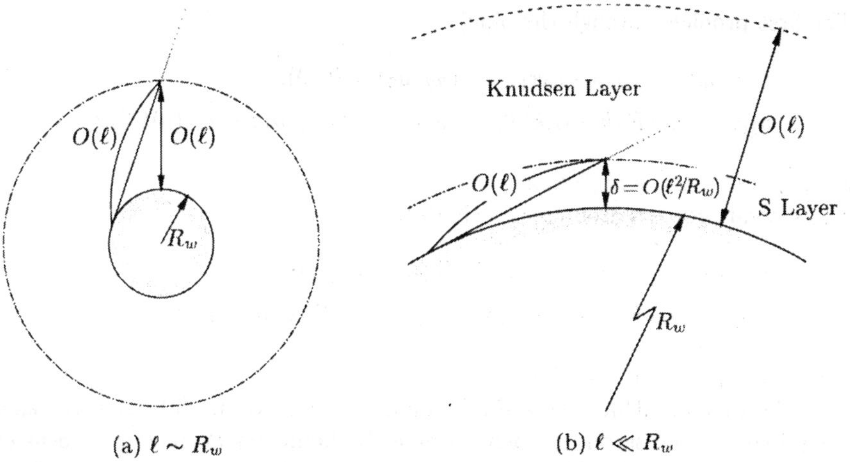

Figure 3.2. Transition region and the region where the discontinuity of the velocity distribution function is appreciable. (a) $\ell \sim R_w$ and (b) $\ell \ll R_w$. The discontinuity is tangent to the boundary and at the velocity of the molecules leaving the boundary along this line. The discontinuity decays with distance from the boundary owing to molecular collisions; it is appreciable on the solid line with length $O(\ell)$, and negligible farther away on the vanishing dotted line. For $\ell \sim R_w$, in panel (a), the discontinuity extends to a region, shown by the chain line, of $O(\ell)$ from the boundary. For small ℓ/R_w, in panel (b), the discontinuity is in a thin layer, shown by the chain line, with thickness $\delta = O(\ell^2/R_w)$; the region, bounded by the dashed line, of $O(\ell)$ from the boundary is the Knudsen layer.

thin layer (or the second boundary layer) at the bottom of the Knudsen layer is called the S layer, which was first found in the analysis of a flow induced around a circular cylinder with a temperature gradient along it by Sone [1973]. The numerical demonstration of the S layer around a spherical body in a uniform flow is given in Sone & Takata [1992], where the correction for $u_i n_i$ is shown to be of higher order than that for other variables. This note on the influence of the discontinuity of the velocity distribution function in the Knudsen-layer analysis applies to that in the following chapters.

Finally, two examples are given for comparison of flows over concave and convex boundaries: one is the flow induced through a circular pipe with a temperature gradient along its axis, and the other is the flow induced around a circular cylinder with a temperature gradient along its axis. Let the axis of the pipe or cylinder of radius L be at $(X_2, X_3) = (0, 0)$ [or $(x_2, x_3) = (0, 0)$]; the gas is in the region $X_2^2 + X_3^2 < L^2$ (or $x_2^2 + x_3^2 < 1$) for the first problem and in $X_2^2 + X_3^2 > L^2$ (or $x_2^2 + x_3^2 > 1$) for the second problem; the temperature distribution of the pipe or cylinder is given by $T_w = T_0(1 + c_0 X_1/L)$ [or $\tau_w = c_0 x_1$]. According to the asymptotic theory developed in this chapter, the steady behavior of the gas for small Knudsen numbers (or $k \ll 1$) is obtained easily as follows:

(The first problem through the pipe)

$$\tau = c_0 x_1, \quad \omega = -c_0 x_1, \quad (u_2, u_3) = (0, 0),$$
$$u_{1G} = -c_0(K_1 k - a_5 k^2/2), \quad u_{1K} = -c_0[Y_1(\eta)k - Y_{a5}(\eta)k^2]/2,$$

where $\eta = [1 - (x_2^2 + x_3^2)^{1/2}]/k$.
(The second problem around the cylinder)

$$\tau = c_0 x_1, \quad \omega = -c_0 x_1, \quad (u_2, u_3) = (0, 0),$$
$$u_{1G} = -c_0(K_1 k + a_5 k^2/2), \quad u_{1K} = -c_0[Y_1(\eta)k + Y_{a5}(\eta)k^2]/2,$$

where $\eta = [(x_2^2 + x_3^2)^{1/2} - 1]/k$.

In the first problem, where the boundary is concave, no S-layer correction is required. In the second problem, where the boundary is convex, an S-layer correction is required. The S layer is studied in Sone [1973] for the BKW equation with the diffuse reflection. According to it, the following correction (say, $u_{S2}k^2$) of the order of k^2 is required to u_1:

$$u_{S2}k^2 = -\frac{c_0 k^2}{\pi}\left[(2K_1 + Y_1(0) - 1)\, g_1(y) + g_2(y)\right],$$

where y is an S-layer variable defined by

$$y = \frac{\eta}{k},$$

and g_1 and g_2 are expressed as

$$g_1(y) = T_4(\sqrt{2y}) - 2T_2(\sqrt{2y}) - y\int_{\sqrt{2y}}^{\infty} \frac{T_0(t)}{t}\mathrm{d}t,$$
$$g_2(y) = T_6(\sqrt{2y}) - 3T_4(\sqrt{2y}) - y\int_{\sqrt{2y}}^{\infty} \frac{T_2(t)}{t}\mathrm{d}t,$$

with the aid of the Abramowitz function $T_n(t)$ (Abramowitz & Stegun [1972]) defined by

$$T_n(t) = \int_0^{\infty} s^n \exp\left(-s^2 - \frac{t}{s}\right)\mathrm{d}s.$$

The function $T_n(t)$ decays rapidly with t:

$$T_n(t) \sim \left(\frac{\pi}{3}\right)^{1/2}\left(\frac{t}{2}\right)^{n/3} \exp\left[-3\left(\frac{t}{2}\right)^{2/3}\right] \quad (t \gg 1).$$

Thus, the correction u_{S2} decays rapidly with y and is appreciable in the thin layer $y = O(1)$ or $(X_2^2 + X_3^2)^{1/2} - L = O(\ell_0^2/L)$.

Incidentally, the differences of the signs of the second terms of u_{1G} and u_{1K} in the two problems are due to the difference of the signs of the curvatures on the convex and concave boundaries.

3.8 Force and mass and energy transfers on a closed body

Consider the total force, total moment of force, and total mass and energy transfers on a closed (or bounded) body in a gas. These global quantities can be obtained only by the knowledge of the Grad–Hilbert part as shown below (Sone [1984c]). The body may be the condensed phase of the gas; there may be other bodies in the gas, and the domain may be enclosed by a boundary.

Proposition 3.1. *The force FB_i and the moment MO_i of force (around the origin) acting on a closed body in a gas are expressed by the Grad–Hilbert part P_{ijG} of the stress tensor as follows:*

$$FB_i = -p_0 L^2 \int_S P_{ijG}\widehat{n}_j \mathrm{d}S, \tag{3.85}$$

$$MO_i = -p_0 L^3 \int_S \epsilon_{ilk} x_l P_{kjG}\widehat{n}_j \mathrm{d}S, \tag{3.86}$$

where the surface S of integration is a closed surface enclosing only the body under consideration, $\widehat{n}_i$ is the outward unit normal vector to the surface S, $\mathrm{d}S$ is the surface element (in the nondimensional x_i space) for integration, and ϵ_{ijk} is Eddington's epsilon (Footnote 13 in Section 2.10). The "outward" means outward with respect to the domain enclosed by S.

Proof. The force FB_i and the moment MO_i are given by the same formulas as Eqs. (3.85) and (3.86) with replacement of P_{ijG} by the full stress tensor P_{ij} ($= P_{ijG} + P_{ijK}$) and with the surface S of integration taken on the body [see Eq. (2.3b)].[30] The Grad–Hilbert part P_{ijG} as well as the full stress tensor P_{ij} satisfies the conservation equation (2.86) ($\partial P_{ij}/\partial x_j = 0$ and $\partial P_{ijG}/\partial x_j = 0$). From these relations, we have $\partial \epsilon_{ilk} x_l P_{kj}/\partial x_j = 0$ and $\partial \epsilon_{ilk} x_l P_{kjG}/\partial x_j = 0$. Owing to these solenoidal relations, the surface S of integration of the force and moment formulas with the full stress or with the Grad–Hilbert stress can be transformed continuously in the domain in the gas. The surface S being taken outside the Knudsen layer, the formulas with the full stress and those with the Grad–Hilbert stress agree. ■

Proposition 3.2. *The mass MF and the energy ET transferred to a closed body in a gas per unit time are expressed by the Grad–Hilbert parts u_{iG} and Q_{iG} of the (nondimensional) flow velocity and heat-flow vector as follows:*

$$MF = -\rho_0 (2RT_0)^{1/2} L^2 \int_S u_{iG}\widehat{n}_i \mathrm{d}S, \tag{3.87}$$

$$ET = -p_0 (2RT_0)^{1/2} L^2 \int_S (Q_{iG} + \frac{5}{2}u_{iG})\widehat{n}_i \mathrm{d}S, \tag{3.88}$$

[30] In this section, the Knudsen-layer part means the remainder of the Hilbert part, including the S layer.

where the definitions of S, dS, and $\widehat{n}_i$ are the same as in Proposition 3.1.

Proof. With the aid of the solenoidal relations (2.85) and (2.87) for the full and Grad–Hilbert mass and heat fluxes ($\partial u_i/\partial x_i = 0$, $\partial u_{iG}/\partial x_i = 0$, $\partial Q_i/\partial x_i = 0$, $\partial Q_{iG}/\partial x_i = 0$), the proof goes in the same way as in Proposition 3.1. ■

Lemma. *Let $g(\boldsymbol{x})$ be three times continuously differentiable function in a shell domain containing a closed surface (say, ∂B_0). Then, the following relations hold:*

$$\int_{\partial B_0} \left(\frac{\partial^2 g}{\partial x_i \partial x_j} - \frac{\partial^2 g}{\partial x_k^2}\delta_{ij} \right) \widehat{n}_j dS = 0,$$

$$\int_{\partial B_0} \epsilon_{ihk} x_h \left(\frac{\partial^2 g}{\partial x_k \partial x_j} - \frac{\partial^2 g}{\partial x_l^2}\delta_{kj} \right) \widehat{n}_j dS = 0.$$

Proof. Extend the function $g(\boldsymbol{x})$ to the whole domain enclosed by ∂B_0 without losing its smoothness, and then apply Gauss's divergence theorem to the above integrals. In the process, it is noted that the relation $\partial \epsilon_{ilk} x_l \psi_{kj}/\partial x_j = 0$ holds for any function ψ_{ij} for which $\partial \psi_{ij}/\partial x_j = 0$ and $\psi_{ij} = \psi_{ji}$. ■

As for the contribution of the thermal stress and "pressure stress", we have the following proposition.

Proposition 3.3. *The thermal stress $\partial^2 \tau_G/\partial x_i \partial x_j$ and the "pressure stress" $\partial^2 P_G/\partial x_i \partial x_j$ in the stress tensor P_{ijG} do not contribute to the force and the moment of force acting on a closed body. That is,*

$$\int_{\partial B} \frac{\partial^2 \tau_G}{\partial x_i \partial x_j} \widehat{n}_j dS = 0, \quad \int_{\partial B} \epsilon_{ilk} x_l \frac{\partial^2 \tau_G}{\partial x_k \partial x_j} \widehat{n}_j dS = 0, \tag{3.89a}$$

$$\int_{\partial B} \frac{\partial^2 P_G}{\partial x_i \partial x_j} \widehat{n}_j dS = 0, \quad \int_{\partial B} \epsilon_{ilk} x_l \frac{\partial^2 P_G}{\partial x_k \partial x_j} \widehat{n}_j dS = 0, \tag{3.89b}$$

where ∂B is the body surface.

Proof. From the Stokes set (3.18)–(3.19c) of equations,

$$\partial^2 \tau_G/\partial x_i^2 = 0, \quad \partial^2 P_G/\partial x_i^2 = 0.$$

Then, the tensors in the integrands in Eqs. (3.89a) and (3.89b) are solenoidal [$\partial^3 \tau_G/\partial x_i \partial x_j^2 = 0$, $\partial(\epsilon_{ilk} x_l \partial^2 \tau_G/\partial x_k \partial x_j)/\partial x_j = 0$, etc.]. Thus, the surface ∂B of integration can be deformed continuously in the gas (say, to a surface S inside the gas). With the harmonic property of τ_G and P_G, the four integrals in the statement of Proposition 3.3 are rewritten as

$$\int_S \left(\frac{\partial^2 \tau_G}{\partial x_i \partial x_j} - \frac{\partial^2 \tau_G}{\partial x_k^2}\delta_{ij} \right) \widehat{n}_j dS, \quad \int_S \epsilon_{ihk} x_h \left(\frac{\partial^2 \tau_G}{\partial x_j \partial x_k} - \frac{\partial^2 \tau_G}{\partial x_l^2}\delta_{jk} \right) \widehat{n}_j dS,$$

$$\int_S \left(\frac{\partial^2 P_G}{\partial x_i \partial x_j} - \frac{\partial^2 P_G}{\partial x_k^2}\delta_{ij} \right) \widehat{n}_j dS, \quad \int_S \epsilon_{ihk} x_h \left(\frac{\partial^2 P_G}{\partial x_j \partial x_k} - \frac{\partial^2 P_G}{\partial x_l^2}\delta_{jk} \right) \widehat{n}_j dS.$$

With the aid of the above lemma, these integrals vanish. (A different proof is given in Sone [1984c].) ■

3.9 Viscosity and thermal conductivity

Up to now we have discussed the problem in nondimensional variables and have not given the formulas of viscosity and thermal conductivity [see Eq. (2.15) for their definitions in classical gas dynamics], though we have mentioned them. They are simply obtained by rewriting the formulas for the stress tensor and heat-flow vector in the dimensional variables. That is, the viscosity μ and the thermal conductivity λ are expressed by the mean free path as follows:

$$\mu = \frac{\gamma_1 mRT_0}{B_0} = \frac{\sqrt{\pi}}{2}\frac{\gamma_1 p_0 \ell_0}{(2RT_0)^{1/2}} = \frac{\gamma_1 m(RT_0)^{1/2}}{4\sqrt{\pi}d_m^2}, \tag{3.90}$$

$$\lambda = \frac{5}{2}\frac{\gamma_2 mR(RT_0)}{B_0} = \frac{5\sqrt{\pi}}{4}\frac{\gamma_2 Rp_0\ell_0}{(2RT_0)^{1/2}} = \frac{5\gamma_2 mR(RT_0)^{1/2}}{8\sqrt{\pi}d_m^2}. \tag{3.91}$$

For molecules with infinite range of intermolecular potential, the extended definition of B_0 explained in Footnote 10 in Section 2.9 should be used.[31] From the last formulas of Eqs. (3.90) and (3.91), the viscosity and the thermal conductivity of a hard-sphere gas are both proportional to the square root of the temperature of the gas, since γ_1 and γ_2 are constants independent of T_0. The second formulas in the above equations are applicable to the BKW equation.[32] From Eqs. (3.90) and (3.91), the viscosity and thermal conductivity are related as

$$\frac{5R\mu}{2\lambda} = \frac{\gamma_1}{\gamma_2}. \tag{3.92}$$

The quantity $5R\mu/2\lambda$ is called the Prandtl number Pr. That is,

$$\mathrm{Pr} = \frac{5R\mu}{2\lambda} = \frac{\gamma_1}{\gamma_2}. \tag{3.93}$$

For example,

$$\mathrm{Pr} = 0.66069446 \text{ (hard-sphere)}, \quad \mathrm{Pr} = 1 \text{ (BKW)}. \tag{3.94}$$

From the relation (3.90) of the viscosity and mean free path, the three parameters: Knudsen number Kn, Reynolds number Re, and Mach number Ma,

---

[31]The viscosity μ and the thermal conductivity λ are independent of the choice of the extended B_0, which has some arbitrariness. That is, γ_1 and γ_2 defined in Eq. (3.33) are, respectively, proportional to $B(\zeta)$ and $A(\zeta)$ defined by Eqs. (3.26) and (3.25); $A(\zeta)$ and $B(\zeta)$ are inversely proportional to B_0, since $\widehat{B}$ in the operator $\mathcal{L}$ is so by definition [see Eqs. (2.41c) and (2.58b)].

[32]The viscosity μ and thermal conductivity λ for BKW is further deformed as $\mu = RT/A_c$ and $\lambda = 5R^2T/2A_c$. Thus, they are proportional to the temperature T. Sometimes, A_c is taken as a function of T so as to match the temperature dependence of μ and λ with that of the gas (e.g., a hard-sphere gas) under consideration. However, the collision frequency of a gas generally depends on molecular speed, but it does not for the BKW model (modified or not). Thus, in this choice of A_c, the consistency that holds between the cutoff Maxwell molecule and the BKW model of constant A_c is not maintained.

the latter two of which are defined by $\mathrm{Re} = UL\rho_0\mu^{-1}$ and $\mathrm{Ma} = U(\gamma RT_0)^{-1/2}$ [U: the characteristic flow speed in the system, γ: the specific-heat ratio (5/3 for a monatomic gas)] and are important parameters in the classical gas dynamics, are found not to be independent, i.e.,

$$\mathrm{Ma} = \left(\frac{3\pi}{40}\right)^{1/2} \gamma_1 \,\mathrm{Re}\,\mathrm{Kn}, \tag{3.95}$$

which is sometimes called the von Karman relation (von Karman [1963]).

In the linearized theory, we neglected the nonlinear terms of the perturbed velocity distribution function ϕ [thus, u_i (or the Mach number), the temperature perturbation τ, etc.] but retained the quantities of the order of $\phi\mathrm{Kn}^n$ [thus, $u_i\mathrm{Kn}^n$ (or $\mathrm{Ma}\mathrm{Kn}^n$), $\tau\mathrm{Kn}^n$, etc.], where n is a positive integer. This means that $\mathrm{Ma} \ll \mathrm{Kn}^n$ or $\mathrm{Re} \ll \mathrm{Kn}^n \ll 1$. The Stokes set of equations in Section 3.2 reflects this situation.

3.10 Summary of the asymptotic theory

Solving the Stokes set of equations (3.18)–(3.19c) under the slip boundary conditions on a simple boundary or on an interface given in Section 3.5 from the zero-th order, we obtain the fluid-dynamic parts of the density, flow velocity, and temperature of the gas and, on the basis of this information, their Knudsen-layer corrections from the formulas there. Further, we can obtain the stress tensor and heat-flow vector with the aid of Eqs. (3.30a)–(3.31d) and (3.68)–(3.73) or (3.81)–(3.84). Thus the problem for the steady behavior of a slightly rarefied gas is reduced to solving the Stokes set of equations under the slip boundary condition. That is, a problem for a slightly rarefied gas can be treated with the same ease as the corresponding classical fluid-dynamic problem.

In the linearized problem, which we have treated in this chapter, the fluid-dynamic-type equations (i.e., the Stokes set of equations) remain the same form when we advance the degree of approximation. Thus, the effect of gas rarefaction for the macroscopic variables, such as density, flow velocity, and temperature, enters only through the slip boundary condition. The thermal creep flow and thermal-stress slip flow, which appear, respectively, in the first- and the second-order slip conditions, are flows particular to a rarefied gas and will be discussed in more detail in Sections 3.11.1 and 3.11.2. The boundary conditions in Eqs. (3.75a) and (3.75c) are those to be used in the continuum gas dynamics (or classical gas dynamics) on an interface of a gas and its condensed phase.

3.11 Applications

3.11.1 Thermal creep flow and thermal transpiration

On a simple boundary of nonuniform temperature in a slightly rarefied gas, the slip condition (3.60a) requires a flow proportional to the temperature gradient as the boundary condition of the Stokes set of equations. Thus, a global flow is induced in a gas around a body with nonuniform temperature. The flow, particular to a rarefied gas, has been known for a long time as the thermal creep flow (e.g., Kennard [1938]), but its rigorous formulation and accurate analysis on the basis of the Boltzmann equation are relatively new (Sone [1966b], Ohwada, Sone & Aoki [1989a]). For example, consider a gas ($X_2 > 0$) over a plane wall at $X_2 = 0$ with a uniform temperature gradient $\mathrm{d}T_w/\mathrm{d}X_1$ in the direction of the X_1 axis. Then, from the Stokes set of equations in Section 3.2 and the slip boundary condition in Section 3.5.1, we can easily obtain the velocity and temperature fields as follows:[33]

$$v_1 = -\left(\frac{\pi R T_0}{2}\right)^{1/2} \frac{\ell_0}{T_0}\frac{\mathrm{d}T_w}{\mathrm{d}X_1}\left[K_1 + \frac{1}{2}Y_1(2X_2/\sqrt{\pi}\ell_0)\right], \quad v_2 = v_3 = 0,$$

$$T = \frac{\mathrm{d}T_w}{\mathrm{d}X_1}X_1 + T_0,$$

where T_0 is the temperature at $X_1 = 0$. The flow is in the direction of the temperature gradient of the wall[34] and uniform except in the neighborhood of the plane wall of the order of the mean free path ℓ_0, and the temperature is uniform with respect to X_2. The thermal creep flow is already encountered in Section 3.7, where a flow along the outside of a circular cylinder with a temperature gradient along it and a flow through a circular pipe with a temperature gradient along it are compared in connection with the S layer. The physical mechanism of the thermal creep flow is given as follows.

Consider a gas at rest over a plane wall with a temperature gradient along it. Take a small area $\mathrm{d}S$ on the wall and estimate the momentum transferred to $\mathrm{d}S$ from the gas. The molecules impinging on $\mathrm{d}S$ come from various directions directly (or without molecular collisions) over a distance of the order of the mean free path, keeping the property of their origins. The average speed of molecules arriving from the hotter region is larger than those from the colder region. Thus the momentum transferred to $\mathrm{d}S$ by the molecules impinging on it has a component in the direction opposite to the temperature gradient on the wall. (A little more precisely, the number of molecules in the hotter region is less than that in the colder region, and the mass fluxes from the two regions balance in a gas at rest. The momentum transfer, to which the effect of molecular velocity is amplified, has a component in the direction opposite to the temperature gradient.) The contribution of the molecules leaving the wall to the tangential component of the momentum transfer is nothing in the case of the diffuse reflection. (Generally except in the specular reflection, the velocity distribution function of the molecules leaving the wall, characterized by the wall condition, is qualitatively different from that of the impinging molecules, formed by collisions of surrounding gas molecules. Thus the two contributions are generally different.) Thus, a momentum in the opposite direction to the

[33] One can take $T_0/|\mathrm{d}T_w/\mathrm{d}X_1|$ as the characteristic length L.

[34] This is the case of a hard-sphere gas or the BKW model, for which K_1 is negative.

temperature gradient is transferred to $\mathrm{d}S$ from the gas. As its reaction, the gas is subject to a force in the direction of the temperature gradient, and a flow is induced in that direction. In a gas in motion, on the other hand, a momentum in the direction of motion is transferred to the wall or $\mathrm{d}S$. Thus, a steady flow is established when the two contributions of momentum transfer balance. Obviously from the above explanation, the boundary (or the qualitative difference of the velocity distribution functions of the molecules impinging on the boundary and those leaving there) plays an essential role in inducing the thermal creep flow.

The thermal creep flow is demonstrated and its relative magnitude (with respect to the temperature gradient of the boundary and the pressure of the gas) is measured by simple experiments in Sone [1991b] and Sone, Sawada & Hirano [1994]. In the former experiment, a flow induced around a vertical plate heated at its lower edge is observed by a small windmill. The natural convection (an upward flow), which is generally seen around a heated body at atmospheric condition, is overcome by the thermal creep flow (a downward flow) as the pressure decreases, and the rotation of the windmill reverses at some pressure.

It has been known, as thermal transpiration, for a long time that a flow is induced through a channel or pipe with a temperature gradient along it in the direction of the gradient (see, e.g., Maxwell [1879], Knudsen [1910a], Kennard [1938], Loeb [1961], Sone [2000a]). The flow for small Knudsen numbers through a straight pipe of an arbitrary cross section with a weak temperature gradient is easily obtained with the aid of the asymptotic theory. Its leading term for small Knudsen numbers is the thermal creep flow of a uniform velocity over the cross section with a Knudsen-layer correction in the neighborhood of the pipe wall. The leading term of the mass flux M_T through the pipe is given by

$$M_T = \frac{c_T \ell p S}{T_w \sqrt{RT_w}} \frac{\mathrm{d}T_w}{\mathrm{d}X}, \tag{3.96}$$

where T_w, $\mathrm{d}T_w/\mathrm{d}X$, p, ℓ, S, R, and c_T are, respectively, the temperature of the pipe, its temperature gradient along the pipe, the pressure of the gas, the mean free path of the gas molecules, the cross sectional area of the pipe, the specific gas constant, and a constant depending on the molecular model. The mass flux is proportional to the cross sectional area. This is a good contrast to the flow induced by a pressure gradient along the pipe (or Poiseuille flow). The leading term of the mass flux M_P of the flow for small Knudsen numbers is given by

$$M_P = -\frac{c_P p S^2}{\mu R T_w} \frac{\mathrm{d}p}{\mathrm{d}X}, \tag{3.97}$$

where $\mathrm{d}p/\mathrm{d}X$, μ, and c_P are, respectively, the pressure gradient in the gas, its viscosity, and a constant depending on the shape of the cross section of the pipe. The mass flux is in the opposite direction to the pressure gradient and proportional to the square of the cross sectional area. An application of the difference between the two flows about the dependence on S will be discussed in Section 3.11.6.

3.11.2 Thermal-stress slip flow

Consider a gas over a simple boundary of a uniform temperature, where no thermal creep flow is induced. Then only the $(\partial G_{iG0}/\partial x_j)n_i t_j$ term can be nonzero in the second-order slip condition (3.61a). For uniform boundary temperature, the temperature gradient $\partial \tau_{G0}/\partial x_i$ is normal to the boundary, and $(\partial G_{iG0}/\partial x_j)n_i t_j$ or $-(\partial^2 \tau_{G0}/\partial x_i \partial x_j)n_i t_j$ on the boundary is rewritten as

$$\frac{\partial^2 \tau_{G0}}{\partial x_i \partial x_j} n_i t_j = t_j \frac{\partial}{\partial x_j}\left(n_i \frac{\partial \tau_{G0}}{\partial x_i}\right),$$

because the derivative of n_i along the boundary is parallel to the boundary. When the temperature gradient $n_i \partial \tau_{G0}/\partial x_i$ is not uniform over the boundary or the isothermal surface (τ_{G0} = const) is not parallel to the boundary, a flow is induced in the direction or in the opposite direction to the gradient of the distance of the isothermal surface, depending on whether the temperature of the boundary is lower or higher than that of the surrounding gas.[35] The flow is called the thermal-stress slip flow. Several examples of the flow have been studied (Sone [1971, 1972, 1974], Sone & Tanaka [1980]). One may naturally think that the flow is a kind of thermal creep flow due to the modified temperature field induced by the first-order temperature jump (3.60c); but this is in the opposite direction to the thermal-stress slip flow, and therefore the thermal-stress slip flow is a new type of flow.

For example, consider a gas between two noncoaxial circular cylinders with different uniform temperatures T_1 and T_2 (Sone & Tanaka [1980]). Then, no thermal creep flow is induced because of uniform temperature on each cylinder. The isothermal surfaces being not parallel between the noncoaxial cylinders, the thermal-stress slip flow is induced between the cylinders as shown in Fig. 3.3, where streamlines are shown with arrows indicating the direction of flow when the inner cylinder is heated ($T_1 > T_2$).

As is seen in Section 3.3, the thermal stress $\partial^2 \tau_{G0}/\partial x_i \partial x_j$ appears with advance of approximation [see Eq. (3.30c)]. The average speed of molecules that arrive at a point under consideration depends on the direction from which they come directly (or without molecular collision) over a distance of the order of the mean free path, when the temperature gradient is nonuniform. This introduces a momentum flux resulting in the thermal stress.[36] The stress in the linearized problem does not make any contribution to the Stokes set of equations. Take a small cube in a gas with one face of the cube on an isothermal surface. The thermal stress over the surface of the cube balances by itself because of its absence in the Stokes set, and no flow is induced by the thermal stress. Let the face on the isothermal surface be replaced by a solid wall where the diffuse reflection takes place. Then, the tangential momentum flux through the surface is reduced by half, and the balance of the momentum flux (or thermal stress)

[35] This is the case of a hard-sphere gas or the BKW model, for which a_4 in Eq. (3.61a) is positive.

[36] The form of stress in a two-dimensional case can easily be understood by considering the representative molecules from four different directions.

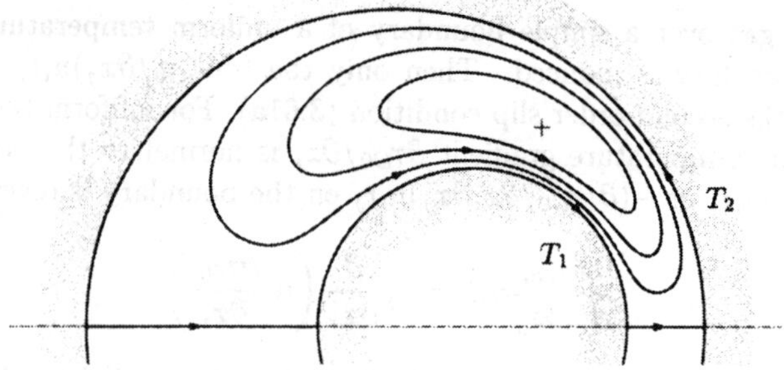

Figure 3.3. Thermal-stress slip flow. The direction of flow is shown by arrows on streamlines when the temperature T_1 of the inner cylinder is higher than the temperature T_2 of the outer cylinder.

over the cube is violated. Thus, the thermal-stress slip flow is induced. As in the case of the thermal creep flow, the boundary (or the qualitative difference of the velocity distribution function of the molecules impinging on the boundary and that of the molecules leaving there) plays an essential role in inducing the thermal-stress slip flow.

Incidentally, the flow between two noncoaxial cylinders is studied for various Knudsen numbers numerically on the basis of the BKW equation in Aoki, Sone & Yano [1989]. The overall flow pattern is similar to that in Fig. 3.3 for intermediate Knudsen numbers, but the difference of local velocity profile shows the shift of the region of acceleration of the flow. The flow vanishes in the limit Kn$\to\infty$. The disappearance of flow induced by a temperature field in the limit Kn$\to\infty$ is proved in a general situation in Sone [1984a, 1985].

3.11.3 Nonlinear thermal-stress flow

In the framework of the linearized theory, the thermal creep flow and the thermal-stress slip flow are the only flows, up to the second order of the Knudsen number, induced by a temperature field. The former is of the first order of the Knudsen number and the latter is of the second order. When the temperature variation in a gas is not small and its nonlinear terms are not negligible, the situation is different (see Section 5.6.1 for a detailed discussion). That is, the stress tensor contains nonlinear terms of the temperature variation, and the nonlinear thermal stress does not balance by itself over a control surface in the gas in contrast to the linearized problem, and as a result a flow is induced when the isothermal surfaces are not parallel. The flow was found by Kogan, Galkin & Fridlender [1976] and is called the nonlinear thermal-stress flow. It is of the first order of the Knudsen number. In this flow, in contrast to the thermal creep and thermal-stress slip flows, the boundary plays only an indirect role to form such a temperature field.

In a gas between two noncoaxial circular cylinders with different uniform temperatures considered in the preceding subsection (Section 3.11.2), the nonlinear thermal-stress flow is also induced if the temperature difference between the cylinders is so large that nonlinear terms of temperature gradients are not negligible. Its flow pattern, including the direction of flow, is similar to that in Fig. 3.3; the detailed flow structure is, of course, different because the driving force is strong where the isothermal surfaces are dense and strongly nonuniform, whereas in the thermal-stress slip flow the driving force is on the boundary.

3.11.4 Thermal edge flow

We have discussed flows induced by temperature fields for small Knudsen numbers on the basis of the asymptotic theory, and found the thermal creep flow and the nonlinear thermal-stress flow at the first order of the Knudsen number and the thermal-stress slip flow at the second order of the Knudsen number.[37] In the asymptotic analysis developed in this chapter or to be developed in the following chapters, we assumed or assume that the shape of the boundary is smooth, i.e., the Knudsen number based on the radius of curvature of the boundary is small. Thus the case where the boundary has a sharp edge is excluded. Here we will examine the possibility of a temperature-induced flow due to the sharp edge of the boundary.

Consider a uniformly heated plate in a gas, where the mean free path is much smaller than the size of the plate. In the neighborhood of the plate far away (on the scale of the mean free path) from its edges, the temperature of the gas is uniform along the plate, and the isothermal surfaces are parallel. There is no cause of a flow along the plate. Near the edges of the plate, however, the isothermal surfaces are sharply curved. The temperatures T_A at A and T_B at B in Fig. 3.4, where A and B are at about one mean free path ℓ_1 away from one of the edges, are, respectively, roughly estimated to be $T_1[1-(\alpha_0+\alpha_1)\ell_1^{1/2}]$ and $T_1[1-(\alpha_0-\alpha_1)\ell_1^{1/2}]$, where T_1 is the temperature of the plate, and α_0 and α_1 are some constants ($\alpha_0>\alpha_1>0$ for a heated plate).[38] The character of the group of molecules impinging on the edge region is roughly represented by the combination of two groups: one from A and the other from B. The situation is similar to that over a plane wall with the temperature gradient $(T_B-T_A)/\sqrt{2}\ell_1 \sim \alpha_1 T_1 \ell_1^{-1/2}$, where the thermal creep flow is induced. Thus a flow proportional to the square root of the mean free path ($\sim \ell_1^{1/2}$) is induced near the edge, and the flow direction is from A to B. In fact, the flow is found numerically (Aoki, Sone & Masukawa [1995]) and experimentally (Sone & Yoshimoto [1997]). In Aoki, Sone & Masukawa [1995], a rarefied gas around a uniformly heated (or cooled) plate in a rectangular vessel is considered in the absence of an external force (Fig. 3.5a), and a steady flow induced in the

[37] Only a brief outline of nonlinear thermal-stress flow is given in the preceding subsection, since the reader has not yet been prepared. A complete discussion will be given in Section 5.6.1.

[38] The temperature field is estimated by the solution of the heat-conduction equation around the edge, although it is not guaranteed that the equation describes the field exactly even in the continuum limit (see Section 5.6).

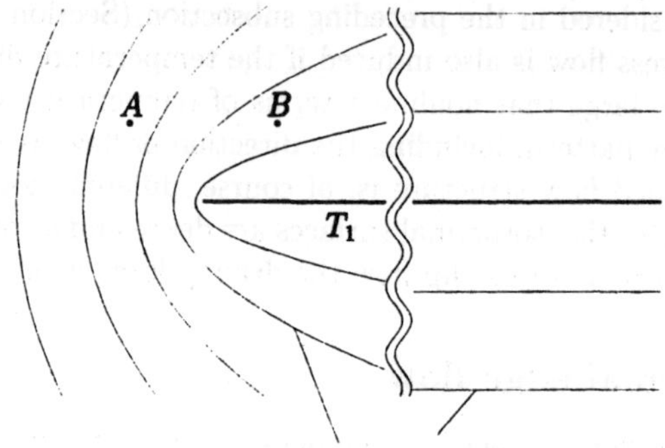

Figure 3.4. The temperature field near a uniformly heated plate. The points A and B are about one mean free path away from the edge of the plate. Far away (on the scale of the mean free path) from the edge of the plate, the temperature of the gas is uniform along the plate. Near the edge, on the other hand, the isothermal surfaces are sharply curved, and thus the temperature is nonuniform along the plate. The molecules impinging on the edge region from the left side is roughly represented by those from A and the molecules from the right side by those from B, where the gas is hotter than at A when the plate is heated. The situation is similar to that over a nonuniformly heated plate, where the thermal creep flow is induced.

gas is investigated for a wide range of Knudsen number numerically by the standard direct-simulation Monte-Carlo method (e.g., Bird [1994], Sone & Aoki [1994]). The computation is done for a hard-sphere gas and the diffuse-reflection boundary condition. The flow field of a quarter region in the vessel is shown in Figs. 3.5b.1–b.4.[39] The Knudsen number in these figures is defined by the mean free path at the average density over the domain of the gas and the plate width as the reference length. The flow is strongest at an intermediate Knudsen number. As the Knudsen number becomes smaller, the flow becomes weaker on the whole and tends to be localized near the edge of the plate. The localized flow remains fairly strong even for small Knudsen numbers.

The experiment (Sone & Yoshimoto [1997]) is performed in a cylindrical vacuum chamber (250-mm diameter, 300-mm height) of a glass bell jar on a steel base, where the pressure can be controlled from the atmospheric pressure down to several pascals (Fig. 3.6). A simple rectangular copper plate (about 100×200 mm) of 0.2-mm thickness with heaters along the shorter edges is set vertically in the chamber, with the longer edge vertical or horizontal. The flows induced around the longer edges are detected by small windmills (5- or 7-mm radius) made of plastic film (Fig. 3.7). In the following experiments, the temperature of the plate is about 44 °C when heated and the room temperature is around 20 °C. The uniformity of the temperature over the plate is kept with its maximum difference 1–3 °C. According to the experiment in Sone, Sawada

[39]This excerpt is from Sone & Yoshimoto [1997] for the convenience of comparison with an experiment explained later.

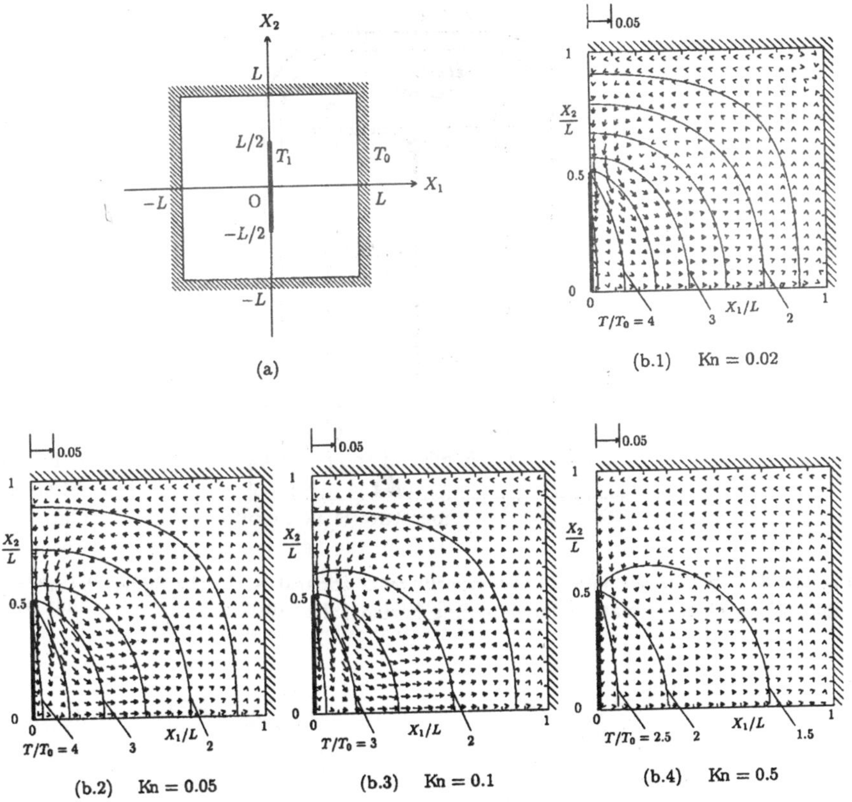

Figure 3.5. Flows induced around an edge of a heated plane plate at temperature T_1 in a rectangular box at T_0. (a) Geometry and (b.1)–(b.4) Flow patterns ($T_1/T_0 = 5$). The arrows indicate the flow velocity v_i at their starting points; the scale 0.05 of $(v_i^2)^{1/2}/(2RT_0)^{1/2}$ is shown by the arrow on the left shoulder of each figure. The solid lines — indicate isothermal lines; the difference of T/T_0 between neighboring isothermal lines is 0.5.

& Hirano [1994], this difference is too small for the thermal creep flow to be observed by the windmill.

In the first experiment with the longer edges horizontal, vertical flows around the two horizontal edges are observed. The positions of the windmill are shown in Fig. 3.8. At the atmospheric pressure, where only upward natural convection is expected, the windmill rotates 30 rpm in the direction caused by an upward flow at position A in Fig. 3.8 and is at rest at position B. The upward flow at A is attributed to natural convection. The speed of rotation (rpm) of the windmill at pressure 10 Pa is shown in Fig. 3.8, where the rotation caused by an upward flow is taken positive. The flows at positions A, a, and b in Fig. 3.8 are downwards, opposite from natural convection. At 5 Pa, the speed of rotation of the windmill is −65 rpm at position a, where it is −25 rpm at 10 Pa. Naturally,

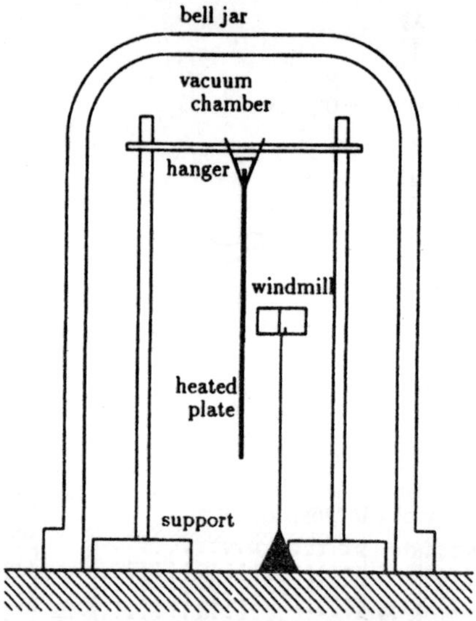

Figure 3.6. Schematic view of the experimental apparatus. In the figure the set of a windmill and heated plate is for detecting horizontal wind.

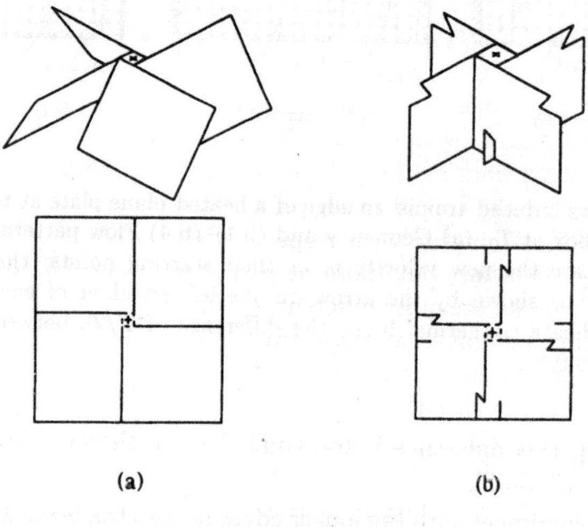

Figure 3.7. The vanes of windmills. (a) Vanes detecting a vertical wind; (b) Vanes detecting a horizontal wind. The vanes are made of a square sheet of transparent plastic film. The size of the square is 14 × 14 mm in panel (a) and 10 × 10 mm in panel (b).

when the plate is unheated, the windmill stands still at all positions for any pressure. Incidentally, the mean free path of air at room temperature is about 0.65 mm at 10 Pa.

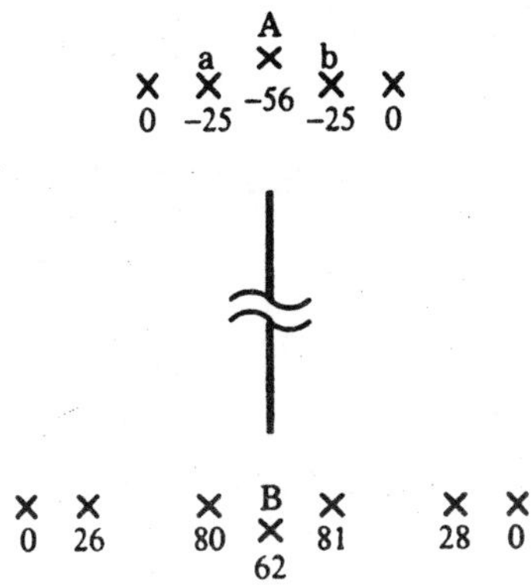

Figure 3.8. The side view of the position of the windmill in the detection of vertical flows around the heated plate and the speed of rotation (rpm) of the windmill at 10 Pa. The windmill is on the vertical plane passing the plate perpendicularly at the middle point along its horizontal edge. The position of the windmill on the plane is marked by ×, and the letters a, b, etc. above a cross are the symbols for reference in the text; the crosses are distributed on lattice points of the lattice size 10 mm. They are shown in proportion to the real geometry. The speed of rotation (rpm) of the windmill at 10 Pa is shown below the cross, where the positive sign corresponds to an upward flow.

In the second experiment with the longer edges vertical, horizontal flows around the vertical edges are observed. The speeds of rotation of the windmill at 10 Pa at various points in the gas are shown in Fig. 3.9, where clockwise rotation viewed from the top is taken positive. The data at positions (e.g., a and d in Fig. 3.9) symmetric with respect to the vertical center line of the plate are roughly the same, but those at positions (e.g., a and b) symmetric with respect to the plate are quite different. This difference does not mean that the flow speed differs so much at these two points, because the configuration of the vanes of the windmill is not symmetric with respect to the direction of rotation. In fact, the speeds of rotation of the vanes antisymmetric to the original ones are, respectively, 50 and -146 rpm at the symmetric positions a and b in Fig. 3.9, where 113 and -55 rpm are obtained with the original vanes. For other pressures, the result is as follows. At atmospheric pressure the windmill stands still at all positions; at 5 and 20 Pa, the speeds are, respectively, 223 and 53 rpm at position a, where the speed is 113 rpm at 10 Pa. Naturally, when the plate is unheated, the windmill stands still at all positions for any pressure.

The flows shown in the numerical computation and the experiment are compatible with the flow explained at the beginning of this subsection. We will call the flow the thermal edge flow. One may find a similarity of the thermal edge flow to the thermal-stress slip flow or the nonlinear thermal-stress flow between two noncoaxial circular cylinders explained in Sections 3.11.2 and 3.11.3, but the direction of flow is opposite.

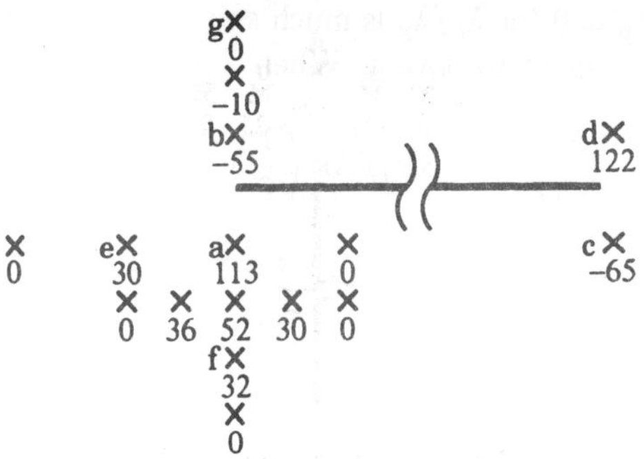

Figure 3.9. The plan view of the position of the windmill in the detection of horizontal flows around the heated plate and the speed of rotation (rpm) of the windmill at 10 Pa. The height of the windmill is at the middle point of the plate. The horizontal position of the windmill is marked by ×, and the letters a, b, etc. to the left of a cross are the symbols for reference in the text; the crosses are distributed on lattice points of the lattice size 10 mm. They are shown in proportion to the real geometry. The speed of rotation (rpm) of the windmill at 10 Pa is shown below the cross, where clockwise rotation viewed from the top is taken positive.

3.11.5 Thermophoresis

Consider a spherical particle of radius L in a slightly rarefied gas at rest and at a uniform pressure p_0; there is a small temperature gradient [say, $(\partial T/\partial X_1)_\infty$] in the gas, that is, the temperature T is given by $T = T_0 + (\partial T/\partial X_1)_\infty X_1$ and $(L/T_0)(\partial T/\partial X_1)_\infty \ll 1$. The particle is assumed to be a uniform solid particle, and the heat flux in the particle is given by Fourier's law of uniform heat conductivity λ_p; thus, the temperature field in the particle is determined by the Laplace equation. The steady behavior of the gas disturbed by the presence of the particle is of our interest. The gas being slightly rarefied and the temperature gradient being small, the asymptotic theory in this chapter can be applied to the analysis of the behavior of the gas. On the boundary, in addition to the slip boundary condition in Section 3.5.1, the law of conservation of energy flux through the surface of the particle should be satisfied. In this subsection, the thermal conductivity of the gas is denoted by λ_g with the subscript g for clearness.

Here we give only the results for the nontrivial leading-order terms in the Knudsen number (or k) of the temperature and velocity fields without derivation (see Sone [1972], Sone & Aoki [1977b, 1981], Ohwada & Sone [1992]). The velocity field of the order of unity vanishes. For convenience of description of the results, we introduce the spherical coordinate system (r, θ, φ) with its origin at the center of the spherical particle and with the half line $\theta = 0$ along the positive X_1 axis. The r, θ, and φ components of the flow velocity are denoted, respectively, by v_r, v_θ, and v_φ. For simplicity of expression, we give the results separately for the two cases: the ratio $\lambda_g/\lambda_p = O(1)$ (or λ_g/λ_p is much larger than k) and $\lambda_g/\lambda_p = 0$ (or λ_g/λ_p is much smaller than k), although the unified expression is not difficult to obtain. When $\lambda_g/\lambda_p = O(1)$,[40]

$$\frac{v_r}{\sqrt{2RT_0}} = -3K_1\Gamma\Lambda k\left(\frac{1}{\hat{r}} - \frac{1}{\hat{r}^3}\right)\cos\theta,$$

$$\frac{v_\theta}{\sqrt{2RT_0}} = \frac{3}{2}\Gamma\Lambda k\left[K_1\left(\frac{1}{\hat{r}} + \frac{1}{\hat{r}^3}\right) + Y_1((\hat{r}-1)k^{-1})\right]\sin\theta,$$

$$v_\varphi = 0,$$

$$\frac{p}{p_0} - 1 = -3K_1\gamma_1\Gamma\Lambda\frac{k^2}{\hat{r}^2}\cos\theta,$$

$$\frac{T}{T_0} - 1 = \Gamma\left(\hat{r} + \frac{3\Lambda - 1}{\hat{r}^2}\right)\cos\theta \quad (\hat{r} > 1),$$

$$\frac{T}{T_0} - 1 = 3\Gamma\Lambda\hat{r}\cos\theta \quad (\hat{r} \le 1),$$

where

$$\hat{r} = r/L, \quad \Gamma = \frac{L}{T_0}\left(\frac{\partial T}{\partial X_1}\right)_\infty, \quad \Lambda = \frac{\lambda_g/\lambda_p}{2\lambda_g/\lambda_p + 1}.$$

From the above flow field, the force F_{Ti} acting on the particle is obtained as

$$F_{T1} = \frac{48\pi K_1}{5}\frac{\Pr L^2}{\sqrt{2RT_0}}\left(\frac{\lambda_g/\lambda_p}{2\lambda_g/\lambda_p + 1}\right)\lambda_g\left(\frac{\partial T}{\partial X_1}\right)_\infty k, \quad F_{T2} = F_{T3} = 0.$$

When $\lambda_g/\lambda_p = 0$,[41]

$$\frac{v_r}{\sqrt{2RT_0}} = -3a_4\Gamma k^2\left(\frac{1}{\hat{r}} - \frac{1}{\hat{r}^3}\right)\cos\theta,$$

$$\frac{v_\theta}{\sqrt{2RT_0}} = \frac{3}{2}\Gamma k^2\left[a_4\left(\frac{1}{\hat{r}} + \frac{1}{\hat{r}^3}\right) + 2Y_{a4}((\hat{r}-1)k^{-1})\right]\sin\theta,$$

$$v_\varphi = 0,$$

$$\frac{p}{p_0} - 1 = -3a_4\gamma_1\Gamma k^3\frac{1}{\hat{r}^2}\cos\theta,$$

$$\frac{T}{T_0} - 1 = \Gamma\left(\hat{r} - \frac{1}{\hat{r}^2}\right)\cos\theta \quad (\hat{r} > 1),$$

$$\frac{T}{T_0} - 1 = 0, \quad (\hat{r} \le 1).$$

The force F_{Ti} acting on the particle is given by

$$F_{T1} = \frac{48\pi a_4}{5}\frac{\Pr L^2}{\sqrt{2RT_0}}\lambda_g\left(\frac{\partial T}{\partial X_1}\right)_\infty k^2, \quad F_{T2} = F_{T3} = 0.$$

[40] For p, the Knudsen-layer correction is not shown.
[41] See Footnote 40.

When the thermal conductivity of the gas is comparable to that of the particle $[\lambda_g/\lambda_p = O(1)]$, the part of the sphere facing the hotter side of the gas is heated, and the part facing the colder side is cooled. Therefore, a thermal creep flow, of the first order of k, is induced from the colder side to the hotter. As a result, the particle is subject to a force in the direction opposite to the imposed temperature gradient. On the other hand, when the thermal conductivity of the gas is negligible to that of the particle $(\lambda_g/\lambda_p = 0)$, the temperature of the particle is uniform and a thermal creep flow is not induced, but a thermal-stress slip flow, of the second order of k, is induced in the direction from the hotter region to the colder. As a result, the particle is subject to a force in the direction of the imposed temperature gradient. If the particle is set free in the gas, it drifts in the direction or in the opposite direction to the imposed temperature gradient depending on $\lambda_g/\lambda_p = 0$ or $\lambda_g/\lambda_p = O(1)$. This phenomenon, which is of interest in aerosol science, is called thermophoresis,[42] and the force acting on the particle is called thermal force. Thermophoresis of a spherical particle of arbitrary thermal conductivity is studied for the whole range of the Knudsen number in Takata & Sone [1995]. Various references are found in Loyalka [1983], Bakanov [1991], Kogan [1992], Sone [2000a]. Interesting results of nonlinear effects, which are not discussed here, are found in Kogan [1992].

3.11.6 Knudsen compressor

Engineering application, to such as a pumping system or flow control, of a flow induced by a temperature field in a rarefied gas, such as thermal transpiration (Section 3.11.1), has been of interest for a long time because of the absence of a mechanical (or moving) part, and more attention has been paid anew to it recently, for example, in relation to micromechanical systems (Knudsen [1910a], [1910b], Huber [1995], Pham-van-Diep, Keeley, Muntz & Weaver [1995], Sone, Waniguchi & Aoki [1996], Vargo & Muntz [1997], Hudson & Bartel [1999], Sone & Sato [2000], Aoki, Sone, Takata, Takahashi & Bird [2001], Sone, Fukuda, Hokazono & Sugimoto [2001] etc.). The basic idea of the pumping system is as follows. When two reservoirs are joined by a pipe with a temperature gradient, a flow is induced in the direction of the temperature gradient of the pipe, and a gas is pumped from the reservoir of lower temperature into the reservoir of higher temperature. With increase of the pressure in the reservoir of higher temperature, an inverse flow is driven by the pressure difference of the two reservoirs. The two flows are counterbalanced at some pressure difference and the pumping flow vanishes. In order to obtain a stronger pumping effect (a larger flow speed at a given pressure difference or a larger pressure difference maintained between the two reservoirs at a given pumping state), we have to make the thermal transpiration stronger or the pressure-driven flow weaker, that is, to make the temperature gradient larger or the pipe longer. In either way, the temperature difference between the two reservoirs should be kept larger. This is not practical. Thus, some cascade system should be devised.

[42]The case where a particle drifts in the direction of the imposed temperature gradient is,

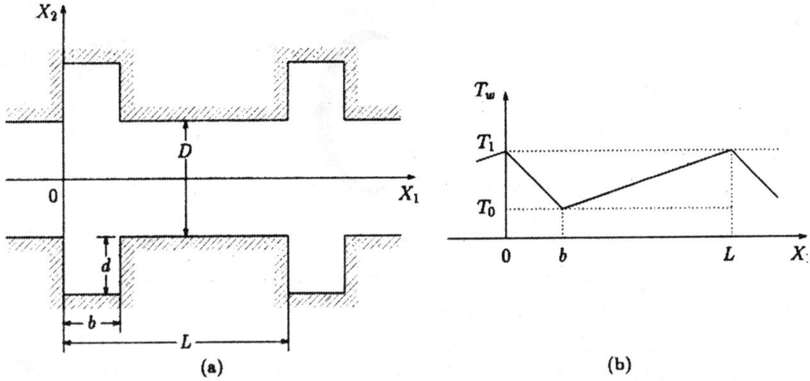

Figure 3.10. Channel configuration and its wall temperature distribution.

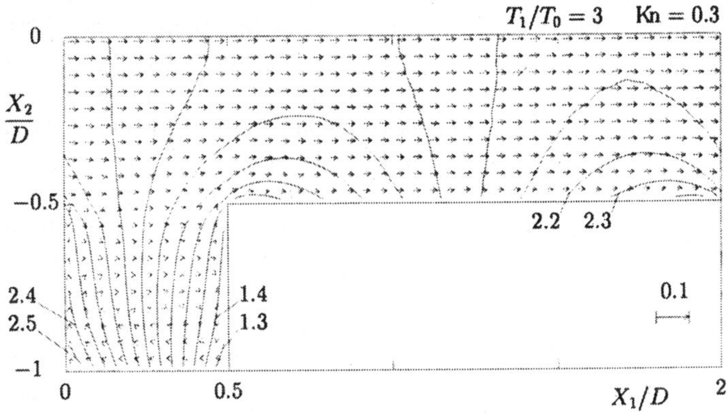

Figure 3.11. One-way flow induced in the channel with ditches ($L/D = 2, b/D = 1/2$, $d/D = 1/2$, $T_1/T_0 = 3$, Kn $= 0.3$). The arrows indicate the flow velocity v_i at their starting points; the scale 0.1 of $(v_i^2)^{1/2}/(2RT_0)^{1/2}$ is shown by the arrow near the right lower corner in the figure. The solid curves indicate the isothermal lines $T/T_0 = 1.3$, $1.4, \ldots, 2.5$. Here the Knudsen number Kn is defined by Kn $= \ell_{av}/D$, where ℓ_{av} is the mean free path of the equilibrium state at rest with temperature T_0 and the average density in the channel.

One of the basic problems of the cascade system is to investigate the possibility of a one-way flow through an infinite pipe under a periodic condition along the pipe or through a finite pipe with its two ends at an equal condition (equal temperature and pressure). In the case of a straight pipe with uniform cross section, a one-way flow cannot be induced by devising the temperature distribution along the pipe. This is shown mathematically by Golse [unpub-

---

specifically, called negative thermophoresis.

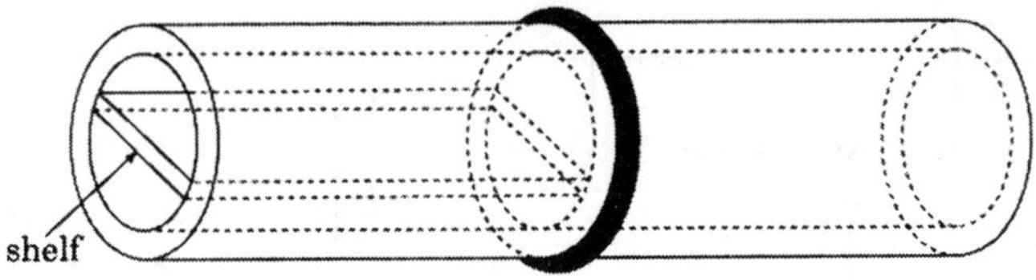

Figure 3.12. A pipe with a shelf put up partially.

lished] (see Appendix A.4), numerically in Sone, Waniguchi & Aoki [1996] and Aoki, Sone, Takata, Takahashi & Bird [2001], and experimentally in Sone & Sato [2000].[43] In these numerical and experimental works, one-way flows are shown to be induced by devising the shape of a pipe. That is, in Sone, Waniguchi & Aoki [1996], a rarefied gas in a channel between two parallel walls with ditches dug periodically [Fig. 3.10 (a)] and with a saw-like temperature distribution in harmony with the ditches [Fig. 3.10 (b)] is considered, and the periodic solution of the system is obtained numerically by the direct-simulation Monte Carlo method (e.g., Bird [1994], Sone & Aoki [1994]) for various sets of the parameters of the system. Thus, a one-way flow is shown to be induced without average pressure and temperature gradients along it (e.g., Fig. 3.11). A flow is induced in the $-X_1$ direction over the bottom of the ditches and a flow is induced in the X_1 direction over the rest (or main part) of the walls. Some of the momentum transmitted or to be transmitted to the gas from the bottom wall of the ditches in the $-X_1$ direction is absorbed by the side wall of the ditches, and a one-way flow is established in the X_1 direction as in the figure. In Aoki, Sone, Takata, Takahashi & Bird [2001], the problem is extended to the case of a circular pipe with ditches. One-way flow is more effectively induced in the channel than in the pipe. An experiment related to this circular pipe problem is carried out by Sone & Sato [2000], where one-way flows are demonstrated to be induced in one period of the pipe with ditches with its two ends at an equal condition (pressure and temperature). Here, we introduce another simple way to induce a one-way flow and estimate the mass flux for small Knudsen numbers with the aid of the asymptotic theory.

Consider a straight uniform pipe with a heater at the middle position along the pipe. A shelf or shelves are put up in a half part of the pipe on one side of the heater (Part S, say), where the passage of the pipe is divided into several passages by the shelf or shelves (Fig. 3.12), and no shelf in the other half of the pipe (Part N, say). In each part of the pipe, thermal transpiration flow is induced towards the heater. The mass fluxes of the two flows are equal, because

[43]The examination is not complete: The mathematical proof is done on the basis of the linearized Boltzmann equation, and thus, a one-way flow due to a nonlinear effect is not excluded; the numerical computation and experiment are done only for limited cases, though they are done in a nonlinear region.

the mass flux of thermal transpiration is proportional to the cross sectional area of the passage [see Eq. (3.96)].[44] Encounter of the two flows increases the pressure of the middle part of the pipe, and a flow away from the heater is driven in each part of the pipe by the pressure gradient induced. The mass flux through Part S is smaller than that through Part N because the flux is proportional to the square of the cross sectional area [see Eq. (3.97)]. Thus, a one-way flow is induced from Part S to Part N. To estimate the flow simply, we take a rectangular pipe with Part S divided into $n \times n$ equal passages.

Let the common temperature at the two ends of the pipe, the temperature at the heater, the common pressure of the gas at its two ends, the pressure of the gas at the heater, and the half length of the pipe (or the length of Part S or N) be, respectively, denoted by T_O, T_{HT}, p_0, p_{HT}, and L. From Eqs. (3.96) and (3.97), the mass flux M_S through Part S is given by

$$M_S = AS\frac{(T_{HT} - T_O)}{L} - n^2 B\left(\frac{S}{n^2}\right)^2 \frac{(p_{HT} - p_0)}{L}, \tag{3.98}$$

where $A = c_T \ell p_0 T_O^{-1}(RT_O)^{-1/2} = 2\sqrt{2/\pi}c_T\mu_O/\gamma_1 T_O$ [μ_O: the viscosity at temperature T_O; see Eq. (3.90)], $B = c_P p_0/\mu_O R T_O$, and the mass flux is taken to be positive for the flux from Part S to Part N. The mass flux M_N in Part N is given by

$$M_N = -AS\frac{(T_{HT} - T_O)}{L} + BS^2\frac{(p_{HT} - p_0)}{L}. \tag{3.99}$$

The two fluxes should be equal ($M_S = M_N = M$; M : the mass flux through the pipe) because of mass conservation. Therefore, the mass flux through the pipe M is given by

$$M = \left(\frac{n^2 - 1}{n^2 + 1}\right)\frac{AS(T_{HT} - T_O)}{L}. \tag{3.100}$$

The $(n^2 - 1)/(n^2 + 1)$ in Eq. (3.100) is the ratio of the mass flux in the present system to that in the system without Part N. That is, the fraction $2/(n^2 + 1)$ of the mass flux in the latter system is reduced by joining the inverse-temperature-gradient Part N to keep the two ends of pipe at an equal temperature. For more shelves, one-way flow is more effectively induced, but the factor fairly quickly approaches unity (e.g., it is $15/17$ for $n = 4$). Too many shelves do not increase the flux effectively.

Another problem of the cascade system is to investigate the pressure difference (or ratio) induced between the two ends of a pipe by blocking a one-way flow through the pipe, keeping the temperature of the two ends equal. The problem (or the pressure problem) was studied as early as 1910 by Knudsen [1910b]. In Sone, Waniguchi & Aoki [1996] and Aoki, Sone, Takata, Takahashi & Bird [2001], the pressure problem for the system of channel or pipe with ditches also is studied. In contrast to the one-way flow, the pressure difference is more effectively induced in the pipe than in the channel.

[44] The Knudsen number based on the size of the divided passage is also assumed to be small.

The pressure problem for the S+N system with its ends closed can be estimated in a similar way to the one-way flow. When the mass flux is blocked ($M_S = 0$ and $M_N = 0$), the pressure gradient induced in each part of the pipe is, according to Eqs. (3.98) and (3.99), given by

$$\frac{(p_{HT} - p_0)}{L} = \frac{A}{B}\frac{n^2}{S}\frac{(T_{HT} - T_O)}{L} \qquad \text{(in Part S)},$$

$$\frac{(p_M - p_{HT})}{L} = -\frac{A}{B}\frac{1}{S}\frac{(T_{HT} - T_O)}{L} \qquad \text{(in Part N)},$$

where p_0 and p_M are, respectively, the pressures at the ends of Part S and Part N at temperature T_O. Thus, the difference $p_M - p_0$ of the pressures at the two ends of the pipe is

$$p_M - p_0 = \frac{A}{B}\frac{(n^2 - 1)}{S}(T_{HT} - T_O), \tag{3.101}$$

$$\frac{A}{B} = \frac{2\sqrt{2}c_T R \mu_O^2}{\sqrt{\pi} c_P \gamma_1}\frac{1}{p_0}.$$

The pressure difference increases indefinitely with the number of the shelves. The effect of the shelf appears quite differently in the one-way flow and in the pressure difference. The flat plane shelves parallel to the pipe axis and their negligible thickness do not affect the thermal transpiration flow, but the pressure-driven flow is reduced to vanish by their resistance with their number. Thus, the pressure difference required to block the thermal transpiration in Part S increases indefinitely.

When the pressure p_1 at the end of Part N is less than the value p_M given by Eq. (3.101), the gas is pumped from Part S to Part N. Let the mass flux at this condition be m_F. Then, by a similar discussion, the relation between p_1 and m_F is given as

$$\begin{aligned} \frac{m_F}{M} &= 1 - \frac{p_1 - p_0}{p_M - p_0} \\ &= 1 - C p_0 (p_1 - p_0), \end{aligned} \tag{3.102}$$

$$C = \frac{\sqrt{\pi} c_P \gamma_1 S}{2\sqrt{2} c_T R (n^2 - 1) \mu_O^2 (T_{HT} - T_O)}.$$

This is the relation in a general operating pumping state.

Consider a cascade system consisting of equal pipe units made of Part N and Part S studied above. Let the pressures of the entrance and exit of the s-th stage be, respectively, p_{s-1} and p_s. From Eq. (3.102), the relation between p_s and p_{s-1} is given by

$$p_s = p_{s-1} + \frac{1}{C}\left(1 - \frac{m_F}{M}\right)\frac{1}{p_{s-1}}, \tag{3.103}$$

where the quantities M and C, as well as m_F, are independent of the stage of the units, since the temperature is periodic. This formula is applicable only when the variation of pressure over one stage, i.e., $(1 - m_F/M)/Cp_{s-1}$, is small.

The above estimates are rough approximations: The discussion is based on the formulas for an infinitely long pipe with small temperature and pressure gradients, and the end effects at the joints of Part S and Part N, as well as at the system ends, are neglected; the passages cannot generally be divided into passages with equal cross section, and thus the shape factor c_P is not common; the formulas of mass fluxes are the leading contributions for a slightly rarefied gas. For many shelves, the Knudsen number based on the width of each passage in Part S can be moderate or large, and more general formulas should be used when the Knudsen number in Part N is small but finite.[45] Thus, the results (3.100)–(3.103) have a limited application: For example, besides small Knudsen numbers in Part S, the temperature variation $(T_{HT} - T_O)/T_O$ and the pressure increase over a unit, i.e., $(1 - m_F/M)/Cp_{s-1}$, should be small, though the pressure variation over many units may be large, and each pipe unit should be slender. However, the discussion clarifies the mechanism of induction of one-way flow and pressure difference.

The one-way flow in the above system (or a pipe with a shelf or shelves put up partially) is easily demonstrated experimentally in the following way (Sone, Fukuda, Hokazono & Sugimoto [2001]). The pipe systems shown in Fig. 3.13 are prepared. The inner diameters of the pipes are 11.5 mm; the length of Part S or N is 65 mm. A system is put in the vacuum chamber shown in Fig. 3.6 in Section 3.11.4, and a flow induced through the pipe is observed by windmills set at the entrances of the pipe. The speeds of rotation of the windmills are shown in Fig. 3.14 when the temperature of the pipe at the heater is 95 °C and that at the entrances is 28 °C. The direction of the flow is from Part S to Part N. Naturally no one-way flow is induced in the symmetric system.

Recently, a cascade pumping system consisting of ten revised units of the above experiment was devised and an experiment was carried out to obtain the data of the pumping effect of the system (the maximum pressure ratio attained between two reservoirs joined with the system and the mass flux versus the pressure ratio between the reservoirs) (Sone & Sugimoto [2001]). The Part N is a pipe of inner diameter 15 mm and length 15 mm, and the Part S is a bundle of 18 pipes of inner diameter 1.6 mm and length 15 mm (Fig. 3.15).[46] The unit makes use of two effects: the effects of variation of the cross section and division of the passage. The system is joined with two reservoirs: one is a vacuum chamber of the bell jar used in the experiment in Section 3.11.4, where the pumping system is put and the pressure is kept constant, and the other is a steel tank of 8×10^6 mm^3 (Fig. 3.16). First the external rotary vacuum pump is operated to keep the whole system (the bell jar, the pumping system, and the

[45]When the Knudsen number of Part N is not small, the difference of the Knudsen number is important even for $n = 2$. The flow pattern and mass flux through a pipe for finite Knudsen numbers are discussed for various cross sections in Sone & Hasegawa [1987] and Hasegawa & Sone [1988]. A software package on Windows 95, 98, and 2000 by which the velocity profiles and mass fluxes of thermal transpiration and Poiseuille flow through a channel between two parallel plates or through a circular pipe at an arbitrary Knudsen number are given in seconds is available (Sone, Itakura & Handa [1998]).

[46]The estimate obtained, with the aid of the references given in Footnote 45, by generalizing the foregoing discussion leading to Eq. (3.103) is useful for the design of the system.

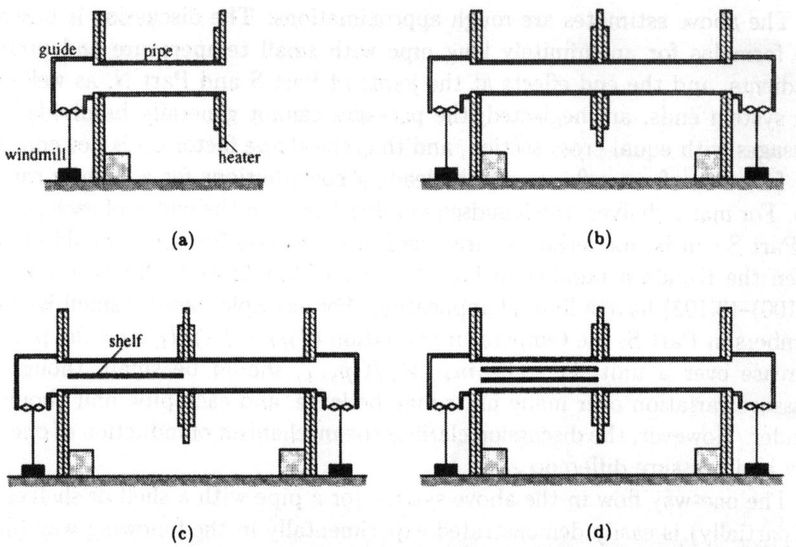

Figure 3.13. The pipe systems. (a) System only of Part N, (b) Symmetric system, (c) System of Part S with one shelf, and (d) System of Part S with two shelves.

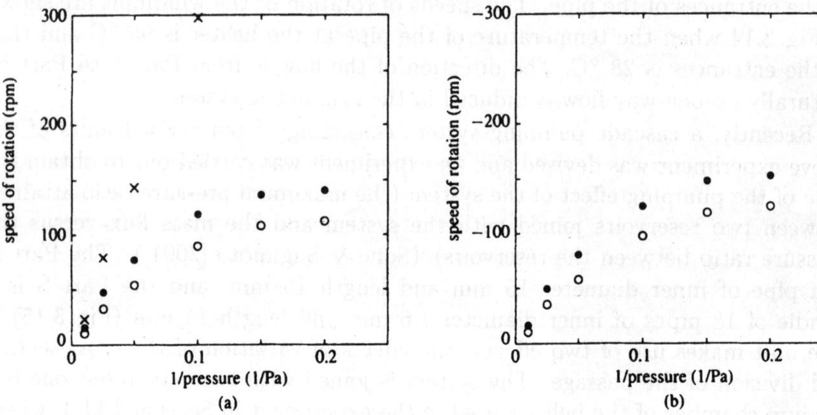

Figure 3.14. The speed of rotation of the windmills. (a) The windmill on the left side in Fig. 3.13 and (b) the windmill on the right side in Fig. 3.13. The symbol × indicates the data of the system (a) in Fig. 3.13, the symbol ∘ those of the system (c), and the symbol • those of the system (d). The positive value corresponds to an upward flow and the negative to a downward flow.

tank) at a desired constant pressure. Keeping the bell jar at this pressure, we put on the heater and observe the pressure variation in the tank. During the initial state before heating, the pressure in the tank is higher than that in the bell jar by $0 \sim 0.3$ Pa, which may be attributed to desorption of a gas from the

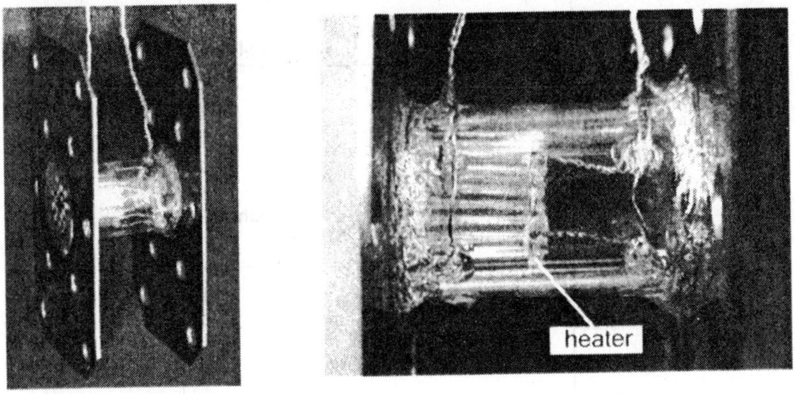

(a) (b)

Figure 3.15. A unit of a pipe (Part S + Part N). (a) Overall view and (b) closer view. The plates at both ends of the pipe, copper plates of thickness 1 mm, play multi-roles: They are joints to the next units and the supports, which are fixed to the base of a thick copper plate; the supports serve to keep the temperature of these ends of the pipe at a constant temperature close to the room temperature. The wires are the leads to the heater in the middle of the pipe.

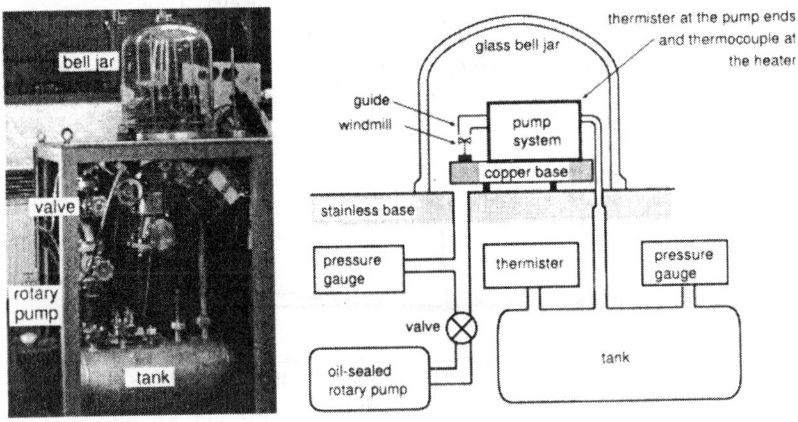

(a) (b)

Figure 3.16. Experimental system. (a) Real system and (b) explanatory diagram. The windmill is used to visualize a flow. The experimental data in the text and in the following figures are those when the windmill is removed.

tank and pumping system. The results are shown in Figs. 3.17 and 3.18. With the bell jar at 20 Pa, for example, the pressure of the tank is reduced from 20 Pa to 11 Pa in 200 sec and to 8 Pa in 500 sec after the heater is put on. The latter is the minimum pressure attained in the tank. There is a difference of

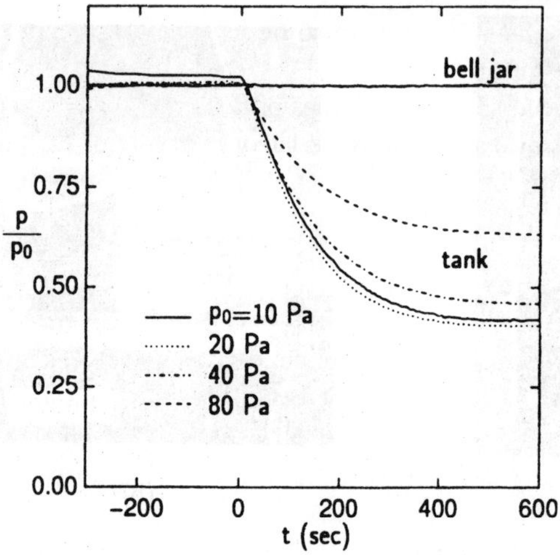

Figure 3.17. Time variation of the pressure p in the bell jar and in the tank. The t is the time (sec) after the heater is put on and p_0 is the pressure in the bell jar at time $t = 0$.

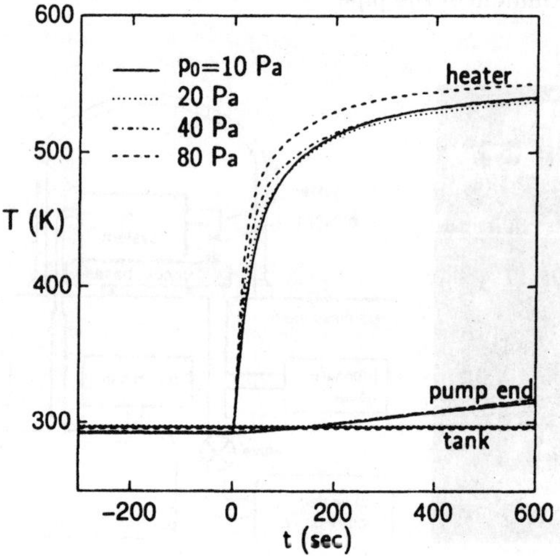

Figure 3.18. Time variation of the temperature T (K) at the heater, an end of the pumping system, and the tank. (See the figure caption of Fig. 3.17.) The temperature difference of the two ends of the pumping system is less than 1 K.

temperatures between the pumping-system ends (or the bell jar) and the tank about 20 K at 600 sec owing to the heater on the pumping system in the bell jar. In a long connecting pipe (700 mm) with curved and narrow parts between the pumping system and the tank, this temperature difference is not large enough to induce appreciable thermal transpiration. In view of ease of control as well as its simple structure, the system can also be used as a valve in a vacuum system as well as a vacuum pump.

We have introduced several examples of flows induced by temperature fields that can be studied by the asymptotic theory in Sections 3.11.1–3.11.6. More general description of the phenomena, not limited to small Knudsen numbers, is found in Kogan, Galkin & Fridlender [1976], Kogan [1992], and Sone [2000a].

3.11.7 Negative-temperature-gradient phenomenon

Consider a gas between its two parallel condensed phases with different uniform temperatures T_1 and T_2 ($T_2 > T_1$) (or two-surface problem of evaporation and condensation). Let $p_w^{(1)} = p_s(T_1)$ and $p_w^{(2)} = p_s(T_2)$, where $p_s(T)$ is the saturated gas pressure at temperature T. The $p_s(T)$ is a fairly rapidly increasing function of temperature T determined by the Clausius–Clapeyron relation (see, e.g., Reif [1965], Landau & Lifshitz [1963]). Thus, $p_w^{(2)} > p_w^{(1)}$. Assuming that the temperature difference of the condensed phases is small [$(T_2 - T_1)/T_1 \ll 1$], we discuss the behavior of the gas in the continuum limit on the basis of the asymptotic theory of the linearized Boltzmann equation developed in this chapter. The notations for dimensionless quantities are the same as in the foregoing sections, i.e., those defined in Section 2.10. Let $(\tau^{(1)}, P_w^{(1)})$ and $(\tau^{(2)}, P_w^{(2)})$ be, respectively, (τ_w, P_w) corresponding to temperatures T_1 and T_2. Let the x_1 coordinate be normal to the condensed phases and let the condensed phase at temperature T_1 and that at T_2 be set, respectively, at $x_1 = 0$ and $x_1 = 1$.

According to Eqs. (3.18) and (3.19a), $P_{G0} = \text{const}$ and $u_{1G0} = \text{const}$, since the quantities may be assumed to depend only on x_1. From the boundary condition (3.75c) at each condensed phase, these constants are determined as

$$P_{G0} = (P_w^{(1)} + P_w^{(2)})/2, \qquad u_{1G0} = (P_w^{(2)} - P_w^{(1)})/2C_4^*. \tag{3.104}$$

Here, u_{1G0} is negative (or evaporation at $x_1 = 1$ and condensation at $x_1 = 0$), since C_4^* is negative.[47] Corresponding to condensation at $x_1 = 0$ and evaporation at $x_1 = 1$, there are a temperature jump $\tau_{G0} - \tau^{(1)} = (d_4^*/2C_4^*)(P_w^{(2)} - P_w^{(1)})$ at $x_1 = 0$ and a drop $\tau_{G0} - \tau^{(2)} = -(d_4^*/2C_4^*)(P_w^{(2)} - P_w^{(1)})$ at $x_1 = 1$. With these conditions at the condensed phases, τ_{G0}, the solution of Eq. (3.19c) with $m = 0$, obviously varies linearly in x_1 from $\tau^{(1)} + (d_4^*/2C_4^*)(P_w^{(2)} - P_w^{(1)})$ to $\tau^{(2)} - (d_4^*/2C_4^*)(P_w^{(2)} - P_w^{(1)})$ as

$$\tau_{G0} = \tau^{(1)} + \frac{d_4^*}{2C_4^*}(P_w^{(2)} - P_w^{(1)}) + \left[(\tau^{(2)} - \tau^{(1)}) - \frac{d_4^*}{C_4^*}(P_w^{(2)} - P_w^{(1)})\right] x_1. \tag{3.105}$$

If the condition $(P_w^{(2)} - P_w^{(1)})/(\tau^{(2)} - \tau^{(1)}) > C_4^*/d_4^*$ (= 4.6992 for a hard-sphere gas, = 4.7723 for the BKW model) or

$$\frac{T_1}{p_s(T_1)}\left[\frac{p_s(T_2) - p_s(T_1)}{T_2 - T_1}\right] > \frac{C_4^*}{d_4^*}, \tag{3.106}$$

[47]The sign is for a hard-sphere gas or the BKW model.

is fulfilled, which is so for many kinds of gases,[48] then τ_{G0} at $x_1 = 1$ is smaller than τ_{G0} at $x_1 = 0$, and therefore the temperature gradient in the gas is opposite to that imposed between the two condensed phases. This phenomenon, first pointed out by Pao [1971], is called negative- (or inverted-) temperature-gradient phenomenon, and is also studied from the point of the irreversible thermodynamics (Hermans & Beenakker [1986]). As is obviously seen from the above analysis, the negative-temperature-gradient phenomenon is due to the temperature jump condition (3.75c), where the temperature drop occurs on an evaporating condensed phase (or a jump on a condensing condensed phase). This is understood in the following way (Sone & Onishi [1978]).

In a gas several mean free paths from a condensed phase (simply, in a gas, for simplicity in this paragraph), the velocity distribution of the gas molecules is established by intermolecular collisions and is close to a Maxwellian with the macroscopic velocity $v_i n_i$ as its center. On the other hand, the velocity distribution on the condensed phase is drastically different from that in the gas and is split into two parts of different features: The velocity distribution of the molecules leaving the condensed phase is determined by the condition of the condensed phase [or the half of the Maxwellian at rest with pressure $p_s(T_w)$ and temperature T_w], but that of the molecules incoming on the condensed phase is subject to the condition of the condensed phase only by intermolecular collisions with molecules leaving the condensed phase and retains the feature of the velocity distribution in a gas (or the negative $\xi_i n_i$ part of the Maxwellian in the gas deformed by intermolecular collision). The temperature of the gas on the condensed phase is determined by the width of this distribution. In an evaporating flow, the center of the distribution is shifted to the positive $\xi_i n_i$ side. Thus, the half width of the distribution is narrower than that of the Maxwellian with the temperature T_w of the condensed phase by the order of the speed of the evaporation. That is, the temperature of the gas on the condensed phase is lower than that of the condensed phase. The gas with lower temperature than that of the condensed phase flows downstream expanding to a state in a gas.

The condition on temperature difference $(T_2 - T_1)/T_1 \ll \mathrm{Kn}^n$ for the linear theory can be loosened. When $(T_2 - T_1)/T_1 = O(\mathrm{Kn})$, the weakly nonlinear theory (Chapter 4) can be applied. It can be easily seen, without solving the problem explicitly, from the one-dimensional simplified forms of Eqs. (4.30), (4.31a), and (4.31c) and their boundary condition (4.68c), that the negative-temperature-gradient condition $(P_w^{(2)} - P_w^{(1)})/(\tau^{(2)} - \tau^{(1)}) > C_4^*/d_4^*$ remains unchanged and that the temperature profile is deformed from a linear profile owing to the convection effect of mass flow (nonlinear effect).[49] When $(T_2 - T_1)/T_1 = O(1)$ or the Mach number of the flow is of the order of unity, the nonlinear theory in Chapter 7 can be applied. The convection effect is so strong that the temperature field becomes uniform except in the Knudsen layer (Section 7.8.1).

[48]The data of the saturated gas pressure at various temperatures for various gases are found at http://webbook.nist.gov.

[49]The solution for the BKW equation, which is a simple application of Sone [1978] or Onishi & Sone [1979], is given in Aoki & Cercignani [1983].

This two-surface problem is studied in more detail including the case of arbitrary Knudsen number (Sone & Onishi [1978], Sone, Ohwada & Aoki [1991], Aoki & Masukawa [1994]) or experimentally (Shankar & Deshpande [1990]).

Chapter 4

Weakly Nonlinear Theory – Finite Reynolds Numbers

According to the discussion in Section 3.9, the asymptotic theory of the linearized Boltzmann equation developed in the last chapter (Chapter 3) is applicable only to the case where the Reynolds number is very small. In the present chapter, we extend the asymptotic theory, according to Sone [1971], so as to be applicable to the case where the Reynolds number takes a finite value.

When the Reynolds number is finite (or of the order of unity), the Mach number is a quantity of the same order as the Knudsen number according to the von Karman relation (3.95). Thus, when the Knudsen number is small, the Mach number is also a small quantity, but its nonlinear terms are the same-order quantities as the corresponding nonlinear terms in the Knudsen number and should be retained in the asymptotic analysis for small Knudsen numbers, that is, the asymptotic analysis has to be carried out on the basis of the original nonlinear Boltzmann equation. Furthermore, the condition that the Mach number is of the same order as the Knudsen number has to be taken into account. The Mach number is one of the scales that indicate the deviation of the system from a uniform equilibrium state at rest. Extending this restriction to that on the velocity distribution function, we develop the asymptotic theory for small Knudsen numbers. That is, in this chapter, we consider the case where the deviation of the velocity distribution function from a uniform equilibrium state at rest is of the order of the Knudsen number and investigate the asymptotic behavior for small Knudsen numbers of the time-independent boundary-value problem of the Boltzmann equation. This restriction on the velocity distribution function is a little weaker than that in the linearized theory, where the deviation is much smaller than the Knudsen number. The result derived is, however, decisively different. In terms of the macroscopic parameters, the assumption on the velocity distribution function requires that the (nondimensional) temperature and density variations, as well as the Mach number, are of the order of the Knudsen number. Elimination of the temperature and density condition is discussed in the next chapter (Chapter 5), from which a very interesting fact showing the incompleteness of the classical gas dynamics will come out.

4.1 Problem

Consider the time-independent boundary-value problem of the Boltzmann equation in a general domain, where no external force is acting on the gas [or F_i in Eq. (2.5) is zero]. Geometrical configuration of the boundary of the domain is arbitrary except that its shape is smooth enough. We investigate the asymptotic behavior of the solution of the boundary-value problem for small Knudsen numbers under the assumption that the deviation of the system from a uniform equilibrium state at rest is of the order of the Knudsen number. In view of this assumption, we use the notation introduced in Section 2.10.

The basic equation is Eq. (2.58a) with $\partial/\partial\hat{t} = 0$, i.e.,

$$\zeta_i \frac{\partial \phi}{\partial x_i} = \frac{1}{k}[\mathcal{L}(\phi) + \mathcal{J}(\phi, \phi)]. \tag{4.1}$$

The boundary condition is, symbolically,

$$\phi = \phi_w \quad (\zeta_i n_i > 0), \tag{4.2}$$

where n_i is the unit normal vector to the boundary, pointed to the gas region, and the condition $u_{wi} n_i = 0$, required in a steady system, is used. For example, for the diffuse reflection [Eq. (2.79a) with (2.79b) and $\alpha = 1$],

$$\phi(x_i, \zeta_i) = \phi_e(\check{\sigma}_w, u_{wi}, \tau_w) \quad (\zeta_j n_j > 0), \tag{4.3a}$$

$$\check{\sigma}_w = -2\left(\frac{\pi}{1+\tau_w}\right)^{1/2} \int_{\zeta_j n_j < 0} \zeta_j n_j E(\zeta)\phi(x_i, \zeta_i)\mathbf{d}\boldsymbol{\zeta} + \left(\frac{1}{1+\tau_w}\right)^{1/2} - 1, \tag{4.3b}$$

and for the complete condensation condition [Eq. (2.81) with $\alpha_c = 1$],

$$\phi(x_i, \zeta_i) = \phi_e(\omega_w, u_{wi}, \tau_w) \quad (\zeta_j n_j > 0), \tag{4.4}$$

where ϕ_e is the perturbed local Maxwellian defined by Eq. (2.71). For the moment, the explicit form of the boundary condition is not specified. The boundary may be a simple boundary or an interface of the gas and its condensed phase.

The asymptotic behavior of the solution ϕ of the boundary-value problem (4.1) with (4.2) for small k ($k \ll 1$) is studied under the assumption that $\phi = O(k)$ on the basis of Sone [1971].

4.2 S solution

As in Section 3.2, putting aside the boundary condition, we look for a moderately varying solution of Eq. (4.1), whose length scale of variation is of the order of the characteristic length L of the system $[\partial\phi/\partial x_i = O(\phi)]$, in a power series of k:

$$\phi_S = \phi_{S1}k + \phi_{S2}k^2 + \cdots , \tag{4.5}$$

where the series starts from the first order of k, since ϕ is assumed to be $O(k)$, and the component function ϕ_{Sm} is a quantity of the order of unity; the subscript S is attached for discrimination of the solution or expansion under the assumptions $\phi = O(k)$ and $\partial\phi/\partial x_i = O(\phi)$. This solution or expansion will be called the S solution or expansion.

The relation between the macroscopic variables and the distribution function is the same as Eqs. (2.60a)–(2.60f) except for the subscript S:

$$\omega_S = \int \phi_S E \mathbf{d}\boldsymbol{\zeta}, \tag{4.6a}$$

$$(1+\omega_S)u_{iS} = \int \zeta_i \phi_S E \mathbf{d}\boldsymbol{\zeta}, \tag{4.6b}$$

$$\frac{3}{2}(1+\omega_S)\tau_S = \int \left(\zeta_i^2 - \frac{3}{2}\right) \phi_S E \mathbf{d}\boldsymbol{\zeta} - (1+\omega_S)u_{iS}^2, \tag{4.6c}$$

$$P_S = \omega_S + \tau_S + \omega_S\tau_S, \tag{4.6d}$$

$$P_{ijS} = 2\int \zeta_i\zeta_j\phi_S E \mathbf{d}\boldsymbol{\zeta} - 2(1+\omega_S)u_{iS}u_{jS}, \tag{4.6e}$$

$$Q_{iS} = \int \zeta_i\zeta_j^2\phi_S E \mathbf{d}\boldsymbol{\zeta} - \frac{5}{2}u_{iS} - u_{jS}P_{ijS} - \frac{3}{2}P_S u_{iS} - (1+\omega_S)u_{iS}u_{jS}^2. \tag{4.6f}$$

Corresponding to the expansion (4.5), the macroscopic variables ω_S, u_{iS}, τ_S, etc. are also expanded in k:

$$h_S = h_{S1}k + h_{S2}k^2 + \cdots , \tag{4.7}$$

where h represents ω, u_i, τ, etc., and the component function h_{Sm} is a quantity of the order of unity. The component function h_{Sm} of the macroscopic variable h_S is related to the component function ϕ_{Sr} $(r \le m)$ of the velocity distribution function ϕ_S as follows:

$$\omega_{S1} = \int \phi_{S1} E \mathbf{d}\boldsymbol{\zeta}, \tag{4.8a}$$

$$u_{iS1} = \int \zeta_i \phi_{S1} E \mathbf{d}\boldsymbol{\zeta}, \tag{4.8b}$$

$$\frac{3}{2}\tau_{S1} = \int \left(\zeta_i^2 - \frac{3}{2}\right) \phi_{S1} E \mathbf{d}\boldsymbol{\zeta}, \tag{4.8c}$$

$$P_{S1} = \omega_{S1} + \tau_{S1}, \tag{4.8d}$$

$$P_{ijS1} = 2\int \zeta_i\zeta_j\phi_{S1} E \mathbf{d}\boldsymbol{\zeta}, \tag{4.8e}$$

$$Q_{iS1} = \int \zeta_i\zeta_j^2\phi_{S1} E \mathbf{d}\boldsymbol{\zeta} - \frac{5}{2}u_{iS1}, \tag{4.8f}$$

$$\omega_{S2} = \int \phi_{S2} E \mathbf{d}\boldsymbol{\zeta}, \tag{4.9a}$$

$$u_{iS2} = \int \zeta_i \phi_{S2} E \mathbf{d}\boldsymbol{\zeta} - \omega_{S1} u_{iS1}, \tag{4.9b}$$

$$\frac{3}{2}\tau_{S2} = \int \left(\zeta_i^2 - \frac{3}{2}\right) \phi_{S2} E \mathbf{d}\boldsymbol{\zeta} - u_{iS1}^2 - \frac{3}{2}\omega_{S1}\tau_{S1}, \tag{4.9c}$$

$$P_{S2} = \omega_{S2} + \tau_{S2} + \omega_{S1}\tau_{S1}, \tag{4.9d}$$

$$P_{ijS2} = 2\int \zeta_i \zeta_j \phi_{S2} E \mathbf{d}\boldsymbol{\zeta} - 2u_{iS1}u_{jS1}, \tag{4.9e}$$

$$Q_{iS2} = \int \zeta_i \zeta_j^2 \phi_{S2} E \mathbf{d}\boldsymbol{\zeta} - \frac{5}{2}u_{iS2} - u_{jS1}P_{ijS1} - \frac{3}{2}P_{S1}u_{iS1}, \tag{4.9f}$$

. . . .

These formulas are more complicated than those, Eqs. (3.7a)–(3.7f), in the Grad–Hilbert expansion, because of nonlinearity of Eqs. (4.6b)–(4.6f).

Substituting the expansion (4.5) into the Boltzmann equation (4.1) and arranging the same order quantities in k, we obtain the following series of integral equations for ϕ_{Sm}:

$$\mathcal{L}(\phi_{S1}) = 0, \tag{4.10}$$

$$\mathcal{L}(\phi_{Sm}) = \zeta_i \frac{\partial \phi_{Sm-1}}{\partial x_i} - \sum_{r=1}^{m-1} \mathcal{J}(\phi_{Sm-r}, \phi_{Sr}) \quad (m = 2, 3, \ldots). \tag{4.11}$$

These are similar to Eqs. (3.8) and (3.9) in the linear theory; the difference is the $\mathcal{J}$ terms in the inhomogeneous term.[1] Their corresponding homogeneous equation $\mathcal{L}(\phi) = 0$ has five independent solutions 1, ζ_i, and ζ_i^2 [see Eq. (2.61c)].

In view of Eqs. (4.8a)–(4.8c), the solution ϕ_{S1} of Eq. (4.10) is given by

$$\phi_{S1} = \omega_{S1} + 2\zeta_i u_{iS1} + \left(\zeta_i^2 - \frac{3}{2}\right)\tau_{S1}. \tag{4.12}$$

This is the first term of the expansion of the perturbed Maxwellian $\phi_e(\omega_S, u_{iS}, \tau_S)$ $(= \phi_{eS})$ in terms of k:

$$\phi_e(\omega_S, u_{iS}, \tau_S) = \phi_{eS1}k + \phi_{eS2}k^2 + \cdots, \tag{4.13}$$

---

[1]The collision term of the BKW equation is not in a bilinear form, and its expansion is more complicated than that shown here. The second term on the right-hand side is more complicated and should be taken as a symbol of the terms of expansion at this order as a whole. This comment applies to the whole discussion relating to expansion of the collision integral in the other parts of this monograph. See also Footnote 12 in Section 2.9.

where

$$\phi_{eS1} = \omega_{S1} + 2\zeta_i u_{iS1} + \left(\zeta_i^2 - \frac{3}{2}\right)\tau_{S1}, \tag{4.14a}$$

$$\begin{aligned}\phi_{eS2} = {} & \omega_{S2} + 2\zeta_i u_{iS2} + \left(\zeta_i^2 - \frac{3}{2}\right)\tau_{S2} \\ & + 2\zeta_i\omega_{S1}u_{iS1} + \left(\zeta_j^2 - \frac{3}{2}\right)\left(\frac{2}{3}u_{iS1}^2 + \omega_{S1}\tau_{S1}\right) \\ & + 2\left(\zeta_i\zeta_j - \frac{1}{3}\zeta_k^2\delta_{ij}\right)u_{iS1}u_{jS1} + 2\zeta_i\left(\zeta_j^2 - \frac{5}{2}\right)u_{iS1}\tau_{S1} \\ & + \left(\frac{1}{2}\zeta_i^2\zeta_j^2 - \frac{5}{2}\zeta_i^2 + \frac{15}{8}\right)\tau_{S1}^2. \end{aligned} \tag{4.14b}$$

As in the Grad–Hilbert expansion, for the inhomogeneous equation (4.11) to have a solution, its inhomogeneous term should satisfy the solvability condition

$$\int g\left[\zeta_i\frac{\partial\phi_{Sm-1}}{\partial x_i} - \sum_{r=1}^{m-1}\mathcal{J}(\phi_{Sm-r},\phi_{Sr})\right]E\mathbf{d}\boldsymbol{\zeta} = 0, \tag{4.15a}$$

where

$$g = 1,\ \zeta_j,\ \text{or}\ \zeta_j^2,$$

since $\int g\mathcal{L}(\phi)E\mathbf{d}\boldsymbol{\zeta} = 0$ for any ϕ [Eq. (2.61b)]. The $\sum$ term in Eq. (4.15a) can be shown to vanish with the aid of Eq. (2.63). Thus, the solvability condition (4.15a) is reduced to

$$\int g\zeta_i\frac{\partial\phi_{Sm-1}}{\partial x_i}E\mathbf{d}\boldsymbol{\zeta} = 0 \quad (m = 2, 3, \ldots). \tag{4.15b}$$

This condition shows that the two inhomogeneous terms: $\zeta_i\partial\phi_{Sm-1}/\partial x_i$ and $-\sum_{r=1}^{m-1}\mathcal{J}(\phi_{Sm-r},\phi_{Sr})$ can be treated independently to obtain the solution of the integral equation (4.11).

Then, the solution of Eq. (4.11) is uniquely expressed in the form

$$\phi_{Sm} = c_0 + c_i\zeta_i + c_4\left(\zeta_j^2 - \frac{3}{2}\right) + \hat{\phi}_{Sm}, \tag{4.16}$$

where $\hat{\phi}_{Sm}$ is the particular solution of Eq. (4.11) orthogonal to g:

$$\int g\hat{\phi}_{Sm}E\mathbf{d}\boldsymbol{\zeta} = 0, \tag{4.17}$$

and c_0, c_i, and c_4 are

$$c_0 = \omega_{Sm}, \tag{4.18a}$$

$$c_i = 2\left(u_{iSm} + \sum_{r=1}^{m-1} \omega_{Sr} u_{iSm-r}\right), \tag{4.18b}$$

$$c_4 = \tau_{Sm} + \sum_{r=1}^{m-1}\left[\omega_{Sr}\tau_{Sm-r} + \frac{2}{3}\left(u_{iSr}u_{iSm-r} + \sum_{h=1}^{m-1-r} \omega_{Sr}u_{iSh}u_{iSm-r-h}\right)\right]. \tag{4.18c}$$

These relations between (c_0, c_i, c_4) and [ω_{Sr}, u_{iSr}, τ_{Sr} $(r \leq m)$] are obtained by substituting Eq. (4.16) into the expanded forms of Eqs. (4.6a)–(4.6c), e.g., Eqs. (4.8a)–(4.8c) and Eqs. (4.9a)–(4.9c), where there is no contribution of $\hat{\phi}_{Sm}$ owing to the orthogonal condition (4.17).

More explicitly, in view of Eqs. (4.12) and (4.14a), the equation for ϕ_{S2} is

$$\mathcal{L}(\phi_{S2}) = \zeta_i \frac{\partial \phi_{eS1}}{\partial x_i} - \mathcal{J}(\phi_{eS1}, \phi_{eS1}). \tag{4.19}$$

The solvability condition (4.15b) for Eq. (4.19) is given by

$$\int g\left[\zeta_i \frac{\partial \omega_{S1}}{\partial x_i} + 2\zeta_i\zeta_j \frac{\partial u_{jS1}}{\partial x_i} + \zeta_i\left(\zeta^2 - \frac{3}{2}\right)\frac{\partial \tau_{S1}}{\partial x_i}\right] E d\boldsymbol{\zeta} = 0,$$

from which

$$\frac{\partial u_{iS1}}{\partial x_i} = 0, \tag{4.20}$$

$$\frac{\partial P_{S1}}{\partial x_i} = 0. \tag{4.21}$$

The two relations in the solvability condition with $g = 1$ and $g = \zeta_i^2$ degenerate into the same equation (4.20).

The solution ϕ_{S2} is expressed by the sum

$$\phi_{S2} = \phi_a + \phi_b,$$

where ϕ_a and ϕ_b are, respectively, the solutions of the equations

$$\mathcal{L}(\phi_a) = -\mathcal{J}(\phi_{eS1}, \phi_{eS1}), \tag{4.22}$$

$$\mathcal{L}(\phi_b) = \zeta_i \frac{\partial \phi_{eS1}}{\partial x_i}, \tag{4.23}$$

where, with the aid of Eqs. (4.20) and (4.21),

$$\zeta_i \frac{\partial \phi_{eS1}}{\partial x_i} = 2\left(\zeta_i\zeta_j - \frac{1}{3}\zeta^2\delta_{ij}\right)\frac{\partial u_{jS1}}{\partial x_i} + \zeta_i\left(\zeta^2 - \frac{5}{2}\right)\frac{\partial \tau_{S1}}{\partial x_i}.$$

From the expanded form in k of Eq. (2.72) with $\phi_e = \phi_{eS}$,

$$\mathcal{L}(\phi_{eS2}) + \mathcal{J}(\phi_{eS1}, \phi_{eS1}) = 0.$$

Therefore,

$$\phi_a = \phi_{eS2}.$$

Noting that ϕ_S and ϕ_{eS} take the common values ω_S, u_{iS}, and τ_S for the perturbed density, velocity, and temperature by definition, we find that ϕ_b is the solution of the equation (4.23) satisfying the orthogonal relation[2]

$$\int g\phi_b E d\boldsymbol{\zeta} = 0. \tag{4.24}$$

With the above form of the inhomogeneous term, the solution ϕ_b, which satisfies Eq. (4.24), is expressed by a linear combination of $\partial u_{jS1}/\partial x_i$ and $\partial \tau_{S1}/\partial x_i$, whose coefficients are functions of ζ_i. According to the general discussion in Appendix A.2, ϕ_b is expressed in the form

$$\phi_b = -\left(\zeta_i\zeta_j - \frac{1}{3}\zeta^2\delta_{ij}\right) B(\zeta)\frac{\partial u_{jS1}}{\partial x_i} - \zeta_i A(\zeta)\frac{\partial \tau_{S1}}{\partial x_i},$$

where $A(\zeta)$ and $B(\zeta)$ are, respectively, defined by Eqs. (3.25) and (3.26).[3] Thus, the solution ϕ_{S2} is expressed in the form

$$\phi_{S2} = \phi_{eS2} - \zeta_i\zeta_j B(\zeta)\frac{\partial u_{jS1}}{\partial x_i} - \zeta_i A(\zeta)\frac{\partial \tau_{S1}}{\partial x_i}. \tag{4.25}$$

With Eq. (4.25) for ϕ_{S2}, the solvability condition (4.15b) for $m = 3$ is given by

$$\int g\left(\zeta_i\frac{\partial \phi_{eS2}}{\partial x_i} - \zeta_i\zeta_j\zeta_k B(\zeta)\frac{\partial^2 u_{jS1}}{\partial x_i \partial x_k} - \zeta_i\zeta_j A(\zeta)\frac{\partial^2 \tau_{S1}}{\partial x_i \partial x_j}\right) E d\boldsymbol{\zeta} = 0. \tag{4.26}$$

From the condition (4.26) with $g = 1$,

$$\frac{\partial u_{iS2}}{\partial x_i} + \frac{\partial \omega_{S1} u_{iS1}}{\partial x_i} = 0, \tag{4.27}$$

with the aid of the subsidiary condition on $A(\zeta)$ in Eq. (3.25). From the condition (4.26) with $g = \zeta_l$,

$$\frac{1}{2}\frac{\partial P_{S2}}{\partial x_l} + \frac{\partial u_{lS1} u_{jS1}}{\partial x_j} - \frac{\gamma_1}{2}\frac{\partial^2 u_{lS1}}{\partial x_j^2} = 0, \tag{4.28}$$

---

[2]The condition (4.24) is equivalent to the orthogonal condition (4.17). That is, the first five terms on the right-hand side of Eq. (4.14b) correspond to the first three terms on the right-hand side of Eq. (4.16) with Eqs. (4.18a)–(4.18c); and each of the last three terms on the right-hand side of Eq. (4.14b) satisfies the orthogonal condition (4.17).

[3]The subsidiary condition on $A(\zeta)$ in Eq. (3.25) corresponds to the orthogonal condition (4.24).

where γ_1 is defined by Eq. (3.33). From the condition (4.26) with $g = \zeta_l^2$,

$$\frac{5}{2}\left(\frac{\partial u_{iS2}}{\partial x_i} + \frac{\partial \omega_{S1} u_{iS1}}{\partial x_i}\right) + \frac{5}{2}\frac{\partial u_{iS1}\tau_{S1}}{\partial x_i} - \frac{5\gamma_2}{4}\frac{\partial^2 \tau_{S1}}{\partial x_i^2} = 0,$$

where γ_2 is defined by Eq. (3.33). With the aid of Eqs. (4.20) and (4.27), this equation is reduced to

$$u_{iS1}\frac{\partial \tau_{S1}}{\partial x_i} = \frac{\gamma_2}{2}\frac{\partial^2 \tau_{S1}}{\partial x_i^2}. \tag{4.29}$$

We can proceed with the analysis in a similar way. From the solvability conditions, we obtain a series of equations that are to be satisfied by the component functions of the expansion of the macroscopic variables.

4.3 Fluid-dynamic-type equations

Here we arrange the equations obtained from the series of the solvability conditions in a series of sets of equations that determine the component functions of the expansion of the macroscopic variables from the lowest order:

$$\frac{\partial P_{S1}}{\partial x_i} = 0, \tag{4.30}$$

$$\frac{\partial u_{iS1}}{\partial x_i} = 0, \tag{4.31a}$$

$$u_{jS1}\frac{\partial u_{iS1}}{\partial x_j} = -\frac{1}{2}\frac{\partial P_{S2}}{\partial x_i} + \frac{1}{2}\gamma_1\frac{\partial^2 u_{iS1}}{\partial x_j^2}, \tag{4.31b}$$

$$u_{jS1}\frac{\partial \tau_{S1}}{\partial x_j} = \frac{1}{2}\gamma_2\frac{\partial^2 \tau_{S1}}{\partial x_j^2}, \tag{4.31c}$$

$$\frac{\partial u_{jS2}}{\partial x_j} = -u_{jS1}\frac{\partial \omega_{S1}}{\partial x_j}, \tag{4.32a}$$

$$\begin{aligned} u_{jS1}\frac{\partial u_{iS2}}{\partial x_j} &+ (\omega_{S1}u_{jS1} + u_{jS2})\frac{\partial u_{iS1}}{\partial x_j} \\ &= -\frac{1}{2}\frac{\partial}{\partial x_i}\left[P_{S3} + \frac{1}{6}(4\gamma_3 - \gamma_1\gamma_2)\frac{\partial^2 \tau_{S1}}{\partial x_j^2}\right] \\ &\quad + \frac{1}{2}\gamma_1\frac{\partial^2 u_{iS2}}{\partial x_j^2} + \frac{1}{2}\gamma_4\frac{\partial}{\partial x_j}\left[\tau_{S1}\left(\frac{\partial u_{iS1}}{\partial x_j} + \frac{\partial u_{jS1}}{\partial x_i}\right)\right], \end{aligned} \tag{4.32b}$$

$$\begin{aligned} u_{jS1}\frac{\partial \tau_{S2}}{\partial x_j} &+ (\omega_{S1}u_{jS1} + u_{jS2})\frac{\partial \tau_{S1}}{\partial x_j} - \frac{2}{5}u_{jS1}\frac{\partial P_{S2}}{\partial x_j} \\ &= \frac{1}{5}\gamma_1\left(\frac{\partial u_{iS1}}{\partial x_j} + \frac{\partial u_{jS1}}{\partial x_i}\right)^2 + \frac{1}{2}\frac{\partial^2}{\partial x_j^2}\left(\gamma_2\tau_{S2} + \frac{1}{2}\gamma_5{\tau_{S1}}^2\right), \end{aligned} \tag{4.32c}$$

where γ_n is a constant (nondimensional transport coefficient, e.g., nondimensional viscosity and thermal conductivity) related to the collision integral [γ_1, γ_2, and γ_3 are defined in Eq. (3.33); γ_4 and γ_5 will be defined in Eq. (4.37)].

First, P_{S1} is a constant from Eq. (4.30), and then the component functions ω_{Sm}, u_{iSm}, τ_{Sm}, and P_{Sm+1} ($m = 1, 2, \ldots$) are determined successively from the series of sets of equations, together with the equation of state, Eqs. (4.8d), (4.9d), etc., that is, ω_{S1}, u_{iS1}, τ_{S1}, and P_{S2} are determined from Eqs. (4.31a)–(4.31c); ω_{S2}, u_{iS2}, τ_{S2}, and P_{S3} from Eqs. (4.32a)–(4.32c). The staggered combination is due to the degeneracy of the two relations of the solvability condition (4.15b) with $g = 1$ and $g = \zeta_j^2$ for $m = 2$ to the single equation (4.20). The degrees of these differential systems remain unchanged if the level of approximation is advanced, in contrast to the Chapman–Enskog expansion (Chapman & Cowling [1952]).[4]

The five relations of the solvability condition (4.15b) with $g = 1$, ζ_i, and ζ_i^2 at each order correspond to the component equations of the expansion of the conservation equations (2.75)–(2.77) (with $\partial/\partial\hat{t} = 0$) of the corresponding order. Equation (4.30) is the momentum equation at the order of k, where the nonlinear terms of flow velocity are absent because they are of higher order. In other words, the pressure variation of the order of k must vanish for a flow field with Mach number of the order of k to be established.[5] The next set of equations (4.31a)–(4.31c), which determines ω_{S1}, u_{iS1}, τ_{S1}, and P_{S2} is the Navier–Stokes set of equations for an incompressible fluid.[6] It may be noted that the velocity distribution function ϕ_{S1} at the order of k is Maxwellian and that the parameter functions ω_{S1}, u_{iS1}, and τ_{S1} determining the Maxwellian ϕ_{S1} are governed by "the incompressible Navier–Stokes set of equations".[7]

---

[4]The structure of the Chapman–Enskog expansion is clearly and briefly explained in Grad [1958].

[5]This is not special. For a gas in the continuum limit, where the small Mach number condition applies, this condition always holds.

[6]Strictly speaking, Eq. (4.31c) is a little different from the energy equation in the Navier–Stokes set of equations for an incompressible fluid [$v_i\partial\rho/\partial X_i = 0$ (or $u_i\partial\omega/\partial x_i = 0$)] under the present situation with a small velocity ($u_i \ll 1$ or small Mach number), a small temperature variation ($\tau \ll 1$), and a finite Reynolds number. In the latter, the convection term $u_{jS1}\partial\tau_{S1}/\partial x_j$ should be replaced by $(3/5)u_{jS1}\partial\tau_{S1}/\partial x_j$. This difference comes from the following fact: the convection term is, precisely speaking, the combination of the two terms, convection of internal energy and work done by pressure; the latter is of higher order in the incompressible fluid under the present situation. However, if we consider an incompressible fluid with the thermal conductivity multiplied by 3/5 (or with the internal energy multiplied by 5/3), the set (4.30)–(4.31c) is the same as the incompressible Navier–Stokes set. For economy of words, we will, hereafter, call the set of equations "the Navier–Stokes set of equations for an incompressible fluid" or "the incompressible Navier–Stokes set of equations" with the quotation mark.

[7]Sometimes the relation between the velocity distribution function and the fluid-dynamic equations is misunderstood in the following way: the velocity distribution function is Maxwellian, and therefore the stress consists only of pressure and the heat flow vanishes, from which the Euler set of equations is derived. This cursory discussion is not correct. This is not special to the present case, where the Mach number of a flow is of the order of k, and is found in various cases. We have already encountered the Stokes set of equations in Chapter 3 and will see other cases in the following chapters. The case in Chapter 5 is especially important.

To see the feature of the set of equations (4.32a)–(4.32c) for ω_{S2}, u_{iS2}, τ_{S2}, and P_{S3}, take the Navier–Stokes set of equations for a compressible fluid, Eqs. (2.11)–(2.13) with Eq. (2.15) and $F_i = 0$, where the viscosity and thermal conductivity are replaced by the mean free path with the aid of Eqs. (3.90) and (3.91), and the other variables are rewritten in nondimensional form by the transformation introduced in Eq. (2.57).[8] This set of equations being expanded in k under the assumption that the perturbed quantities are of the order of k, a series of the sets of equations that determine the component functions ω_m, u_{im}, τ_m, and P_m of the expansion of the macroscopic variables ω, u_i, τ, and P, i.e., the equation for P_1 and the set of equations for (ω_m, u_{im}, τ_m, and P_{m+1}) ($m = 1, 2, \ldots$), is obtained. The first equation for P_1 and the second set of equations for ω_1, u_{i1}, τ_1, and P_2 are, respectively, the same as Eq. (4.30) and the set of equations (4.31a)–(4.31c).[9] The set of equations for ω_2, u_{i2}, τ_2, and P_3 is different from the set of equations (4.32a)–(4.32c) at the point that the term multiplied by γ_3 in Eq. (4.32b) is absent. The term proportional to γ_3 is the contribution of thermal stress due to an effect of gas rarefaction.

The functions ω_{Sr}, u_{iSr}, τ_{Sr}, and P_{Sr} being obtained up to the m-th order from the fluid-dynamic-type equations (4.30)–(4.32c), etc., the velocity distribution function ϕ_{Sm} is given by these functions by Eq. (4.16) with Eqs. (4.17)–(4.18c) or more explicitly for $m = 1$ and $m = 2$ by Eqs. (4.12) and (4.25). The explicit expressions are listed here again for convenience of reference:

$$\phi_{S1} = \phi_{eS1}, \tag{4.33a}$$

$$\phi_{S2} = \phi_{eS2} - \frac{1}{2}\zeta_i\zeta_j B(\zeta)\left(\frac{\partial u_{iS1}}{\partial x_j} + \frac{\partial u_{jS1}}{\partial x_i}\right) - \zeta_i A(\zeta)\frac{\partial \tau_{S1}}{\partial x_i}, \tag{4.33b}$$

where

$$\phi_{eS1} = \omega_{S1} + 2\zeta_i u_{iS1} + \left(\zeta_j^2 - \frac{3}{2}\right)\tau_{S1}, \tag{4.34a}$$

$$\begin{aligned}\phi_{eS2} = {} & \omega_{S2} + 2\zeta_i u_{iS2} + \left(\zeta_j^2 - \frac{3}{2}\right)\tau_{S2} \\ & + 2\zeta_i\omega_{S1}u_{iS1} + \left(\zeta_j^2 - \frac{3}{2}\right)\left(\frac{2}{3}u_{iS1}^2 + \omega_{S1}\tau_{S1}\right) \\ & + 2\left(\zeta_i\zeta_j - \frac{1}{3}\zeta_k^2\delta_{ij}\right)u_{iS1}u_{jS1} + 2\zeta_i\left(\zeta_j^2 - \frac{5}{2}\right)u_{iS1}\tau_{S1} \\ & + \left(\frac{1}{2}\zeta_i^2\zeta_j^2 - \frac{5}{2}\zeta_i^2 + \frac{15}{8}\right)\tau_{S1}^2.\end{aligned} \tag{4.34b}$$

[8] The local values p, T, and ℓ should be used for p_0, T_0, and ℓ_0 in the formulas (3.90) and (3.91).

[9] In view of Footnote 6 in this subsection, the leading set of equations of the small-Mach-number (and small-temperature-variation) expansion of the compressible Navier–Stokes set of equations is not identical to the incompressible Navier–Stokes set of equations for the same situation, in contrast to rather prevailing understanding. It is probably due to the fact that the energy equation is not paid attention to when mentioning an incompressible fluid in many cases.

From the velocity distribution function ϕ_S, the component functions P_{ijSm} and Q_{iSm} of the stress tensor P_{ijS} and the heat-flow vector Q_{iS} are obtained as follows:

$$P_{ijS1} = P_{S1}\delta_{ij}, \tag{4.35a}$$

$$P_{ijS2} = P_{S2}\delta_{ij} - \gamma_1\left(\frac{\partial u_{iS1}}{\partial x_j} + \frac{\partial u_{jS1}}{\partial x_i}\right), \tag{4.35b}$$

$$P_{ijS3} = P_{S3}\delta_{ij} - \gamma_1\left(\frac{\partial u_{iS2}}{\partial x_j} + \frac{\partial u_{jS2}}{\partial x_i} - \frac{2}{3}\frac{\partial u_{kS2}}{\partial x_k}\delta_{ij}\right)$$
$$- \gamma_4\tau_{S1}\left(\frac{\partial u_{iS1}}{\partial x_j} + \frac{\partial u_{jS1}}{\partial x_i}\right) + \gamma_3\left(\frac{\partial^2 \tau_{S1}}{\partial x_i \partial x_j} - \frac{1}{3}\frac{\partial^2 \tau_{S1}}{\partial x_k^2}\delta_{ij}\right), \tag{4.35c}$$

$$Q_{iS1} = 0, \tag{4.36a}$$

$$Q_{iS2} = -\frac{5}{4}\gamma_2\frac{\partial \tau_{S1}}{\partial x_i}, \tag{4.36b}$$

$$Q_{iS3} = -\frac{5}{4}\gamma_2\frac{\partial \tau_{S2}}{\partial x_i} - \frac{5}{4}\gamma_5\tau_{S1}\frac{\partial \tau_{S1}}{\partial x_i} + \frac{1}{2}\gamma_3\frac{\partial^2 u_{iS1}}{\partial x_j^2}, \tag{4.36c}$$

where nondimensional transport coefficients γ_1, γ_2, and γ_3 are defined in Eq. (3.33), and γ_4 and γ_5 are defined as follows. Functions $C(\zeta)$, $D(\zeta)$, and $G(\zeta)$ of ζ are first defined from $A(\zeta)$ and $B(\zeta)$ by the equations

$$2\mathcal{J}\left(\zeta^2 - \frac{3}{2}, \zeta_i\zeta_j B(\zeta)\right) = \zeta_i\zeta_j C(\zeta) + \delta_{ij}D(\zeta),$$

$$2\mathcal{J}\left(\zeta^2 - \frac{3}{2}, \zeta_i A(\zeta)\right) = \zeta_i G(\zeta).$$

The functions expressed by the collision integrals on the left-hand sides of the above equations are spherically symmetric (Section 2.9), and therefore the forms on their right-hand sides are legitimate (Appendix B). With $A(\zeta)$, $B(\zeta)$, $C(\zeta)$, and $G(\zeta)$, the constants γ_4 and γ_5 are defined by the equations

$$\left.\begin{aligned} \gamma_4 &= -\tfrac{5}{2}\gamma_1 + I_8(B) + \tfrac{1}{2}I_6(BC), \\ \gamma_5 &= -6\gamma_2 + 2I_8(A) + 2I_4(AG), \end{aligned}\right\} \tag{4.37}$$

where I_n is defined by Eq. (3.34). For a hard-sphere gas,[10]

$$\gamma_4 = \gamma_1/2 = 0.635021, \qquad \gamma_5 = \gamma_2/2 = 0.961142, \tag{4.38}$$

and for the BKW model,

$$\gamma_4 = \gamma_5 = 1. \tag{4.39}$$

[10]Formulas in Appendix A.8 are used for the derivation of the relation between γ_1 and γ_4 and that between γ_2 and γ_5 from Eq. (4.37). Their relations in the general molecular model are obtained in Chapters 5 and 6. The meaning of the relations is explained in the last paragraph of the present section.

The functions C, D, and G, as well as A and B, and thus γ_4 and γ_5, as well as γ_1 and γ_2, generally depend on the reference temperature T_0 (see Footnote 11 in Section 2.9).

It is noted that the stress P_{ijS3} and the heat flow Q_{iS3} at the third order are expressed only by P_{S3} and the derivatives of u_{iSm} and τ_{Sm} with $m \leq 2$. That is, the stress and heat flow are obtained up to the third order by analysis of the sets of fluid-dynamic-type equations to the second order [or Eqs. (4.30)–(4.32c)]. The stress and heat-flow formulas are obtained without solving the third-order distribution function ϕ_{S3} explicitly in a similar way to the derivation of $P_{ijG3} - P_{G3}\delta_{ij}$ and Q_{iG3} in Eqs. (3.37) and (3.38) with the aid of the self-adjoint property (2.62) of $\mathcal{L}(*)$ as follows:

$$P_{ijS3} - P_{S3}\delta_{ij} = 2\int \left(\zeta_i\zeta_j - \frac{1}{3}\zeta^2\delta_{ij}\right)\phi_{S3}E\mathbf{d}\boldsymbol{\zeta} + \cdots \quad \text{[see (4.6e)]}$$

$$= -\int \left(\zeta_i\zeta_j - \frac{1}{3}\zeta^2\delta_{ij}\right)B(\zeta)\left[\zeta_k\frac{\partial\phi_{S2}}{\partial x_k} - 2\mathcal{J}(\phi_{S1},\phi_{S2})\right]E\mathbf{d}\boldsymbol{\zeta} + \cdots,$$

and

$$Q_{iS3} = \int \zeta_i\left(\zeta_j^2 - \frac{5}{2}\right)\phi_{S3}E\mathbf{d}\boldsymbol{\zeta} + \cdots \quad \text{[see (4.6f)]}$$

$$= -\int \zeta_iA(\zeta)\left[\zeta_j\frac{\partial\phi_{S2}}{\partial x_j} - 2\mathcal{J}(\phi_{S1},\phi_{S2})\right]E\mathbf{d}\boldsymbol{\zeta} + \cdots,$$

where the parts shown by the dots $\cdots$ are expressed by ω_{Sr}, u_{iSr}, τ_{Sr}, and P_{Sr} $(r \leq 2)$. Using Eq. (4.33b) in each of the last expressions, we obtain the desired expressions. These third-order formulas are useful when discussing a force or heat flow on a closed (or bounded) body (Section 4.8).

In the formulas of stress (4.35a)–(4.35c), the terms containing γ_1 correspond to the viscous stress in the classical gas dynamics, the term with γ_4 reflects the variation of the viscosity owing to the local variation of temperature (or τ_{S1}),[11] and the term with γ_3 is the thermal stress. In the formulas of heat flow (4.36a)–(4.36c), the terms with γ_2 correspond to the heat flow given by Fourier's law in classical gas dynamics, the term with γ_5 reflects the variation of thermal conductivity owing to the local variation of temperature (or τ_{S1}), and the term with γ_3, where $\partial^2 u_{iS1}/\partial x_j^2$ can also be written as $\partial(\partial u_{iS1}/\partial x_j + \partial u_{jS1}/\partial x_i)/\partial x_j$ (shear gradient) owing to Eq. (4.31a), is not reduced to the form of pressure gradient, in contrast to Eq. (3.31c) or (3.31d), because of the convection term in Eq. (4.31b).

[11] According to the discussion in Section 3.9, the viscosity μ and the thermal conductivity λ depend on the temperature of the gas, e.g., both are proportional to the square root of temperature for a hard-sphere gas and to the temperature for the BKW model. The variations due to the local temperature variation should be taken into account in the slightly nonlinear theory, in contrast to the linear theory. The relations $\gamma_4 = \gamma_1/2$, $\gamma_5 = \gamma_2/2$ (hard-sphere) and $\gamma_4 = \gamma_1$, $\gamma_5 = \gamma_2$ (BKW) reflect the above-mentioned temperature dependence of the two coefficients.

4.4 Knudsen-layer analysis

The component function ϕ_{Sm} of the velocity distribution function ϕ_S is expressed by a polynomial of the component functions ω_{Sr}, u_{iSr}, τ_{Sr}, and P_{Sr} $(r \leq m)$ and their derivatives, whose coefficients are functions of the molecular velocity ζ_i of special forms.[12] Thus, by the same reason as for the Grad–Hilbert solution ϕ_G in Chapter 3, the S solution ϕ_S does not have freedom enough to be fitted to the kinetic boundary condition (4.2). Thus, we introduce a correction to the S solution in the neighborhood of the boundary by allowing a sharp variation of the velocity distribution function normal to the boundary. That is, we put the solution ϕ of the boundary-value problem (4.1) and (4.2) in the form

$$\phi = \phi_S + \phi_K, \tag{4.40}$$

where ϕ_K is assumed to have a similar behavior $[kn_i\partial\phi_K/\partial x_i = O(\phi_K)]$ to ϕ_K in Section 3.4, that is, ϕ_K is appreciable only in a thin layer, with thickness of the order of the mean free path, adjacent to the boundary and decays very rapidly in the layer in the direction normal to the boundary. The ϕ_S is called the fluid-dynamic part, and ϕ_K is the Knudsen-layer part (or correction). Substituting Eq. (4.40) into the Boltzmann equation (4.1) and taking into account that the S solution ϕ_S itself satisfies the Boltzmann equation (4.1), then we find that the equation for ϕ_K is given as[13]

$$\zeta_i \frac{\partial \phi_K}{\partial x_i} = \frac{1}{k}[\mathcal{L}(\phi_K) + 2\mathcal{J}(\phi_S, \phi_K) + \mathcal{J}(\phi_K, \phi_K)]. \tag{4.41}$$

Here the same Knudsen-layer variables (η, χ_1, χ_2) as in Section 3.4 are introduced:

$$x_i = k\eta n_i(\chi_1, \chi_2) + x_{wi}(\chi_1, \chi_2), \tag{4.42}$$

where x_{wi} is the boundary surface, η is a stretched coordinate normal to the boundary, χ_1 and χ_2 are (unstretched) coordinates within a parallel surface $\eta = \text{const}$, and the normal vector n_i is a function of χ_1 and χ_2. Then, Eq. (4.41) is rewritten as

$$\zeta_i n_i \frac{\partial \phi_K}{\partial \eta} = \mathcal{L}(\phi_K) + 2\mathcal{J}(\phi_S, \phi_K) + \mathcal{J}(\phi_K, \phi_K) - k\zeta_i \left(\frac{\partial \chi_1}{\partial x_i}\frac{\partial \phi_K}{\partial \chi_1} + \frac{\partial \chi_2}{\partial x_i}\frac{\partial \phi_K}{\partial \chi_2} \right). \tag{4.43}$$

The Knudsen-layer correction ϕ_K is also expanded in a power series of k:

$$\phi_K = \phi_{K1}k + \phi_{K2}k^2 + \cdots, \tag{4.44}$$

where $\phi_{Km} = O(1)$ in contrast to the same symbol ϕ_{Km} in Section 3.4. Equation (4.40) with the expansions (4.5) and (4.44) is substituted into Eq. (4.43)

---

[12] More explicitly, ϕ_{Sm} consists of a linear form of ω_{Sm}, u_{iSm}, and τ_{Sm}, i.e., $\omega_{Sm} + 2\zeta_i u_{iSm} + (\zeta_i^2 - 3/2)\tau_{Sm}$, and the polynomials of the lower-order ω_{Sr}, u_{iSr}, and τ_{Sr} $(r < m)$ and their derivatives, of the $(m-r)$-th degree at the highest, with coefficients of functions of ζ_i of special forms [see Eqs. (4.33a) and (4.33b)].

[13] See Footnotes 12 in Section 2.9 and 1 in Section 4.2 for the second and the third terms in the brackets on the right-hand side in the case of the BKW equation.

and the same-order terms in k are arranged. In Eq. (4.43), the S solution ϕ_S appears only as the product with the Knudsen-layer correction ϕ_K, and the latter is assumed to vanish faster than any inverse power of η. Thus, we can apply the following expansion of ϕ_S in a power series of η for ϕ_S in the Knudsen-layer equation (4.43):

$$\begin{aligned}\phi_S &= (\phi_S)_0 + \left(\frac{\partial \phi_S}{\partial x_i}\right)_0 n_i k\eta + \cdots \\ &= (\phi_{S1})_0 k + \left[(\phi_{S2})_0 + \left(\frac{\partial \phi_{S1}}{\partial x_i}\right)_0 n_i \eta\right] k^2 + \cdots,\end{aligned} \tag{4.45}$$

where the parentheses with subscript 0, i.e., $(\quad)_0$, indicate that the quantities in the parentheses $(\quad)$ are evaluated at $\eta = 0$. (This convention is applied in the following discussion.) In this series expansion, the Knudsen-layer variable η always appears in the form $k\eta$ and this leads to reshuffling the order of k in the expansion of ϕ_S. With this reordering, we obtain a series of one-dimensional (homogeneous or inhomogeneous) linearized Boltzmann equations for ϕ_{Km} as follows:

$$\zeta_i n_i \frac{\partial \phi_{K1}}{\partial \eta} = \mathcal{L}(\phi_{K1}), \tag{4.46}$$

$$\begin{aligned}\zeta_i n_i \frac{\partial \phi_{K2}}{\partial \eta} &= \mathcal{L}(\phi_{K2}) + 2\mathcal{J}((\phi_{S1})_0, \phi_{K1}) + \mathcal{J}(\phi_{K1}, \phi_{K1}) \\ &\quad - \zeta_i \left[\left(\frac{\partial \chi_1}{\partial x_i}\right)_0 \frac{\partial \phi_{K1}}{\partial \chi_1} + \left(\frac{\partial \chi_2}{\partial x_i}\right)_0 \frac{\partial \phi_{K1}}{\partial \chi_2}\right],\end{aligned} \tag{4.47}$$

$$\cdots\cdots\cdots\cdots$$

Corresponding to Eq. (4.40), the Knudsen-layer part (ω_K, u_{iK}, τ_K, etc.) of a macroscopic variable (ω, u_i, τ, etc.) is defined as the remainder, i.e.,

$$h = h_S + h_K, \tag{4.48}$$

where h represents ω, u_i, τ, etc. Equations (4.40) and (4.48) being substituted into Eqs. (2.60a)–(2.60f), h_K is expressed by moments of ϕ_S and ϕ_K.[14] For example,[15]

$$\omega_K = \int \phi_K E \mathrm{d}\boldsymbol{\zeta}, \tag{4.49a}$$

$$(1 + \omega_S + \omega_K) u_{iK} = \int \zeta_i \phi_K E \mathrm{d}\boldsymbol{\zeta} - \omega_K u_{iS}, \tag{4.49b}$$

$$\begin{aligned}\frac{3}{2}(1 + \omega_S + \omega_K)\tau_K &= \int \left(\zeta_i^2 - \frac{3}{2}\right) \phi_K E \mathrm{d}\boldsymbol{\zeta} - (1 + \omega_S)(u_{iK}^2 + 2u_{iS}u_{iK}) \\ &\quad - \omega_K (u_{iS}^2 + u_{iK}^2 + 2u_{iS}u_{iK} + \frac{3}{2}\tau_S).\end{aligned} \tag{4.49c}$$

---

[14] The Knudsen layer part h_K depends on ϕ_S as well as ϕ_K, since Eqs. (2.60b)–(2.60f) are nonlinear.

[15] The following expressions are simplified by the relations (4.6a)–(4.6f).

The Knudsen-layer part h_K is also expanded in k:

$$h_K = h_{K1}k + h_{K2}k^2 + \cdots. \tag{4.50}$$

In the Knudsen layer, the following reshuffled form of the expansion of h_S, corresponding to Eq. (4.45), is applicable:

$$h_S = (h_{S1})_0 k + \left[(h_{S2})_0 + \left(\frac{\partial h_{S1}}{\partial x_i}\right)_0 n_i\eta\right] k^2 + \cdots. \tag{4.51}$$

Substituting the expansions (4.50) and (4.51) into Eqs. (4.49a)–(4.49c), etc., the component function h_{Km} is related to ϕ_{Sr-1} and ϕ_{Kr} ($r \le m$) as follows:

$$\omega_{K1} = \int \phi_{K1} E \mathbf{d}\boldsymbol{\zeta}, \tag{4.52a}$$

$$u_{iK1} = \int \zeta_i \phi_{K1} E \mathbf{d}\boldsymbol{\zeta}, \tag{4.52b}$$

$$\frac{3}{2}\tau_{K1} = \int \left(\zeta_i^2 - \frac{3}{2}\right) \phi_{K1} E \mathbf{d}\boldsymbol{\zeta}, \tag{4.52c}$$

$$P_{K1} = \omega_{K1} + \tau_{K1}, \tag{4.52d}$$

$$P_{ijK1} = 2\int \zeta_i\zeta_j \phi_{K1} E \mathbf{d}\boldsymbol{\zeta}, \tag{4.52e}$$

$$Q_{iK1} = \int \zeta_i \left(\zeta_j^2 - \frac{5}{2}\right) \phi_{K1} E \mathbf{d}\boldsymbol{\zeta}, \tag{4.52f}$$

$$\omega_{K2} = \int \phi_{K2} E \mathbf{d}\boldsymbol{\zeta}, \tag{4.53a}$$

$$u_{iK2} = \int \zeta_i \phi_{K2} E \mathbf{d}\boldsymbol{\zeta} - (\omega_{S1})_0 u_{iK1} - (u_{iS1})_0 \omega_{K1} - \omega_{K1} u_{iK1}, \tag{4.53b}$$

$$\begin{aligned}\frac{3}{2}\tau_{K2} = {} & \int \left(\zeta_i^2 - \frac{3}{2}\right) \phi_{K2} E \mathbf{d}\boldsymbol{\zeta} - 2(u_{iS1})_0 u_{iK1} - u_{iK1}^2 \\ & - \frac{3}{2}[(\omega_{S1})_0 \tau_{K1} + \omega_{K1}(\tau_{S1})_0 + \omega_{K1}\tau_{K1}],\end{aligned} \tag{4.53c}$$

$$P_{K2} = \omega_{K2} + \tau_{K2} + (\omega_{S1})_0 \tau_{K1} + \omega_{K1}(\tau_{S1})_0 + \omega_{K1}\tau_{K1}, \tag{4.53d}$$

$$P_{ijK2} = 2\int \zeta_i\zeta_j \phi_{K2} E \mathbf{d}\boldsymbol{\zeta} - 2[(u_{iS1})_0 u_{jK1} + u_{iK1}(u_{jS1})_0 + u_{iK1}u_{jK1}], \tag{4.53e}$$

$$\begin{aligned}Q_{iK2} = {} & \int \zeta_i\zeta_j^2 \phi_{K2} E \mathbf{d}\boldsymbol{\zeta} - \frac{5}{2}u_{iK2} - (u_{jS1})_0 P_{ijK1} - u_{jK1}(P_{ijS1})_0 - u_{jK1}P_{ijK1} \\ & - \frac{3}{2}[(P_{S1})_0 u_{iK1} + P_{K1}(u_{iS1})_0 + P_{K1}u_{iK1}],\end{aligned} \tag{4.53f}$$

$$\cdots\cdots\cdots\cdots.$$

From Eq. (4.2), the boundary condition for ϕ_{Km} at $\eta = 0$ is given as

$$\phi_{Km} = \phi_{wm} - \phi_{Sm} \qquad (\zeta_i n_i > 0), \tag{4.54}$$

where ϕ_{wm} is the expansion coefficient of ϕ_w (see Footnote 7 in Chapter 3):

$$\phi_w = \phi_{w1}k + \phi_{w2}k^2 + \cdots . \tag{4.55}$$

For the diffuse reflection, the above condition is reduced to

$$\phi_{K1} = \check{\sigma}_{w1} - \omega_{S1} + 2\zeta_i(u_{wi1} - u_{iS1}) + \left(\zeta_i^2 - \frac{3}{2}\right)(\tau_{w1} - \tau_{S1}) \quad (\zeta_i n_i > 0), \tag{4.56a}$$

$$\begin{aligned}\check{\sigma}_{w1} &= -\frac{1}{2}\tau_{w1} - 2\sqrt{\pi}\int_{\zeta_i n_i<0} \zeta_i n_i E(\zeta)(\phi_{S1} + \phi_{K1})\mathbf{d}\zeta \\ &= -\frac{1}{2}(\tau_{w1} - \tau_{S1}) + \omega_{S1} - \sqrt{\pi}u_{iS1}n_i - 2\sqrt{\pi}\int_{\zeta_i n_i<0} \zeta_i n_i E(\zeta)\phi_{K1}\mathbf{d}\zeta,\end{aligned} \tag{4.56b}$$

$$\begin{aligned}\phi_{K2} = {}& \check{\sigma}_{w2} - \omega_{S2} + 2\zeta_i(u_{wi2} - u_{iS2}) + \left(\zeta_i^2 - \frac{3}{2}\right)(\tau_{w2} - \tau_{S2}) \\ & + \phi_{e2}(\check{\sigma}_{w1}, u_{wi1}, \tau_{w1}) - \phi_{e2}(\omega_{S1}, u_{iS1}, \tau_{S1}) + \zeta_i\zeta_j B(\zeta)\frac{\partial u_{iS1}}{\partial x_j} + \zeta_i A(\zeta)\frac{\partial \tau_{S1}}{\partial x_i} \\ & (\zeta_i n_i > 0),\end{aligned} \tag{4.57a}$$

$$\begin{aligned}\check{\sigma}_{w2} = {}& -\frac{1}{2}\tau_{w2} + \frac{3}{8}\tau_{w1}^2 + \sqrt{\pi}\tau_{w1}\int_{\zeta_j n_j<0} \zeta_j n_j E(\zeta)(\phi_{S1} + \phi_{K1})\mathbf{d}\zeta \\ & - 2\sqrt{\pi}\int_{\zeta_j n_j<0} \zeta_j n_j E(\zeta)(\phi_{S2} + \phi_{K2})\mathbf{d}\zeta,\end{aligned} \tag{4.57b}$$

$\cdots\cdots\cdots\cdots ,$

where $\phi_{e2}(\check{\sigma}_{w1}, u_{wi1}, \tau_{w1})$ and $\phi_{e2}(\omega_{S1}, u_{iS1}, \tau_{S1})$ are, respectively, the second-order component functions of the expansion of Maxwellian $\phi_e(\omega, u_i, \tau)$ with $\omega = k\check{\sigma}_{w1}$, $u_i = ku_{wi1}$, $\tau = k\tau_{w1}$ and with $\omega = k\omega_{S1}$, $u_i = ku_{iS1}$, $\tau = k\tau_{S1}$ [see Eq. (4.13)]. More explicitly, for example, $\phi_{e2}(\omega_{S1}, u_{iS1}, \tau_{S1})$ is given by Eq. (4.14b) with $\omega_{S2} = 0$, $u_{iS2} = 0$, $\tau_{S2} = 0$. For the complete condensation condition,

$$\phi_{K1} = \omega_{w1} - \omega_{S1} + 2\zeta_i(u_{wi1} - u_{iS1}) + \left(\zeta_i^2 - \frac{3}{2}\right)(\tau_{w1} - \tau_{S1}) \quad (\zeta_i n_i > 0), \tag{4.58}$$

$$\begin{aligned}\phi_{K2} = {}& \omega_{w2} - \omega_{S2} + 2\zeta_i(u_{wi2} - u_{iS2}) + \left(\zeta_i^2 - \frac{3}{2}\right)(\tau_{w2} - \tau_{S2}) \\ & + \phi_{e2}(\omega_{w1}, u_{wi1}, \tau_{w1}) - \phi_{e2}(\omega_{S1}, u_{iS1}, \tau_{S1}) + \zeta_i\zeta_j B(\zeta)\frac{\partial u_{iS1}}{\partial x_j} + \zeta_i A(\zeta)\frac{\partial \tau_{S1}}{\partial x_i} \\ & (\zeta_i n_i > 0),\end{aligned} \tag{4.59}$$

$\cdots\cdots\cdots\cdots ,$

where $\phi_{e2}(\omega_{w1}, u_{wi1}, \tau_{w1})$ is given by $\phi_{e2}(\check{\sigma}_{w1}, u_{wi1}, \tau_{w1})$ with $\check{\sigma}_{w1} = \omega_{w1}$.

Since the Knudsen layer ϕ_K is introduced as the correction to ϕ_S in the neighborhood of the boundary, the condition at infinity is

$$\phi_{Km} \to 0 \quad \text{as } \eta \to \infty, \tag{4.60}$$

where the decay is assumed to be faster than any inverse power of η. This rapid decay is required for obtaining the solution in the sum of the two series.[16]

The analysis of the Knudsen layer is reduced to that of the one-dimensional half-space boundary-value problem of the linearized Boltzmann equation, i.e., Eqs. (4.46), (4.47), etc. and the boundary condition (4.54) [more explicitly, for example, the diffuse reflection condition (4.56a), (4.57a), etc. and the complete condensation condition (4.58), (4.59), etc.] and the condition (4.60) at infinity. This problem is practically the same problem as that explained in Section 3.4; the formal difference is the additional nonlinear terms of the perturbed quantities, for example, the second and third terms on the right-hand side of inhomogeneous Boltzmann equation (4.47), ϕ_{e2} terms in Eqs. (4.57a) and (4.59), the second and third terms on the right-hand side of Eq. (4.57b). The discussion goes parallel to that in Section 3.4 with the aid of Bardos's theorem. Thus, the Knudsen-layer solution ϕ_{Km} is determined uniquely together with the relations among the undetermined boundary values of the macroscopic variables ω_{Sm}, u_{iSm}, and τ_{Sm} of the S solution ϕ_{Sm}; further, the solution ϕ_{Km} decays exponentially as $\eta \to \infty$, which is assumed for the construction of the solution ϕ as the sum of ϕ_S and ϕ_K. On a general locally isotropic simple solid boundary with a finite diffuse reflection part, the velocity slip $u_{iSm} - u_{wim}$ and the temperature jump $\tau_{Sm} - \tau_{wm}$ are expressed by the lower-order macroscopic variables and their derivatives; on an interface of the gas and its locally isotropic condensed phase with a finite complete condensation part, the tangential velocity slip $(u_{iSm} - u_{jSm}n_j n_i) - u_{wim}$ is expressed by the lower-order macroscopic variables and their derivatives, and two linear combinations of $\omega_{Sm} - \omega_{wm}$, $\tau_{Sm} - \tau_{wm}$, and $u_{iSm}n_i$ are expressed by the lower-order macroscopic variables and their derivatives. These relations serve as the boundary condition for the fluid-dynamic-type equations (4.30)–(4.32c), etc. on a simple boundary or on an interface of the gas and its condensed phase.

4.5 Slip condition and Knudsen layer

Here, the slip boundary condition and the Knudsen-layer correction for the macroscopic variables on a simple boundary and on an interface of a gas and its condensed phase are summarized. The formulas apply to a general locally isotropic boundary (Footnote 18 in Chapter 3), where the reflection law is expressed by Eq. (2.22) with a finite diffuse reflection part or by Eq. (2.25) with a finite complete condensation part.

[16] The rapid decay is required for application of the expansion (4.45) in Eq. (4.43). See also the comment below Eq. (3.52).

4.5.1 On a simple solid boundary

On a simple solid boundary, the slip boundary condition for the fluid-dynamic-type equations and the Knudsen-layer correction for the macroscopic variables are given as follows (Sone [1971], Sone & Aoki [1987]):

$$u_{iS1} - u_{wi1} = 0, \tag{4.61a}$$

$$\tau_{S1} - \tau_{w1} = 0, \tag{4.61b}$$

$$u_{iK1} = 0, \quad \omega_{K1} = \tau_{K1} = 0, \tag{4.61c}$$

$$\begin{bmatrix} (u_{iS2} - u_{wi2})t_i \\ u_{iK2}t_i \end{bmatrix} = S_{ijS1}n_i t_j \begin{bmatrix} k_0 \\ Y_0(\eta) \end{bmatrix} + G_{iS1}t_i \begin{bmatrix} K_1 \\ \frac{1}{2}Y_1(\eta) \end{bmatrix}, \tag{4.62a}$$

$$\begin{bmatrix} u_{iS2}n_i \\ u_{iK2}n_i \end{bmatrix} = 0, \tag{4.62b}$$

$$\begin{bmatrix} \tau_{S2} - \tau_{w2} \\ \omega_{K2} \\ \tau_{K2} \end{bmatrix} = -G_{iS1}n_i \begin{bmatrix} d_1 \\ \Omega_1(\eta) \\ \Theta_1(\eta) \end{bmatrix}, \tag{4.62c}$$

$$S_{ijSm} = -\left(\frac{\partial u_{iSm}}{\partial x_j} + \frac{\partial u_{jSm}}{\partial x_i}\right), \quad G_{iSm} = \frac{\partial \tau_{Sm}}{\partial x_i}.$$

Here, (i) As already explained, u_{wim} and τ_{wm} are the component functions of the expansions of u_{wi} and τ_w, where $(2RT_0)^{1/2}u_{wi}$ $(u_{wi}n_i = 0)$ and $T_0(1+\tau_w)$ are, respectively, the velocity and temperature of the boundary [see Eqs. (2.57) and Footnote 7 in Chapter 3]:

$$u_{wi} = u_{wi1}k + u_{wi2}k^2 + \cdots, \qquad \tau_w = \tau_{w1}k + \tau_{w2}k^2 + \cdots. \tag{4.63}$$

(ii) The t_i is a unit vector tangential to the boundary.
(iii) The quantities with the subscript S are evaluated on the boundary.
(iv) The slip coefficients k_0, K_1, and d_1 and the Knudsen-layer functions $Y_0(\eta)$, $Y_1(\eta)$, $\Omega_1(\eta)$, and $\Theta_1(\eta)$ are the same as those in the linear theory.

The Knudsen-layer parts of the stress tensor P_{ij} and the heat-flow vector Q_i are

$$P_{ijK1} = 0, \tag{4.64}$$

$$P_{ijK2} = -\frac{3}{2}G_{kS1}n_k[\Omega_1(\eta) + \Theta_1(\eta)](\delta_{ij} - n_i n_j), \tag{4.65}$$

$$Q_{iK1} = 0, \tag{4.66}$$

$$Q_{iK2}n_i = 0, \quad Q_{iK2}t_i = -S_{ijS1}t_i n_j H_A(\eta) - G_{iS1}t_i H_B(\eta), \tag{4.67}$$

where $H_A(\eta)$ and $H_B(\eta)$ are the same as those in the linear theory, and the quantities with the subscript S are evaluated on the boundary.

The slip boundary condition and the Knudsen-layer correction up to the second order of k are essentially the same as those (up to the first order in k) in the linear theory.

4.5.2 On an interface of a gas and its condensed phase

On an interface of a gas and its condensed phase, the boundary condition for the fluid-dynamic-type equations and the Knudsen-layer correction for the macroscopic variables are given as follows (Onishi & Sone [1979], Sone & Aoki [1987], Sone [1991a]):[17]

$$(u_{iS1} - u_{wi1})t_i = 0, \tag{4.68a}$$

$$u_{iK1} = 0, \tag{4.68b}$$

$$\begin{bmatrix} P_{S1} - P_{w1} \\ \tau_{S1} - \tau_{w1} \\ \omega_{K1} \\ \tau_{K1} \end{bmatrix} = u_{iS1}n_i \begin{bmatrix} C_4^* \\ d_4^* \\ \Omega_4^*(\eta) \\ \Theta_4^*(\eta) \end{bmatrix}, \tag{4.68c}$$

$$\begin{bmatrix} (u_{iS2} - u_{wi2})t_i \\ u_{iK2}t_i \end{bmatrix} = S_{ijS1}n_it_j \begin{bmatrix} k_0 \\ Y_0(\eta) \end{bmatrix} + G_{iS1}t_i \begin{bmatrix} K_1 \\ \frac{1}{2}Y_1(\eta) \end{bmatrix}$$
$$+t_j\frac{\partial}{\partial x_j}(u_{iS1}n_i) \begin{bmatrix} K_2 \\ Y_{K2}(\eta) \end{bmatrix}, \tag{4.69a}$$

$$u_{iK2}n_i = -(u_{iS1}n_i)^2\Omega_4^*(\eta), \tag{4.69b}$$

$$\begin{bmatrix} P_{S2} - P_{w2} \\ \tau_{S2} - \tau_{w2} \\ \omega_{K2} \\ \tau_{K2} \end{bmatrix} = u_{iS2}n_i \begin{bmatrix} C_4^* \\ d_4^* \\ \Omega_4^*(\eta) \\ \Theta_4^*(\eta) \end{bmatrix} - G_{iS1}n_i \begin{bmatrix} C_1 \\ d_1 \\ \Omega_1(\eta) \\ \Theta_1(\eta) \end{bmatrix}$$
$$-S_{ijS1}n_in_j \begin{bmatrix} C_6 \\ d_6 \\ \Omega_6(\eta) \\ \Theta_6(\eta) \end{bmatrix} - 2\bar{\kappa}u_{iS1}n_i \begin{bmatrix} C_7 \\ d_7 \\ \Omega_7(\eta) \\ \Theta_7(\eta) \end{bmatrix}$$
$$+(u_{iS1}n_i)^2 \begin{bmatrix} C_8 \\ d_8 \\ \Omega_8(\eta) \\ \Theta_8(\eta) \end{bmatrix} + \tau_{w1}u_{iS1}n_i \begin{bmatrix} C_9 \\ d_9 \\ \Omega_9(\eta) \\ \Theta_9(\eta) \end{bmatrix}$$
$$+P_{w1}u_{iS1}n_i \begin{bmatrix} C_{10} \\ d_{10} \\ \Omega_{10}(\eta) \\ \Theta_{10}(\eta) \end{bmatrix}. \tag{4.69c}$$

Here, (i) See note (i) after Eq. (4.62c) for u_{wim} and τ_{wm}; the P_{wm} is the component function of the expansion of P_w, where $p_0(1 + P_w)$ is the saturated gas

[17] As explained in Footnote 15 in Section 3.4, the formulas for $u_{iK2}n_i$ in Eq. (4.69b) and those for $P_{ijK2}n_i$ and $Q_{iK2}n_i$ in the forthcoming Eqs. (4.73a), (4.73b), and (4.75) can be obtained in a simple way directly from Eq. (4.47). In this process there is no contribution from $\mathcal{L}(\phi_{K2}) + 2\mathcal{J}((\phi_{S1})_0, \phi_{K1}) + \mathcal{J}(\phi_{K1}, \phi_{K1})$ owing to the general relations (2.61b) and (2.63).

pressure at temperature $T_0(1+\tau_w)$ [see Eqs. (2.57) and Footnote 7 in Chapter 3]:

$$P_w = P_{w1}k + P_{w2}k^2 + \cdots . \tag{4.70}$$

(ii) The $\bar{\kappa}/L$ is the mean curvature of the boundary [see Eq. (3.62) and the note (iii) after it].
(iii) The quantities with the subscript S are evaluated on the interface.
(iv) The slip (jump) coefficients C_1, C_4^*, d_4^*, K_2, C_6, d_6, C_7, and d_7, as well as k_0, K_1, and d_1, and the Knudsen-layer functions $Y_{K2}(\eta)$, $\Omega_4^*(\eta)$, $\Theta_4^*(\eta)$, $\Omega_6(\eta)$, $\Theta_6(\eta)$, $\Omega_7(\eta)$, and $\Theta_7(\eta)$, as well as $Y_0(\eta)$, $Y_1(\eta)$, $\Omega_1(\eta)$, and $\Theta_1(\eta)$, are the same as those in the linear theory. The remaining coefficients C_8, d_8, C_9, d_9, C_{10}, and d_{10} and the Knudsen-layer functions $\Omega_8(\eta)$, $\Theta_8(\eta)$, $\Omega_9(\eta)$, $\Theta_9(\eta)$, $\Omega_{10}(\eta)$, and $\Theta_{10}(\eta)$, which first appeared here, are obtained for the BKW equation under the complete condensation condition (Onishi & Sone [1979]).[18] The new slip coefficients for BKW are

$$\left.\begin{aligned} C_8 &= 2.320074, & C_9 &= 1.066019, & C_{10} &= C_4^*, \\ d_8 &= -0.0028315, & d_9 &= -0.223375, & d_{10} &= 0. \end{aligned}\right\} \tag{4.71}$$

The leading terms of the slip condition and the Knudsen-layer correction are essentially the same as those in the linear theory.

The Knudsen-layer parts of the stress tensor P_{ij} and the heat-flow vector Q_i are given by

$$P_{ijK1} = \frac{3}{2}u_{kS1}n_k[\Omega_4^*(\eta) + \Theta_4^*(\eta)](\delta_{ij} - n_in_j), \tag{4.72}$$

$$P_{ijK2}n_in_j = 3\bar{\kappa}u_{iS1}n_i \int_\infty^\eta [\Omega_4^*(\eta_0) + \Theta_4^*(\eta_0)]\mathrm{d}\eta_0, \tag{4.73a}$$

$$P_{ijK2}n_it_j = -\frac{3}{2}t_j\frac{\partial u_{iS1}n_i}{\partial x_j}\int_\infty^\eta [\Omega_4^*(\eta_0) + \Theta_4^*(\eta_0)]\mathrm{d}\eta_0, \tag{4.73b}$$

$$Q_{iK1} = 0, \tag{4.74}$$

$$Q_{iK2}n_i = (u_{kS1}n_k)^2\left[\Omega_4^*(\eta) - \frac{3}{2}\Theta_4^*(\eta)\right], \tag{4.75}$$

where $u_{iS1}n_i$ is evaluated on the boundary.

4.6 Determination of macroscopic variables

The process of obtaining the component functions of the S expansion of the macroscopic variables from the equations and their boundary conditions given in Sections 4.3 and 4.5 is similar to that explained in Section 3.6. However, we will repeat the explanation to avoid confusion in replacing variables and

[18] In Onishi & Sone [1979], $\int_0^\eta [1+\omega(\eta_0)]\mathrm{d}\eta_0$ instead of η is used as the Knudsen-layer variable. The difference affects only Ω_8 and Θ_8.

equations. The discussion is constructed separately for a simple boundary and for an interface of a gas and its condensed phase.

First we start with the discussion of the case of a simple boundary. From Eq. (4.30), P_{S1} is uniform, i.e.,

$$P_{S1} = P_1.$$

In an open-(or unbounded-)domain problem, where the pressure is specified, for example, at infinity, the constant P_1 is determined by the pressure there. In a closed-(or bounded-)domain problem, where the total mass in the domain is specified (see Footnote 27 in Section 3.6), P_1 cannot be determined at this stage. Now consider the set of equations (4.31a)–(4.31c), where the variables u_{iS1}, τ_{S1}, and P_{S2} are contained. The variables u_{iS1}, τ_{S1}, and P_{S2} are determined by this set of equations under the boundary conditions (4.61a) and (4.61b) leaving an undetermined integration constant (say, P_2) in P_{S2}. From Eq. (4.8d), ω_{S1} is expressed by P_1 and τ_{S1} as

$$\omega_{S1} = P_1 - \tau_{S1}.$$

At this stage the constant P_1, if undetermined previously, is determined, for example, from the total mass condition with the aid of ω_{S1}. In the next step, the variables u_{iS2}, τ_{S2}, ω_{S2}, and P_{S3} and the constant P_2 (if undetermined) are determined by Eqs. (4.32a)–(4.32c) with Eq. (4.9d) under the boundary conditions (4.62a)–(4.62c) leaving an undetermined integration constant (say, P_3) in P_{S3}, where the determination of P_2 and ω_{S2} are similar to that of P_1 and ω_{S1}. In this way, we can determine the macroscopic variables successively.

For a boundary of an interface of a gas and its condensed phase, the process is similar but there is some difference for determination of the integration constant in P_{Sm}. From Eq. (4.30), P_{S1} is uniform, i.e.,

$$P_{S1} = P_1.$$

The constant P_1 is determined by the condition at infinity in an open-domain problem or undetermined at this stage in a closed-domain problem. From the set of equations (4.31a)–(4.31c) and the boundary conditions (4.68a) and (4.68c), the variables u_{iS1}, τ_{S1}, and P_{S2} are determined leaving an undetermined integration constant (say, P_2), in the result of which the (determined or undetermined) parameter P_{S1} (or P_1) is included through the boundary condition (4.68c). The constant P_1 cannot be arbitrary, because it is determined by the condition at infinity in an open-domain problem, or because the quantity $u_{iS1}n_i$ integrated over the boundary must be zero from Eq. (4.31a) in a closed-domain problem. Then ω_{S1} is determined by Eq. (4.8d). In the next step, the variables u_{iS2}, τ_{S2}, ω_{S2}, and P_{S3} and the constant P_2 (if undetermined) are determined by Eqs. (4.32a), (4.32b), and (4.32c) with Eq. (4.9d) under the boundary conditions (4.69a) and (4.69c), leaving an undetermined integration constant (say, P_3) in P_{S3}, where P_2 and ω_{S2} are determined in a similar way to P_1 and ω_{S1}. In this way, we can determine the macroscopic variables of the fluid-dynamic part successively.

4.7 Rarefaction effect

The equations (4.31a)–(4.31c) are "the Navier–Stokes set of equations for an incompressible fluid".[19] Their boundary conditions on a simple solid boundary are the nonslip conditions (4.61a) and (4.61b). This set of equations and boundary conditions is what is used in the classical gas dynamics. Equation (4.68a) and the first two relations in Eq. (4.68c) are the boundary conditions for "the Navier–Stokes set" on an interface of the gas and its condensed phase. There is a temperature jump as well as a pressure jump when evaporation or condensation is taking place. This fact is not well known in the field of classical gas dynamics. At the next order, from the discussion in Section 4.3, the fluid-dynamic-type equations contain a non-Navier–Stokes term or the term due to thermal stress. However, this additional term to the slightly compressible Navier–Stokes equations can be incorporated into the pressure term by introducing the new variable P^*_{S3} (a modified pressure) defined by

$$P^*_{S3} = P_{S3} + \frac{2}{3}\gamma_3 \frac{\partial^2 \tau_{S1}}{\partial x_i^2}, \tag{4.76}$$

instead of P_{S3}. With this replacement, Eqs. (4.32a)–(4.32c) agree apparently with the next-order set of equations after "the incompressible Navier–Stokes set" in the expansion of the Navier–Stokes set of equations for a compressible gas by the Mach number. Further, the slip conditions in Eqs. (4.61a)–(4.62c) or in Eqs. (4.68a)–(4.69c) do not contain P_{S3}. Thus, we have the following proposition:

Proposition 4.1. *Except for the Knudsen-layer correction, the macroscopic variables of a slightly rarefied gas are obtained correctly up to the second order of the Knudsen number (i.e., up to ω_{S2}, u_{iS2}, τ_{S2}, etc.) by solving the Navier–Stokes sets of equations for a slightly compressible fluid (or with the second-order quantities of the Mach number and temperature variation retained) under the slip boundary conditions in Eqs. (4.61a)–(4.62c) or in Eqs. (4.68a)–(4.69c). The effect of rarefaction of the gas comes in only through the slip boundary conditions.*

4.8 Force and mass and energy transfers on a closed body

We will give the extension of the propositions on the force and the moment of force or the mass and energy transfers on a closed (or bounded) body in the linear theory (Section 3.8) to the present weakly nonlinear case (Sone & Aoki [1987]).

---

[19] See Footnote 6 in Section 4.3 for the meaning of the quotation mark. It may be recalled that the case of a small Mach number, a small temperature variation, and a finite Reynolds number is considered here, when the comparison with the Navier–Stokes set of equations for an incompressible fluid is made.

Consider the total force, total moment of force, and total mass and energy transfers on a closed (or bounded) body in a gas. These global quantities can be obtained only by the knowledge of the fluid-dynamic part as follows. The body may be the condensed phase of the gas; there may be other bodies in the gas, and the domain may be enclosed by a boundary.

Proposition 4.2. *The force FB_i and the moment MO_i of force (around the origin) acting on a closed body in a gas are expressed by the fluid-dynamic part ψ_{ijS} [$= P_{ijS}+2(1+\omega_S)u_{iS}u_{jS}$] of the (nondimensional) momentum flux tensor as follows:*

$$FB_i = -p_0 L^2 \int_S \psi_{ijS}\widehat{n}_j \mathrm{d}S, \tag{4.77}$$

$$MO_i = -p_0 L^3 \int_S \epsilon_{ilk} x_l \psi_{kjS}\widehat{n}_j \mathrm{d}S, \tag{4.78}$$

where the surface S of integration is a closed surface enclosing only the body under consideration, $\widehat{n}_i$ is the outward unit normal vector to the surface S, $\mathrm{d}S$ is the surface element (in the nondimensional x_i space) for integration, and ϵ_{ijk} is Eddington's epsilon (Footnote 13 in Section 2.10). The "outward" means outward with respect to the domain enclosed by S.

Proof. The force FB_i and the moment MO_i of force (around the origin) acting on a body (say, B) in a gas are expressed by the (nondimensional) momentum flux tensor ψ_{ij} [$= P_{ij} + 2(1+\omega)u_i u_j$] as follows:

$$FB_i = -p_0 L^2 \int_{\partial B} \psi_{ij}\widehat{n}_j \mathrm{d}S, \tag{4.79}$$

$$MO_i = -p_0 L^3 \int_{\partial B} \epsilon_{ilk} x_l \psi_{kj}\widehat{n}_j \mathrm{d}S, \tag{4.80}$$

where ∂B is the surface of the body B. The surface ∂B of integration in Eq. (4.79) can be transformed continuously in the gas, since $\partial\psi_{ij}/\partial x_j = 0$ [Eq. (2.76)] in the gas. Thus, moving ∂B to a surface (say ∂B_1) outside the Knudsen layer, we have

$$FB_i = -p_0 L^2 \int_{\partial B_1} \psi_{ijS}\widehat{n}_j \mathrm{d}S, \tag{4.81}$$

because the Knudsen-layer part ψ_{ijK} vanishes there. The S solution being also a solution of the Boltzmann equation (4.1), $\partial\psi_{ijS}/\partial x_j = 0$. Thus, the surface ∂B_1 in Eq. (4.81) can be continuously transformed freely into a surface S in the gas. It can be shown easily from the relations $\partial\psi_{ij}/\partial x_j = 0$ and $\psi_{ij} = \psi_{ji}$ that $\partial\epsilon_{ilk}x_l\psi_{kj}/\partial x_j = 0$. With the aid of this condition, the statement on MO_i can be shown similarly to that on FB_i. ■

Corollary. *On a simple boundary, FB_i/p_0L^2 and MO_i/p_0L^3 are expressed only by P_{ijS} up to the third order of k.*

Proof. On a simple solid boundary, $u_{iS1}\widehat{n}_i = u_{iS2}\widehat{n}_i = 0$ [see Eqs. (4.61a) and (4.62b)]. Take S in the formulas (4.77) and (4.78) on the boundary (or ∂B), and apply the above conditions to the formulas (4.77) and (4.78). ∎

Proposition 4.3. *The mass MF and energy ET transferred to a closed body in a gas per unit time are expressed by the fluid-dynamic part* ψ_{iS}^m $[=(1+\omega_S)u_{iS}]$ *of the (nondimensional) mass-flux vector and that* ψ_{iS}^e $[=\int \zeta_i\zeta_j^2\phi_S E \mathrm{d}\boldsymbol{\zeta} = \frac{5}{2}u_{iS} + u_{jS}P_{ijS} + \frac{3}{2}P_S u_{iS} + (1+\omega_S)u_{iS}u_{jS}^2 + Q_{iS}]$ *of the (nondimensional) energy-flux vector as follows:*

$$MF = -\rho_0(2RT_0)^{1/2}L^2\int_S \psi_{iS}^m \widehat{n}_i \mathrm{d}S, \tag{4.82}$$

$$ET = -p_0(2RT_0)^{1/2}L^2\int_S \psi_{iS}^e \widehat{n}_i \mathrm{d}S, \tag{4.83}$$

where the definitions of S, $\mathrm{d}S$, *and* $\widehat{n}_i$ *are the same as in Proposition 4.2.*

Proof. The mass flux ψ_i^m $[=(1+\omega)u_i]$ and the energy flux ψ_i^e $[=\int \zeta_i\zeta_j^2\phi E\mathrm{d}\boldsymbol{\zeta} = (5/2)u_i + u_jP_{ij} + (3/2)Pu_i + (1+\omega)u_iu_j^2 + Q_i]$ and their fluid-dynamic parts ψ_{iS}^m and ψ_{iS}^e are all solenoidal [see Eqs. (2.75) and (2.77)]:

$$\frac{\partial \psi_i^m}{\partial x_i} = 0, \qquad \frac{\partial \psi_{iS}^m}{\partial x_i} = 0,$$
$$\frac{\partial \psi_i^e}{\partial x_i} = 0, \qquad \frac{\partial \psi_{iS}^e}{\partial x_i} = 0.$$

With the aid of these conditions, the proof can be carried out in a similar way to that in Proposition 4.2. ∎

Now we consider the contribution of the thermal stress. First rewrite the expression of P_{ijS3} given by Eq. (4.35c) using P_{S3}^* defined by Eq. (4.76) as follows:

$$P_{ijS3} = P_{S3}^*\delta_{ij} - \gamma_1\left(\frac{\partial u_{iS2}}{\partial x_j} + \frac{\partial u_{jS2}}{\partial x_i} - \frac{2}{3}\frac{\partial u_{kS2}}{\partial x_k}\delta_{ij}\right)$$
$$- \gamma_4\tau_{S1}\left(\frac{\partial u_{iS1}}{\partial x_j} + \frac{\partial u_{jS1}}{\partial x_i}\right) + \gamma_3\left(\frac{\partial^2 \tau_{S1}}{\partial x_i \partial x_j} - \frac{\partial^2\tau_{S1}}{\partial x_k^2}\delta_{ij}\right).$$

We will call this expression the stress in the P_{S3}^* (or modified pressure) system. In this system the thermal stress (non-Newtonian stress) may be taken as

$$\gamma_3\left(\frac{\partial^2 \tau_{S1}}{\partial x_i \partial x_j} - \frac{\partial^2\tau_{S1}}{\partial x_k^2}\delta_{ij}\right). \tag{4.84}$$

Proposition 4.4. *The thermal stress (non-Newtonian stress) in the* P_{S3}^* *system contributes neither to the force nor to the moment of force on a closed body in a gas.*

Proof. The contributions of the non-Newtonian stress in the P_{S3}^* system to FB_i and MO_i are, respectively, proportional to the integrals

$$\int_{\partial B} \left(\frac{\partial^2 \tau_{S1}}{\partial x_i \partial x_j} - \frac{\partial^2 \tau_{S1}}{\partial x_k^2} \delta_{ij} \right) \widehat{n}_j \mathrm{d}S,$$

$$\int_{\partial B} \epsilon_{ihk} x_h \left(\frac{\partial^2 \tau_{S1}}{\partial x_k \partial x_j} - \frac{\partial^2 \tau_{S1}}{\partial x_l^2} \delta_{kj} \right) \widehat{n}_j \mathrm{d}S.$$

Transforming the surface ∂B continuously from the body surface into the gas, which is allowed because of the tensors in the integrands being solenoidal, and applying the lemma in Section 3.8, we obtain the proposition. ∎

From Propositions 4.1 and 4.4, we have

Proposition 4.5. *Under the condition of Proposition 4.1, the force and the moment of force on a closed body can be obtained up to the third order of k in the expressions* FB_i/p_0L^2 and MO_i/p_0L^3 *by the classical gas dynamic calculation where the slip condition is taken into account. Here we mean by the classical gas dynamic calculation that the force or the moment of force is calculated with the aid of the Newtonian stress from the flow velocity and pressure fields obtained by solving the Navier–Stokes set of equations under the slip boundary condition.*

4.9 Summary of the asymptotic theory and a comment on a time-dependent problem

In this chapter we discussed the asymptotic behavior of a gas for small Knudsen numbers in the case where the deviation from a uniform equilibrium state at rest was of the order of the Knudsen number. The systematic process of derivation and the formulas (up to the second order of the Knudsen number) of the fluid-dynamic-type equations and their associated slip boundary conditions were presented. It should be noted that the fluid-dynamic-type equations that are to be used with the slip conditions (4.62a)–(4.62c) or (4.69a) and (4.69c) are not the Navier–Stokes set of equations but contain the thermal-stress term. By the discussions in Sections 4.7 and 4.8, it is found that the behavior of a gas up to this order can be treated by the Navier–Stokes set with the replacement (4.76). Incidentally, in the case where the temperature variation in a gas is not small (Chapter 5), the thermal-stress term cannot be included in the pressure term, and therefore the fluid-dynamic-type equations that should be used with the (leading) slip condition contain a term that is not found in the Navier–Stokes set of equations. This introduces an interesting result particular to a rarefied gas. [Moreover, a serious result about the behavior of a gas in the continuum limit (or incompleteness of the classical gas dynamics) is derived there.]

As mentioned at the beginning of this chapter, we cannot neglect the nonlinear effect for any small Mach number, if it is of the order of the Knudsen number. This is also shown by the numerical analysis of the BKW equation in Sone, Kataoka, Ohwada, Sugimoto & Aoki [1994]. In some infinite-domain problems, the variation of the velocity distribution function becomes more and more moderate in the far field or the length scale of variation becomes larger and larger there, which means that the effective Knudsen number becomes smaller and smaller. Thus, the nonlinear effect cannot be neglected for any small Mach number, and the linearized Boltzmann equation does not give a uniformly valid solution over the whole domain, which corresponds to the Stokes paradox in classical gas dynamics. This is studied in Cercignani [1968], Sone [1978], Onishi & Sone [1983], Sone, Kataoka, Ohwada, Sugimoto & Aoki [1994]. Some simple examples are given in Sections 4.10.1 and 4.10.2.

The fluid-dynamic-type equations derived by the S expansion are a series of the Navier–Stokes-type sets of equations with "the incompressible Navier–Stokes set of equations" as their leading set, and do not have the inconvenience that the degree of differentiation increases with the progress of approximation, which is encountered in the Chapman–Enskog expansion. Further the assumption that the velocity distribution function depends on the space variables only through the five macroscopic variables and their derivatives is unnecessary, and it comes out as a result. The difficulty of the Chapman–Enskog expansion (Chapman & Cowling [1952]) in the boundary-value problem is discussed in Cercignani [1988]. Incidentally, a non-well-posed equation is derived from this expansion (see Sone [1984b]).

The S expansion is a solution of the Boltzmann equation in a power series of k under the assumptions that $\phi = O(k)$ and $\partial\phi/\partial x_i = O(\phi)$. This is a Hilbert expansion (Hilbert [1912]) (see Section 6.2 or 7.2) with the additional constraint $\phi = O(k)$. In the framework of the Hilbert expansion, the S expansion, which is motivated by physical consideration, corresponds to the case where the Maxwellian distribution at the leading term of the Hilbert expansion is a uniform state at rest [a state with a uniform density and temperature (thus, a uniform pressure) and zero flow velocity]. These macroscopic variables at the trivial state obviously satisfy the Euler set of equations, which governs the macroscopic variables of the leading term of the Hilbert expansion. The leading nontrivial behavior of the gas is determined by the next-order term of the Hilbert expansion. The velocity distribution function up to this order is still Maxwellian and the macroscopic variables satisfy "the incompressible Navier–Stokes set of equations". Thus, the S expansion is a special case of the Hilbert expansion, but the correct understanding of the asymptotic behavior of a gas for small Knudsen numbers is not obtained by such a general discussion that the Hilbert expansion gives only the Euler set, but requires more detailed discussion. This is clearer and more important in the situation discussed in the next chapter, where the leading Maxwellian is in a state at rest but not in a uniform state. In Chapter 6, the Hilbert expansion where the length scale of variation of the velocity distribution function differs in directions is discussed, and the (compressible or incompressible) viscous (or Prandtl) boundary-layer set of equations is derived as the leading set of equations for the macroscopic variables. Special cases are not simply special cases. They cannot be understood by superficial general discussions.[20]

In this monograph, the time-independent behavior of a gas is exclusively considered. The extension of the analyses deriving fluid-dynamic-type equations in this monograph to a time-dependent problem is straightforward, that is, to retain $\mathrm{Sh}\partial\phi/\partial\hat{t}$ term in Eq. (2.58a) or (2.82) or $\mathrm{Sh}\partial\hat{f}/\partial\hat{t}$ term in Eq. (2.40a) and to carry out a similar expansion analysis. What is to be noted is the time scale t_0 of variation of the velocity distribution function. There are three important cases: Sh, defined in Eq. (2.41a), is of the order of unity, Sh is of the order of k, and Sh is of the order of Ma [Ma: Mach number (Section 3.9)]. In view of the definitions of the Strouhal number Sh and the mean free time $\bar{\tau}_c$ defined just before Eq. (2.19), the condition of the first case corresponds to $\bar{\tau}_c/t_0 = \mathrm{Sh}k = O(k)$, which is naturally conceived by the expansion of small k. The time scale t_0 of variation of the velocity distribution function is $t_0 = O(L/(2RT_0)^{1/2})$, that is, it is the time for the sound wave to proceed over the distance L (or the length scale of variation of the velocity distribution function), since $(2RT_0)^{1/2}$ is of the order of the sound speed. Thus, the first case describes the behavior of wave propagation. This case is discussed in Grad [1963a] and Bardos, Golse & Levermore [1998, 2000] (see also Bouchut, Golse & Pulvirenti [2000]). The second case [$\mathrm{Sh} = O(k)$ or $\bar{\tau}_c/t_0 = O(k^2)$] describes much slower variation or viscous diffusion because the time scale t_0 is expressed in terms of kinematic viscosity μ/ρ_0 as $t_0 = O(L^2\rho_0/\mu)$. Some examples in this case are discussed in Sone [1964, 1965, 1968] and the mathematical theory is developed in Bardos, Golse & Levermore [1993, 1998]. In the third case [$\mathrm{Sh} = O(\mathrm{Ma})$ or $\bar{\tau}_c/t_0 = O(k\mathrm{Ma})$], the time scale t_0 is of the order of L/U (U: a characteristic flow speed), i.e., the time required to traverse the distance L by the gas flow. In view of the von Karman relation (3.95), the third time scale is comparable to the second one when the Reynolds number is finite (the case related to this chapter and Chapter 5); it is much longer than the second when the Reynolds number is small (the case related to Chapter 3); it is comparable to the first one when the Mach number is of the order unity (thus, the Reynolds number is large) (the case related to Chapter 6).

The extension of the S expansion to a time-dependent problem under the first condition [or $\mathrm{Sh} = O(1)$] leads to a linear system. The leading set of fluid-dynamic-type equations is given by the linear conservation equations (2.85)–(2.87) with replacement: $\omega = \omega_{S1}$, $u_i = u_{iS1}$, $P = P_{S1}$, $P_{ij} = P_{S1}\delta_{ij}$, and $Q_i = 0$. The nonlinear terms and the viscosity and thermal conductivity of the gas first appear in the next-order equations. The variables u_{iS1} and P_{S1} are determined by the acoustic equations:

$$2\mathrm{Sh}\frac{\partial u_{iS1}}{\partial\hat{t}} + \frac{\partial P_{S1}}{\partial x_i} = 0, \qquad \frac{3}{5}\mathrm{Sh}\frac{\partial P_{S1}}{\partial\hat{t}} + \frac{\partial u_{iS1}}{\partial x_i} = 0. \tag{4.85}$$

The other variables ω_{S1} and τ_{S1} are derived from P_{S1} by the equations

$$\frac{\partial\omega_{S1}}{\partial\hat{t}} = \frac{3}{5}\frac{\partial P_{S1}}{\partial\hat{t}}, \qquad \tau_{S1} = P_{S1} - \omega_{S1}.$$

[20] Misleading discussions of the asymptotic behavior of a gas for small Knudsen numbers based on superficial understanding of the Chapman–Enskog expansion are prevalent.

The same equations with replacement of the subscript $S1$ by $G0$ are derived for the extension of the Grad–Hilbert expansion in Chapter 3.

The extension of the S expansion to a time-dependent problem under the second condition [or $\mathrm{Sh} = O(k)$] retains the feature of the time-independent equations. That is, Eqs. (4.30)–(4.31c) are subject to the change with additional terms $(\mathrm{Sh}/k)\partial u_{iS1}/\partial \hat{t}$ and $(\mathrm{Sh}/k)[\partial \tau_{S1}/\partial \hat{t} - (2/5)\partial P_{S1}/\partial \hat{t}]$ to the left-hand sides of Eqs. (4.31b) and (4.31c), respectively. That is,

$$\frac{\partial P_{S1}}{\partial x_i} = 0, \tag{4.86a}$$

$$\frac{\partial u_{iS1}}{\partial x_i} = 0, \tag{4.86b}$$

$$\frac{\mathrm{Sh}}{k}\frac{\partial u_{iS1}}{\partial \hat{t}} + u_{jS1}\frac{\partial u_{iS1}}{\partial x_j} = -\frac{1}{2}\frac{\partial P_{S2}}{\partial x_i} + \frac{1}{2}\gamma_1\frac{\partial^2 u_{iS1}}{\partial x_j^2}, \tag{4.86c}$$

$$\frac{\mathrm{Sh}}{k}\left(\frac{\partial \tau_{S1}}{\partial \hat{t}} - \frac{2}{5}\frac{\partial P_{S1}}{\partial \hat{t}}\right) + u_{jS1}\frac{\partial \tau_{S1}}{\partial x_j} = \frac{1}{2}\gamma_2\frac{\partial^2 \tau_{S1}}{\partial x_j^2}. \tag{4.86d}$$

The generalized equations are the leading set of equations derived from the compressible Navier–Stokes set of equations by the small-Mach-number (and small-temperature variation) expansion under the conditions that the Reynolds number Re $(= UL\rho_0/\mu)$ and the parameter $L^2\rho_0/\mu t_0$ are both finite. The extended form (4.86c) is the momentum equation of the Navier–Stokes set of equations for an incompressible fluid, but the extended form (4.86d) is different from the energy equation of the incompressible Navier–Stokes set, that is, in the latter the time-derivative term is given by $(3\mathrm{Sh}/5k)\partial \tau_{S1}/\partial \hat{t}$ and the other term $u_{iS1}\partial \tau_{S1}/\partial x_i$ on the left-hand side in the extended form (4.86d) should be replaced by $(3/5)u_{iS1}\partial \tau_{S1}/\partial x_i$.[21] An example showing the qualitative difference between the two systems will be shown in Section 4.10.3. The leading set of equations for the corresponding extension of the Grad–Hilbert expansion in Chapter 3 is obtained from the above set of equations by eliminating the nonlinear terms and replacing the subscript Sm by $Gm-1$.

In this way, we can understand the fluid-dynamic behavior of the solution of the Boltzmann equation on the basis of the Hilbert expansion by specifying the situation of our interest and ordering the size of variables correctly.

[21] In view of the above discussion, the small-Mach-number (and small-temperature-variation) expansion of the compressible Navier–Stokes set of equations for an ideal gas with a finite Reynolds number and with Sh being of the order of the Mach number does not give, as its leading set of equations, the Navier–Stokes set of equations for an incompressible fluid $[\partial\rho/\partial t + v_i\partial\rho/\partial X_i = 0$ (or $\mathrm{Sh}\partial\omega/\partial\hat{t} + u_i\partial\omega/\partial x_i = 0$)] under the same situation. The difference is due to the fact that the work done by pressure is of higher order in the incompressible fluid, as in the time-independent case. If the boundary condition on P_{S1} is time-independent, the $\partial P_{S1}/\partial\hat{t}$ term vanishes because of Eq. (4.86a). Then the difference between the two is reduced to the difference of the thermal conductivity as in the time-independent case explained in Footnote 6 in Section 4.3. If not, the relation is not so simple.

4.10 Applications

4.10.1 Half-space problem of evaporation and condensation

We will discuss the half-space problem of weak evaporation and condensation. Consider a semi-infinite expanse ($X_1 > 0$) of a gas bounded by its plane condensed phase at $X_1 = 0$. The plane condensed phase is at rest and is kept at a uniform temperature T_w. The saturated gas pressure at this temperature (or T_w) is p_w. The state of the gas far away from the condensed phase is a uniform state with pressure p_∞, temperature T_∞, and velocity $(v_{1\infty}, v_{2\infty}, 0)$. We discuss the steady behavior of the gas in the case where $(p_\infty - p_w)/p_w$, $(T_\infty - T_w)/T_w$, and $v_{i\infty}/\sqrt{2RT_w}$, denoted by P_∞, τ_∞, and $u_{i\infty}$ respectively, are small quantities of the same order, that is, τ_∞ and $u_{i\infty}$ are $O(|P_\infty|)$. The notations defined in Section 2.10 are used here with p_w and T_w as the reference quantities p_0 and T_0.

The problem does not contain a geometric characteristic length, and the mean free path is the only quantity with dimension of length. Thus, apparently, the solution has a length scale of variation of the order of the mean free path. However, we will show that the solution can have a different length scale $\sqrt{\pi}\ell_w/2|P_\infty|$ of variation (or a slowly varying structure in the far field), where ℓ_w is the mean free path in the equilibrium state at rest with pressure p_w and temperature T_w. We take $\sqrt{\pi}\ell_w/2|P_\infty|$ as the reference length L. Then k defined by Eq. (2.41b) is given by $|P_\infty|$. In the present situation the perturbed velocity distribution function ϕ, defined in Eq. (2.57), may be considered to be of the same order as the small quantity $|P_\infty|$ [or ϕ is of the order of k ($\ll 1$)].[22] This is the case that is studied in this chapter, and we can, therefore, apply the results of the asymptotic theory.

In the present one-dimensional problem, the governing equations (4.30)–(4.31c) are reduced to

$$\frac{\mathrm{d}P_{S1}}{\mathrm{d}x_1} = 0, \tag{4.87}$$

$$\frac{\mathrm{d}u_{1S1}}{\mathrm{d}x_1} = 0, \tag{4.88a}$$

$$u_{1S1}\frac{\mathrm{d}u_{iS1}}{\mathrm{d}x_1} = -\frac{1}{2}\frac{\mathrm{d}P_{S2}}{\mathrm{d}x_1}\delta_{i1} + \frac{1}{2}\gamma_1\frac{\mathrm{d}^2u_{iS1}}{\mathrm{d}x_1^2}, \tag{4.88b}$$

$$u_{1S1}\frac{\mathrm{d}\tau_{S1}}{\mathrm{d}x_1} = \frac{1}{2}\gamma_2\frac{\mathrm{d}^2\tau_{S1}}{\mathrm{d}x_1^2}. \tag{4.88c}$$

[22] In the situation where the length scale of variation is $O(\sqrt{\pi}\ell_w/2|P_\infty|)$ and the size of ϕ is $O(|P_\infty|)$, the nonlinear term $\mathcal{J}(\phi,\phi)/k$ of the Boltzmann equation (4.1) is of the same order [or $O(|P_\infty|)$] as its differential term $\zeta_i\partial\phi/\partial x_i$. Thus, nonlinear effect cannot be neglected even for small ϕ in the far field. The problem is studied from this view point in Sone [1978].

The boundary condition on the condensed phase at $x_1 = 0$ is, from Eqs. (4.68a) and (4.68c),

$$P_{S1} = C_4^* u_{1S1}, \tag{4.89a}$$

$$\tau_{S1} = d_4^* u_{1S1}, \tag{4.89b}$$

$$u_{2S1} = u_{3S1} = 0. \tag{4.89c}$$

The condition at infinity ($x_1 \to \infty$) is

$$P_{S1} \to P_\infty/|P_\infty|, \quad \tau_{S1} \to \tau_\infty/|P_\infty|, \tag{4.90a}$$

$$u_{1S1} \to u_{1\infty}/|P_\infty|, \quad u_{2S1} \to u_{2\infty}/|P_\infty|, \quad u_{3S1} \to 0. \tag{4.90b}$$

Obviously, it can be taken that $u_{3S1} = 0$. From Eqs. (4.87) and (4.88a), $P_{S1} =$ const and $u_{1S1} =$ const. Then, from Eq. (4.88b), $P_{S2} =$ const. From the condition on P_{S1} at infinity,

$$P_{S1} = P_\infty/|P_\infty|. \tag{4.91}$$

The normal velocity u_{1S1} (= const) is determined by Eqs. (4.89a) and (4.91) as

$$u_{1S1} = \frac{1}{C_4^*}\frac{P_\infty}{|P_\infty|},$$

and the velocity u_{1S1} can be made to satisfy the condition (4.90b) at infinity if and only if the boundary data $P_\infty/|P_\infty|$ and $u_{1\infty}/|P_\infty|$ are related as

$$u_{1\infty} = \frac{1}{C_4^*}P_\infty. \tag{4.92}$$

That is, the rate of evaporation or condensation is determined by the pressure difference P_∞.

The tangential velocity u_{2S1} and temperature τ_{S1} are determined in two different forms depending on the sign of P_∞. When $P_\infty \le 0$,[23] the following u_{2S1} and τ_{S1} are the only solutions u_{2S1} and τ_{S1} of Eqs. (4.88b) with $i = 2$ and (4.88c) that satisfy the conditions (4.89b) and (4.89c) and are finite at infinity:

$$u_{2S1} = 0,$$

$$\tau_{S1} = \frac{d_4^*}{C_4^*}\frac{P_\infty}{|P_\infty|},$$

for which the boundary data must satisfy the relations

$$u_{2\infty} = 0, \quad \tau_\infty = \frac{d_4^*}{C_4^*}P_\infty.$$

[23] The constant C_4^*, which is negative for a hard-sphere gas, the BKW model, etc., is assumed to be generally negative.

When $P_\infty > 0$, the solutions u_{2S1} and τ_{S1} of Eqs. (4.88b) with $i = 2$ and (4.88c) that satisfy the boundary conditions on the condensed phase and at infinity [or Eqs. (4.89b), (4.89c), (4.90a), and (4.90b)] are given as

$$u_{2S1} = \frac{u_{2\infty}}{|P_\infty|}\left[1 - \exp\left(\frac{2}{\gamma_1 C_4^*}x_1\right)\right],$$

$$\tau_{S1} = \frac{1}{|P_\infty|}\left[\tau_\infty + \left(\frac{d_4^*}{C_4^*}P_\infty - \tau_\infty\right)\exp\left(\frac{2}{\gamma_2 C_4^*}x_1\right)\right].$$

The Knudsen-layer corrections are given in the following forms common to both cases $P_\infty \leq 0$ and $P_\infty > 0$:

$$u_{iK1} = 0, \quad \omega_{K1} = u_{1S1}\Omega_4^*(x_1/|P_\infty|), \quad \tau_{K1} = u_{1S1}\Theta_4^*(x_1/|P_\infty|).$$

To summarize the results in dimensional variables, the leading terms of the expansion are as follows:

$$\frac{p - p_\infty}{p_w} = \frac{1}{C_4^*}\left(\frac{p_\infty}{p_w} - 1\right)\left[\Omega_4^*(2X_1/\sqrt{\pi}\ell_w) + \Theta_4^*(2X_1/\sqrt{\pi}\ell_w)\right], \tag{4.93a}$$

$$\frac{v_1}{\sqrt{2RT_w}} = \frac{1}{C_4^*}\left(\frac{p_\infty}{p_w} - 1\right), \tag{4.93b}$$

and when $p_\infty/p_w \leq 1$,

$$v_2 = 0, \quad v_3 = 0, \quad \frac{T}{T_w} - 1 = \frac{1}{C_4^*}\left(\frac{p_\infty}{p_w} - 1\right)\left[d_4^* + \Theta_4^*(2X_1/\sqrt{\pi}\ell_w)\right], \tag{4.94}$$

or when $p_\infty/p_w > 1$,

$$v_2 = v_{2\infty}\left[1 - \exp\left(\frac{4(p_\infty/p_w - 1)}{\sqrt{\pi}\gamma_1 C_4^*}\frac{X_1}{\ell_w}\right)\right], \qquad v_3 = 0, \tag{4.95a}$$

$$\frac{T - T_\infty}{T_w} = \left[\frac{d_4^*}{C_4^*}\left(\frac{p_\infty}{p_w} - 1\right) - \frac{T_\infty}{T_w} + 1\right]\exp\left(\frac{4(p_\infty/p_w - 1)}{\sqrt{\pi}\gamma_2 C_4^*}\frac{X_1}{\ell_w}\right) + \frac{1}{C_4^*}\left(\frac{p_\infty}{p_w} - 1\right)\Theta_4^*(2X_1/\sqrt{\pi}\ell_w). \tag{4.95b}$$

For the solution to exist, the boundary data must satisfy the condition

$$\frac{v_{1\infty}}{\sqrt{2RT_w}} = \frac{1}{C_4^*}\left(\frac{p_\infty}{p_w} - 1\right), \tag{4.96}$$

and in addition, the two conditions

$$v_{2\infty} = 0, \quad \frac{T_\infty}{T_w} - 1 = \frac{d_4^*}{C_4^*}\left(\frac{p_\infty}{p_w} - 1\right), \tag{4.97}$$

when $p_\infty/p_w \leq 1$. The condition (4.96) is the relation between the rate of evaporation or condensation and the pressure ratio.

When $p_\infty/p_w \leq 1$, v_1 is positive or zero (an evaporating flow), and when $p_\infty/p_w > 1$, v_1 is negative (a condensing flow). The feature of flow is qualitatively different in the two cases. The former solution is determined by giving only the pressure ratio p_∞/p_w, but the latter is determined by giving three parameters p_∞/p_w, T_∞/T_w, and $v_{2\infty}/\sqrt{2RT_w}$. The solution of the evaporating flow is practically the same as that of the linearized problem,[24] but the solution of the condensing flow, which contains exponential terms, is decisively different from the solution of the linearized problem, which is given by the common formulas to that of the evaporating flow. The difference is decisive in the far field. The nonlinear convection terms $u_{1S1}\mathrm{d}u_{iS1}/\mathrm{d}x_1$ and $u_{1S1}\mathrm{d}\tau_{S1}/\mathrm{d}x_1$ allow a slowly varying structure, other than the Knudsen layer, in the condensing flow, and the nonlinear effect is not negligible, however small $|p_\infty - p_w|/p_w$, $|T_\infty - T_w|/T_w$, and $|v_{2\infty}|/\sqrt{2RT_w}$ may be (Stokes paradox: see Section 4.9). The range in the parameter space $(p_\infty/p_w, T_\infty/T_w, v_{1\infty}/\sqrt{2RT_w}, v_{2\infty}/\sqrt{2RT_w})$ where a solution exists abruptly expands from a (one-dimensional) line [Eqs. (4.96) and (4.97)] to a three-dimensional hyperplane [Eq. (4.96)] as $p_\infty/p_w - 1$ passes zero from negative to positive values. (These results were first derived for the BKW equation in Sone [1978].)

In the present problem, Eq. (4.88b) with $i = 1$ does not participate in determining u_{iS1} and works only to determine P_{S2}. We will encounter a similar situation in Sections 4.10.2, 4.10.3, and 5.7. This is a common feature of one-dimensional problems with small Mach numbers.

4.10.2 Evaporation and condensation around a cylindrical or spherical condensed phase

Consider a cylindrical or spherical condensed phase (radius L) surrounded by an infinite expanse of its vapor gas at rest with pressure p_∞ and temperature T_∞ far from the condensed phase. The condensed phase is at rest and is kept at a uniform temperature T_w. The saturated gas pressure at T_w is denoted by p_w. We are interested here in the time-independent behavior of the gas under the following conditions: $|T_\infty - T_w|/T_w \ll 1$ and $|p_\infty - p_w|/p_w \ll 1$; the Knudsen number Kn $(= \ell_w/L = 2k/\sqrt{\pi})$, where ℓ_w is the mean free path in the equilibrium state at rest with pressure p_w and temperature T_w, is small, i.e., $k \ll 1$; and these small quantities, i.e., $|T_\infty - T_w|/T_w$, $|p_\infty - p_w|/p_w$, and Kn, are of the same order, i.e., $O(k)$. The behavior of the gas can easily be obtained with the aid of the asymptotic theory in this chapter. Here only the results of the leading term with respect to these small quantities are presented without derivation.

First the results of evaporation and condensation for a cylindrical condensed phase are given as follows:[25]

[24]The solution of the linearized problem is not given, but it is easily obtained with the aid of the asymptotic theory in the preceding chapter, if necessary. That is, the solution of the linearized problem is given by Eqs. (4.93a)–(4.94), (4.96), and (4.97) commonly for $p_\infty/p_w \leq 1$ and for $p_\infty/p_w > 1$.

[25]The result for the BKW equation is given in Onishi & Sone [1983].

I. Case $p_\infty/p_w \le 1$:

$$\frac{p-p_\infty}{p_w} = \frac{1}{C_4^*}\left(\frac{p_\infty}{p_w}-1\right)\left[\Omega_4^*(\eta)+\Theta_4^*(\eta)\right], \tag{4.98a}$$

$$\frac{v_r}{\sqrt{2RT_w}} = \frac{1}{C_4^*}\left(\frac{p_\infty}{p_w}-1\right)\frac{L}{r}, \tag{4.98b}$$

$$\frac{T}{T_w}-1 = \frac{1}{C_4^*}\left(\frac{p_\infty}{p_w}-1\right)\left[d_4^*+\Theta_4^*(\eta)\right], \tag{4.98c}$$

where r is the distance from the axis of the cylinder, v_r is the radial component of the flow velocity v_i in the cylindrical coordinate system and

$$\eta = \frac{1}{k}\left(\frac{r}{L}-1\right).$$

In this case, the solution exists only when the temperature ratio T_∞/T_w is related to the pressure ratio p_∞/p_w as

$$\frac{T_\infty}{T_w}-1 = \frac{d_4^*}{C_4^*}\left(\frac{p_\infty}{p_w}-1\right). \tag{4.99}$$

II. Case $p_\infty/p_w > 1$:

$$\frac{p-p_\infty}{p_w} = \frac{1}{C_4^*}\left(\frac{p_\infty}{p_w}-1\right)\left[\Omega_4^*(\eta)+\Theta_4^*(\eta)\right], \tag{4.100a}$$

$$\frac{v_r}{\sqrt{2RT_w}} = \frac{1}{C_4^*}\left(\frac{p_\infty}{p_w}-1\right)\frac{L}{r}, \tag{4.100b}$$

$$\begin{aligned}\frac{T}{T_w}-1 = {} & \frac{d_4^*}{C_4^*}\left(\frac{p_\infty}{p_w}-1\right)\exp\left(\frac{2}{\gamma_2 C_4^*}\frac{p_\infty-p_w}{p_w k}\ln\left(\frac{r}{L}\right)\right) \\ & + \left(\frac{T_\infty}{T_w}-1\right)\left[1-\exp\left(\frac{2}{\gamma_2 C_4^*}\frac{p_\infty-p_w}{p_w k}\ln\left(\frac{r}{L}\right)\right)\right] \\ & + \frac{1}{C_4^*}\left(\frac{p_\infty}{p_w}-1\right)\Theta_4^*(\eta).\end{aligned} \tag{4.100c}$$

When $p_\infty/p_w \le 1$, v_r is positive or zero (an evaporating flow), and when $p_\infty/p_w > 1$, v_r is negative (a condensing flow). The flow is qualitatively different in the two cases as in the half-space problem in Section 4.10.1; the former solution is determined only by the pressure ratio p_∞/p_w, but the latter is determined by the two parameters p_∞/p_w and T_∞/T_w. In the present cylinder problem as well as the half-space problem, the conduction of heat, expressed by the right-hand side of Eq. (4.31c), is strong enough to warm up (or cool down) the gas at infinity to the value on the condensed phase. In a condensing flow, its convection effect, expressed by the left-hand side of Eq. (4.31c), works to confine the conduction and establishes a structure, which is expressed by the exponential terms in Eq. (4.100c). According to the linear theory in Chapter 3, where the situation that the perturbation from a uniform state at rest is much smaller than the Knudsen number [or $|T_\infty - T_w|/T_w$ and $|p_\infty - p_w|/p_w \ll k \ll 1$] is considered and where the convection is absent [see Eq. (3.19c)], the results of the cylinder problem are given by Eqs. (4.98a)–(4.98c) and (4.99), irrespective of $p_\infty/p_w \leq 1$ or $p_\infty/p_w > 1$. This result for the condensation is obtained formally by putting $(p_\infty - p_w)/p_w k = 0$ in the exponential functions in Eq. (4.100c). However, this process is not legitimate in the far field, because the argument of the exponential function, i.e., $(|p_\infty - p_w|/p_w k)\ln(r/L)$, is not small in the far field even for small $|p_\infty - p_w|/p_w k$. The nonlinear effect cannot be neglected in the case $v_r < 0$, however small $|p_\infty - p_w|/p_w$ and $|T_\infty - T_w|/T_w$ may be (Stokes paradox: see Section 4.9).

Now we give the results for a spherical condensed phase.

$$\frac{p - p_\infty}{p_w} = \frac{1}{C_4^*}\left(\frac{p_\infty}{p_w} - 1\right)[\Omega_4^*(\eta) + \Theta_4^*(\eta)], \tag{4.101a}$$

$$\frac{v_r}{\sqrt{2RT_w}} = \frac{1}{C_4^*}\left(\frac{p_\infty}{p_w} - 1\right)\left(\frac{L}{r}\right)^2, \tag{4.101b}$$

$$\begin{aligned}\frac{T}{T_w} - 1 = \left(\frac{T_\infty}{T_w} - 1\right) + \left[\frac{d_4^*}{C_4^*}\left(\frac{p_\infty}{p_w} - 1\right) - \left(\frac{T_\infty}{T_w} - 1\right)\right]\frac{1 - \exp(-c_A L/r)}{1 - \exp(-c_A)} \\ + \frac{1}{C_4^*}\left(\frac{p_\infty}{p_w} - 1\right)\Theta_4^*(\eta),\end{aligned} \tag{4.101c}$$

$$c_A = \frac{2}{\gamma_2}\frac{v_r r^2}{\sqrt{2RT_w}L^2 k} = \frac{2}{\gamma_2 C_4^*}\frac{p_\infty - p_w}{p_w k}, \tag{4.101d}$$

where r is the distance from the center of the sphere, v_r is the radial component of the flow velocity v_i in the spherical coordinate system and

$$\eta = \frac{1}{k}\left(\frac{r}{L} - 1\right).$$

When $p_\infty/p_w < 1$, v_r is positive (an evaporating flow), and when $p_\infty/p_w > 1$, v_r is negative (a condensing flow). The solution is given by a single set of formulas. The feature of the solution is common to the two cases; the solution is determined by the two parameters p_∞/p_w and T_∞/T_w. The corresponding solution by the linear theory in Chapter 3 is given by putting $c_A = 0$ in Eqs. (4.101a)–(4.101c). The solution of the linear theory is also determined by the two parameters p_∞/p_w and T_∞/T_w. In the spherical case, the conduction effect is not strong enough to warm up (or cool down) the gas at infinity to the temperature on the condensed phase, and the temperature at infinity can be specified at our disposal, because the heat has to be supplied to (or extracted from) a wider region compared with the cylindrical case. The convection effect gives a structure, expressed by the exponential functions in Eq. (4.101c), to the conduction temperature field. Its effect is not so strong to warm up (or cool down) the gas at infinity in an evaporating flow because of faster decay of v_r as $r \to \infty$. The solution by the linear theory is valid for small $(p_\infty - p_w)/p_w k$ in contrast to that in the cylinder problem.

4.10.3 The difference of the temperature field for the S expansion and for the incompressible Navier–Stokes set in a time-dependent problem

Here we show an example showing the difference between the leading-order solution of the S expansion and that of the incompressible Navier–Stokes set of equations in time-dependent problems, which is briefly discussed in Section 4.9. This difference is also the difference between the leading-order solution of the small-Mach-number expansion of the compressible Navier–Stokes set and that of the incompressible set.

Consider a gas between two parallel plane walls at rest at $x_1 = 0$ and $x_1 = 1$. The wall at $x_1 = 1$ is a simple boundary, and the one at $x_1 = 0$ is the condensed phase of the gas. The temperature of the simple wall is given by T_0 (or $\tau_w = 0$), and that of the condensed phase by $T_0[1 + k\tau_C(\hat{t})]$, where $\tau_C(\hat{t})$ is a given moderately varying function of $\hat{t}$ with the reference time t_0 for which Sh/k $[= L/t_0(2RT_0)^{1/2}k]$ is of the order of unity. As the initial state at $\hat{t} = 0$, the uniform equilibrium state at rest at temperature T_0 and pressure p_0, where p_0 is the saturated gas pressure at temperature T_0, is taken.

The extended equations (4.86a)–(4.86d) for the corresponding time-dependent problem, discussed in Section 4.9, are reduced to the following forms for the present one-dimensional case:

$$\frac{\partial P_{S1}}{\partial x_1} = 0, \tag{4.102}$$

$$\frac{\partial u_{1S1}}{\partial x_1} = 0, \tag{4.103a}$$

$$\frac{\mathrm{Sh}}{k}\frac{\partial u_{1S1}}{\partial \hat{t}} + u_{1S1}\frac{\partial u_{1S1}}{\partial x_1} = -\frac{1}{2}\frac{\partial P_{S2}}{\partial x_1} + \frac{1}{2}\gamma_1\frac{\partial^2 u_{1S1}}{\partial x_1^2}, \tag{4.103b}$$

$$\frac{\mathrm{Sh}}{k}\left(\frac{\partial \tau_{S1}}{\partial \hat{t}} - \frac{2}{5}\frac{\partial P_{S1}}{\partial \hat{t}}\right) + u_{1S1}\frac{\partial \tau_{S1}}{\partial x_1} = \frac{1}{2}\gamma_2\frac{\partial^2 \tau_{S1}}{\partial x_1^2}, \tag{4.103c}$$

where the tangential velocities u_{2S1} and u_{3S1} are consistently taken to be zero because of the symmetry. The boundary conditions on the condensed phase at $x_1 = 0$ are, from Eq. (4.68c),[26]

$$\begin{bmatrix} P_{S1} - P_C(\hat{t}) \\ \tau_{S1} - \tau_C(\hat{t}) \end{bmatrix} = u_{1S1}\begin{bmatrix} C_4^* \\ d_4^* \end{bmatrix} \quad \text{at } x_1 = 0, \tag{4.104}$$

[26]The boundary conditions given in Section 4.5.2 are derived for the time-independent problem. However, for the slow variation considered here, the time-dependent term does not contribute to the Knudsen-layer equation (4.46).

and the conditions on the simple boundary at $x_1 = 1$ are, from Eqs. (4.61a) and (4.61b),

$$u_{1S1} = 0, \quad \tau_{S1} = 0 \quad \text{at } x_1 = 1, \tag{4.105}$$

where $p_0[1+kP_C(\hat{t})]$ is the saturated gas pressure at temperature $T_0[1+k\tau_C(\hat{t})]$. The initial condition at $\hat{t} = 0$ is[27]

$$u_{1S1} = 0, \quad \tau_{S1} = 0, \quad P_{S1} = 0. \tag{4.106}$$

On the other hand, the incompressible Navier–Stokes set of equations for the corresponding situation is given by

$$\frac{\partial P_{N1}}{\partial x_1} = 0, \tag{4.107}$$

$$\frac{\partial u_{1N1}}{\partial x_1} = 0, \tag{4.108a}$$

$$\frac{\mathrm{Sh}}{k}\frac{\partial u_{1N1}}{\partial \hat{t}} + u_{1N1}\frac{\partial u_{1N1}}{\partial x_1} = -\frac{1}{2}\frac{\partial P_{N2}}{\partial x_1} + \frac{1}{2}\gamma_1\frac{\partial^2 u_{1N1}}{\partial x_1^2}, \tag{4.108b}$$

$$\frac{3\mathrm{Sh}}{5k}\frac{\partial \tau_{N1}}{\partial \hat{t}} + \frac{3}{5}u_{1N1}\frac{\partial \tau_{N1}}{\partial x_1} = \frac{1}{2}\gamma_2\frac{\partial^2 \tau_{N1}}{\partial x_1^2}, \tag{4.108c}$$

where the subscript N is attached for discrimination. The boundary conditions are[28]

$$\begin{bmatrix} P_{N1} - P_C(\hat{t}) \\ \tau_{N1} - \tau_C(\hat{t}) \end{bmatrix} = u_{1N1}\begin{bmatrix} C_4^* \\ d_4^* \end{bmatrix} \quad \text{at } x_1 = 0, \tag{4.109}$$

and

$$u_{1N1} = 0, \quad \tau_{N1} = 0 \quad \text{at } x_1 = 1. \tag{4.110}$$

The initial condition at $\hat{t} = 0$ is

$$u_{1N1} = 0, \quad \tau_{N1} = 0, \quad P_{N1} = 0. \tag{4.111}$$

First consider the solution of the S expansion. From Eqs. (4.103a) and (4.105)

$$u_{S1} = 0.$$

Then, the boundary condition (4.104) at $x_1 = 0$ is reduced to

$$P_{S1} = P_C(\hat{t}), \quad \tau_{S1} = \tau_C(\hat{t}). \tag{4.112}$$

From Eq. (4.102) with the former relation in this boundary condition, the pressure P_{S1} in the gas is determined as

$$P_{S1} = P_C(\hat{t}). \tag{4.113}$$

---

[27]If the variations of the boundary data $\tau_C(\hat{t})$ and $P_C(\hat{t})$ start smoothly from their initial values $\tau_C(0) = 0$ and $P_C(0) = 0$ and are moderate as functions of $\hat{t}$ (or on the time scale of $L^2\rho_0/\mu$), the effects of the initial layer (Grad [1963a]) and the acoustic region (Section 4.9) can be neglected.

[28]The boundary conditions derived for the S system are used here.

Then the temperature field is determined by the following heat-conduction equation with a source term:

$$\frac{\mathrm{Sh}}{k}\frac{\partial \tau_{S1}}{\partial \hat{t}} = \frac{1}{2}\gamma_2 \frac{\partial^2 \tau_{S1}}{\partial x_1^2} + \frac{2\mathrm{Sh}}{5k}\frac{\mathrm{d}P_C(\hat{t})}{\mathrm{d}\hat{t}}.$$

The solution of the equation satisfying the boundary and initial conditions (4.105), (4.106) and (4.112) is expressed in the following form (Carslaw & Jaeger [1959]):

$$\begin{aligned}\tau_{S1} = \frac{2}{5}P_C(\hat{t}) &+ \int_0^{\hat{t}} \left(\frac{\mathrm{d}\tau_C(\hat{t})}{\mathrm{d}\hat{t}} - \frac{2}{5}\frac{\mathrm{d}P_C(\hat{t})}{\mathrm{d}\hat{t}}\right)_{\hat{t}=\hat{t}'} \mathcal{H}(x_1, \gamma_2(\hat{t}-\hat{t}')/(2\mathrm{Sh}/k))\mathrm{d}\hat{t}' \\ &- \frac{2}{5}\int_0^{\hat{t}} \left(\frac{\mathrm{d}P_C(\hat{t})}{\mathrm{d}\hat{t}}\right)_{\hat{t}=\hat{t}'} \mathcal{H}(1-x_1, \gamma_2(\hat{t}-\hat{t}')/(2\mathrm{Sh}/k))\mathrm{d}\hat{t}',\end{aligned}$$

where the function $\mathcal{H}(x,t)$ is the solution of the following simpler initial- and boundary-value problem of the heat-conduction equation:[29]

$$\begin{aligned}&\frac{\partial \mathcal{H}}{\partial t} = \frac{\partial^2 \mathcal{H}}{\partial x^2}, \\ &\mathcal{H}(0,t) = 1, \quad \mathcal{H}(1,t) = 0, \\ &\mathcal{H}(x,0) = 0.\end{aligned}$$

The solution of the incompressible Navier–Stokes system is similarly obtained in the form

$$\begin{aligned}u_{N1} &= 0, \quad P_{N1} = P_C(\hat{t}), \\ \tau_{N1} &= \int_0^{\hat{t}} \left(\frac{\mathrm{d}\tau_C(\hat{t})}{\mathrm{d}\hat{t}}\right)_{\hat{t}=\hat{t}'} \mathcal{H}(x_1, \gamma_2(\hat{t}-\hat{t}')/(6\mathrm{Sh}/5k))\mathrm{d}\hat{t}'.\end{aligned}$$

The temperature field of the incompressible Navier–Stokes set is independent of the saturated gas pressure $P_C(\hat{t})$, which is determined by the Clausius–Clapeyron relation (e.g., Reif [1965], Landau & Lifshitz [1987]).

---

[29] If the simple wall is at infinity ($x_1 = \infty$) instead of at $x_1 = 1$, the corresponding function $\mathcal{H}(x,t)$ satisfying $\mathcal{H}(\infty,t) = 0$, instead of $\mathcal{H}(1,t) = 0$, is given by

$$\mathcal{H}(x,t) = \frac{2}{\sqrt{\pi}}\int_{x/2\sqrt{t}}^{\infty} \exp(-s^2)\mathrm{d}s.$$

Chapter 5

Nonlinear Theory I – Finite Temperature Variations and Ghost Effect

In the preceding chapter, we discussed the case where the deviation of the system from a uniform equilibrium state at rest is of the order of the Knudsen number and derived a series of fluid-dynamic-type equations with "the incompressible Navier–Stokes set of equations"[1] as the leading set of equations. In this system, the variation of temperature of the gas (compared with the average temperature of the system) as well as the flow Mach number is limited to be a small quantity of the order of the Knudsen number. In the present chapter, we try to eliminate the restriction on temperature variation. The result is not just an extension of the formulas but contains a very important fact concerning the behavior of a gas in the continuum limit. That is, the classical gas dynamics is found to be incomplete to describe the behavior of a gas in the continuum limit, in contradiction with the purpose of the classical gas dynamics. This problem is discussed in the second half of this chapter.

5.1 Problem

Consider a slightly rarefied gas in a steady state in an arbitrary domain except that its boundary shape is smooth. No external force is acting on the gas [or F_i in Eq. (2.5) is zero]. The flow speed is small and its Mach number is of the order of the Knudsen number. We will investigate the asymptotic behavior of this system for small Knudsen numbers. That is, the problem is the asymptotic analysis for small Knudsen and Mach numbers of the time-independent boundary-value problem of the Boltzmann equation in a general domain under the condition

[1] See Footnote 6 in Section 4.3 for the meaning of the quotation mark.

that the Mach number is of the same order as the Knudsen number. In this chapter, we use the notation introduced in Section 2.9.

The basic equation is Eq. (2.40a) with $\partial/\partial\hat{t} = 0$ and $\hat{F}_i = 0$, i.e.,

$$\zeta_i \frac{\partial \hat{f}}{\partial x_i} = \frac{1}{k}\hat{J}(\hat{f}, \hat{f}). \tag{5.1}$$

The boundary condition is, symbolically,

$$\hat{f} = \hat{f}_w \quad (\zeta_i n_i > 0), \tag{5.2}$$

where n_i is the unit normal vector to the boundary, pointed to the gas region, and the condition $\hat{v}_{wi} n_i = 0$, required in a steady system, is used. For example, for the diffuse reflection [Eq. (2.55a) with (2.55b) and $\alpha = 1$],

$$\hat{f}(x_i, \zeta_i) = \frac{\hat{\sigma}_w}{(\pi \hat{T}_w)^{3/2}} \exp\left(-\frac{(\zeta_i - \hat{v}_{wi})^2}{\hat{T}_w}\right) \quad (\zeta_j n_j > 0), \tag{5.3a}$$

$$\hat{\sigma}_w = -2\left(\frac{\pi}{\hat{T}_w}\right)^{1/2} \int_{\zeta_j n_j < 0} \zeta_j n_j \hat{f}(x_i, \zeta_i) \mathbf{d}\boldsymbol{\zeta}, \tag{5.3b}$$

and for the complete condensation condition [Eq. (2.56a) with $\alpha_c = 1$],

$$\hat{f}(x_i, \zeta_i) = \frac{\hat{p}_w}{\pi^{3/2} \hat{T}_w^{5/2}} \exp\left(-\frac{(\zeta_i - \hat{v}_{wi})^2}{\hat{T}_w}\right) \quad (\zeta_j n_j > 0). \tag{5.4}$$

The boundary parameters $\hat{v}_{wi}$, $\hat{T}_w$, and $\hat{p}_w$ generally depend on k, and they are assumed to be expanded in power series of k; Corresponding to the assumption of vanishing flow velocity in the continuum limit, the series of $\hat{v}_{wi}$ starts from the term of the order of k, that is,

$$\hat{v}_{wi} = \hat{v}_{wi1} k + \cdots, \tag{5.5a}$$

$$\hat{T}_w = \hat{T}_{w0} + \hat{T}_{w1} k + \cdots, \tag{5.5b}$$

$$\hat{p}_w = \hat{p}_{w0} + \hat{p}_{w1} k + \cdots. \tag{5.5c}$$

For the moment, the explicit form of the boundary condition is not specified. The boundary may be a simple boundary or an interface of the gas and its condensed phase.

In the following sections, the asymptotic behavior of the solution $\hat{f}(x_i, \zeta_i)$ of the boundary-value problem (5.1) with (5.2) for small k (or $k \ll 1$) is studied under the assumption that

$$\int \zeta_i \hat{f} \mathbf{d}\boldsymbol{\zeta} = O(k), \tag{5.6}$$

on the basis of the asymptotic analysis by Sone & Wakabayashi [1988] for the BKW equation, which is later generalized to the standard Boltzmann equation in Sone, Aoki, Takata, Sugimoto & Bobylev [1996]. The assumption $\phi = O(k)$ on the velocity distribution function in the preceding chapter (Chapter 4) is weakened here to the assumption (5.6) on a moment of the velocity distribution function.

5.2 SB solution

As in the preceding two chapters, putting aside the boundary condition, we look for a moderately varying solution of Eq. (5.1), whose length scale of variation is of the order of the characteristic length L of the system $[\partial\hat{f}/\partial x_i = O(\hat{f})]$, in a power series of k:

$$\hat{f}_{SB} = \hat{f}_{SB0} + \hat{f}_{SB1}k + \hat{f}_{SB2}k^2 + \cdots, \tag{5.7}$$

where the subscript *SB* is attached to discriminate the moderately varying solution satisfying the condition (5.6). This type of solution (or expansion) will be called SB solution (or expansion). The condition (5.6) is reduced to the following condition on the component function $\hat{f}_{SB0}$ of the expansion (5.7):

$$\int \zeta_i \hat{f}_{SB0} \mathbf{d}\boldsymbol{\zeta} = 0. \tag{5.8}$$

The relation between the macroscopic variables and the velocity distribution function is the same as Eqs. (2.48a)–(2.48f) except for the subscript *SB*:

$$\hat{\rho}_{SB} = \int \hat{f}_{SB} \mathbf{d}\boldsymbol{\zeta}, \tag{5.9a}$$

$$\hat{\rho}_{SB}\hat{v}_{iSB} = \int \zeta_i \hat{f}_{SB} \mathbf{d}\boldsymbol{\zeta}, \tag{5.9b}$$

$$\frac{3}{2}\hat{\rho}_{SB}\hat{T}_{SB} = \int (\zeta_i - \hat{v}_{iSB})^2 \hat{f}_{SB} \mathbf{d}\boldsymbol{\zeta}, \tag{5.9c}$$

$$\hat{p}_{SB} = \hat{\rho}_{SB}\hat{T}_{SB}, \tag{5.9d}$$

$$\hat{p}_{ijSB} = 2\int (\zeta_i - \hat{v}_{iSB})(\zeta_j - \hat{v}_{jSB}) \hat{f}_{SB} \mathbf{d}\boldsymbol{\zeta}, \tag{5.9e}$$

$$\hat{q}_{iSB} = \int (\zeta_i - \hat{v}_{iSB})(\zeta_j - \hat{v}_{jSB})^2 \hat{f}_{SB} \mathbf{d}\boldsymbol{\zeta}. \tag{5.9f}$$

Corresponding to the expansion (5.7), the macroscopic variable $\hat{h}_{SB}$, where $\hat{h}$ represents $\hat{\rho}$, $\hat{v}_i$, $\hat{T}$, etc., is also expanded in k:

$$\hat{h}_{SB} = \hat{h}_{SB0} + \hat{h}_{SB1}k + \hat{h}_{SB2}k^2 + \cdots. \tag{5.10}$$

The component function $\hat{h}_{SBm}$ is related to the component function of the velocity distribution function as follows:

$$\hat{\rho}_{SB0} = \int \hat{f}_{SB0}\mathbf{d}\boldsymbol{\zeta}, \tag{5.11a}$$

$$\hat{\rho}_{SB0}\hat{v}_{iSB0} = \int \zeta_i \hat{f}_{SB0}\mathbf{d}\boldsymbol{\zeta} = 0, \tag{5.11b}$$

$$\frac{3}{2}\hat{\rho}_{SB0}\hat{T}_{SB0} = \int {\zeta_i}^2 \hat{f}_{SB0}\mathbf{d}\boldsymbol{\zeta}, \tag{5.11c}$$

$$\hat{p}_{SB0} = \hat{\rho}_{SB0}\hat{T}_{SB0}, \tag{5.11d}$$

$$\hat{p}_{ijSB0} = 2\int \zeta_i\zeta_j \hat{f}_{SB0}\mathbf{d}\boldsymbol{\zeta}, \tag{5.11e}$$

$$\hat{q}_{iSB0} = \int \zeta_i{\zeta_j}^2 \hat{f}_{SB0}\mathbf{d}\boldsymbol{\zeta}, \tag{5.11f}$$

$$\hat{\rho}_{SB1} = \int \hat{f}_{SB1}\mathbf{d}\boldsymbol{\zeta}, \tag{5.12a}$$

$$\hat{\rho}_{SB0}\hat{v}_{iSB1} = \int \zeta_i \hat{f}_{SB1}\mathbf{d}\boldsymbol{\zeta}, \tag{5.12b}$$

$$\frac{3}{2}\hat{\rho}_{SB0}\hat{T}_{SB1} = \int \zeta_i^2 \hat{f}_{SB1}\mathbf{d}\boldsymbol{\zeta} - \frac{3}{2}\hat{\rho}_{SB1}\hat{T}_{SB0}, \tag{5.12c}$$

$$\hat{p}_{SB1} = \hat{\rho}_{SB0}\hat{T}_{SB1} + \hat{\rho}_{SB1}\hat{T}_{SB0}, \tag{5.12d}$$

$$\hat{p}_{ijSB1} = 2\int \zeta_i\zeta_j \hat{f}_{SB1}\mathbf{d}\boldsymbol{\zeta}, \tag{5.12e}$$

$$\hat{q}_{iSB1} = \int \zeta_i\zeta_j^2 \hat{f}_{SB1}\mathbf{d}\boldsymbol{\zeta} - \frac{3}{2}\hat{\rho}_{SB0}\hat{T}_{SB0}\hat{v}_{iSB1} - \hat{p}_{ijSB0}\hat{v}_{jSB1}, \tag{5.12f}$$

$$\cdots\cdots\cdots\cdots,$$

where the condition (5.8) is used.

Now return to obtaining the SB solution. Substituting Eq. (5.7) into the Boltzmann equation (5.1) and arranging the same-order terms in k, we obtain a series of integral equations for the component function $\hat{f}_{SBm}$:[2]

$$\hat{J}(\hat{f}_{SB0}, \hat{f}_{SB0}) = 0, \tag{5.13}$$

$$2\hat{J}(\hat{f}_{SB0}, \hat{f}_{SBm}) = \zeta_i \frac{\partial \hat{f}_{SBm-1}}{\partial x_i} - \sum_{r=1}^{m-1} \hat{J}(\hat{f}_{SBr}, \hat{f}_{SBm-r}) \quad (m \geq 1), \tag{5.14}$$

where the $\sum$ term is absent when $m = 1$.

The solution $\hat{f}_{SB0}$ of the integral equation (5.13) satisfying the condition (5.8) is given by

$$\hat{f}_{SB0} = \frac{\hat{\rho}_{SB0}}{(\pi\hat{T}_{SB0})^{3/2}} \exp\left(-\frac{\zeta_i^2}{\hat{T}_{SB0}}\right), \tag{5.15}$$

[2] For the BKW equation, the collision integral on the left-hand side in Eq. (5.14) is taken as the collision integral linearized around the Maxwellian $\hat{f}_{SB0}$ (see Footnotes 12 in Section 2.9 and 26 in Appendix A.7). See Footnote 1 in Section 4.2 for the meaning of the terms under $\sum$ sign on the right-hand side.

where the relations (5.11a) and (5.11c) are used. The solution (5.15) is incomplete to determine $\hat{f}_{SB0}$, because the spatial variations of the parameter functions $\hat{\rho}_{SB0}$ and $\hat{T}_{SB0}$ are not specified. With this $\hat{f}_{SB0}$, the equation (5.14) is the inhomogeneous linear integral equation for $\hat{f}_{SBm}$ $(m \geq 1)$. The homogeneous equation corresponding to Eq. (5.14), i.e.,

$$\hat{J}(\hat{f}_{SB0}, \hat{g}) = 0, \tag{5.16}$$

has five independent solutions, i.e.,

$$\hat{g} = \hat{f}_{SB0}g, \quad g = 1, \quad \zeta_i, \quad \zeta_j^2, \tag{5.17}$$

which is seen from Eq. (2.61c) with Eqs. (2.59a) and (2.59b) (or from the relations $g' + g'_* = g + g_*$ and $\hat{f}'_{SB0}\hat{f}'_{SB0*} = \hat{f}_{SB0}\hat{f}_{SB0*}$). From the general relation (2.44), the integral $\int g\hat{J}(\hat{f}_{SB0}, \hat{f}_{SBm})\mathbf{d}\boldsymbol{\zeta}$ of the left-hand side of Eq. (5.14) vanishes. Therefore, the inhomogeneous term of the integral equation (5.14) must satisfy the following relation (solvability condition) for Eq. (5.14) to have a solution:

$$\int g\left(\zeta_i \frac{\partial \hat{f}_{SBm-1}}{\partial x_i} - \sum_{r=1}^{m-1} \hat{J}(\hat{f}_{SBr}, \hat{f}_{SBm-r})\right)\mathbf{d}\boldsymbol{\zeta} = 0. \tag{5.18}$$

With the general relation (2.44) of the collision integral, the above solvability condition is reduced to

$$\int g\zeta_i \frac{\partial \hat{f}_{SBm-1}}{\partial x_i}\mathbf{d}\boldsymbol{\zeta} = 0. \tag{5.19}$$

The solvability condition (5.19) being satisfied, the solution of the integral equation (5.14) is expressed in the form:

$$\hat{f}_{SBm} = \hat{f}_{SB0}(c_{0m} + c_{im}\zeta_i + c_{4m}\zeta_i^2) + \hat{\psi}_{SBm}, \tag{5.20}$$

where $\hat{\psi}_{SBm}$ is the particular solution satisfying the orthogonal relation

$$\int g\hat{\psi}_{SBm}\mathbf{d}\boldsymbol{\zeta} = 0, \tag{5.21}$$

and

$$c_{0m} = \frac{5\hat{\rho}_{SBm}}{2\hat{\rho}_{SB0}} - \frac{3\hat{p}_{SBm}}{2\hat{p}_{SB0}} - \frac{(\hat{\rho}_{SB}\hat{v}_{jSB}^2)_m}{\hat{p}_{SB0}}, \tag{5.22a}$$

$$c_{im} = \frac{2(\hat{\rho}_{SB}\hat{v}_{iSB})_m}{\hat{p}_{SB0}}, \tag{5.22b}$$

$$c_{4m} = \frac{1}{\hat{T}_{SB0}}\left[\frac{\hat{p}_{SBm}}{\hat{p}_{SB0}} - \frac{\hat{\rho}_{SBm}}{\hat{\rho}_{SB0}} + \frac{2(\hat{\rho}_{SB}\hat{v}_{jSB}^2)_m}{3\hat{p}_{SB0}}\right]. \tag{5.22c}$$

The notation $(\cdots)_m$ indicates the m-th order component function of the SB expansion, for example $(\hat{\rho}_{SB}\hat{v}_{iSB}^2)_3 = 2\hat{\rho}_{SB0}\hat{v}_{iSB1}\hat{v}_{iSB2} + \hat{\rho}_{SB1}\hat{v}_{iSB1}^2$.

More explicitly, the inhomogeneous term of Eq. (5.14) for $m = 1$ is

$$\zeta_i \frac{\partial \hat{f}_{SB0}}{\partial x_i} = \zeta_i \left[\frac{1}{\hat{\rho}_{SB0}} \frac{\partial \hat{\rho}_{SB0}}{\partial x_i} + \frac{1}{\hat{T}_{SB0}} \frac{\partial \hat{T}_{SB0}}{\partial x_i} \left(\frac{\zeta_j^2}{\hat{T}_{SB0}} - \frac{3}{2} \right) \right] \hat{f}_{SB0}. \tag{5.23}$$

The two relations for $g = 1$ and $g = \zeta_i^2$ in the solvability condition (5.19) for $m = 1$ are reduced to identities, and the relation for $g = \zeta_i$ is

$$\frac{\partial \hat{p}_{SB0}}{\partial x_i} = 0. \tag{5.24}$$

Then, the inhomogeneous term (5.23) is reduced to

$$\zeta_i \frac{\partial \hat{f}_{SB0}}{\partial x_i} = \frac{\zeta_i}{\hat{T}_{SB0}} \frac{\partial \hat{T}_{SB0}}{\partial x_i} \left(\frac{\zeta_j^2}{\hat{T}_{SB0}} - \frac{5}{2} \right) \hat{f}_{SB0}. \tag{5.25}$$

With this inhomogeneous term, the solution $\hat{f}_{SB1}$ is found to be of the form (see Appendix A.7):

$$\begin{aligned} \hat{f}_{SB1} &= \hat{f}_{SB0} \left[\frac{\hat{p}_{SB1}}{\hat{p}_{SB0}} + \frac{2\zeta_i \hat{\rho}_{SB0} \hat{v}_{iSB1}}{\hat{p}_{SB0}} + \left(\frac{\zeta^2}{\hat{T}_{SB0}} - \frac{5}{2} \right) \frac{\hat{T}_{SB1}}{\hat{T}_{SB0}} \right. \\ &\qquad \left. - \frac{\zeta_i \mathcal{A}(\zeta/\hat{T}_{SB0}^{1/2}, \hat{T}_{SB0})}{\hat{T}_{SB0}^{1/2} \hat{p}_{SB0}} \frac{\partial \hat{T}_{SB0}}{\partial x_i} \right] \\ &= \frac{\hat{p}_{SB0}}{\hat{T}_{SB0}^{3/2}} E(\tilde{\zeta}) \left[\frac{\hat{p}_{SB1}}{\hat{p}_{SB0}} + \frac{2\hat{T}_{SB0}^{1/2} \hat{\rho}_{SB0} \hat{v}_{iSB1}}{\hat{p}_{SB0}} \tilde{\zeta}_i + \frac{\hat{T}_{SB1}}{\hat{T}_{SB0}} \left(\tilde{\zeta}^2 - \frac{5}{2} \right) \right. \\ &\qquad \left. - \frac{1}{\hat{p}_{SB0}} \frac{\partial \hat{T}_{SB0}}{\partial x_i} \tilde{\zeta}_i \mathcal{A}(\tilde{\zeta}, \hat{T}_{SB0}) \right], \end{aligned} \tag{5.26}$$

$$\tilde{\zeta}_i = \frac{\zeta_i}{\hat{T}_{SB0}^{1/2}}, \quad \tilde{\zeta} = \sqrt{\tilde{\zeta}_i^2}, \quad E(\tilde{\zeta}) = \frac{1}{\pi^{3/2}} \exp(-\tilde{\zeta}^2),$$

where $\mathcal{A}(\tilde{\zeta}, \hat{T}_{SB0})$ is a function defined by Eq. (A.84) in Appendix A.5. The definition of the symbol $\tilde{\zeta}_i$ differs in Chapters 5, 6, and 7 [see Eqs. (A.109a)–(A.109d)], but this will not introduce confusion.

From this $\hat{f}_{SB1}$, the first component of the inhomogeneous term of Eq. (5.14) for $m = 2$ is

$$\begin{aligned} \zeta_i \frac{\partial \hat{f}_{SB1}}{\partial x_i} &= \hat{f}_{SB0}(I + II) \\ &= \frac{\hat{p}_{SB0}}{\hat{T}_{SB0}^{3/2}} E(\tilde{\zeta})(I + II), \end{aligned} \tag{5.27}$$

where

$$
\begin{aligned}
I = & \left(\frac{\hat{T}_{SB0}^{1/2}}{\hat{p}_{SB0}}\frac{\partial \hat{p}_{SB1}}{\partial x_i}\right)\tilde{\zeta}_i + \left(\frac{2}{\hat{p}_{SB0}}\frac{\partial \hat{p}_{SB0}\hat{v}_{jSB1}}{\partial x_i}\right)\tilde{\zeta}_i\tilde{\zeta}_j \\
& + \hat{T}_{SB0}^{1/2}\left[\frac{\hat{p}_{SB1}}{\hat{T}_{SB0}\hat{p}_{SB0}}\frac{\partial \hat{T}_{SB0}}{\partial x_i} + \frac{\partial}{\partial x_i}\left(\frac{\hat{T}_{SB1}}{\hat{T}_{SB0}}\right)\right]\tilde{\zeta}_i\left(\tilde{\zeta}^2 - \frac{5}{2}\right) \\
& + \left(\frac{2\hat{p}_{SB0}\hat{v}_{jSB1}}{\hat{p}_{SB0}}\frac{\partial \hat{T}_{SB0}}{\partial x_i}\right)\tilde{\zeta}_i\tilde{\zeta}_j\left(\tilde{\zeta}^2 - \frac{5}{2}\right) \\
& + \left(\frac{\hat{T}_{SB1}}{\hat{T}_{SB0}^{3/2}}\frac{\partial \hat{T}_{SB0}}{\partial x_i}\right)\tilde{\zeta}_i\left(\tilde{\zeta}^4 - 6\tilde{\zeta}^2 + \frac{25}{4}\right), \\
II = & -\left(\frac{1}{\hat{p}_{SB0}\hat{T}_{SB0}^{1/2}}\frac{\partial \hat{T}_{SB0}}{\partial x_i}\frac{\partial \hat{T}_{SB0}}{\partial x_j}\right) \\
& \times \tilde{\zeta}_i\tilde{\zeta}_j\left[\left(\tilde{\zeta}^2 - 3\right)\mathcal{A}(\tilde{\zeta}, \hat{T}_{SB0}) - \frac{1}{2}\tilde{\zeta}\frac{\partial \mathcal{A}(\tilde{\zeta}, \hat{T}_{SB0})}{\partial \tilde{\zeta}} + \hat{T}_{SB0}\frac{\partial \mathcal{A}(\tilde{\zeta}, \hat{T}_{SB0})}{\partial \hat{T}_{SB0}}\right] \\
& - \left(\frac{\hat{T}_{SB0}^{1/2}}{\hat{p}_{SB0}}\frac{\partial^2 \hat{T}_{SB0}}{\partial x_i \partial x_j}\right)\tilde{\zeta}_i\tilde{\zeta}_j\mathcal{A}(\tilde{\zeta}, \hat{T}_{SB0}).
\end{aligned}
$$

With this inhomogeneous term, the solvability condition (5.19) for $m = 2$ gives the following three equations:[3]

$$\frac{\partial \hat{p}_{SB0}\hat{v}_{iSB1}}{\partial x_i} = 0, \tag{5.28}$$

$$\frac{\partial \hat{p}_{SB1}}{\partial x_i} = 0, \tag{5.29}$$

$$\hat{p}_{SB0}\hat{v}_{iSB1}\frac{\partial \hat{T}_{SB0}}{\partial x_i} = \frac{1}{2}\frac{\partial}{\partial x_i}\left(\hat{\gamma}_2(\hat{T}_{SB0})\hat{T}_{SB0}^{1/2}\frac{\partial \hat{T}_{SB0}}{\partial x_i}\right), \tag{5.30}$$

where $\hat{\gamma}_2(\hat{T}_{SB0})$, related to thermal conductivity, is a nondimensional function of $\hat{T}_{SB0}$ defined by Eq. (A.99) in Appendix A.5; for example,

$$\hat{\gamma}_2(\hat{T}_{SB0}) = 1.922284066 \quad \text{for a hard-sphere gas,}$$

$$\hat{\gamma}_2(\hat{T}_{SB0}) = \hat{T}_{SB0}^{1/2} \quad \text{for the BKW model.}$$

---

[3]The solvability condition (5.19) is, separately for each g, equivalent to

$$\int \bar{g}\tilde{\zeta}_i \frac{\partial \hat{f}_{SBm-1}}{\partial x_i}\mathbf{d}\tilde{\boldsymbol{\zeta}} = 0, \quad \bar{g} = 1,\ \tilde{\zeta}_i,\ \tilde{\zeta}^2 \quad \text{for each } \bar{g},$$

which is simpler for manipulation. For $\tilde{\zeta}_i$ in Eqs. (A.109b)–(A.109d), the condition $\int g\zeta_i G(\boldsymbol{\zeta})\mathbf{d}\boldsymbol{\zeta} = 0$ is equivalent to $\int \bar{g}\tilde{\zeta}_i G(\boldsymbol{\zeta})\mathbf{d}\tilde{\boldsymbol{\zeta}} = 0$ as a whole. The latter is convenient for manipulation of the solvability condition in Chapters 6 and 7.

Now putting $\hat{f}_{SB2}$ in the form

$$\begin{aligned}\hat{f}_{SB2} &= \hat{f}_{SB0}\phi_2(x_i, \tilde{\zeta}_i) \\ &= \frac{\hat{\rho}_{SB0}}{\hat{T}_{SB0}^{3/2}} E(\tilde{\zeta})\phi_2(x_i, \tilde{\zeta}_i),\end{aligned} \tag{5.31}$$

we arrange the integral equation for $\hat{f}_{SB2}$ [or Eq. (5.14) for $m = 2$] as the equation for $\phi_2(x_i, \tilde{\zeta}_i)$. The collision integral $\hat{J}(\hat{f}_{SB1}, \hat{f}_{SB1})$ in the inhomogeneous term, as well as $\hat{J}(\hat{f}_{SB0}, \hat{f}_{SB2})$, in Eq. (5.14) for $m = 2$ is arranged with the aid of formulas in Appendixes A.7 and A.8, and then the whole inhomogeneous term is further arranged with the aid of the solvability conditions (5.24) and (5.28)–(5.30). Thus, we obtain the equation for $\phi_2(x_i, \tilde{\zeta}_i)$ [or $\phi_2(\tilde{\zeta}_i)$ for short] in the following form:[4]

$$\begin{aligned}
&\mathcal{L}_{\hat{T}_{SB0}}(\phi_2(\tilde{\zeta}_i)) \\
&= -\frac{1}{2}\left(\frac{\hat{T}_{SB1}}{\hat{T}_{SB0}}\right)^2 \mathcal{L}_{\hat{T}_{SB0}}(\tilde{\zeta}^2(1-\tilde{\zeta}^2)) - \frac{2\hat{T}_{SB0}^{1/2}\hat{\rho}_{SB0}\hat{v}_{iSB1}}{\hat{p}_{SB0}}\frac{\hat{T}_{SB1}}{\hat{T}_{SB0}}\mathcal{L}_{\hat{T}_{SB0}}(\tilde{\zeta}_i(1-\tilde{\zeta}^2)) \\
&\quad - \left(\frac{\hat{T}_{SB0}^{1/2}\hat{\rho}_{SB0}}{\hat{p}_{SB0}}\right)^2 \hat{v}_{iSB1}\hat{v}_{jSB1}\mathcal{L}_{\hat{T}_{SB0}}(\delta_{ij} - 2\tilde{\zeta}_i\tilde{\zeta}_j) \\
&\quad + \frac{\hat{\rho}_{SB0}\hat{v}_{iSB1}\hat{T}_{SB0}^{1/2}}{\hat{p}_{SB0}^2}\frac{\partial \hat{T}_{SB0}}{\partial x_j}\mathcal{L}_{\hat{T}_{SB0}}\left(\frac{\tilde{\zeta}_i\tilde{\zeta}_j}{\tilde{\zeta}}\frac{\partial \mathcal{A}(\tilde{\zeta}, \hat{T}_{SB0})}{\partial \tilde{\zeta}} - (2\tilde{\zeta}_i\tilde{\zeta}_j - \delta_{ij})\mathcal{A}(\tilde{\zeta}, \hat{T}_{SB0})\right) \\
&\quad + \frac{1}{\hat{p}_{SB0}}\frac{\hat{T}_{SB1}}{\hat{T}_{SB0}}\frac{\partial \hat{T}_{SB0}}{\partial x_i} \\
&\quad \times \mathcal{L}_{\hat{T}_{SB0}}\left(\tilde{\zeta}_i\left[\frac{1}{2}\tilde{\zeta}\frac{\partial \mathcal{A}(\tilde{\zeta}, \hat{T}_{SB0})}{\partial \tilde{\zeta}} - \left(\tilde{\zeta}^2 - 3\right)\mathcal{A}(\tilde{\zeta}, \hat{T}_{SB0}) - \hat{T}_{SB0}\frac{\partial \mathcal{A}(\tilde{\zeta}, \hat{T}_{SB0})}{\partial \hat{T}_{SB0}}\right]\right) \\
&\quad + \frac{1}{\hat{p}_{SB0}}\frac{\partial \hat{T}_{SB1}}{\partial x_i}\underbrace{\tilde{\zeta}_i\left(\tilde{\zeta}^2 - \frac{5}{2}\right)} + 2\left(\frac{\hat{T}_{SB0}^{1/2}}{\hat{p}_{SB0}}\frac{\partial \hat{v}_{jSB1}}{\partial x_i}\right)\underbrace{\left(\tilde{\zeta}_i\tilde{\zeta}_j - \frac{\tilde{\zeta}^2}{3}\delta_{ij}\right)} \\
&\quad - \frac{1}{3\hat{p}_{SB0}^2}\left(\frac{\partial \hat{T}_{SB0}}{\partial x_i}\right)\left(\frac{\partial \hat{T}_{SB0}}{\partial x_j}\right)\Big[\frac{3}{2}\underbrace{\text{Iha1}} + \delta_{ij}\underbrace{\text{Iha2}} + \delta_{ij}\underbrace{\text{Iha3}} + \delta_{ij}\underbrace{\text{Iha4}} \\
&\qquad + 3\mathcal{J}_{\hat{T}_{SB0}}(\tilde{\zeta}_i\mathcal{A}(\tilde{\zeta}, \hat{T}_{SB0}), \tilde{\zeta}_j\mathcal{A}(\tilde{\zeta}, \hat{T}_{SB0}))\Big] \\
&\quad - \frac{\hat{T}_{SB0}}{3\hat{p}_{SB0}^2}\frac{\partial^2 \hat{T}_{SB0}}{\partial x_i \partial x_j}(3\underbrace{\text{Ihb1}} + \delta_{ij}\underbrace{\text{Ihb2}}),
\end{aligned} \tag{5.32}$$

where the operators $\mathcal{L}_{\hat{T}_{SB0}}(\phi(\tilde{\zeta}_i))$ and $\mathcal{J}_{\hat{T}_{SB0}}(\phi(\tilde{\zeta}_i), \psi(\tilde{\zeta}_i))$ are defined by

$$\mathcal{L}_{\hat{T}_{SB0}}(\phi(\tilde{\zeta}_i)) = \int E_*(\phi' + \phi'_* - \phi - \phi_*)\hat{B}_{\hat{T}_{SB0}}\,\mathrm{d}\Omega(\boldsymbol{\alpha})\mathrm{d}\tilde{\boldsymbol{\zeta}}_*,$$

[4] For the BKW equation, $3\mathcal{J}_{\hat{T}_{SB0}}(\tilde{\zeta}_i\mathcal{A}(\tilde{\zeta}, \hat{T}_{SB0}), \tilde{\zeta}_j\mathcal{A}(\tilde{\zeta}, \hat{T}_{SB0}))$ in the term before the last of the inhomogeneous terms happens to be absent (see also Footnote 1 in Section 4.2).

$$\mathcal{J}_{\hat{T}_{SB0}}(\phi(\tilde{\zeta}_i),\psi(\tilde{\zeta}_i))$$
$$=\frac{1}{2}\int E_*(\phi'\psi'_*+\phi'_*\psi'-\phi\psi_*-\phi_*\psi)\hat{B}_{\hat{T}_{SB0}}\,\mathrm{d}\Omega(\boldsymbol{\alpha})\,\mathrm{d}\tilde{\boldsymbol{\zeta}}_*,$$

with

$$\hat{B}_{\hat{T}_{SB0}}=\hat{B}_{\hat{T}_{SB0}}(|\alpha_i(\tilde{\zeta}_{i*}-\tilde{\zeta}_i)|/|\tilde{\zeta}_{k*}-\tilde{\zeta}_k|,\,|\tilde{\zeta}_{i*}-\tilde{\zeta}_i|)$$
$$=\frac{\hat{B}(|\alpha_i(\tilde{\zeta}_{i*}-\tilde{\zeta}_i)|/|\tilde{\zeta}_{k*}-\tilde{\zeta}_k|,\,|\tilde{\zeta}_{i*}-\tilde{\zeta}_i|\hat{T}_{SB0}^{1/2})}{\hat{T}_{SB0}^{1/2}},$$

$$\phi=\phi(\tilde{\zeta}_i),\ \phi_*=\phi(\tilde{\zeta}_{i*}),\ \phi'=\phi(\tilde{\zeta}'_i),\ \phi'_*=\phi(\tilde{\zeta}'_{i*}),\quad E_*=E(\tilde{\zeta}_*),$$

$$\tilde{\zeta}'_i=\tilde{\zeta}_i+\alpha_k(\tilde{\zeta}_{k*}-\tilde{\zeta}_k)\alpha_i,\quad \tilde{\zeta}'_{i*}=\tilde{\zeta}_{i*}-\alpha_k(\tilde{\zeta}_{k*}-\tilde{\zeta}_k)\alpha_i,$$

and Iha1 etc. in the inhomogeneous terms are

$$\mathrm{Iha1}=\left(\tilde{\zeta}_i\tilde{\zeta}_j-\frac{1}{3}\tilde{\zeta}^2\delta_{ij}\right)$$
$$\times\left(2\left(\tilde{\zeta}^2-3\right)\mathcal{A}(\tilde{\zeta},\hat{T}_{SB0})-\tilde{\zeta}\frac{\partial\mathcal{A}(\tilde{\zeta},\hat{T}_{SB0})}{\partial\tilde{\zeta}}+2\hat{T}_{SB0}\frac{\partial\mathcal{A}(\tilde{\zeta},\hat{T}_{SB0})}{\partial\hat{T}_{SB0}}\right),$$

$$\mathrm{Iha2}=\tilde{\zeta}^2\left(\tilde{\zeta}^2-\frac{7}{2}\right)\mathcal{A}(\tilde{\zeta},\hat{T}_{SB0})-\frac{1}{2}\tilde{\zeta}^3\frac{\partial\mathcal{A}(\tilde{\zeta},\hat{T}_{SB0})}{\partial\tilde{\zeta}},$$

$$\mathrm{Iha3}=\frac{1}{2}\left(\tilde{\zeta}^2\mathcal{A}(\tilde{\zeta},\hat{T}_{SB0})-\frac{5}{2}\hat{\gamma}_2(\hat{T}_{SB0})\left(\tilde{\zeta}^2-\frac{3}{2}\right)\right),$$

$$\mathrm{Iha4}=\hat{T}_{SB0}\tilde{\zeta}^2\frac{\partial\mathcal{A}(\tilde{\zeta},\hat{T}_{SB0})}{\partial\hat{T}_{SB0}}-\frac{5}{2}\hat{T}_{SB0}\frac{\mathrm{d}\hat{\gamma}_2(\hat{T}_{SB0})}{\mathrm{d}\hat{T}_{SB0}}\left(\tilde{\zeta}^2-\frac{3}{2}\right),$$

and

$$\mathrm{Ihb1}=\left(\tilde{\zeta}_i\tilde{\zeta}_j-\frac{1}{3}\tilde{\zeta}^2\delta_{ij}\right)\mathcal{A}(\tilde{\zeta},\hat{T}_{SB0}),$$

$$\mathrm{Ihb2}=\tilde{\zeta}^2\mathcal{A}(\tilde{\zeta},\hat{T}_{SB0})-\frac{5}{2}\hat{\gamma}_2(\hat{T}_{SB0})\left(\tilde{\zeta}^2-\frac{3}{2}\right)$$
$$=2\mathrm{Iha3}.$$

Here, each of the inhomogeneous terms marked by $\underbrace{***}$ as well as the terms expressed by operator $\mathcal{L}_{\hat{T}_{SB0}}$ satisfies the solvability condition.

The solution of the integral equation (5.32) is expressed in the following form:

$$
\begin{aligned}
&\phi_2(\tilde{\zeta}_i)\\
&= \frac{\hat{\rho}_{SB2}}{\hat{\rho}_{SB0}} + \frac{2(\hat{\rho}_{SB1}\hat{v}_{iSB1} + \hat{\rho}_{SB0}\hat{v}_{iSB2})\hat{T}_{SB0}^{1/2}}{\hat{p}_{SB0}}\tilde{\zeta}_i\\
&\quad + \left[\frac{\hat{p}_{SB2}}{\hat{p}_{SB0}} - \frac{\hat{\rho}_{SB2}}{\hat{\rho}_{SB0}} + \frac{2\hat{\rho}_{SB0}\hat{v}_{iSB1}^2}{3\hat{p}_{SB0}}\right]\left(\tilde{\zeta}^2 - \frac{3}{2}\right)\\
&\quad + \frac{1}{2}\left(\frac{\hat{T}_{SB1}}{\hat{T}_{SB0}}\right)^2 \underleftrightarrow{\left(\tilde{\zeta}^4 - 5\tilde{\zeta}^2 + \frac{15}{4}\right)} + \frac{2\hat{T}_{SB0}^{1/2}\hat{\rho}_{SB0}\hat{v}_{iSB1}}{\hat{p}_{SB0}}\frac{\hat{T}_{SB1}}{\hat{T}_{SB0}}\tilde{\zeta}_i \underleftrightarrow{\left(\tilde{\zeta}^2 - \frac{5}{2}\right)}\\
&\quad + 2\left(\frac{\hat{T}_{SB0}^{1/2}\hat{\rho}_{SB0}}{\hat{p}_{SB0}}\right)^2 \hat{v}_{iSB1}\hat{v}_{jSB1} \underleftrightarrow{\left(\tilde{\zeta}_i\tilde{\zeta}_j - \frac{\tilde{\zeta}^2}{3}\delta_{ij}\right)}\\
&\quad + \frac{\hat{\rho}_{SB0}\hat{v}_{iSB1}\hat{T}_{SB0}^{1/2}}{\hat{p}_{SB0}^2}\frac{\partial \hat{T}_{SB0}}{\partial x_j}\left(\frac{\tilde{\zeta}_i\tilde{\zeta}_j}{\tilde{\zeta}}\frac{\partial \mathcal{A}(\tilde{\zeta},\hat{T}_{SB0})}{\partial \tilde{\zeta}} - (2\tilde{\zeta}_i\tilde{\zeta}_j - \delta_{ij})\mathcal{A}(\tilde{\zeta},\hat{T}_{SB0})\right)\\
&\quad + \frac{1}{\hat{p}_{SB0}}\frac{\hat{T}_{SB1}}{\hat{T}_{SB0}}\frac{\partial \hat{T}_{SB0}}{\partial x_i}\\
&\qquad \times \tilde{\zeta}_i\left[\frac{1}{2}\tilde{\zeta}\frac{\partial \mathcal{A}(\tilde{\zeta},\hat{T}_{SB0})}{\partial \tilde{\zeta}} - \left(\tilde{\zeta}^2 - 3\right)\mathcal{A}(\tilde{\zeta},\hat{T}_{SB0}) - \hat{T}_{SB0}\frac{\partial \mathcal{A}(\tilde{\zeta},\hat{T}_{SB0})}{\partial \hat{T}_{SB0}}\right]\\
&\quad - \frac{1}{\hat{p}_{SB0}}\frac{\partial \hat{T}_{SB1}}{\partial x_i}\tilde{\zeta}_i\mathcal{A}(\tilde{\zeta},\hat{T}_{SB0}) - \left(\frac{\hat{T}_{SB0}^{1/2}}{\hat{p}_{SB0}}\frac{\partial \hat{v}_{jSB1}}{\partial x_i}\right)\left(\tilde{\zeta}_i\tilde{\zeta}_j - \frac{\tilde{\zeta}^2}{3}\delta_{ij}\right)\mathcal{B}(\tilde{\zeta},\hat{T}_{SB0})\\
&\quad + \frac{1}{2\hat{p}_{SB0}^2}\left(\frac{\partial \hat{T}_{SB0}}{\partial x_i}\right)\left(\frac{\partial \hat{T}_{SB0}}{\partial x_j}\right)\left[\left(\tilde{\zeta}_i\tilde{\zeta}_j - \frac{\tilde{\zeta}^2}{3}\delta_{ij}\right)\mathcal{B}_2 + \delta_{ij}\mathcal{N}^A\right]\\
&\quad + \frac{\hat{T}_{SB0}}{\hat{p}_{SB0}^2}\frac{\partial^2 \hat{T}_{SB0}}{\partial x_i \partial x_j}\left[\left(\tilde{\zeta}_i\tilde{\zeta}_j - \frac{\tilde{\zeta}^2}{3}\delta_{ij}\right)\mathcal{B}_1 + \delta_{ij}\mathcal{N}^B\right],
\end{aligned}
\tag{5.33}
$$

where $\mathcal{B}(\tilde{\zeta},\hat{T}_{SB0})$, $\mathcal{B}_1(\tilde{\zeta},\hat{T}_{SB0})$, $\mathcal{B}_2(\tilde{\zeta},\hat{T}_{SB0})$, $\mathcal{N}^A(\tilde{\zeta},\hat{T}_{SB0})$, and $\mathcal{N}^B(\tilde{\zeta},\hat{T}_{SB0})$ are defined in Appendix A.5. The first six terms on the right-hand side are the second-order terms of the local Maxwellian. The terms marked by $\underleftrightarrow{***}$ are obtained by modifying the obvious solutions known from the forms of the inhomogeneous terms expressed by operator $\mathcal{L}_{\hat{T}_{SB0}}$ with the solutions $(1, \zeta_i, \zeta_j^2)$ of the corresponding homogeneous equation in order for the orthogonal condition (5.21) to be satisfied.

We proceed with the analysis in a similar way. Then, from the solvability condition (5.19) for $m = 3$, we obtain the following equations:

$$
\frac{\partial \hat{\rho}_{SB0}\hat{v}_{iSB2}}{\partial x_i} + \frac{\partial \hat{\rho}_{SB1}\hat{v}_{iSB1}}{\partial x_i} = 0, \tag{5.34}
$$

$$
\begin{aligned}
\hat{\rho}_{SB0}\hat{v}_{jSB1}\frac{\partial \hat{v}_{iSB1}}{\partial x_j} = & -\frac{1}{2}\frac{\partial \hat{p}_{SB2}}{\partial x_i} \\
& + \frac{1}{2}\frac{\partial}{\partial x_j}\left[\hat{\gamma}_1(\hat{T}_{SB0})\hat{T}_{SB0}^{1/2}\left(\frac{\partial \hat{v}_{iSB1}}{\partial x_j} + \frac{\partial \hat{v}_{jSB1}}{\partial x_i} - \frac{2}{3}\frac{\partial \hat{v}_{kSB1}}{\partial x_k}\delta_{ij}\right)\right] \\
& - \frac{1}{2\hat{p}_{SB0}}\frac{\partial}{\partial x_j}\left\{\hat{\gamma}_7(\hat{T}_{SB0})\left[\frac{\partial \hat{T}_{SB0}}{\partial x_i}\frac{\partial \hat{T}_{SB0}}{\partial x_j} - \frac{1}{3}\left(\frac{\partial \hat{T}_{SB0}}{\partial x_k}\right)^2\delta_{ij}\right]\right\} \\
& - \frac{1}{2\hat{p}_{SB0}}\frac{\partial}{\partial x_j}\left[\hat{\gamma}_3(\hat{T}_{SB0})\hat{T}_{SB0}\left(\frac{\partial^2 \hat{T}_{SB0}}{\partial x_i \partial x_j} - \frac{1}{3}\frac{\partial^2 \hat{T}_{SB0}}{\partial x_k^2}\delta_{ij}\right)\right],
\end{aligned} \tag{5.35}
$$

$$
\begin{aligned}
& \hat{\rho}_{SB0}\hat{v}_{iSB1}\frac{\partial \hat{T}_{SB1}}{\partial x_i} + (\hat{\rho}_{SB0}\hat{v}_{iSB2} + \hat{\rho}_{SB1}\hat{v}_{iSB1})\frac{\partial \hat{T}_{SB0}}{\partial x_i} \\
& \quad = \frac{1}{2}\frac{\partial}{\partial x_i}\left(\hat{\gamma}_2(\hat{T}_{SB0})\hat{T}_{SB0}^{1/2}\frac{\partial \hat{T}_{SB1}}{\partial x_i} + \hat{T}_{SB1}\frac{\mathrm{d}\hat{\gamma}_2(\hat{T}_{SB0})\hat{T}_{SB0}^{1/2}}{\mathrm{d}\hat{T}_{SB0}}\frac{\partial \hat{T}_{SB0}}{\partial x_i}\right),
\end{aligned} \tag{5.36}
$$

where $\hat{\gamma}_1(\hat{T}_{SB0})$, $\hat{\gamma}_3(\hat{T}_{SB0})$, and $\hat{\gamma}_7(\hat{T}_{SB0})$, related to transport coefficients, are nondimensional functions of $\hat{T}_{SB0}$ defined in Eqs. (A.99) and (A.102) in Appendix A.5; for example, for a hard-sphere gas,

$$
\hat{\gamma}_1(\hat{T}_{SB0}) = 1.270042427, \quad \hat{\gamma}_3(\hat{T}_{SB0}) = 1.947906335, \quad \hat{\gamma}_7(\hat{T}_{SB0}) = 0.189201,
$$

and for the BKW model,

$$
\hat{\gamma}_1(\hat{T}_{SB0}) = \hat{T}_{SB0}^{1/2}, \quad \hat{\gamma}_3(\hat{T}_{SB0}) = \hat{T}_{SB0}, \quad \hat{\gamma}_7(\hat{T}_{SB0}) = \hat{T}_{SB0}.
$$

Now, at the stage of the solvability condition (5.19) for $m = 3$, the equations that determine the component functions of the macroscopic variables at the leading order are lined up. From Eqs. (5.24) and (5.29), which are required for the flow velocity $\hat{v}_i$ to be a small quantity of the order of k, $\hat{p}_{SB0}$ and $\hat{p}_{SB1}$ are constants (say, $\hat{p}_0$ and $\hat{p}_1$):

$$
\hat{p}_{SB0} = \hat{p}_0, \quad \hat{p}_{SB1} = \hat{p}_1, \tag{5.37}
$$

from which

$$
\hat{\rho}_{SB0} = \frac{\hat{p}_0}{\hat{T}_{SB0}}, \quad \hat{\rho}_{SB1} = \frac{\hat{p}_1 - \hat{\rho}_{SB0}\hat{T}_{SB1}}{\hat{T}_{SB0}}, \tag{5.38}
$$

with the aid of the equations of state (5.11d) and (5.12d). Equations (5.28), (5.30), and (5.35), which are derived from the solvability condition (5.19) for ($m = 2$, $g = 1$ and ζ_i^2) and ($m = 3$, $g = \zeta_i$), contain the component functions $\hat{\rho}_{SB0}$, $\hat{T}_{SB0}$, $\hat{v}_{iSB1}$, and $\hat{p}_{SB2}$, but from Eq. (5.38), they are the equations for $\hat{T}_{SB0}$, $\hat{v}_{iSB1}$, and $\hat{p}_{SB2}$. Generally, the set of equations derived from the solvability condition (5.19) for ($m = s + 2$, $g = 1$ and ζ_i^2) and ($m = s + 3$, $g = \zeta_i$)

contains the functions $\hat{\rho}_{SBs}$, $\hat{T}_{SBs}$, $\hat{v}_{iSBs+1}$, and $\hat{p}_{SBs+2}$ as well as the functions $\hat{\rho}_{SBr}$, $\hat{T}_{SBr}$, $\hat{v}_{iSBr+1}$, and $\hat{p}_{SBr+2}$ $(r \leq s-1)$ encountered in the equations at the previous stages. Thus, with the aid of the expanded form of the equation of state (5.9d), the staggered combination of functions $\hat{\rho}_{SBs}$, $\hat{T}_{SBs}$, $\hat{v}_{iSBs+1}$, and $\hat{p}_{SBs+2}$ is determined consistently and successively from the lowest order by the rearranged sets of equations given by the solvability condition (5.19).

The set of equations for $\hat{\rho}_{SB0}$, $\hat{T}_{SB0}$, $\hat{v}_{iSB1}$, and $\hat{p}_{SB2}$ has a striking feature. That is, the leading temperature field $\hat{T}_{SB0}$ is determined together with the next-order velocity component $\hat{v}_{iSB1}$. This is an important result related to the incompleteness of the classical gas dynamics, which will be discussed in detail in Section 5.6. The component function $\hat{f}_{SBm}$ of the velocity distribution function is determined by the macroscopic variables $\hat{\rho}_{SBs}$, $\hat{T}_{SBs}$, $\hat{v}_{iSBs}$, and $\hat{p}_{SBs}$ $(s \leq m)$. The leading component function $\hat{f}_{SB0}$ is the Maxwellian at rest with the parameters $\hat{\rho}_{SB0}$ and $\hat{T}_{SB0}$, i.e.,

$$\hat{f}_{SB0} = \frac{\hat{\rho}_{SB0}}{(\pi \hat{T}_{SB0})^{3/2}} \exp\left(-\frac{\zeta_i^2}{\hat{T}_{SB0}}\right). \tag{5.39}$$

However, the parameter $\hat{T}_{SB0}$ is not determined by the Euler set of equations. We have already seen this type of example for the parameters in ϕ_{G0} in Chapter 3 and for those in ϕ_{S1} in Chapter 4. Furthermore, in the present case it is determined together with the higher order variable $\hat{v}_{iSB1}$.

From $\hat{f}_{SBm}$ obtained [Eqs. (5.15), (5.26), and (5.31) with (5.33)], the component functions $\hat{p}_{ijSBm}$ and $\hat{q}_{iSBm}$ of the stress tensor and heat-flow vector are easily obtained as follows:

$$\hat{p}_{ijSB0} = \hat{p}_{SB0}\delta_{ij}, \tag{5.40a}$$

$$\hat{p}_{ijSB1} = \hat{p}_{SB1}\delta_{ij}, \tag{5.40b}$$

$$\begin{aligned}\hat{p}_{ijSB2} = {} & \hat{p}_{SB2}\delta_{ij} - \hat{\gamma}_1(\hat{T}_{SB0})\hat{T}_{SB0}^{1/2}\left(\frac{\partial \hat{v}_{iSB1}}{\partial x_j} + \frac{\partial \hat{v}_{jSB1}}{\partial x_i} - \frac{2}{3}\frac{\partial \hat{v}_{kSB1}}{\partial x_k}\delta_{ij}\right) \\ & + \frac{\hat{\gamma}_7(\hat{T}_{SB0})}{\hat{p}_{SB0}}\left[\frac{\partial \hat{T}_{SB0}}{\partial x_i}\frac{\partial \hat{T}_{SB0}}{\partial x_j} - \frac{1}{3}\left(\frac{\partial \hat{T}_{SB0}}{\partial x_k}\right)^2\delta_{ij}\right] \\ & + \frac{\hat{\gamma}_3(\hat{T}_{SB0})\hat{T}_{SB0}}{\hat{p}_{SB0}}\left(\frac{\partial^2 \hat{T}_{SB0}}{\partial x_i \partial x_j} - \frac{1}{3}\frac{\partial^2 \hat{T}_{SB0}}{\partial x_k^2}\delta_{ij}\right),\end{aligned} \tag{5.40c}$$

$$\hat{q}_{iSB0} = 0, \tag{5.41a}$$

$$\hat{q}_{iSB1} = -\frac{5}{4}\hat{\gamma}_2(\hat{T}_{SB0})\hat{T}_{SB0}^{1/2}\frac{\partial \hat{T}_{SB0}}{\partial x_i}, \tag{5.41b}$$

$$\hat{q}_{iSB2} = -\frac{5}{4}\left(\hat{\gamma}_2(\hat{T}_{SB0})\hat{T}_{SB0}^{1/2}\frac{\partial \hat{T}_{SB1}}{\partial x_i} + \hat{T}_{SB1}\frac{\mathrm{d}\hat{\gamma}_2(\hat{T}_{SB0})\hat{T}_{SB0}^{1/2}}{\mathrm{d}\hat{T}_{SB0}}\frac{\partial \hat{T}_{SB0}}{\partial x_i}\right). \tag{5.41c}$$

The term with the factor $\hat{\gamma}_1$ in $\hat{p}_{ijSB2}$ is the viscous stress, due to the first-order velocity field $\hat{v}_{iSB1}$, given by Newton's law, and the terms with the factor

$\hat{\gamma}_2$ in $\hat{q}_{iSB1}$ and $\hat{q}_{iSB2}$ are the heat flow by Fourier's law. The $\hat{\gamma}_1\hat{T}_{SB0}^{1/2}$ and $\hat{\gamma}_2\hat{T}_{SB0}^{1/2}$ are, respectively, the (nondimensional) viscosity and thermal conductivity of the gas, and $\hat{T}_{SB1}\mathrm{d}\hat{\gamma}_2\hat{T}_{SB0}^{1/2}/\mathrm{d}\hat{T}_{SB0}$ in $\hat{q}_{iSB2}$ is due to the temperature dependence of the thermal conductivity (see Appendix A.6). The third and fourth terms in $\hat{p}_{ijSB2}$, as a whole, are called thermal stress, and are the source of Kogan's flow (Kogan, Galkin & Fridlender [1976]) to be explained in Section 5.6.1. This stress is related to the incompleteness of the classical gas dynamics discussed in Section 5.6.

The nondimensional transport coefficients $\hat{\gamma}_1$, $\hat{\gamma}_2$, $\hat{\gamma}_3$, and $\hat{\gamma}_7$ generally depend on the reference temperature T_0 besides $\hat{T}_{SB0}$ (see Footnote 13 in Appendix A.2). Their dimensional forms are naturally independent of T_0, but it is not clear from the nondimensional forms, since their dependences on T_0 and $\hat{T}_{SB0}$ are not obtained explicitly. This is clarified in Appendix A.6.

5.3 Fluid-dynamic-type equations

In the preceding section we found that the set of equations for $\hat{T}_{SB0}$, $\hat{v}_{iSB1}$, and $\hat{p}_{SB2}$ has to be solved to obtain the behavior of the gas (here, $\hat{T}_{SB0}$) in the continuum limit (or the limit where the Knudsen number tends to zero). Now we first list the equations for $\hat{T}_{SB0}$, $\hat{v}_{iSB1}$, and $\hat{p}_{SB2}$ for convenience of reference and then discuss their forms. The set of equations for $\hat{T}_{SB0}$, $\hat{v}_{iSB1}$, and $\hat{p}_{SB2}$ is

$$\frac{\partial \hat{\rho}_{SB0}\hat{v}_{iSB1}}{\partial x_i} = 0, \tag{5.42}$$

$$\begin{aligned}\hat{\rho}_{SB0}\hat{v}_{jSB1}\frac{\partial \hat{v}_{iSB1}}{\partial x_j} =& -\frac{1}{2}\frac{\partial \hat{p}_{SB2}}{\partial x_i}\\ &+\frac{1}{2}\frac{\partial}{\partial x_j}\left[\hat{\gamma}_1(\hat{T}_{SB0})\hat{T}_{SB0}^{1/2}\left(\frac{\partial \hat{v}_{iSB1}}{\partial x_j}+\frac{\partial \hat{v}_{jSB1}}{\partial x_i}-\frac{2}{3}\frac{\partial \hat{v}_{kSB1}}{\partial x_k}\delta_{ij}\right)\right]\\ &-\frac{1}{2\hat{p}_{SB0}}\frac{\partial}{\partial x_j}\left\{\hat{\gamma}_7(\hat{T}_{SB0})\left[\frac{\partial \hat{T}_{SB0}}{\partial x_i}\frac{\partial \hat{T}_{SB0}}{\partial x_j}-\frac{1}{3}\left(\frac{\partial \hat{T}_{SB0}}{\partial x_k}\right)^2\delta_{ij}\right]\right\}\\ &-\frac{1}{2\hat{p}_{SB0}}\frac{\partial}{\partial x_j}\left[\hat{\gamma}_3(\hat{T}_{SB0})\hat{T}_{SB0}\left(\frac{\partial^2 \hat{T}_{SB0}}{\partial x_i\partial x_j}-\frac{1}{3}\frac{\partial^2 \hat{T}_{SB0}}{\partial x_k^2}\delta_{ij}\right)\right],\end{aligned} \tag{5.43}$$

$$\hat{\rho}_{SB0}\hat{v}_{iSB1}\frac{\partial \hat{T}_{SB0}}{\partial x_i} = \frac{1}{2}\frac{\partial}{\partial x_i}\left(\hat{\gamma}_2(\hat{T}_{SB0})\hat{T}_{SB0}^{1/2}\frac{\partial \hat{T}_{SB0}}{\partial x_i}\right), \tag{5.44}$$

where $\hat{\rho}_{SB0}$ is expressed with $\hat{T}_{SB0}$ by Eq. (5.38) as

$$\hat{\rho}_{SB0} = \frac{\hat{p}_0}{\hat{T}_{SB0}}. \tag{5.45}$$

In the above set of equations (5.42)–(5.44), the variable $\hat{\rho}_{SB0}$ is retained without replacing it by $\hat{p}_0/\hat{T}_{SB0}$ to keep its resemblance to the classical gas dynamic set of equations. The set of equations being compared with the set of compressible Navier–Stokes equations and the order of expansion being ignored, the main formal differences are additional thermal stress terms in the equation (5.43) and the relation (5.45) between $\hat{\rho}_{SB0}$ and $\hat{T}_{SB0}$ instead of the relation among $\hat{\rho}_{SB0}$, $\hat{T}_{SB0}$, and $\hat{p}_{SB2}$ corresponding to the equation of state. The role of the pressure term $\partial\hat{p}_{SB2}/\partial x_i$ in the momentum equation (5.43) is the same as that of the pressure term in the incompressible Navier–Stokes equations. That is, the set of equations (5.42)–(5.44) has a similar character to the incompressible Navier–Stokes set of equations in the way the solution is determined.[5]

A part of the thermal-stress terms in Eq. (5.43) is the third-order differential terms. The third-order equation generally requires more boundary conditions than the slip conditions for the Grad–Hilbert solution in Section 3.5 or for the S solution in Section 4.5. However, Eq. (5.43) is of the third order only in its appearance. That is, the third-order differential term (the fourth term on the right-hand side) of Eq. (5.43) can be transformed in the following way:

$$
\begin{aligned}
&\frac{\partial}{\partial x_j}\left[\hat{\gamma}_3\hat{T}_{SB0}\left(\frac{\partial^2\hat{T}_{SB0}}{\partial x_i\partial x_j}-\frac{1}{3}\frac{\partial^2\hat{T}_{SB0}}{\partial x_k^2}\delta_{ij}\right)\right]\\
&=\frac{\partial}{\partial x_j}\left[\frac{\partial}{\partial x_i}\left(\hat{\gamma}_3\hat{T}_{SB0}\frac{\partial\hat{T}_{SB0}}{\partial x_j}\right)-\frac{\mathrm{d}\hat{\gamma}_3\hat{T}_{SB0}}{\mathrm{d}\hat{T}_{SB0}}\frac{\partial\hat{T}_{SB0}}{\partial x_i}\frac{\partial\hat{T}_{SB0}}{\partial x_j}\right]\\
&\quad-\frac{1}{3}\frac{\partial}{\partial x_i}\left(\hat{\gamma}_3\hat{T}_{SB0}\frac{\partial^2\hat{T}_{SB0}}{\partial x_k^2}\right)\\
&=\frac{\partial}{\partial x_i}\frac{\partial}{\partial x_j}\left(\hat{\gamma}_3\hat{T}_{SB0}\frac{\partial\hat{T}_{SB0}}{\partial x_j}\right)-\frac{\partial}{\partial x_j}\left(\frac{\mathrm{d}\hat{\gamma}_3\hat{T}_{SB0}}{\mathrm{d}\hat{T}_{SB0}}\frac{\partial\hat{T}_{SB0}}{\partial x_i}\frac{\partial\hat{T}_{SB0}}{\partial x_j}\right)\\
&\quad-\frac{1}{3}\frac{\partial}{\partial x_i}\frac{\partial}{\partial x_k}\left(\hat{\gamma}_3\hat{T}_{SB0}\frac{\partial\hat{T}_{SB0}}{\partial x_k}\right)+\frac{1}{3}\frac{\partial}{\partial x_j}\left[\frac{\mathrm{d}\hat{\gamma}_3\hat{T}_{SB0}}{\mathrm{d}\hat{T}_{SB0}}\left(\frac{\partial\hat{T}_{SB0}}{\partial x_k}\right)^2\delta_{ij}\right]\\
&=\frac{2}{3}\frac{\partial}{\partial x_i}\frac{\partial}{\partial x_k}\left(\hat{\gamma}_3\hat{T}_{SB0}\frac{\partial\hat{T}_{SB0}}{\partial x_k}\right)\\
&\quad-\frac{\partial}{\partial x_j}\left\{\frac{\mathrm{d}\hat{\gamma}_3\hat{T}_{SB0}}{\mathrm{d}\hat{T}_{SB0}}\left[\frac{\partial\hat{T}_{SB0}}{\partial x_i}\frac{\partial\hat{T}_{SB0}}{\partial x_j}-\frac{1}{3}\left(\frac{\partial\hat{T}_{SB0}}{\partial x_k}\right)^2\delta_{ij}\right]\right\}.
\end{aligned}
$$

With the following new quantities $\hat{p}^*_{SB2}$ (a modified pressure) and Γ_7:

$$\hat{p}^*_{SB2}=\hat{p}_{SB2}+\frac{2}{3\hat{p}_{SB0}}\frac{\partial}{\partial x_k}\left(\hat{\gamma}_3(\hat{T}_{SB0})\hat{T}_{SB0}\frac{\partial\hat{T}_{SB0}}{\partial x_k}\right), \tag{5.46}$$

$$\Gamma_7(\hat{T}_{SB0})=\frac{\mathrm{d}\hat{\gamma}_3(\hat{T}_{SB0})\hat{T}_{SB0}}{\mathrm{d}\hat{T}_{SB0}}-\hat{\gamma}_7(\hat{T}_{SB0}), \tag{5.47}$$

[5]The process of solution under the boundary conditions to be derived in the next section will be explained in Section 5.5.

Eq. (5.43) is rewritten in the following second-order equation:

$$\hat{\rho}_{SB0}\hat{v}_{jSB1}\frac{\partial\hat{v}_{iSB1}}{\partial x_j} = -\frac{1}{2}\frac{\partial\hat{p}^*_{SB2}}{\partial x_i} + \frac{1}{2}\frac{\partial}{\partial x_j}\left[\hat{\gamma}_1(\hat{T}_{SB0})\hat{T}^{1/2}_{SB0}\left(\frac{\partial\hat{v}_{iSB1}}{\partial x_j} + \frac{\partial\hat{v}_{jSB1}}{\partial x_i} - \frac{2}{3}\frac{\partial\hat{v}_{kSB1}}{\partial x_k}\delta_{ij}\right)\right] + \frac{1}{2\hat{p}_{SB0}}\frac{\partial}{\partial x_j}\left\{\Gamma_7(\hat{T}_{SB0})\left[\frac{\partial\hat{T}_{SB0}}{\partial x_i}\frac{\partial\hat{T}_{SB0}}{\partial x_j} - \frac{1}{3}\left(\frac{\partial\hat{T}_{SB0}}{\partial x_k}\right)^2\delta_{ij}\right]\right\}. \tag{5.48}$$

The thermal-stress term (or the third term on the right-hand side) in Eq. (5.48) can be further reduced to the first order with the aid of Eq. (5.44). With the new modified pressure $\hat{p}^\dagger_{SB2}$ defined by

$$\begin{aligned}\hat{p}^\dagger_{SB2} &= \hat{p}_{SB2} + \frac{2}{3\hat{p}_{SB0}}\frac{\partial}{\partial x_k}\left(\hat{\gamma}_3(\hat{T}_{SB0})\hat{T}_{SB0}\frac{\partial\hat{T}_{SB0}}{\partial x_k}\right) - \frac{\Gamma_7(\hat{T}_{SB0})}{6\hat{p}_{SB0}}\left(\frac{\partial\hat{T}_{SB0}}{\partial x_k}\right)^2 \\ &= \hat{p}^*_{SB2} - \frac{\Gamma_7(\hat{T}_{SB0})}{6\hat{p}_{SB0}}\left(\frac{\partial\hat{T}_{SB0}}{\partial x_k}\right)^2,\end{aligned} \tag{5.49}$$

Eq. (5.48) is rewritten in the following form with the first-order thermal-stress term:

$$\hat{\rho}_{SB0}\hat{v}_{jSB1}\frac{\partial\hat{v}_{iSB1}}{\partial x_j} = -\frac{1}{2}\frac{\partial\hat{p}^\dagger_{SB2}}{\partial x_i} + \frac{1}{2}\frac{\partial}{\partial x_j}\left[\hat{\gamma}_1\hat{T}^{1/2}_{SB0}\left(\frac{\partial\hat{v}_{iSB1}}{\partial x_j} + \frac{\partial\hat{v}_{jSB1}}{\partial x_i} - \frac{2}{3}\frac{\partial\hat{v}_{kSB1}}{\partial x_k}\delta_{ij}\right)\right] + \left[\frac{\Gamma_7}{\hat{\gamma}_2\hat{T}^{1/2}_{SB0}}\frac{\hat{v}_{jSB1}}{\hat{T}_{SB0}}\frac{\partial\hat{T}_{SB0}}{\partial x_j} + \frac{\hat{\gamma}_2^2\hat{T}_{SB0}}{4\hat{p}_{SB0}}\frac{\mathrm{d}\Gamma_7/\hat{\gamma}_2^2\hat{T}_{SB0}}{\mathrm{d}\hat{T}_{SB0}}\left(\frac{\partial\hat{T}_{SB0}}{\partial x_j}\right)^2\right]\frac{\partial\hat{T}_{SB0}}{\partial x_i}, \tag{5.50}$$

where $\hat{\gamma}_1 = \hat{\gamma}_1(\hat{T}_{SB0})$, $\hat{\gamma}_2 = \hat{\gamma}_2(\hat{T}_{SB0})$, and $\Gamma_7 = \Gamma_7(\hat{T}_{SB0})$. The $\partial\hat{p}^*_{SB2}/\partial x_i$ in Eq. (5.48) and $\partial\hat{p}^\dagger_{SB2}/\partial x_i$ in Eq. (5.50) play the same role as $\partial\hat{p}_{SB2}/\partial x_i$ in Eq. (5.43) (or the pressure term of the incompressible Navier–Stokes equations) and they are eliminated by taking the curl (or rotation) of Eqs. (5.48) and (5.50) respectively.

Incidentally,

$$\Gamma_7(\hat{T}_{SB0}) = 1.758705 \text{ (a hard-sphere gas)}, \quad \Gamma_7(\hat{T}_{SB0}) = \hat{T}_{SB0} \text{ (BKW)}.$$

5.4 Knudsen layer and slip condition

The velocity distribution function of the SB solution has a similar structure as a function with respect to the molecular velocity ζ_i to that of the Grad–Hilbert solution in Chapter 3 or that of the S solution in Chapter 4, and therefore does not have enough freedom to be fitted to the kinetic boundary condition (5.2). Thus, we introduce a correction to the SB solution in the neighborhood of the boundary by allowing a sharp variation of the velocity distribution function normal to the boundary. That is, we put the solution $\hat{f}$ of the boundary-value problem (5.1) and (5.2) in the form

$$\hat{f} = \hat{f}_{SB} + \hat{f}_K, \tag{5.51}$$

where $\hat{f}_K$ is assumed to have a similar behavior $[kn_i\partial\hat{f}_K/\partial x_i = O(\hat{f}_K)]$ to ϕ_K in Section 3.4 or 4.4, that is, $\hat{f}_K$ is appreciable only in a thin layer, with thickness of the order of the mean free path, adjacent to the boundary and decays very rapidly in the layer in the direction normal to the boundary. The $\hat{f}_{SB}$ is called the fluid-dynamic part, and $\hat{f}_K$ is the Knudsen-layer part (or correction). Substituting Eq. (5.51) into the Boltzmann equation (5.1) and taking into account that the SB solution $\hat{f}_{SB}$ itself satisfies the Boltzmann equation (5.1), then we find that the equation for $\hat{f}_K$ is given as[6]

$$\zeta_i \frac{\partial \hat{f}_K}{\partial x_i} = \frac{1}{k}[2\hat{J}(\hat{f}_{SB}, \hat{f}_K) + \hat{J}(\hat{f}_K, \hat{f}_K)]. \tag{5.52}$$

Here the same Knudsen-layer variables (η, χ_1, χ_2) as in Section 3.4 are introduced:

$$x_i = k\eta n_i(\chi_1, \chi_2) + x_{wi}(\chi_1, \chi_2), \tag{5.53}$$

where x_{wi} is the boundary surface, η is a stretched coordinate normal to the boundary, χ_1 and χ_2 are (unstretched) coordinates within a parallel surface $\eta =$ const, and the normal vector n_i is a function of χ_1 and χ_2. Then, Eq. (5.52) is rewritten as

$$\zeta_i n_i \frac{\partial \hat{f}_K}{\partial \eta} = 2\hat{J}(\hat{f}_{SB}, \hat{f}_K) + \hat{J}(\hat{f}_K, \hat{f}_K) - k\zeta_i \left(\frac{\partial \chi_1}{\partial x_i}\frac{\partial \hat{f}_K}{\partial \chi_1} + \frac{\partial \chi_2}{\partial x_i}\frac{\partial \hat{f}_K}{\partial \chi_2} \right). \tag{5.54}$$

We can proceed with the analysis in a similar way to that in Section 3.4 or 4.4. However, before continuing the straightforward analysis, some discussion is made to save manipulations. The leading term of the SB expansion has been shown to be the Maxwellian

$$\hat{f}_{SB0} = \frac{\hat{\rho}_{SB0}}{(\pi \hat{T}_{SB0})^{3/2}} \exp\left(-\frac{\zeta_i^2}{\hat{T}_{SB0}}\right). \tag{5.55}$$

Taking into account the condition (2.23c) on a simple boundary or the condition (2.26c) on an interface of a gas and its condensed phase, we can easily see that

---

[6]For the BKW equation, $2\hat{J}(\hat{f}_{SB}, \hat{f}_K) + \hat{J}(\hat{f}_K, \hat{f}_K)$ on the right-hand side should be understood as

$$\hat{J}(\hat{f}_{SB} + \hat{f}_K, \hat{f}_{SB} + \hat{f}_K) - \hat{J}(\hat{f}_{SB}, \hat{f}_{SB}),$$

(see Footnote 12 in Section 2.9).

the boundary condition can be satisfied at the leading order [or $O(1)$] by taking

$$\hat{T}_{SB0} = \hat{T}_{w0}, \tag{5.56}$$

on a simple boundary or

$$\hat{T}_{SB0} = \hat{T}_{w0}, \tag{5.57a}$$

$$\hat{p}_{SB0} = \hat{p}_{w0}, \tag{5.57b}$$

on an interface. In this chapter, the case $\hat{v}_i = O(k)$ is discussed, and thus $\hat{v}_{wi}$ in the conditions (2.23c) and (2.26c) should be taken to be zero at the order of unity (i.e., $\hat{v}_{wi0} = 0$). Two conditions are imposed on an interface instead of one on a simple boundary. The condition (5.57b) should be taken as the condition on the saturated gas pressure $\hat{p}_{w0}$ that $\hat{p}_{w0}$ be uniform, owing to $\hat{p}_{SB0} = \hat{p}_0$ [Eq. (5.37)], for the absence of evaporation and condensation of the order of unity. Thus, under the no-temperature-jump condition [Eq. (5.56) or Eq. (5.57a)], the correction $\hat{f}_K$ to the SB solution $\hat{f}_{SB}$ required to fit the solution $\hat{f}$ to the boundary condition (5.2) is of the order of k.

According to the above discussion, the Knudsen-layer correction $\hat{f}_K$ is expanded in a power series of k starting from the term of the order of k:

$$\hat{f}_K = \hat{f}_{K1}k + \hat{f}_{K2}k^2 + \cdots. \tag{5.58}$$

Substituting the series (5.58) and the reshuffled series of Eq. (5.7), i.e.,

$$\hat{f}_{SB} = (\hat{f}_{SB0})_0 + \left[(\hat{f}_{SB1})_0 + \left(n_i\frac{\partial \hat{f}_{SB0}}{\partial x_i}\right)_0 \eta\right]k + \cdots, \tag{5.59}$$

which is applicable in the Knudsen layer [see the discussion following Eq. (4.45)], into Eq. (5.54), we obtain the series of the Knudsen-layer equations:[7]

$$\zeta_i n_i \frac{\partial \hat{f}_{K1}}{\partial \eta} = 2\hat{J}((\hat{f}_{SB0})_0, \hat{f}_{K1}), \tag{5.60}$$

$$\cdots\cdots\cdots,$$

where the quantities in the parentheses with subscript 0, i.e., $(\cdots)_0$, are evaluated at $\eta = 0$.

The boundary condition for the Knudsen-layer correction $\hat{f}_{Km}$ at $\eta = 0$ is

$$\hat{f}_{Km} = \hat{f}_{wm} - \hat{f}_{SBm}. \tag{5.61}$$

More explicitly, for the diffuse reflection condition on a simple boundary,

$$\hat{f}_{K1} = \hat{f}_{SB0}\left[\frac{\hat{\sigma}_{w1} - \hat{\rho}_{SB1}}{\hat{\rho}_{SB0}} + \frac{2\zeta_i(\hat{v}_{wi1} - \hat{v}_{iSB1})}{\hat{T}_{w0}} + \left(\frac{\zeta_i^2}{\hat{T}_{w0}} - \frac{3}{2}\right)\frac{\hat{T}_{w1} - \hat{T}_{SB1}}{\hat{T}_{w0}} + \frac{\zeta_i \mathcal{A}(\zeta/\hat{T}_{w0}^{1/2}, \hat{T}_{w0})}{\hat{T}_{w0}^{1/2}\hat{p}_{SB0}}\frac{\partial \hat{T}_{SB0}}{\partial x_i}\right] \quad (\zeta_i n_i > 0), \tag{5.62}$$

[7]For the term on the right-hand side, see Footnotes 26 and 28 in Appendix A.

where

$$\frac{\hat{\sigma}_{w1}}{\hat{\rho}_{SB0}} = -\frac{1}{2}\frac{\hat{T}_{w1}}{\hat{T}_{w0}} - \frac{2\sqrt{\pi}\hat{T}_{w0}^{1/2}}{\hat{p}_{SB0}}\int_{\zeta_i n_i<0} \zeta_i n_i(\hat{f}_{SB1}+\hat{f}_{K1})\mathbf{d}\boldsymbol{\zeta}$$
$$= \frac{\hat{\rho}_{SB1}}{\hat{\rho}_{SB0}} - \frac{\hat{T}_{w1}-\hat{T}_{SB1}}{2\hat{T}_{w0}} - \frac{\sqrt{\pi}\hat{v}_{iSB1}n_i}{\hat{T}_{w0}^{1/2}} - \frac{2\sqrt{\pi}\hat{T}_{w0}^{1/2}}{\hat{p}_{SB0}}\int_{\zeta_i n_i<0} \zeta_i n_i \hat{f}_{K1}\mathbf{d}\boldsymbol{\zeta},$$
$$\cdots\cdots\cdots,$$

and for the complete condensation condition on an interface of a gas and its condensed phase,

$$\hat{f}_{K1} = \hat{f}_{SB0}\left[\frac{\hat{p}_{w1}-\hat{p}_{SB1}}{\hat{p}_{w0}} + \frac{2\zeta_i(\hat{v}_{wi1}-\hat{v}_{iSB1})}{\hat{T}_{w0}} + \left(\frac{\zeta_i^2}{\hat{T}_{w0}}-\frac{5}{2}\right)\frac{\hat{T}_{w1}-\hat{T}_{SB1}}{\hat{T}_{w0}} + \frac{\zeta_i \mathcal{A}(\zeta/\hat{T}_{w0}^{1/2},\hat{T}_{w0})}{\hat{T}_{w0}^{1/2}\hat{p}_{w0}}\frac{\partial \hat{T}_{SB0}}{\partial x_i}\right] \quad (\zeta_i n_i > 0), \tag{5.63}$$
$$\cdots\cdots\cdots.$$

The Knudsen-layer parts ($\hat{\rho}_K$, $\hat{v}_{iK}$, $\hat{T}_K$, $\hat{p}_K$, $\hat{p}_{ijK}$, $\hat{q}_{iK}$) of the macroscopic variables ($\hat{\rho}$, $\hat{v}_i$, $\hat{T}$, $\hat{p}$, $\hat{p}_{ij}$, $\hat{q}_i$) are also defined as the remainders ($\hat{\rho}-\hat{\rho}_{SB}$, $\hat{v}_i-\hat{v}_{iSB}$, $\hat{T}-\hat{T}_{SB}$, etc.). Then, they depend on $\hat{f}_{SB}$ as well as $\hat{f}_K$, since the relations between the macroscopic variables and the velocity distribution functions, i.e., Eqs. (2.48a)–(2.48f), are nonlinear. For example,

$$(\hat{\rho}_{SB}+\hat{\rho}_K)\hat{v}_{iK} = \int \zeta_i \hat{f}_K \mathbf{d}\boldsymbol{\zeta} - \hat{v}_{iSB}\hat{\rho}_K. \tag{5.64}$$

Corresponding to the expansion (5.58), the Knudsen-layer parts of the macroscopic variables are also expanded in power series of k:

$$\hat{h}_K = \hat{h}_{K1}k + \hat{h}_{K2}k^2 + \cdots, \tag{5.65}$$

where $\hat{h}$ represents $\hat{\rho}$, $\hat{v}_i$, $\hat{T}$, $\hat{p}$, $\hat{p}_{ij}$, and $\hat{q}_i$. The relations of the component functions of the expansions to $\hat{f}_{Km}$ are given as follows:

$$\hat{\rho}_{K1} = \int \hat{f}_{K1}\mathbf{d}\boldsymbol{\zeta}, \tag{5.66a}$$

$$(\hat{\rho}_{SB0})_0\hat{v}_{iK1} = \int \zeta_i \hat{f}_{K1}\mathbf{d}\boldsymbol{\zeta}, \tag{5.66b}$$

$$\frac{3}{2}(\hat{\rho}_{SB0})_0\hat{T}_{K1} = \int \zeta_j^2 \hat{f}_{K1}\mathbf{d}\boldsymbol{\zeta} - \frac{3}{2}\hat{\rho}_{K1}(\hat{T}_{SB0})_0, \tag{5.66c}$$

$$\hat{p}_{K1} = (\hat{\rho}_{SB0})_0\hat{T}_{K1} + (\hat{T}_{SB0})_0\hat{\rho}_{K1}, \tag{5.66d}$$

$$\hat{p}_{ijK1} = 2\int \zeta_i\zeta_j \hat{f}_{K1}\mathbf{d}\boldsymbol{\zeta}, \tag{5.66e}$$

$$\hat{q}_{iK1} = \int \zeta_i\zeta_j^2 \hat{f}_{K1}\mathbf{d}\boldsymbol{\zeta} - \frac{5}{2}(\hat{p}_{SB0})_0\hat{v}_{iK1}, \tag{5.66f}$$
$$\cdots\cdots\cdots,$$

where the quantities in the parentheses with subscript 0, i.e., $(**)_0$, are evaluated on the boundary.

The boundary-value problem for Knudsen-layer correction $\hat{f}_{K1}$ is reduced to a familiar half-space problem of the linearized Boltzmann equation by introducing the following variables. In place of the Knudsen-layer variable η and the (nondimensional) molecular velocity ζ_i, a space variable $\tilde{\eta}$ and a molecular velocity $\tilde{\zeta}_i$ locally normalized along the boundary are introduced:

$$\tilde{\eta} = \frac{(\hat{\rho}_{SB0})_0}{g_M(\hat{T}_{w0})}\eta, \quad \tilde{\zeta}_i = \frac{\zeta_i}{\hat{T}_{w0}^{1/2}}, \tag{5.67}$$

where a function $g_M(\hat{T}_{w0})$ of $\hat{T}_{w0}$ is introduced, for simplicity of the results for some molecular models, in the following way: if the linearized collision operator $\mathcal{L}_{\hat{T}_{w0}}(*)$, already encountered often and defined in Eq. (A.23), can be reduced to be independent of $\hat{T}_{w0}$ by multiplication of some function of $\hat{T}_{w0}$, then $g_M(\hat{T}_{w0})$ is chosen in such a way that $g_M(\hat{T}_{w0})\mathcal{L}_{\hat{T}_{w0}}(*) = \mathcal{L}_1(*)$ [$= \mathcal{L}(*)$], and if not, $g_M(\hat{T}_{w0}) = 1$. For a hard-sphere gas, $\mathcal{L}_{\hat{T}_{w0}}(*)$ itself is independent of $\hat{T}_{w0}$ or $g_M(\hat{T}_{w0}) = 1$, and for the BKW model, $\hat{T}_{w0}^{1/2}\mathcal{L}_{\hat{T}_{w0}}(*)$ is independent of $\hat{T}_{w0}$ or $g_M(\hat{T}_{w0}) = \hat{T}_{w0}^{1/2}$. Noting that $(\hat{f}_{SB0})_0$ is expressed with the new variables in the form

$$(\hat{f}_{SB0})_0 = \frac{(\hat{\rho}_{SB0})_0}{\hat{T}_{w0}^{3/2}}E(\tilde{\zeta}),$$

we put the distribution function $\hat{f}_{K1}$ in the following form with a function $\phi_{K1}(\tilde{\eta}, \tilde{\zeta}_i)$ of the new variables:

$$\hat{f}_{K1} = \frac{(\hat{\rho}_{SB0})_0}{\hat{T}_{w0}^{3/2}}E(\tilde{\zeta})\phi_{K1}(\tilde{\eta}, \tilde{\zeta}_i).$$

Then the Knudsen-layer equation (5.60) is reduced to the linearized Boltzmann equation (see Appendix A.9)

$$\tilde{\zeta}_i n_i \frac{\partial \phi_{K1}(\tilde{\eta}, \tilde{\zeta}_i)}{\partial \tilde{\eta}} = g_M(\hat{T}_{w0})\mathcal{L}_{\hat{T}_{w0}}(\phi_{K1}(\tilde{\eta}, \tilde{\zeta}_i)), \tag{5.68}$$

where the operator $\mathcal{L}_{\hat{T}_{w0}}(*)$ is defined in Eq. (A.23).

The boundary condition at $\tilde{\eta} = 0$ is, in the case of the diffuse reflection,

$$\begin{aligned}\phi_{K1} = {} & \frac{\hat{\sigma}_{w1} - \hat{\rho}_{SB1}}{\hat{\rho}_{SB0}} + \frac{2\tilde{\zeta}_i(\hat{v}_{wi1} - \hat{v}_{iSB1})}{\hat{T}_{w0}^{1/2}} + \left(\tilde{\zeta}_i^2 - \frac{3}{2}\right)\frac{\hat{T}_{w1} - \hat{T}_{SB1}}{\hat{T}_{w0}} \\ & + \frac{1}{\hat{\rho}_{SB0}}\frac{\partial \hat{T}_{SB0}}{\partial x_i}\tilde{\zeta}_i \mathcal{A}(\tilde{\zeta}, \hat{T}_{w0}) \qquad (\tilde{\zeta}_i n_i > 0),\end{aligned} \tag{5.69}$$

where

$$\frac{\hat{\sigma}_{w1}}{\hat{\rho}_{SB0}} = \frac{\hat{\rho}_{SB1}}{\hat{\rho}_{SB0}} - \frac{\hat{T}_{w1} - \hat{T}_{SB1}}{2\hat{T}_{w0}} - \frac{\sqrt{\pi}\hat{v}_{iSB1}n_i}{\hat{T}_{w0}^{1/2}} - 2\sqrt{\pi}\int_{\tilde{\zeta}_i n_i < 0} \tilde{\zeta}_i n_i \phi_{K1}(\tilde{\zeta}_i)E(\tilde{\zeta})\mathbf{d}\tilde{\zeta},$$

or, in the case of the complete condensation,

$$\phi_{K1} = \frac{\hat{p}_{w1} - \hat{p}_{SB1}}{\hat{p}_{w0}} + \frac{2\tilde{\zeta}_i(\hat{v}_{wi1} - \hat{v}_{iSB1})}{\hat{T}_{w0}^{1/2}} + \left(\tilde{\zeta}_i^2 - \frac{5}{2}\right)\frac{\hat{T}_{w1} - \hat{T}_{SB1}}{\hat{T}_{w0}} + \frac{1}{\hat{p}_{w0}}\frac{\partial \hat{T}_{SB0}}{\partial x_i}\tilde{\zeta}_i \mathcal{A}(\tilde{\zeta}, \hat{T}_{w0}) \qquad (\tilde{\zeta}_i n_i > 0). \tag{5.70}$$

The condition at infinity ($\tilde{\eta} \to \infty$) is that ϕ_{K1} vanishes faster than any inverse power of $\tilde{\eta}$,

$$\phi_{K1} \to 0 \text{ (faster than any inverse power of } \tilde{\eta}) \text{ as } \tilde{\eta} \to \infty.$$

This speed condition is required for application of the expansion (5.59) to the product of the fluid-dynamic part and the Knudsen-layer correction in deriving the Knudsen-layer equation [see also the comment below Eq. (3.52)].

Now the problem is reduced to the half-space problem to which Bardos's theorem (Appendix A.9) applies. From the discussion in Section 3.4 based on Bardos's theorem, the solution exists uniquely, when and only when the boundary data satisfy special conditions: that is, the boundary data $(\hat{v}_{wi1} - \hat{v}_{iSB1})/\hat{T}_{w0}^{1/2}$ and $(\hat{T}_{w1} - \hat{T}_{SB1})/\hat{T}_{w0}$ are related to $(\partial \hat{T}_{SB0}/\partial x_i)/\hat{p}_{SB0}$ for the diffuse reflection and the boundary data $(\hat{p}_{w1} - \hat{p}_{SB1})/\hat{p}_{w0}$, $(\hat{v}_{wj1} - \hat{v}_{jSB1})(\delta_{ij} - n_j n_i)/\hat{T}_{w0}^{1/2}$, and $(\hat{T}_{w1} - \hat{T}_{SB1})/\hat{T}_{w0}$ are related to $(\partial \hat{T}_{SB0}/\partial x_i)/\hat{p}_{w0}$ and $\hat{v}_{iSB1} n_i$ for the complete condensation condition. The process, where the problem is reduced to two independent problems, is described below.

From the isotropic property of the operator $\mathcal{L}_{\hat{T}_{w0}}$ (Section 2.10), the fields expressed by the integrals $\mathcal{L}_{\hat{T}_{w0}}(\varphi(\tilde{\zeta}, \tilde{\zeta}_l n_l))$ and $\mathcal{L}_{\hat{T}_{w0}}(\tilde{\zeta}_j(\delta_{ij} - n_j n_i)\varphi_i(\tilde{\zeta}, \tilde{\zeta}_l n_l))$ are shown to be axially symmetric with respect to the axis n_i [see the explanation below Eq. (2.47)]. Thus, according to Appendix B.4.2, they are of the following forms:

$$\mathcal{L}_{\hat{T}_{w0}}(\varphi(\tilde{\zeta}, \tilde{\zeta}_l n_l)) = g(\tilde{\zeta}, \tilde{\zeta}_l n_l), \tag{5.71a}$$

$$\mathcal{L}_{\hat{T}_{w0}}(\tilde{\zeta}_j(\delta_{ij} - n_j n_i)\varphi_i(\tilde{\zeta}, \tilde{\zeta}_l n_l)) = \tilde{\zeta}_j(\delta_{ij} - n_j n_i) g_i(\tilde{\zeta}, \tilde{\zeta}_l n_l). \tag{5.71b}$$

That is, the symmetry with respect to $\tilde{\zeta}_j(\delta_{ij} - n_j n_i)$ is conserved by the operation of $\mathcal{L}_{\hat{T}_{w0}}$. Taking into account this property and the boundary conditions (5.69) or (5.70), the boundary-value problem for ϕ_{K1} is reduced to two half-space problems with different parity with respect to $\tilde{\zeta}_j(\delta_{ij} - n_j n_i)$. The solution ϕ_{K1} is split into two parts:

$$\phi_{K1} = \varphi_I + \varphi_{II}, \tag{5.72}$$

where φ_I and φ_{II} are of the following special form with respect to $\tilde{\zeta}_i$:

$$\varphi_I = \varphi_I(\tilde{\eta}, \chi_1, \chi_2, \tilde{\zeta}, \tilde{\zeta}_l n_l),$$

$$\varphi_{II} = \tilde{\zeta}_j(\delta_{ij} - n_j n_i)\phi_i(\tilde{\eta}, \chi_1, \chi_2, \tilde{\zeta}, \tilde{\zeta}_l n_l),$$

each of which is determined independently in the following way.

The first part $\varphi_I(\tilde{\eta}, \chi_1, \chi_2, \tilde{\zeta}, \tilde{\zeta}_l n_l)$ is the solution of the following boundary-value problem: The equation is

$$\tilde{\zeta}_i n_i \frac{\partial \varphi_I(\tilde{\eta}, \tilde{\zeta}, \tilde{\zeta}_l n_l)}{\partial \tilde{\eta}} = g_M(\hat{T}_{w0}) \mathcal{L}_{\hat{T}_{w0}}(\varphi_I(\tilde{\eta}, \tilde{\zeta}, \tilde{\zeta}_l n_l)),$$

where the right-hand collision integral depends only on $\tilde{\zeta}$ and $\tilde{\zeta}_l n_l$ because of its axial symmetry property. The boundary condition at $\tilde{\eta} = 0$ is, in the case of the diffuse reflection,

$$\varphi_I = \frac{\hat{\sigma}_{w1} - \hat{\rho}_{SB1}}{\hat{\rho}_{SB0}} - \frac{2\tilde{\zeta}_i n_i \hat{v}_{jSB1} n_j}{\hat{T}_{w0}^{1/2}} + \left(\tilde{\zeta}^2 - \frac{3}{2}\right) \frac{\hat{T}_{w1} - \hat{T}_{SB1}}{\hat{T}_{w0}}$$
$$+ \frac{1}{\hat{p}_{SB0}} \frac{\partial \hat{T}_{SB0}}{\partial x_i} n_i \tilde{\zeta}_j n_j A(\tilde{\zeta}, \hat{T}_{w0}) \qquad (\tilde{\zeta}_i n_i > 0),$$

where

$$\frac{\hat{\sigma}_{w1}}{\hat{\rho}_{SB0}} = \frac{\hat{\rho}_{SB1}}{\hat{\rho}_{SB0}} - \frac{\hat{T}_{w1} - \hat{T}_{SB1}}{2\hat{T}_{w0}} - \frac{\sqrt{\pi}\hat{v}_{iSB1} n_i}{\hat{T}_{w0}^{1/2}} - 2\sqrt{\pi} \int_{\tilde{\zeta}_i n_i < 0} \tilde{\zeta}_i n_i \varphi_I(\tilde{\zeta}, \tilde{\zeta}_l n_l) E(\tilde{\zeta}) \mathbf{d}\tilde{\boldsymbol{\zeta}},$$

or, in the case of the complete condensation,

$$\varphi_I = \frac{\hat{p}_{w1} - \hat{p}_{SB1}}{\hat{p}_{w0}} - \frac{2\tilde{\zeta}_i n_i \hat{v}_{jSB1} n_j}{\hat{T}_{w0}^{1/2}} + \left(\tilde{\zeta}^2 - \frac{5}{2}\right) \frac{\hat{T}_{w1} - \hat{T}_{SB1}}{\hat{T}_{w0}}$$
$$+ \frac{1}{\hat{p}_{w0}} \frac{\partial \hat{T}_{SB0}}{\partial x_i} n_i \tilde{\zeta}_j n_j A(\tilde{\zeta}, \hat{T}_{w0}) \qquad (\tilde{\zeta}_i n_i > 0),$$

and the condition at infinity (as $\tilde{\eta} \to \infty$) is

$$\varphi_I \to 0.$$

The second part $\varphi_{II}(\tilde{\eta}, \chi_1, \chi_2, \tilde{\zeta}_i)$ is the solution of the following boundary-value problem: The equation is

$$\tilde{\zeta}_k n_k \frac{\partial \varphi_{II}(\tilde{\eta}, \tilde{\zeta}_i)}{\partial \tilde{\eta}} = g_M(\hat{T}_{w0}) \mathcal{L}_{\hat{T}_{w0}}(\varphi_{II}(\tilde{\eta}, \tilde{\zeta}_i)),$$

and therefore, with a new linear operator $\tilde{\mathcal{L}}_{\hat{T}_{w0}}(*)$,

$$\tilde{\zeta}_k n_k \frac{\partial(\delta_{ij} - n_j n_i)\phi_i(\tilde{\eta}, \tilde{\zeta}, \tilde{\zeta}_l n_l)}{\partial \tilde{\eta}} = g_M(\hat{T}_{w0}) \tilde{\mathcal{L}}_{\hat{T}_{w0}}((\delta_{ij} - n_j n_i)\phi_i(\tilde{\eta}, \tilde{\zeta}, \tilde{\zeta}_l n_l)),$$

where $\tilde{\mathcal{L}}_{\hat{T}_{w0}}(\phi_i(\tilde{\eta}, \tilde{\zeta}, \tilde{\zeta}_l n_l))$ is a function of $\tilde{\eta}$, $\tilde{\zeta}$, and $\tilde{\zeta}_l n_l$ because of Eq. (5.71b). The boundary condition at $\tilde{\eta} = 0$ is, commonly for the diffuse reflection and the complete condensation (in the latter $\hat{p}_{SB0} = \hat{p}_{w0}$),

$$\varphi_{II} = \frac{2\tilde{\zeta}_j(\delta_{ij} - n_j n_i)(\hat{v}_{wi1} - \hat{v}_{iSB1})}{\hat{T}_{w0}^{1/2}} + \frac{1}{\hat{p}_{SB0}} \frac{\partial \hat{T}_{SB0}}{\partial x_i} \tilde{\zeta}_j(\delta_{ij} - n_j n_i)\mathcal{A}(\tilde{\zeta}, \hat{T}_{w0}) \quad (\tilde{\zeta}_i n_i > 0),$$

or

$$(\delta_{ij} - n_j n_i)\phi_i = (\delta_{ij} - n_j n_i)\left[\frac{2(\hat{v}_{wi1} - \hat{v}_{iSB1})}{\hat{T}_{w0}^{1/2}} + \frac{1}{\hat{p}_{SB0}} \frac{\partial \hat{T}_{SB0}}{\partial x_i}\mathcal{A}(\tilde{\zeta}, \hat{T}_{w0})\right] \quad (\tilde{\zeta}_i n_i > 0),$$

and the condition at infinity (as $\tilde{\eta} \to \infty$) is

$$\varphi_{II} \to 0 \quad \text{or} \quad \phi_i \to 0.$$

The first problem for φ_I is the combination of the temperature-jump problem and the evaporation and condensation problem (Welander [1954], Sone [1966a], Sone & Onishi [1973], Sone, Ohwada & Aoki [1989a, 1989b], etc.) and the second one for φ_{II} is the thermal-creep-flow problem (Sone [1966b], Ohwada, Sone & Aoki [1989a]). Combining the two solutions, we have the following relations for the boundary data and the Knudsen-layer corrections of the macroscopic variables:
For the diffuse reflection condition, the conditions on the boundary data of the fluid-dynamic part are

$$\frac{(\hat{v}_{jSB1} - \hat{v}_{wj1})(\delta_{ij} - n_j n_i)}{\hat{T}_{w0}^{1/2}} = -\frac{\hat{K}_1}{\hat{p}_{SB0}} \frac{\partial \hat{T}_{SB0}}{\partial x_j}(\delta_{ij} - n_j n_i), \tag{5.73a}$$

$$\hat{v}_{jSB1} n_j = 0, \tag{5.73b}$$

$$\frac{\hat{T}_{SB1} - \hat{T}_{w1}}{\hat{T}_{w0}} = \frac{\hat{d}_1}{\hat{p}_{SB0}} \frac{\partial \hat{T}_{SB0}}{\partial x_j} n_j, \tag{5.73c}$$

and the Knudsen-layer corrections for the density, flow velocity, and temperature are

$$\frac{\hat{v}_{jK1}(\delta_{ij} - n_j n_i)}{\hat{T}_{w0}^{1/2}} = -\frac{1}{2\hat{p}_{SB0}}\left(\frac{\partial \hat{T}_{SB0}}{\partial x_j}\right)_0 (\delta_{ij} - n_j n_i)\hat{Y}_1(\tilde{\eta}), \tag{5.74a}$$

$$\hat{v}_{jK1} n_j = 0, \tag{5.74b}$$

$$\frac{\hat{\rho}_{K1}}{(\hat{\rho}_{SB0})_0} = \frac{1}{\hat{p}_{SB0}}\left(\frac{\partial \hat{T}_{SB0}}{\partial x_j}\right)_0 n_j \hat{\Omega}_1(\tilde{\eta}), \tag{5.74c}$$

$$\frac{\hat{T}_{K1}}{\hat{T}_{w0}} = \frac{1}{\hat{p}_{SB0}}\left(\frac{\partial \hat{T}_{SB0}}{\partial x_j}\right)_0 n_j \hat{\Theta}_1(\tilde{\eta}). \tag{5.74d}$$

For the complete condensation condition, the conditions on the boundary data of the fluid-dynamic part are

$$\frac{(\hat{v}_{jSB1}-\hat{v}_{wj1})(\delta_{ij}-n_jn_i)}{\hat{T}_{w0}^{1/2}} = -\frac{\hat{K}_1}{\hat{p}_{w0}}\frac{\partial \hat{T}_{SB0}}{\partial x_j}(\delta_{ij}-n_jn_i), \tag{5.75a}$$

$$\frac{\hat{p}_{SB1}-\hat{p}_{w1}}{\hat{p}_{w0}} = \hat{C}_4^*\frac{\hat{v}_{jSB1}n_j}{\hat{T}_{w0}^{1/2}} + \frac{\hat{C}_1}{\hat{p}_{w0}}\frac{\partial \hat{T}_{SB0}}{\partial x_j}n_j, \tag{5.75b}$$

$$\frac{\hat{T}_{SB1}-\hat{T}_{w1}}{\hat{T}_{w0}} = \hat{d}_4^*\frac{\hat{v}_{jSB1}n_j}{\hat{T}_{w0}^{1/2}} + \frac{\hat{d}_1}{\hat{p}_{w0}}\frac{\partial \hat{T}_{SB0}}{\partial x_j}n_j, \tag{5.75c}$$

and the Knudsen-layer corrections for the density, flow velocity, and temperature are

$$\frac{\hat{v}_{jK1}(\delta_{ij}-n_jn_i)}{\hat{T}_{w0}^{1/2}} = -\frac{1}{2\hat{p}_{w0}}\left(\frac{\partial \hat{T}_{SB0}}{\partial x_j}\right)_0(\delta_{ij}-n_jn_i)\hat{Y}_1(\tilde{\eta}), \tag{5.76a}$$

$$\hat{v}_{jK1}n_j = 0, \tag{5.76b}$$

$$\frac{\hat{\rho}_{K1}}{\hat{\rho}_{w0}} = \frac{(\hat{v}_{jSB1}n_j)_0}{\hat{T}_{w0}^{1/2}}\hat{\Omega}_4^*(\tilde{\eta}) + \frac{1}{\hat{p}_{w0}}\left(\frac{\partial \hat{T}_{SB0}}{\partial x_j}\right)_0 n_j\hat{\Omega}_1(\tilde{\eta}), \tag{5.76c}$$

$$\frac{\hat{T}_{K1}}{\hat{T}_{w0}} = \frac{(\hat{v}_{jSB1}n_j)_0}{\hat{T}_{w0}^{1/2}}\hat{\Theta}_4^*(\tilde{\eta}) + \frac{1}{\hat{p}_{w0}}\left(\frac{\partial \hat{T}_{SB0}}{\partial x_j}\right)_0 n_j\hat{\Theta}_1(\tilde{\eta}). \tag{5.76d}$$

In these formulas, $\hat{K}_1$, $\hat{d}_1$, $\hat{C}_1$, $\hat{C}_4^*$, and $\hat{d}_4^*$, which are called slip or jump coefficients, are functions of $\hat{T}_{w0}$ depending on molecular models, and $\hat{Y}_1(\tilde{\eta})$, $\hat{\Omega}_1(\tilde{\eta})$, $\hat{\Theta}_1(\tilde{\eta})$, $\hat{\Omega}_4^*(\tilde{\eta})$, and $\hat{\Theta}_4^*(\tilde{\eta})$, called Knudsen-layer functions, are functions of $\tilde{\eta}$ and $\hat{T}_{w0}$ depending on molecular models. It may be noted here that these slip coefficients and the Knudsen-layer functions generally depend on the reference temperature T_0, since the operator $\mathcal{L}_{\hat{T}_{w0}}$ depends on T_0 besides the parameter $\hat{T}_{w0}$ (see Footnotes 11 in Section 2.9 and 13 in Appendix A.2). In case $g_M(\hat{T}_{w0})\mathcal{L}_{\hat{T}_{w0}}$ is independent of $\hat{T}_{w0}$, the modified coefficients and functions $\hat{K}_1/g_M(\hat{T}_{w0})$, $\hat{d}_1/g_M(\hat{T}_{w0})$, $\hat{C}_1/g_M(\hat{T}_{w0})$, $\hat{Y}_1(\tilde{\eta})/g_M(\hat{T}_{w0})$, $\hat{\Omega}_1(\tilde{\eta})/g_M(\hat{T}_{w0})$, and $\hat{\Theta}_1(\tilde{\eta})/g_M(\hat{T}_{w0})$ and the coefficients and functions $\hat{C}_4^*$, $\hat{d}_4^*$, $\hat{\Omega}_4^*(\tilde{\eta})$, and $\hat{\Theta}_4^*(\tilde{\eta})$ themselves are independent of $\hat{T}_{w0}$.[8] For a hard-sphere gas and the BKW model, these coefficients and functions are related to the corresponding coefficients and

[8] In this case the linearized Boltzmann equation (5.68) is reduced to the same equation as that in the linear theory in Chapter 3:

$$\tilde{\zeta}_i n_i\frac{\partial \phi_{K1}(\tilde{\eta},\tilde{\zeta}_i)}{\partial \tilde{\eta}} = \mathcal{L}(\phi_{K1}(\tilde{\eta},\tilde{\zeta}_i)).$$

The essential difference is $\mathcal{A}(\tilde{\zeta},\hat{T}_{w0})$ in the boundary condition (5.69) or (5.70). From the definitions of $\mathcal{A}(\tilde{\zeta},\hat{T}_{w0})$ [Eq. (A.84)], $A(\tilde{\zeta})$ [Eq. (3.25)], and $g_M(\hat{T}_{w0})$, the function $\mathcal{A}(\tilde{\zeta},\hat{T}_{w0})$ is given by $A(\tilde{\zeta})g_M(\hat{T}_{w0})$. The statement in the text is simply derived from these facts in the linear system.

functions without the symbol circumflex ˆ in the linear theory in Chapter 3. For a hard-sphere gas,

$$\hat{K}_1 = K_1\ (= -0.6463),\ \hat{d}_1 = d_1\ (= 2.4001),\ \hat{C}_1 = C_1\ (= 1.0947),$$
$$\hat{C}_4^* = C_4^*\ (= -2.1412),\ \hat{d}_4^* = d_4^*\ (= -0.4557),$$

$$\hat{Y}_1(\bar{\eta}) = Y_1(\bar{\eta}),\quad \hat{\Omega}_1(\bar{\eta}) = \Omega_1(\bar{\eta}),\quad \hat{\Theta}_1(\bar{\eta}) = \Theta_1(\bar{\eta}),$$
$$\hat{\Omega}_4^*(\bar{\eta}) = \Omega_4^*(\bar{\eta}),\quad \hat{\Theta}_4^*(\bar{\eta}) = \Theta_4^*(\bar{\eta}),$$

and for the BKW model,

$$\hat{K}_1/\hat{T}_{w0}^{1/2} = K_1\ (= -0.38316),\ \hat{d}_1/\hat{T}_{w0}^{1/2} = d_1\ (= 1.30272),$$
$$\hat{C}_1/\hat{T}_{w0}^{1/2} = C_1\ (= 0.55844),\ \hat{C}_4^* = C_4^*\ (= -2.13204),\ \hat{d}_4^* = d_4^*\ (= -0.44675),$$

$$\hat{Y}_1(\bar{\eta})/\hat{T}_{w0}^{1/2} = Y_1(\bar{\eta}),\ \hat{\Omega}_1(\bar{\eta})/\hat{T}_{w0}^{1/2} = \Omega_1(\bar{\eta}),\ \hat{\Theta}_1(\bar{\eta})/\hat{T}_{w0}^{1/2} = \Theta_1(\bar{\eta}),$$
$$\hat{\Omega}_4^*(\bar{\eta}) = \Omega_4^*(\bar{\eta}),\quad \hat{\Theta}_4^*(\bar{\eta}) = \Theta_4^*(\bar{\eta}).$$

They are exceptionally independent of the reference temperature T_0.

The higher-order Knudsen-layer analysis, giving the corresponding Knudsen-layer corrections and the conditions on the boundary data of the fluid-dynamic part, can be carried out in a similar way, although the S-layer correction (Section 3.7) is required over a convex boundary.

By a similar discussion to that in Section 3.4, the relations (5.73a)–(5.73c) or (5.75a)–(5.75c) for the boundary data of the fluid-dynamic part and the Knudsen-layer corrections (5.74a)–(5.74d) or (5.76a)–(5.76d) are found to be applicable for a general locally isotropic boundary where the reflection law is expressed by Eq. (2.22) with a finite diffuse-reflection part or by Eq. (2.25) with a finite complete condensation part. The differences are the slip coefficients $\hat{K}_1$, $\hat{d}_1$, $\hat{C}_1$, $\hat{C}_4^*$, and $\hat{d}_4^*$ and the Knudsen-layer functions $\hat{Y}_1(\bar{\eta})$, $\hat{\Omega}_1(\bar{\eta})$, $\hat{\Theta}_1(\bar{\eta})$, $\hat{\Omega}_4^*(\bar{\eta})$, and $\hat{\Theta}_4^*(\bar{\eta})$. They depend on the reflection law on the boundary as well as on the molecular model of the gas.

The relations (5.73a)–(5.73c) or (5.75a)–(5.75c) are the conditions on the boundary values of variables of the order of k. Some of them are, however, required to find the behavior of the gas in the limit $k \to 0$, because the fluid-dynamic equations (5.42)–(5.44) with Eq. (5.45) to determine the behavior of the gas in this limit consist of variables of the order of 1 and k. The process by which the variables are determined succesively from the lowest order is explained in the next section.

5.5 Determination of macroscopic variables

Now we have obtained information enough to discuss how the macroscopic variables are determined from the lowest order. We will discuss the determination of the fluid-dynamic parts (the variables with subscript *SB*), since the Knudsen-layer corrections are determined from the fluid-dynamic parts with the aid of simple formulas. The discussion is made separately for a simple boundary and for an interface of a gas and its condensed phase.[9]

First we start with the discussion of the case of a simple boundary. From Eqs. (5.24) and (5.29), as we have seen in Eq. (5.37), $\hat{p}_{SB0}$ and $\hat{p}_{SB1}$ are constants:

$$\hat{p}_{SB0} = \hat{p}_0, \qquad \hat{p}_{SB1} = \hat{p}_1.$$

In an open-(or unbounded-)domain problem, where the pressure is specified, for example, at infinity, the constants $\hat{p}_0$ and $\hat{p}_1$ are determined by the pressure there. In a closed-(or bounded-)domain problem, where the total mass in the domain is specified (see Footnote 27 in Section 3.6), $\hat{p}_0$ and $\hat{p}_1$ cannot be determined at this stage. Now consider the set of equations (5.28), (5.35) [or (5.48) or (5.50)], and (5.30) with Eq. (5.11d), where the variables $\hat{\rho}_{SB0}$, $\hat{T}_{SB0}$, $\hat{v}_{iSB1}$, and $\hat{p}_{SB2}$ are contained. From Eq. (5.11d), $\hat{\rho}_{SB0}$ is expressed by $\hat{p}_{SB0}$ (or $\hat{p}_0$) and $\hat{T}_{SB0}$ [Eq. (5.38)] as

$$\hat{\rho}_{SB0} = \frac{\hat{p}_0}{\hat{T}_{SB0}}. \tag{5.77}$$

Thus we can discuss the set of equations (5.28), (5.35) [or (5.48) or (5.50)], and (5.30) with the variables $\hat{T}_{SB0}$, $\hat{v}_{iSB1}$, and $\hat{p}_{SB2}$ (and a constant $\hat{p}_0$) under the boundary conditions (5.56), (5.73a), and (5.73b). From this system, the variables $\hat{T}_{SB0}$, $\hat{v}_{iSB1}$, and $\hat{p}_{SB2}$ (thus $\hat{\rho}_{SB0}$) are determined with $\hat{p}_0$ as a parameter, leaving an undetermined integration constant (say, $\hat{p}_2$) in $\hat{p}_{SB2}$. Here the variable $\hat{p}_{SB2}$ is in a similar situation to the pressure in the incompressible Navier–Stokes set of equations. At this stage the constant $\hat{p}_0$, if undetermined previously, is determined, for example, from the total mass condition with the aid of $\hat{\rho}_{SB0}$. We have not used the equations (5.34) and (5.36) and the boundary condition (5.73c). These are used at the next order in the same role as the equations (5.28) and (5.30) and the boundary condition (5.56) at the present stage. The constant $\hat{p}_1$, if undetermined, is also determined at the next stage by the total mass condition, for example, in a similar way to $\hat{p}_0$ at the present stage. The constant $\hat{p}_2$ in $\hat{p}_{SB2}$, if undetermined, is fixed at the stage after next. In this way we can determine the macroscopic variables successively.

For the case of an interface of a gas and its condensed phase, the process is similar but there is some difference for determination of the integration constant in $\hat{p}_{SBm}$. The constant $\hat{p}_0$ is determined by the boundary condition (5.57b). From the set of equations (5.28), (5.35) [or (5.48) or (5.50)], and (5.30) with Eq. (5.11d) and the boundary conditions (5.57a), (5.75a), and (5.75b), the variables $\hat{T}_{SB0}$, $\hat{\rho}_{SB0}$, $\hat{v}_{iSB1}$, and $\hat{p}_{SB2}$ are determined with the constant $\hat{p}_1$ in Eq. (5.37) as a parameter, leaving an undetermined integration constant (say, $\hat{p}_2$) in $\hat{p}_{SB2}$; the parameter $\hat{p}_1$ enters the variables $\hat{T}_{SB0}$, $\hat{\rho}_{SB0}$, $\hat{v}_{iSB1}$, and $\hat{p}_{SB2}$ through $\hat{p}_{SB1}$ in the boundary condition (5.75b). The constant $\hat{p}_1$ cannot be arbitrary, because it is determined by the condition at infinity in an open-domain

[9] Note the discussion at the paragraph before the last in the preceding section.

problem, or because the quantity $\hat{\rho}_{SB0}\hat{v}_{iSB1}n_i$ integrated over the boundary must be zero from Eq. (5.28) in a closed-domain problem. We have not used the equations (5.34) and (5.36) and the boundary condition (5.75c). These are used at the next order in the same role as the equations (5.28) and (5.30) and the boundary condition (5.57a) at the present stage. The constant $\hat{p}_2$ in $\hat{p}_{SB2}$ is determined at the next stage in the same way as $\hat{p}_1$ at the present stage, where Eq. (5.34) plays the same role as Eq. (5.28). In this way we can determine the macroscopic variables of the fluid-dynamic part successively. Formally it is consistent, but some notes should be made here. In evaporation and condensation problems, the saturated gas pressure $\hat{p}_w$ is generally an increasing function of temperature $\hat{T}_w$. Thus from the condition (5.57b), which is required for the condition (5.8) to hold, $\hat{p}_{w0}$ and $\hat{T}_{w0}$ should be uniform. In a closed-domain problem this case corresponds to that discussed in Chapter 4. In an open-domain problem, the pressure $\hat{p}_\infty$ should be close to $\hat{p}_{w0}$ (with difference of the order of k), but the temperature $\hat{T}_\infty$ at infinity may be arbitrary, for which the result of the present chapter is required. In this case, the gas is highly undersaturated or supersaturated at infinity. An example is given in Section 5.7.

Finally, the following important results may be noted: The velocity distribution function of the gas molecules is a Maxwellian in the limit $k \to 0$, but the macroscopic variables in the Maxwellian are neither the solution of the Euler set of equations nor that of the Navier–Stokes set of equations. They are determined by a set of equations that contains higher-order gas rarefaction effects, i.e., Eqs. (5.24), (5.29), (5.28), (5.35) [or (5.48) or (5.50)], and (5.30). From these equations, the temperature $\hat{T}_{SB0}$ and density $\hat{\rho}_{SB0}$ in the continuum limit are determined together with the higher-order velocity $\hat{v}_{iSB1}$. That is, the behavior of a gas in the continuum world cannot be discussed only with the quantities appreciable in the world. This is a serious matter to the classical gas dynamics. This subject will be discussed in more detail in the next section.

5.6 Ghost effect: Incompleteness of the system of the classical gas dynamics

5.6.1 Nonlinear thermal-stress flow and inappropriateness of the heat-conduction equation

In this chapter, we have developed the asymptotic theory for small Knudsen numbers of a time-independent boundary-value problem of the Boltzmann equation when the flow velocity v_i vanishes in the limit as the Knudsen number $\to 0$ (or in the continuum limit). The system of equations and boundary conditions that describes the gas in the continuum limit has an interesting feature. It is not the system of equations and boundary conditions that is given by the classical gas dynamics. Furthermore, the flow velocity of the first order of the Knudsen number[10] appears in the equations that determine the temperature and density

[10] There should be no misunderstanding from the foregoing discussions. In our nondimensional form of equations, this flow velocity is the component function (i.e., $\hat{v}_{iSB1}$) at the first

in the continuum limit. We will discuss the system and the result derived from it in this section.

In the classical gas dynamics, the temperature field of a gas at rest is determined by the well-known heat-conduction equation

$$\frac{\partial}{\partial X_i}\lambda\frac{\partial T}{\partial X_i}=0, \tag{5.78}$$

which is obtained by putting the flow velocity v_i to be zero in the Navier–Stokes set of equations (see Section 2.3). According to our asymptotic analysis, the corresponding temperature field $\hat{T}_{SB0}$ is determined by the set of equations (5.42), (5.43), and (5.44) with Eq. (5.45). If the velocity $\hat{v}_{iSB1}$ of the first order of the Knudsen number vanishes, Eq. (5.44) is reduced to the heat-conduction equation

$$\frac{\partial}{\partial x_i}\left(\hat{\gamma}_2(\hat{T}_{SB0})\hat{T}_{SB0}^{1/2}\frac{\partial \hat{T}_{SB0}}{\partial x_i}\right)=0. \tag{5.79}$$

The first-order velocity $\hat{v}_{iSB1}$, however, cannot be zero except for very limited special cases for the following reasons.
(i) First of all, if the boundary is moving at a very small (or infinitesimal) velocity $\hat{v}_{wi}=\hat{v}_{wi1}k$, which cannot be perceived in the world of the continuum limit, then flow $\hat{v}_{iSB1}k$ of the same order is induced because of the boundary condition (5.73a) or (5.75a).
(ii) When the temperature of the boundary is nonuniform, $\hat{v}_{iSB1}$ cannot be zero because of the boundary condition (5.73a) or (5.75a). (Thermal creep flow)
(iii) Even when the boundary temperature is uniform and $\hat{v}_{wi1}=0$, $\hat{v}_{iSB1}$ cannot be generally zero. Let $\hat{v}_{iSB1}$ be zero in Eq. (5.43). Then Eq. (5.43) is reduced to

$$\frac{\partial \hat{p}_{SB2}}{\partial x_i}+\frac{1}{\hat{p}_{SB0}}\frac{\partial}{\partial x_j}\left\{\hat{\gamma}_7(\hat{T}_{SB0})\left[\frac{\partial \hat{T}_{SB0}}{\partial x_i}\frac{\partial \hat{T}_{SB0}}{\partial x_j}-\frac{1}{3}\left(\frac{\partial \hat{T}_{SB0}}{\partial x_k}\right)^2\delta_{ij}\right]\right\}$$
$$+\frac{1}{\hat{p}_{SB0}}\frac{\partial}{\partial x_j}\left[\hat{\gamma}_3(\hat{T}_{SB0})\hat{T}_{SB0}\left(\frac{\partial^2 \hat{T}_{SB0}}{\partial x_i\partial x_j}-\frac{1}{3}\frac{\partial^2 \hat{T}_{SB0}}{\partial x_k^2}\delta_{ij}\right)\right]=0. \tag{5.80}$$

Taking the curl of Eq. (5.80) and using Eq. (5.79), we have

$$\frac{1}{\hat{p}_{SB0}}\left[\frac{\hat{\gamma}_2^2\hat{T}_{SB0}}{2}\frac{\mathrm{d}}{\mathrm{d}\hat{T}_{SB0}}\left(\frac{\hat{\gamma}_7-\mathrm{d}\hat{\gamma}_3\hat{T}_{SB0}/\mathrm{d}\hat{T}_{SB0}}{\hat{\gamma}_2^2\hat{T}_{SB0}}\right)\right]$$
$$\times\left[\frac{\partial \hat{T}_{SB0}}{\partial x_i}\frac{\partial}{\partial x_j}\left(\frac{\partial \hat{T}_{SB0}}{\partial x_l}\right)^2-\frac{\partial \hat{T}_{SB0}}{\partial x_j}\frac{\partial}{\partial x_i}\left(\frac{\partial \hat{T}_{SB0}}{\partial x_l}\right)^2\right]=0.$$

---

order of the expansion of the flow velocity (i.e., $\hat{v}_{iSB}$) in the Knudsen number. This kind of note will not be repeated if no confusion is expected.

The factor

$$\frac{\hat{\gamma}_2^2\hat{T}_{SB0}}{2}\frac{\mathrm{d}}{\mathrm{d}\hat{T}_{SB0}}\left(\frac{\hat{\gamma}_7-\mathrm{d}\hat{\gamma}_3\hat{T}_{SB0}/\mathrm{d}\hat{T}_{SB0}}{\hat{\gamma}_2^2\hat{T}_{SB0}}\right),$$

a function of $\hat{T}_{SB0}$, can vanish only for special values of $\hat{T}_{SB0}$.[11] Thus,

$$\left[\frac{\partial\hat{T}_{SB0}}{\partial x_i}\frac{\partial}{\partial x_j}\left(\frac{\partial\hat{T}_{SB0}}{\partial x_l}\right)^2-\frac{\partial\hat{T}_{SB0}}{\partial x_j}\frac{\partial}{\partial x_i}\left(\frac{\partial\hat{T}_{SB0}}{\partial x_l}\right)^2\right]=0. \tag{5.81}$$

Besides the obvious case of a uniform temperature, the condition (5.81) is satisfied only when the distance between the isothermal surfaces is uniform along them or they are parallel.[12] Furthermore, Eq. (5.79) requires that the mean curvature is uniform on each isothermal surface.[13] The classes of surfaces that satisfy the two conditions are very limited, for example, parallel plane surfaces, coaxial cylinders or concentric spheres. Thus, in a gas around bodies at rest of different uniform temperatures, where the thermal creep flow is absent, a flow is generally induced. The flow was first pointed out by Kogan's group (Kogan, Galkin & Fridlender [1976]) and is now called the nonlinear thermal-stress flow.

According to these considerations, the temperature field of a gas at rest in the continuum limit is not described correctly by the widely used heat-conduction equation [or Eq. (5.78) or (5.79)]. The system of equations of the classical gas dynamics, the system to describe the behavior of a gas in the continuum limit, fails to describe an important class of problems in the continuum limit. The quantity that completes the classical gas dynamic system is given by the flow velocity of the first order of the Knudsen number. If one lives in the world in the continuum limit, one does not know this velocity. Something not existing in the world affects the behavior of the world. It may be called a ghost effect.

---

[11] For a hard-sphere gas, the factor is $(\hat{\gamma}_3-\hat{\gamma}_7)/2\hat{T}_{SB0}$ $(=0.87935/\hat{T}_{SB0})$, and for the BKW model, it is 1/2.

[12] From the condition (5.81), $(\partial\hat{T}_{SB0}/\partial x_l)^2$ is found to be a function of $\hat{T}_{SB0}$ only (see a standard textbook on analysis).

[13] Take a curvilinear coordinate system (s_1, s_2, s_3), where the coordinate lines s_1 and s_2 lie on the isothermal surfaces and s_3 is orthogonal to the isothermal surfaces. The temperature $\hat{T}_{SB0}$ being independent of s_1 and s_2 by definition, the heat-conduction equation (5.79) is reduced to

$$\frac{\partial}{\partial s_3}\left(\frac{S_{12}\hat{\gamma}_2\hat{T}_{SB0}^{1/2}}{\sqrt{g_{33}}}\frac{\partial\hat{T}_{SB0}}{\partial s_3}\right)=0,\quad S_{12}=(g_{11}g_{22}-g_{12}^2)^{1/2},$$

where g_{11}, g_{22}, g_{33}, and g_{12} are the measure factors, by which the length of a line element $\mathrm{d}s$ is expressed as $(\mathrm{d}s)^2=g_{11}(\mathrm{d}s_1)^2+2g_{12}\mathrm{d}s_1\mathrm{d}s_2+g_{22}(\mathrm{d}s_2)^2+g_{33}(\mathrm{d}s_3)^2$ (Brillouin [1938]). The equation is rewritten as

$$\frac{1}{S_{12}}\frac{\partial}{\partial s_3}S_{12}=-\left(\frac{\hat{\gamma}_2\hat{T}_{SB0}^{1/2}}{\sqrt{g_{33}}}\frac{\partial\hat{T}_{SB0}}{\partial s_3}\right)^{-1}\frac{\partial}{\partial s_3}\left(\frac{\hat{\gamma}_2\hat{T}_{SB0}^{1/2}}{\sqrt{g_{33}}}\frac{\partial\hat{T}_{SB0}}{\partial s_3}\right).$$

The g_{33} as well as $\hat{T}_{SB0}$ is independent of s_1 and s_2 because the isothermal surfaces are parallel. Therefore, the left-hand side $S_{12}^{-1}\partial S_{12}/\partial s_3$ is independent of s_1 and s_2. On the other hand, $S_{12}^{-1}\partial S_{12}/\partial s_3$ is twice the mean curvature of the surface s_3 =const (or isothermal surface).

5.6.2 Ambiguity in the continuum world

The system of equations (5.42), (5.43) [or (5.48)], and (5.44) with Eq. (5.45) and boundary conditions (5.56), (5.73a), and (5.73b) [or (5.57a), (5.75a), and (5.75b)], which is obtained from the Boltzmann system of Eqs. (5.1) and (5.3a) [or (5.4)] as the limiting system in the continuum limit, has ambiguity, when problems are considered at Kn = 0 (or in the world at Kn = 0 or in the continuum world). This is not simply because $\hat{v}_{iSB1}$ enters the system, but because the data $\hat{v}_{wi1}$ enter the boundary condition (5.73a) [or (5.75a)]. The governing equations being established, the variable $\hat{v}_{iSB1}$ is only a function that is determined together with $\hat{T}_{SB0}$ definitely if the boundary data to be substituted in the boundary conditions (5.56), (5.73a), and (5.73b) [or (5.57a), (5.75a), and (5.75b)] are specified. In the boundary conditions (5.56), (5.73a), and (5.73b) [or (5.57a), (5.75a), and (5.75b)], only $\hat{T}_{w0}$ appears as the information of the temperature of the boundary, but $\hat{v}_{wi1}$ appears as the information of the velocity of the boundary. The former is defined in the world at Kn = 0 (or in the continuum world), but there is no way to determine (or discriminate) the latter there. Thus, the ghost effects in the situations (ii) and (iii) in Section 5.6.1 do not introduce any ambiguity and the variables $\hat{T}_{SB0}$ and $\hat{p}_{SB0}$ are determined together with $\hat{v}_{iSB1}$. The effect in the situation (i) in Section 5.6.1 cannot be determined since $\hat{v}_{wi1}$ cannot be specified. There is an ambiguity in the continuum world. In real physical problems, Kn is not exactly zero, thus $\hat{v}_{wi1}$ can be specified in principle, but its very accurate determination is required. That is, an error of the order of Kn in the determination of the boundary velocity introduces errors of the order of unity in $\hat{T}_{SB0}$ and $\hat{p}_{SB0}$.

An interesting effect of infinitesimal boundary velocity is seen in a gas between two coaxial circular cylinders. When the inner cylinder is rotating, the bifurcation of solution giving rise to the Taylor roll, in addition to the axially symmetric and uniform cylindrical Couette flow, is well-known in the classical fluid dynamics (Taylor [1923], Chandrasekhar [1961]). When the (relative) temperature difference between the two cylinders is of the order of unity, according to the discussion in this chapter, we have to be careful to discuss the problem on the basis of the classical gas dynamics. Let the inner cylinder be rotating at an infinitesimal velocity (of the order of the Knudsen number). This is the situation discussed in this chapter. Obviously, an axially symmetric and uniform infinitesimal velocity field is obtained from the fluid-dynamic-type equations and their boundary conditions given in Sections 5.3 and 5.4. For this infinitesimal velocity field, the temperature field is governed by the heat-conduction equation and no ghost effect is seen. According to Sone, Handa & Doi [unpublished], there is an axially symmetric but nonuniform solution of Taylor-roll-type infinitesimal velocity that bifurcates from the axially symmetric and uniform solution of infinitesimal velocity. Owing to the ghost effect of this bifurcated infinitesimal velocity field, the corresponding temperature field is axially nonuniform. The axially nonuniform temperature field and the bifurcation curve (the relation between the ratio of the temperatures of the two cylinders and the infinitesimal speed of rotation of the inner cylinder where the bifurcation of solution occurs) depend on the nonlinear thermal stress.

5.6.3 Reflection on the Navier–Stokes set of equations

We have seen the discrepancy of the classical gas dynamic system (or Navier–Stokes system) in describing an important class of problems in a gas in the continuum limit. This discrepancy can be seen from the Navier–Stokes set of equations itself if we look at the set a little bit carefully. The Navier–Stokes set of equations contains the viscosity μ and thermal conductivity λ of a gas, which are generally functions of the temperature of the gas. These quantities are determined by the properties of the molecule constituting the gas. According to a very crude estimate by elementary kinetic theory (Kennard [1938], Vincenti & Kruger [1965]), or as we have seen in Chapters 3–5, they are given as follows:

$$\mu/\rho = f(T)(2RT)^{1/2}\ell, \quad \lambda/\rho = g(T)(2RT)^{1/2}R\ell, \tag{5.82}$$

where ρ is the density of the gas, T is its temperature, R is the specific gas constant, ℓ is the mean free path of the gas molecules (strictly, the mean free path at the equilibrium state at rest with ρ and T), and f and g are nondimensional functions of T, whose functional forms are determined by the property of the gas molecule. (For a hard-sphere gas, $f = \sqrt{\pi}\gamma_1/4 = 0.56277$, $g = 5\sqrt{\pi}\gamma_2/8 = 2.1295$.) The Navier–Stokes set, which is supposed to describe the behavior of the gas in the continuum limit (or of vanishing mean free path), contains the mean free path. One may naturally suspect whether the Navier–Stokes set is really a consistent set for describing the behavior of a gas in the continuum limit or whether some necessary terms may be neglected if the viscosity and thermal conductivity are really necessary to describe the behavior of the gas. We will discuss this possibility within the framework of the Navier–Stokes set.

Let us examine the order of the main terms in the energy equation of the Navier–Stokes set [Eq. (2.13) with Eq. (2.15)]. The energy equation is given as

$$\frac{5}{2}\rho v_i \frac{\partial RT}{\partial X_i} = \cdots + \frac{\partial}{\partial X_i}\lambda\frac{\partial T}{\partial X_i},$$

where only the important terms in our discussion are given explicitly. The left-hand side is the convection term,[14] which is absent for a gas at rest, and the last term on the right-hand side is the heat-conduction term. The orders of the convection term [say, $O(\text{Evec})$] and the conduction term [say, $O(\text{Educ})$] are

$$O(\text{Evec}) = \rho VRT/L, \qquad O(\text{Educ}) = \rho(RT)^{3/2}\ell/L^2,$$

where V and L are, respectively, the characteristic flow speed and the length scale of variation of the gas temperature, and the temperature variation is of the same order as the temperature itself (a fundamental assumption in this chapter). The two terms are comparable when the flow speed V is of the order

[14] More exactly, it is the heat convection term plus the work done by pressure.

of $(RT)^{1/2}\ell/L$. Thus, the flow vanishing as $\ell \to 0$ cannot be neglected in the energy equation.

We have already seen that the thermal creep flow and the nonlinear thermal-stress flow are of this order and that the thermal edge flow (the flow induced around a heated sharp edge) is much faster though localized. Without this information from kinetic theory, we know that the motion of the boundary with speed of the order of $(RT)^{1/2}\ell/L$ induces a flow with speed of the same order. Thus, it is clear that the convection term cannot be neglected in a general situation. For this size of flow velocity, let us estimate the order of the main terms in the momentum equation of the Navier–Stokes set [Eq. (2.12) with Eq. (2.15)][15]

$$\rho v_j \frac{\partial v_i}{\partial X_j} = \cdots + \frac{\partial}{\partial X_j}\mu\left(\frac{\partial v_i}{\partial X_j} + \frac{\partial v_j}{\partial X_i} - \frac{2}{3}\frac{\partial v_k}{\partial X_k}\delta_{ij}\right).$$

The order of the convection term [say, $O(\text{Mvec})$] on the left-hand side and that of the viscous term [say, $O(\text{Visc})$], the last term on the right-hand side, are

$$O(\text{Mvec}) = \rho V^2/L = (\rho RT/L)(\ell/L)^2, \qquad O(\text{Visc}) = (\rho RT/L)(\ell/L)^2.$$

Both are of the second order of the Knudsen number. The original order $\rho(RT)^{1/2}V\ell/L$ of the viscous stress, which is the first order of the Knudsen number for $V = O((RT)^{1/2})$, is degraded to $(\rho RT)(\ell/L)^2$ because of $V = O((RT)^{1/2}\ell/L)$. Thus, the higher-order stress of the order $(\rho RT)(\ell/L)^2$ must be retained in this momentum equation. The thermal stress in Eq. (5.43) corresponds to this stress. This stress is another cause of the flow of the order of $(RT)^{1/2}\ell/L$, resulting in the nonlinear thermal-stress flow. In this way we can also understand the inconsistency of the Navier–Stokes set and the necessity of the convection term in the energy equation and another higher-order term in the momentum equation by examining the Navier–Stokes set of equations.

5.6.4 Illustrative examples

For better understanding of the ghost effect pointed out by the asymptotic theory developed in this chapter, we will show the comparison of three kinds of solutions, solutions of the kinetic equation, the equations derived by the asymptotic theory, and the heat-conduction equation, of the following simple problem given in Sone, Aoki, Takata, Sugimoto & Bobylev [1996].

Consider a rarefied gas between two parallel plane walls at $X_2 = 0$ and $X_2 = L$. Both walls have a common periodic temperature distribution $T_w = T_0(1 - \alpha\cos 2\pi X_1/L)$ (α: a constant) and a common velocity $v_{wi} = (2RT_0)^{1/2}(\beta\text{Kn}, 0, 0)$ (β: a constant) of the order of Kn in their planes (see Fig. 5.1). The analysis is carried out under the following assumptions: (i) the behavior of the gas is described by the BKW equation; (ii) the gas molecules make the diffuse reflection on the walls; (iii) the solution has period L with respect to X_1. The

[15] For a monatomic gas, the bulk viscosity $\mu_B = 0$.

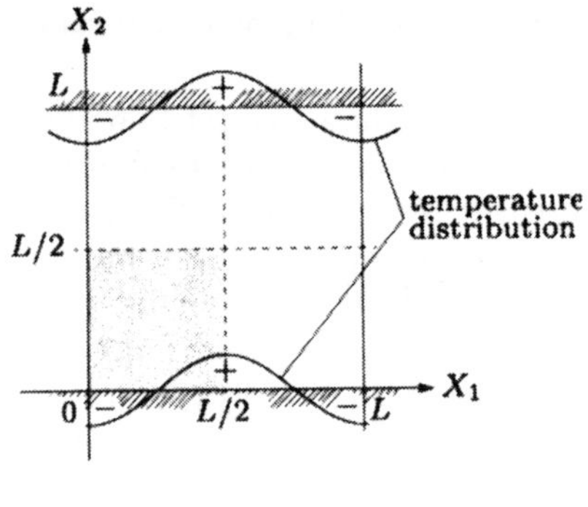

Figure 5.1. Geometry of the examples of the ghost effect.

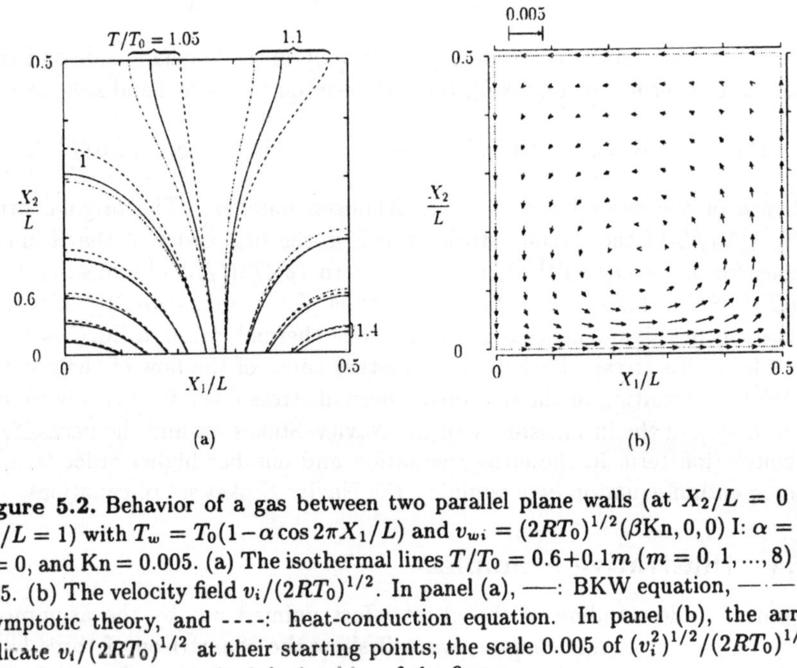

Figure 5.2. Behavior of a gas between two parallel plane walls (at $X_2/L = 0$ and $X_2/L = 1$) with $T_w = T_0(1-\alpha\cos 2\pi X_1/L)$ and $v_{wi} = (2RT_0)^{1/2}(\beta\text{Kn}, 0, 0)$ I: $\alpha = 0.5$, $\beta = 0$, and Kn $= 0.005$. (a) The isothermal lines $T/T_0 = 0.6+0.1m$ ($m = 0, 1, ..., 8$) and 1.05. (b) The velocity field $v_i/(2RT_0)^{1/2}$. In panel (a), —: BKW equation, —·—·—: asymptotic theory, and - - - -: heat-conduction equation. In panel (b), the arrows indicate $v_i/(2RT_0)^{1/2}$ at their starting points; the scale 0.005 of $(v_i^2)^{1/2}/(2RT_0)^{1/2}$ is shown by the arrow on the left shoulder of the figure.

last assumption corresponds to the condition that the average pressure gradient in the X_1-direction is absent. According to the asymptotic theory, the way in which the temperature field deviates from that of the heat-conduction equation is similar for the standard Boltzmann equation and the BKW equation. Thus, the easier BKW equation is used, instead of the standard Boltzmann equation, for the kinetic equation, since very precise computation for very small Knudsen numbers, near the singular point of the Boltzmann equation, is required for the present purpose.

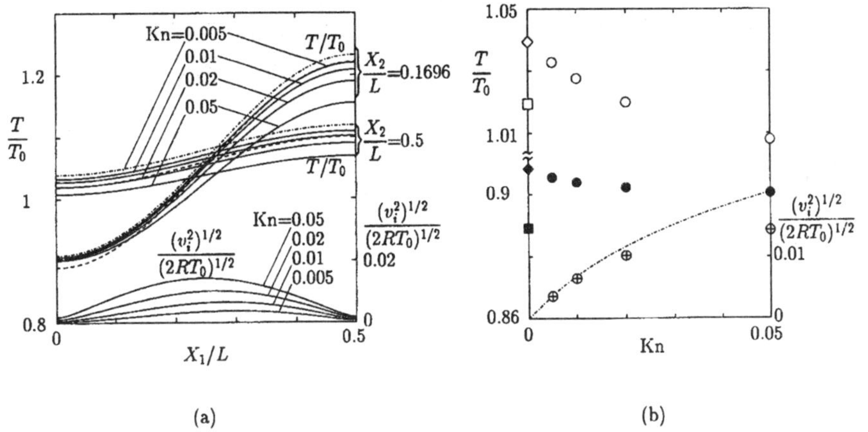

Figure 5.3. Behavior of the solution of the BKW equation as Kn$\to 0$ for the case with $\alpha = 0.5$ and $\beta = 0$. (a) Temperature distributions along $X_2/L = 0.1696$ and $X_2/L = 0.5$ and flow-speed distribution along $X_2/L = 0.0227$ for Kn$= 0.005$, 0.01, 0.02, and 0.05. (b) Temperatures at $(X_1/L, X_2/L) = (0, 0.1696)$ and $(0, 0.5)$ and flow speed at $(X_1/L, X_2/L) = (0.25, 0.0227)$ vs Kn. In panel (a), ——: BKW equation, —·—·—: asymptotic theory, and - - - -: heat-conduction equation. In panel (b), white symbols (○, ◇, □): T/T_0 at $(X_1/L, X_2/L) = (0, 0.5)$, black symbols (●, ♦, ■): T/T_0 at $(X_1/L, X_2/L) = (0, 0.1696)$, and ⊕: $(v_i^2)^{1/2}/(2RT_0)^{1/2}$ at $(X_1/L, X_2/L) = (0.25, 0.0227)$; circles (○, ●, ⊕): BKW equation, diamonds (◇, ♦): asymptotic theory, and squares (□, ■): heat-conduction equation; —·—·—: $|(\hat{v}_{iSB1} + \hat{v}_{iK1})k|$ in the asymptotic theory. Note the difference of the ordinate of T/T_0 from that of $(v_i^2)^{1/2}/(2RT_0)^{1/2}$ in both figures. The solution of the kinetic equation converges to the solution by the asymptotic theory.

The numerical computation is carried out by a finite difference method. The geometry of the problem is simple, and the discontinuity of the velocity distribution function (see Section 3.7), which is a typical feature of a rarefied gas around bodies, is absent, except on the walls. Thus, a standard nonuniform rectangular lattice system is used, and the boundary-value problem is solved by the method of iteration (e.g., Aoki, Sone & Yamada [1990], Sugimoto & Sone [1992], Takata, Sone & Aoki [1993]).

In the examples treated in this section, the average density over the whole field is taken as the reference density ρ_0, and the reference pressure p_0 is, therefore, given by $p_0 = R\rho_0 T_0$ with the average density ρ_0. The first example is the case with $\alpha = 0.5$ and $\beta = 0$ (or the walls are at rest). The isothermal lines and the flow-velocity field at Kn$= 0.005$ in the shaded region of Fig. 5.1 are shown in Figs. 5.2 (a) and 5.2 (b), where the isothermal lines obtained by the asymptotic theory and those by the heat-conduction equation are also shown for comparison. A very small flow ($|\hat{v}_i|$: 0.0041 at most), which is of the same order $[|\hat{v}_i| = O(\text{Kn})]$ as the Knudsen number (Kn$= 0.005$), is induced by the temperature gradient of the wall; the temperature field of the BKW equation deviates from that of the heat-conduction equation by a few percent, which is

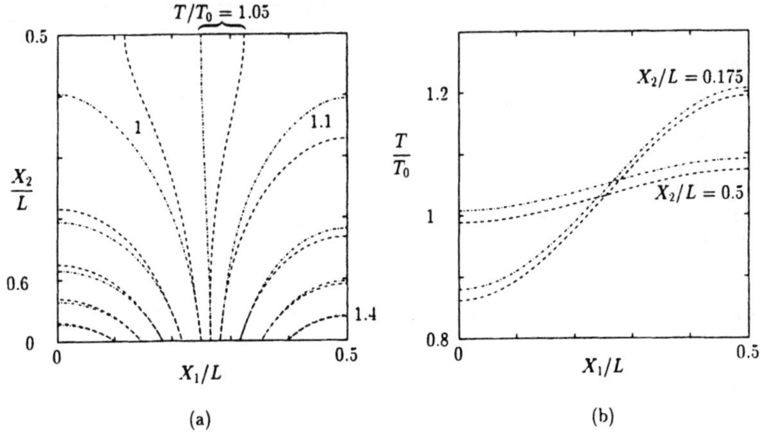

Figure 5.4. The solutions of the asymptotic theory and the heat-conduction equation for a hard-sphere gas for the case with $\alpha = 0.5$ and $\beta = 0$. (a) The isothermal lines $T/T_0 = 0.6 + 0.1m$ ($m = 0, 1, ..., 8$) and 1.05. (b) The temperature distributions along $X_2/L = 0.175$ and $X_2/L = 0.5$. Here, —·—·—: asymptotic theory and - - - -: heat-conduction equation.

much larger than the Knudsen number. From Fig. 5.2 (a) for a special value of the Knudsen number, however, it is not clear where the solution of the BKW equation converges. To make this clear, the temperature distributions along the two lines $X_2/L = 0.1696$ and $X_2/L = 0.5$ and the flow-speed distribution along $X_2/L = 0.0227$ are shown for Kn $= 0.005$, 0.01, 0.02, and 0.05 in Fig. 5.3 (a); the temperatures at $(X_1/L, X_2/L) = (0, 0.1696)$ and $(0, 0.5)$ and the flow speed at $(X_1/L, X_2/L) = (0.25, 0.0227)$ versus Kn are shown in Fig. 5.3 (b). From Figs. 5.3 (a) and 5.3 (b), it is seen that the temperature field of the BKW equation approaches that of the asymptotic theory but not that of the heat-conduction equation as Kn $\to 0$, and that the flow vanishes. On the colder side, near $X_1/L = 0$, where the effective Knudsen number is smaller than the reference Knudsen number Kn because of larger density, the state is closer to the limiting state given by the asymptotic theory. Comparison of the solutions of the asymptotic theory and the heat-conduction equation for a hard-sphere gas is made for $\alpha = 0.5$ and $\beta = 0$ (the same values of the parameters as those in Fig. 5.2 for the BKW equation) in Fig. 5.4 and for $\alpha = 0.8$ and $\beta = 0$ (a larger temperature variation) in Fig. 5.5.

The second example is the case with $\alpha = 0.5$ and $\beta = 1$ (the walls are moving with a speed of the order of Kn). The isothermal lines and the velocity field at Kn $= 0.01$ in a half of the channel (Fig. 5.1) are shown in Figs. 5.6 (a) and 5.6 (b). To see the behavior of the solution of the BKW equation as Kn $\to 0$, the temperature distributions along two lines $X_2/L = 0.2002$ and $X_2/L = 0.5$ and the flow-speed distribution along $X_2/L = 0.0227$ are shown for Kn $= 0.01$, 0.02, 0.05 in Fig. 5.7 (a); the temperatures at $(X_1/L, X_2/L) = (0, 0.2002)$ and $(0, 0.5)$ and the flow speed at $(X_1/L, X_2/L) = (0.3, 0.0227)$ versus Kn are shown

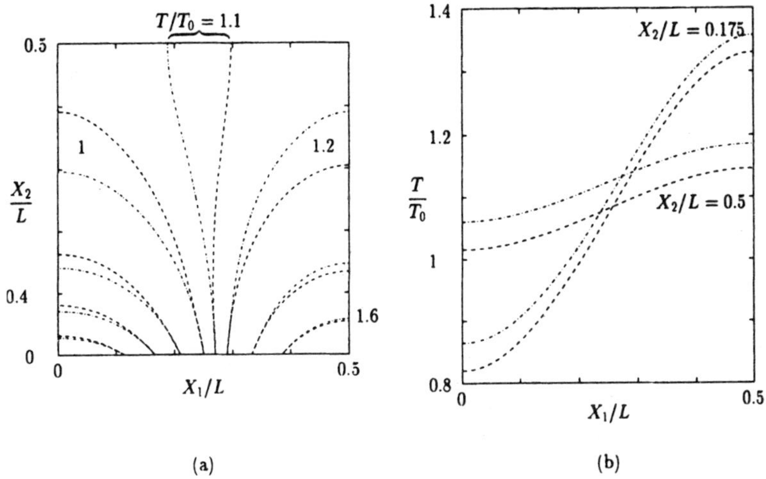

Figure 5.5. The solutions of the asymptotic theory and the heat-conduction equation for a hard-sphere gas in the problem with a larger temperature variation ($\alpha = 0.8$ and $\beta = 0$). (a) The isothermal lines $T/T_0 = 0.4 + 0.2m$ ($m = 0, 1, ..., 6$) and 1.1. (b) The temperature distributions along $X_2/L = 0.175$ and $X_2/L = 0.5$. Here, —·—·—: asymptotic theory and - - - -: heat-conduction equation.

in Fig. 5.7 (b). Also in this case, the temperature field of the BKW equation approaches that of the asymptotic theory and not that of the heat-conduction equation, and the flow vanishes. The difference of the limiting solution from that of the heat-conduction equation is more pronounced; the limiting solution no longer has symmetry with respect to $X_1/L = n/2$ ($n = 0, \pm 1, \ldots$). Comparison of the solutions of the asymptotic theory and the heat-conduction equation for a hard-sphere gas is made in Fig. 5.8.

5.6.5 Supplementary discussion

We have discussed the ghost effect on a temperature field in a gas at rest. The effect is not limited to a temperature field, but also seen in a velocity field. Sone, Takata & Sugimoto [1996] (see also Sone [1997]) considered a gas between two coaxial cylinders made of the condensed phase of the gas and investigated the behavior of the gas in the continuum limit. Let the temperature of the outer cylinder be higher than that of the inner. When both cylinders are at rest, evaporation takes place on the outer cylinder and condensation on the inner, and the flow from the outer cylinder to the inner is a radial isentropic flow. If the outer cylinder is rotating, the evaporating flow has a tangential component, and the pressure of the gas near the outer cylinder increases owing to the centrifugal force by the circumferential gas motion. This pressure rise suppresses the rate of evaporation on the outer cylinder. As the speed of rotation of the outer cylinder increases, the rate of evaporation decreases and finally vanishes at some speed. If the speed of the cylinder is further increased, evaporation or condensation

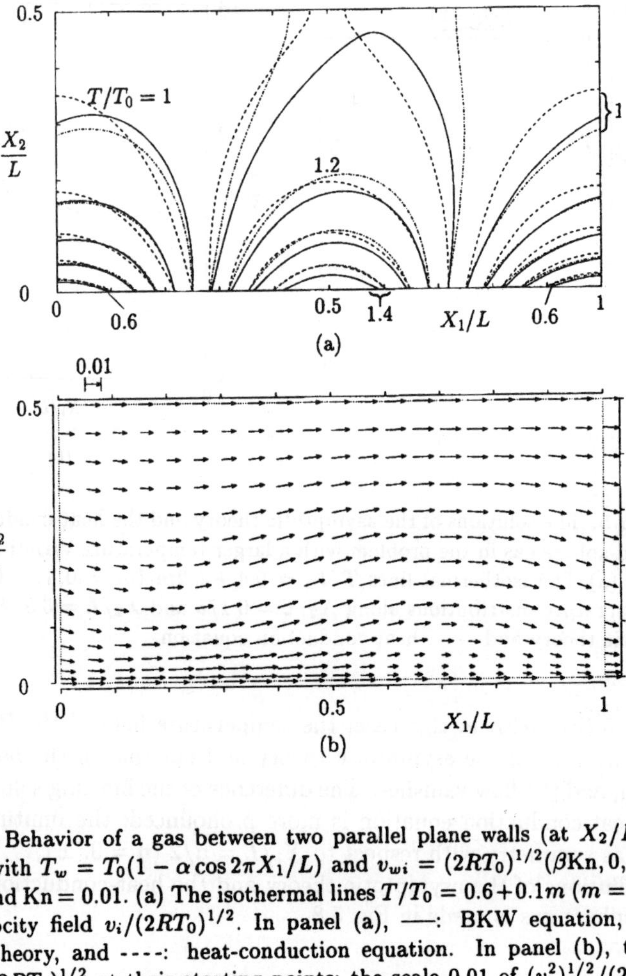

Figure 5.6. Behavior of a gas between two parallel plane walls (at $X_2/L = 0$ and $X_2/L = 1$) with $T_w = T_0(1 - \alpha \cos 2\pi X_1/L)$ and $v_{wi} = (2RT_0)^{1/2}(\beta\text{Kn}, 0, 0)$ II: $\alpha = 0.5$, $\beta = 1$, and Kn $= 0.01$. (a) The isothermal lines $T/T_0 = 0.6 + 0.1m$ $(m = 0, 1, ..., 8)$. (b) The velocity field $v_i/(2RT_0)^{1/2}$. In panel (a), ——: BKW equation, — - — - —: asymptotic theory, and - - - -: heat-conduction equation. In panel (b), the arrows indicate $v_i/(2RT_0)^{1/2}$ at their starting points; the scale 0.01 of $(v_i^2)^{1/2}/(2RT_0)^{1/2}$ is shown by the arrow on the left shoulder of the figure.

on the cylinders remains vanished. The behavior of the gas at this state is not given by the solution without radial velocity of the Navier–Stokes set of equations under the nonslip condition. The circumferential velocity and temperature distributions are affected by the first-order component function of the expansion of the radial velocity in the Knudsen number. The situation where the Mach number of the flow of evaporation or condensation is of the order of unity is the case that will be discussed in Chapter 7. However, at the situation where the radial velocity vanishes in the continuum limit, several fluid-dynamic-type equations derived from the solvability condition degenerate, as we have already seen in other places, and higher-order analysis than that given in Chapter 7 is required to obtain the equations that determine the behavior of the macroscopic variables in the continuum limit. The equations contain the component function

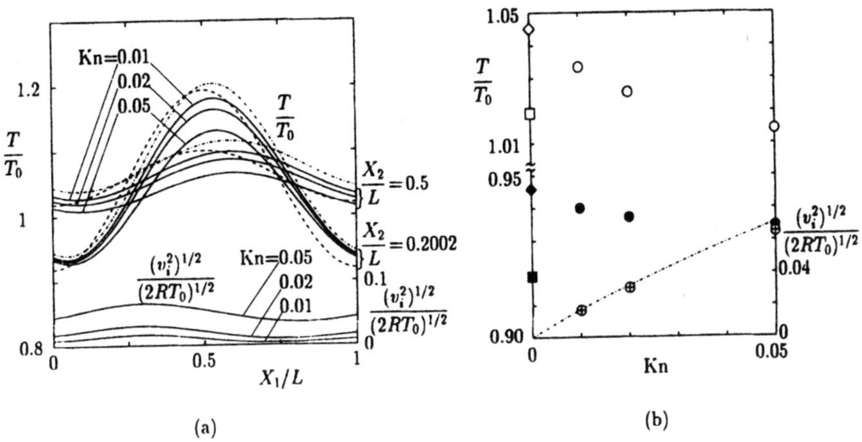

Figure 5.7. Behavior of the solution of the BKW equation as Kn$\to 0$ for the case with $\alpha = 0.5$ and $\beta = 1$. (a) Temperature distributions along $X_2/L = 0.2002$ and $X_2/L = 0.5$ and flow-speed distribution along $X_2/L = 0.0227$ for Kn$= 0.01$, 0.02, and 0.05. (b) Temperatures at $(X_1/L, X_2/L) = (0, 0.2002)$ and $(0, 0.5)$ and flow speed at $(X_1/L, X_2/L) = (0.3, 0.0227)$ vs Kn. In panel (a), ——: BKW equation, —·—·—: asymptotic theory, and ----: heat-conduction equation. In panel (b), white symbols (○, ◇, □): T/T_0 at $(X_1/L, X_2/L) = (0, 0.5)$, black symbols (●, ♦, ■): T/T_0 at $(X_1/L, X_2/L) = (0, 0.2002)$, and ⊕: $(v_i^2)^{1/2}/(2RT_0)^{1/2}$ at $(X_1/L, X_2/L) = (0.3, 0.0227)$; circles (○, ●, ⊕): BKW equation, diamonds (◇, ♦): asymptotic theory, and squares (□, ■): heat-conduction equation; —·—·—: $|(\hat{v}_{iSB1} + \hat{v}_{iK1})k|$ in the asymptotic theory. Note the difference of the ordinate of T/T_0 from that of $(v_i^2)^{1/2}/(2RT_0)^{1/2}$ in both figures. The solution of the kinetic equation converges to the solution of the asymptotic theory.

of the radial velocity at the first order of the Knudsen number as well as the other variables at the leading (or zero-th) order.[16]

In the famous Bénard problem, where a gas under uniform gravity between two parallel plane walls uniformly heated below is considered, the ghost effect and bifurcation appear in the temperature field in a combined way when the temperature difference of the two plates is finite. According to Sone & Doi [unpublished], where the asymptotic theory in this chapter is extended to the system under uniform gravity and the Bénard problem is studied, the field of the gas in the continuum limit is determined in the combination of the temperature and infinitesimal velocity, and the bifurcation of the field is found to exist. In the bifurcated field, the isothermal surfaces are curved in the absence of gas motion. The bifurcated temperature field and bifurcation curve (the relation

[16]This special example will be discussed in Chapter 8 for the case where the speeds of rotation of the cylinders (or their Mach numbers) and the Knudsen number are both small. Its main purpose is to demonstrate the analysis of a marginal case of the general theories in Chapters 3–7 and to show an interesting example of bifurcation of a flow. In this example, degeneracy of a quantity is seen to complicate analysis and to introduce a quite different type of flow field. The ghost effect on the flow is also briefly discussed.

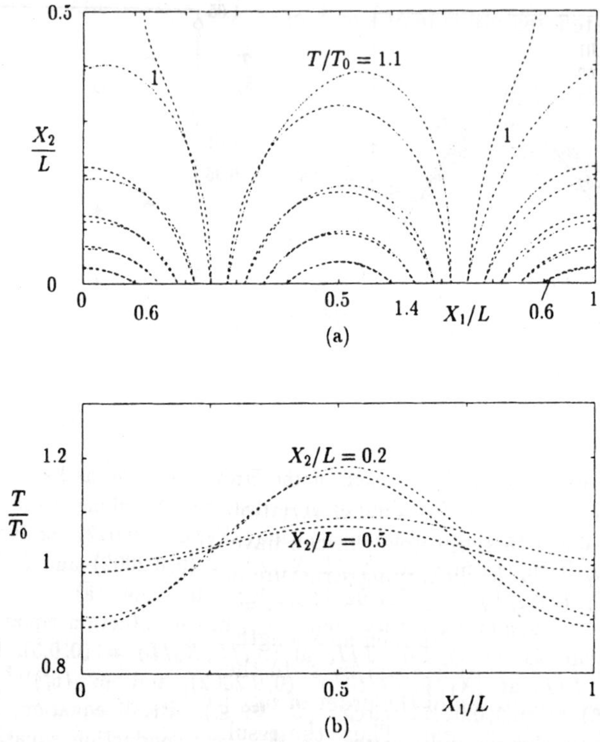

Figure 5.8. The solutions of the asymptotic theory and the heat-conduction equation for a hard-sphere gas for the case with $\alpha = 0.5$ and $\beta = 1$. (a) The isothermal lines $T/T_0 = 0.6 + 0.1m$ ($m = 0, 1, ..., 8$). (b) The temperature distributions along $X_2/L = 0.2$ and $X_2/L = 0.5$. Here, —·—·—: asymptotic theory and - - - -: heat-conduction equation.

between the parameters, the gravity and the ratio of the temperatures of the two walls, where the bifurcation of solution occurs) depend on the nonlinear thermal stress.

The ghost effect appears in a multi-component gas more widely (Takata & Aoki [1999], Takata, Aoki & Muraki [1999], Takata & Aoki [2001]).

5.7 Half-space problem of evaporation and condensation

We will discuss the half-space problem of evaporation and condensation considered in Section 4.10.1 by eliminating one of the assumptions, that is, a small temperature variation. We again describe the problem. Consider a semi-infinite expanse ($X_1 > 0$) of a gas bounded by its plane condensed phase at $X_1 = 0$. The plane condensed phase is at rest and is kept at a uniform temperature T_w.

The saturated gas pressure at this temperature (or T_w) is p_w. The state of the gas far away from the condensed phase is a uniform state with pressure p_∞, temperature T_∞, and velocity $(v_{1\infty}, v_{2\infty}, v_{3\infty})$. We discuss the steady behavior of the gas in the case where $(p_\infty - p_w)/p_w$ and $v_{i\infty}/\sqrt{2RT_w}$ are small quantities of the same order and denoted by $\varepsilon P_{\infty 1}$ and $\varepsilon \hat{v}_{i\infty 1}$ respectively with a small parameter ε. The notation defined in Section 2.9 is used with p_w and T_w as the reference quantities p_0 and T_0. The state at infinity is summarized in this notation as follows:

$$\left.\begin{aligned} \hat{p}_\infty &= p_\infty/p_w = 1 + \varepsilon P_{\infty 1}, \\ \hat{T}_\infty &= T_\infty/T_w, \\ \hat{v}_{i\infty} &= v_{i\infty}/\sqrt{2RT_w} = \varepsilon \hat{v}_{i\infty 1}. \end{aligned}\right\} \tag{5.83}$$

The problem does not contain a geometric characteristic length, and the mean free path is the only quantity with dimension of length. Thus, the solution, apparently, has the length scale of variation of the order of the mean free path. As in Section 4.10.1, the solution can have the different length scale $\sqrt{\pi}\ell_w/2\varepsilon$ of variation (or a slowly varying structure in the far field), where ℓ_w is the mean free path in the equilibrium state at rest with pressure p_w and temperature T_w. We take $\sqrt{\pi}\ell_w/2\varepsilon$ as the reference length L, i.e., $x_1 = 2\varepsilon X_1/\sqrt{\pi}\ell_w$. Then k defined by Eq. (2.41b) is given by ε. In the present situation the flow velocity may be assumed to be of the order of $v_{i\infty}$ [or $\hat{v}_i = O(\varepsilon)$]. This is the case that is studied in this chapter. Thus, the results of the asymptotic theory can be applied.

From the conditions (5.37) and (5.57b) on $\hat{p}_{SB0}$ and $\hat{p}_{SB1}$ and the condition (5.83) at infinity,

$$\hat{p}_{SB0} = 1, \tag{5.84a}$$

$$\hat{p}_{SB1} = P_{\infty 1}, \tag{5.84b}$$

where the condition (5.57b) on $\hat{p}_{SB0}$ at $x_1 = 0$ and the condition (5.83) are compatible.

In the present one-dimensional problem, the governing equations (5.42)–(5.44) are reduced to

$$\frac{\mathrm{d}\hat{p}_{SB0}\hat{v}_{1SB1}}{\mathrm{d}x_1} = 0, \tag{5.85a}$$

$$\begin{aligned} \hat{p}_{SB0}\hat{v}_{1SB1}\frac{\mathrm{d}\hat{v}_{1SB1}}{\mathrm{d}x_1} &= -\frac{1}{2}\frac{\mathrm{d}\hat{p}_{SB2}}{\mathrm{d}x_1} + \frac{2}{3}\frac{\mathrm{d}}{\mathrm{d}x_1}\left(\hat{\gamma}_1(\hat{T}_{SB0})\hat{T}_{SB0}^{1/2}\frac{\mathrm{d}\hat{v}_{1SB1}}{\mathrm{d}x_1}\right) \\ &\quad - \frac{1}{3\hat{p}_{SB0}}\frac{\mathrm{d}}{\mathrm{d}x_1}\left[\hat{\gamma}_7(\hat{T}_{SB0})\left(\frac{\mathrm{d}\hat{T}_{SB0}}{\mathrm{d}x_1}\right)^2 + \hat{\gamma}_3(\hat{T}_{SB0})\hat{T}_{SB0}\frac{\mathrm{d}^2\hat{T}_{SB0}}{\mathrm{d}x_1^2}\right], \end{aligned} \tag{5.85b}$$

$$\hat{p}_{SB0}\hat{v}_{1SB1}\frac{\mathrm{d}\hat{v}_{iSB1}}{\mathrm{d}x_1} = \frac{1}{2}\frac{\mathrm{d}}{\mathrm{d}x_1}\left(\hat{\gamma}_1(\hat{T}_{SB0})\hat{T}_{SB0}^{1/2}\frac{\mathrm{d}\hat{v}_{iSB1}}{\mathrm{d}x_1}\right) \quad (i = 2,\ 3), \tag{5.85c}$$

$$\hat{\rho}_{SB0}\hat{v}_{1SB1}\frac{\mathrm{d}\hat{T}_{SB0}}{\mathrm{d}x_1} = \frac{1}{2}\frac{\mathrm{d}}{\mathrm{d}x_1}\left(\hat{\gamma}_2(\hat{T}_{SB0})\hat{T}_{SB0}^{1/2}\frac{\mathrm{d}\hat{T}_{SB0}}{\mathrm{d}x_1}\right), \tag{5.85d}$$

where $\hat{\rho}_{SB0}$ is, from Eq. (5.38), given by

$$\hat{\rho}_{SB0} = 1/\hat{T}_{SB0}. \tag{5.86}$$

The boundary condition on the condensed phase ($x_1 = 0$) is, from Eqs. (5.57a), (5.75a), and (5.75b),

$$\hat{T}_{SB0} = 1, \tag{5.87a}$$

$$\hat{v}_{2SB1} = \hat{v}_{3SB1} = 0, \tag{5.87b}$$

$$\hat{p}_{SB1} = \hat{C}_4^*(1)\hat{v}_{1SB1} + \hat{C}_1(1)\frac{\mathrm{d}\hat{T}_{SB0}}{\mathrm{d}x_1}, \tag{5.87c}$$

where the argument $\hat{T}_{w0}\,(= 1)$ of $\hat{C}_4^*$ and $\hat{C}_1$ is shown explicitly. The $\hat{C}_4^*(1)$ and $\hat{C}_1(1)$ generally depend on the reference temperature T_0 $(= T_w)$; thus, they are functions of T_w. For a hard-sphere gas and the BKW model, they are exceptionally independent of T_w. The condition at infinity ($x_1 \to \infty$) is, from Eq. (5.83),

$$\hat{T}_{SB0} \to \hat{T}_\infty, \quad \hat{v}_{iSB1} \to \hat{v}_{i\infty1}. \tag{5.88}$$

From Eqs. (5.85a), (5.86), and (5.88), the velocity $\hat{v}_{1SB1}$ is expressed by $\hat{T}_{SB0}$ as

$$\hat{\rho}_{SB0}\hat{v}_{1SB1} = \hat{v}_{1\infty1}/\hat{T}_\infty, \quad \text{or} \quad \hat{v}_{1SB1} = \hat{v}_{1\infty1}\hat{T}_{SB0}/\hat{T}_\infty. \tag{5.89}$$

With Eqs. (5.84b), (5.89), and (5.87a) into the boundary condition (5.87c), the parameters $\hat{T}_\infty$, $P_{\infty1}$, and $\hat{v}_{1\infty1}$ at infinity are related as

$$P_{\infty1} = \hat{C}_4^*(1)\frac{\hat{v}_{1\infty1}}{\hat{T}_\infty} + \hat{C}_1(1)\left(\frac{\mathrm{d}\hat{T}_{SB0}}{\mathrm{d}x_1}\right)_{x_1=0} \tag{5.90}$$

From Eq. (5.85d), (5.87a), and (5.89),

$$x_1 = \frac{\hat{T}_\infty}{2\hat{v}_{1\infty1}}\int_1^{\hat{T}_{SB0}}\frac{\hat{\gamma}_2(t)t^{1/2}}{t - c_0}\mathrm{d}t, \tag{5.91}$$

where c_0 is an integration constant. When $\hat{T}_{SB0}$ approaches c_0 from either side of it, the integral diverges to $+\infty$ for negative $\hat{v}_{1\infty1}$ and to $-\infty$ for positive $\hat{v}_{1\infty1}$. In the latter case, x_1 cannot approach positive infinity for a finite positive value of $\hat{T}_{SB0}$. Thus, it is found that $c_0 = \hat{T}_\infty$ and that when $\hat{v}_{1\infty1} \geq 0$,[17] there is no solution that takes a finite value at infinity, except the obvious solution $\hat{T}_{SB0} = 1$ for the special condition $\hat{T}_\infty = 1$. The only solution of Eqs. (5.85c) and (5.87b) for $\hat{v}_{1\infty1} \geq 0$ and $\hat{T}_\infty = 1$ that is finite at infinity is

$$\hat{v}_{iSB1} = 0 \quad (i = 2,\ 3),$$

[17]The same conclusion as for $\hat{v}_{1\infty1} > 0$ can be easily obtained for $\hat{v}_{1\infty1} = 0$ by direct analysis.

for the special condition

$$\hat{v}_{i\infty 1} = 0 \quad (i = 2,\ 3).$$

For $\hat{v}_{1\infty 1} < 0$, the nonuniform solution can be constructed from Eq. (5.91). Thus, the solution of the problem is given in two different forms depending on the sign of $\hat{v}_{1\infty 1}$ as follows:

When $\hat{v}_{1\infty 1} \geq 0$, the problem has the following uniform solution only when $T_\infty = T_w$ and $\hat{v}_{i\infty 1} = 0$ $(i = 2,3)$:

$$\hat{T}_{SB0} = 1, \quad \hat{v}_{1SB1} = \hat{v}_{1\infty 1}, \quad \hat{v}_{iSB1} = 0 \quad (i = 2,3).$$

Then, the relation (5.90) is reduced to

$$P_{\infty 1} = \hat{C}_4^*(1)\hat{v}_{1\infty 1},$$

or

$$\frac{p_\infty - p_w}{p_w} = \sqrt{\frac{5}{6}}\hat{C}_4^*(1)M_{n\infty}, \tag{5.92}$$

where

$$M_{n\infty} = \frac{v_{1\infty}}{\sqrt{5RT_\infty/3}}.$$

Thus, the parameters $\hat{v}_{i\infty 1}$ $(i = 1,2,3)$, $P_{\infty 1}$, and $\hat{T}_\infty$ in the problem cannot be given independently, and only $\hat{v}_{1\infty 1}$ or $P_{\infty 1}$ can be specified.

When $\hat{v}_{1\infty 1} < 0$, the temperature field $\hat{T}_{SB0}(x_1)$ is obtained from Eq. (5.91) with $c_0 = \hat{T}_\infty$ as the implicit form

$$x_1 = \frac{\hat{T}_\infty}{2\hat{v}_{1\infty 1}} \int_1^{\hat{T}_{SB0}} \frac{\hat{\gamma}_2(t)t^{1/2}}{t - \hat{T}_\infty} dt \qquad \text{when } \hat{T}_\infty \neq 1, \tag{5.93a}$$

and directly from Eq. (5.85d) as the uniform function

$$\hat{T}_{SB0} = 1 \qquad \text{when } \hat{T}_\infty = 1. \tag{5.93b}$$

The solution (5.93a) is reduced for a hard-sphere gas to

$$x_1 = \frac{\hat{\gamma}_2\hat{T}_\infty}{\hat{v}_{1\infty 1}}\left(\hat{T}_{SB0}^{1/2} - 1 + \frac{\hat{T}_\infty^{1/2}}{2}\ln\left|\frac{(\hat{T}_\infty^{1/2} - \hat{T}_{SB0}^{1/2})(\hat{T}_\infty^{1/2} + 1)}{(\hat{T}_\infty^{1/2} - 1)(\hat{T}_\infty^{1/2} + \hat{T}_{SB0}^{1/2})}\right|\right)$$
$$\text{when } \hat{T}_\infty \neq 1,$$

and for the BKW model to

$$x_1 = \frac{\hat{T}_\infty}{2\hat{v}_{1\infty 1}}\left(\hat{T}_{SB0} - 1 + \hat{T}_\infty \ln\left|\frac{\hat{T}_\infty - \hat{T}_{SB0}}{\hat{T}_\infty - 1}\right|\right) \qquad \text{when } \hat{T}_\infty \neq 1.$$

The temperature field $\hat{T}_{SB0}$ is independent of the tangential components $\hat{v}_{2\infty 1}$ and $\hat{v}_{3\infty 1}$.

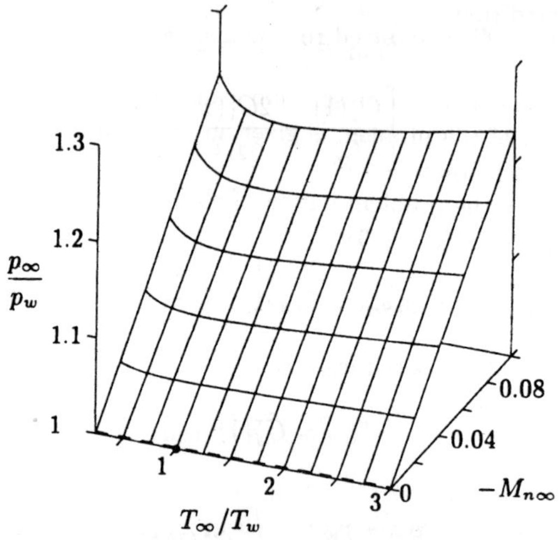

Figure 5.9. The surface (5.95) for a hard-sphere gas in the parameter space $(-M_{n\infty}, p_\infty/p_w, T_\infty/T_w)$. The surface approaches $p_\infty/p_w = 1$ as $M_{n\infty} \to 0_-$. On the line $M_{n\infty} = 0$, $p_\infty/p_w = 1$, the solution exists only at $T_\infty/T_w = 1$. The pressure ratio p_∞/p_w diverges to infinity as T_∞/T_w tends to zero or infinity. In the intermediate range of T_∞/T_w, the variation of p_∞/p_w with respect to T_∞/T_w is mild.

Then, the normal velocity $\hat{v}_{1SB1}$ is determined from $\hat{T}_{SB0}$ by Eq. (5.89). The tangential velocity $\hat{v}_{iSB1}$ $(i = 2, 3)$ is obtained from Eqs. (5.85c), (5.93a), (5.93b), (5.87b), and (5.88) as

$$\hat{v}_{iSB1} = \hat{v}_{i\infty 1}\left[1 - \exp\left(\int_1^{\hat{T}_{SB0}} \frac{\hat{\gamma}_2(t)}{\hat{\gamma}_1(t)(t - \hat{T}_\infty)}\mathrm{d}t\right)\right] \quad \text{when } \hat{T}_\infty \neq 1, \tag{5.94a}$$

$$\hat{v}_{iSB1} = \hat{v}_{i\infty 1}\left[1 - \exp\left(\frac{2\hat{v}_{1\infty 1}}{\hat{\gamma}_1(1)}x_1\right)\right] \quad \text{when } \hat{T}_\infty = 1. \tag{5.94b}$$

For a hard-sphere gas and the BKW model, the tangential velocity $\hat{v}_{iSB1}$ $(i = 2, 3)$ for $\hat{T}_\infty \neq 1$ is reduced to

$$\hat{v}_{iSB1} = \hat{v}_{i\infty 1}\left(1 - \left|\frac{\hat{T}_{SB0} - \hat{T}_\infty}{1 - \hat{T}_\infty}\right|^{\hat{\gamma}_2(1)/\hat{\gamma}_1(1)}\right) \quad \text{when } \hat{T}_\infty \neq 1,$$

since $\hat{\gamma}_2(t)/\hat{\gamma}_1(t)$ is independent of t.

With the aid of the solution (5.93a) or (5.93b), the derivative $\left(\mathrm{d}\hat{T}_{SB0}/\mathrm{d}x_1\right)_{x_1=0}$ in Eq. (5.90) is given as

$$\left(\frac{\mathrm{d}\hat{T}_{SB0}}{\mathrm{d}x_1}\right)_{x_1=0} = -\frac{2\hat{v}_{1\infty 1}}{\hat{\gamma}_2(1)}\left(1 - \frac{1}{\hat{T}_\infty}\right).$$

Then, the relation (5.90) is reduced to

$$P_{\infty 1} = \hat{v}_{1\infty 1}\left[\frac{\hat{C}_4^*(1)}{\hat{T}_\infty} - \frac{2\hat{C}_1(1)}{\hat{\gamma}_2(1)}\left(1 - \frac{1}{\hat{T}_\infty}\right)\right],$$

or

$$\frac{p_\infty - p_w}{p_w} = \sqrt{\frac{5}{6}}M_{n\infty}\left[\left(\hat{C}_4^*(1) + \frac{2\hat{C}_1(1)}{\hat{\gamma}_2(1)}\right)\sqrt{\frac{T_w}{T_\infty}} - \frac{2\hat{C}_1(1)}{\hat{\gamma}_2(1)}\sqrt{\frac{T_\infty}{T_w}}\right], \tag{5.95}$$

where

$$M_{n\infty} = \frac{v_{1\infty}}{\sqrt{5RT_\infty/3}},$$

and a similar note to that on $\hat{C}_4^*(1)$ and $\hat{C}_1(1)$ just after Eq. (5.87c) applies to $\hat{\gamma}_2(1)$. The relation (5.95) required among the parameters in the problem is independent of the tangential components $v_{2\infty}$ and $v_{3\infty}$. It is drawn for a hard-sphere gas in Fig. 5.9.

The solution of the half-space problem of evaporation and condensation is determined together with the parameters p_∞/p_w, $v_{i\infty}/\sqrt{2RT_w}$ $(i = 2, 3)$, and T_∞/T_w by specifying only $v_{1\infty}/\sqrt{2RT_w}$ when $v_{1\infty} \geq 0$, but it is determined together with the parameter p_∞/p_w by specifying $v_{i\infty}/\sqrt{2RT_w}$ $(i = 1, 2, 3)$ and T_∞/T_w when $v_{1\infty} < 0$.

In the present problem, Eq. (5.85b) does not participate in determining $\hat{v}_{1SB1}$ and works only to determine $\hat{p}_{SB2}$. We encountered a similar situation in Sections 4.10.1–4.10.3.

Chapter 6

Nonlinear Theory II – Flow with a Finite Mach Number around a Simple Boundary

Up to now, we have considered the cases where the Mach number of a gas flow is a small quantity much smaller than or of the same order as the Knudsen number. In this chapter a flow with a finite Mach number around a simple boundary is considered. In this case, from the von Karman relation (3.95), the Reynolds number of the flow is very large, and a viscous boundary layer (or Prandtl boundary layer) is expected outside the Knudsen layer. Thus, the asymptotic analysis is developed by introducing an intermediate anisotropic region with thickness of the order of the square root of the Knudsen number. Owing to the presence of this layer, the velocity distribution function and macroscopic variables are developed in a power series of the square root of the Knudsen number, in contrast to the cases discussed in the foregoing chapters, where they are developed in a power series of the Knudsen number. The analysis for a finite Mach number, with this anisotropy, does not cover the analyses in the foregoing chapters, but this is a special case with a very large Reynolds number to be treated separately.

6.1 Problem

Consider a rarefied gas in a general domain except that the shape of the domain is smooth. No external force is acting on the gas [or F_i in Eq. (2.5) is zero]. The boundary of the bodies in the gas is a solid boundary through which there is no mass flux, and the characteristic speed of gas flow compared with the sonic speed is of the order of unity (or the Mach number is of the order of unity). We will investigate the asymptotic behavior of the gas when the mean free path of the gas molecules is small compared with the geometrical characteristic length of the system (or the Knudsen number of the system is small), and try to find the fluid-dynamic-type equations and their associated boundary conditions. We use the notation introduced in Section 2.9.

The basic equation is Eq. (2.40a) with $\partial/\partial\hat{t} = 0$ and $\hat{F}_i = 0$, i.e.,

$$\zeta_i \frac{\partial \hat{f}}{\partial x_i} = \frac{1}{k}\hat{J}(\hat{f}, \hat{f}). \tag{6.1}$$

The boundary condition is, symbolically,

$$\hat{f} = \hat{f}_w \quad (\zeta_i n_i > 0), \tag{6.2}$$

where n_i is the unit normal vector to the boundary, pointed to the gas region, and the condition $\hat{v}_{wi}n_i = 0$, required in a steady system, is used. The condition of a simple boundary is

$$\int \zeta_i n_i \hat{f} \mathbf{d}\boldsymbol{\zeta} = 0,$$

on the boundary. For the diffuse reflection [Eq. (2.55a) with (2.55b) and $\alpha = 1$], for example, the boundary condition (6.2) is expressed by

$$\hat{f}(x_i, \zeta_i) = \frac{\hat{\sigma}_w}{(\pi \hat{T}_w)^{3/2}} \exp\left(-\frac{(\zeta_i - \hat{v}_{wi})^2}{\hat{T}_w}\right) \quad (\zeta_j n_j > 0), \tag{6.3a}$$

$$\hat{\sigma}_w = -2\left(\frac{\pi}{\hat{T}_w}\right)^{1/2} \int_{\zeta_j n_j < 0} \zeta_j n_j \hat{f}(x_i, \zeta_i) \mathbf{d}\boldsymbol{\zeta}, \tag{6.3b}$$

where the parameters $\hat{v}_{wi}$ and $\hat{T}_w$ are, for simplicity, taken to be independent of k in this chapter.

The asymptotic behavior of the solution $\hat{f}(x_i, \zeta_i)$ of the boundary-value problem (6.1) with (6.2) for small k (or $k \ll 1$) is studied without restriction on the size of the variables on the basis of Sone, Bardos, Golse & Sugimoto [2000].[1]

6.2 Hilbert solution

In this section, we consider the case where appreciable variation of the state of gas occurs in the distance of the characteristic length L of the system (or the length scale of variation of the variables is of the order of L), i.e., $\partial\hat{f}/\partial x_i = O(\hat{f})$. The asymptotic solution for small k describing this situation is the well-known Hilbert solution or expansion [Hilbert (1912)], which is obtained by a power series expansion in k under the assumption $\partial\hat{f}/\partial x_i = O(\hat{f})$. In the analysis of the next section, where the adjustment to the boundary condition is considered, we encounter a series expansion in the square root of k, and therefore we will

[1]The problem was first discussed by Darrozes [1969]. However, Darrozes's work, though pioneering, is simple, and thus, the systematic system of formulas, as well as a course of analysis, that is required to understand the theory and to apply it to concrete problems is not given. As mentioned briefly in Chapter 1 and to be explained in this chapter, his main statement disagrees with our result.

rewrite the Hilbert solution in a power series of $\sqrt{k}$. Here we introduce the notation ε:

$$\varepsilon = \sqrt{k}. \tag{6.4}$$

The Hilbert solution (or expansion) $\hat{f}_h$ is now rewritten as

$$\hat{f}_h = \hat{f}_{h0} + \hat{f}_{h1}\varepsilon + \hat{f}_{h2}\varepsilon^2 + \cdots, \tag{6.5}$$

where the subscript h is chosen to discriminate this type of solution.

The relation between the macroscopic variables and the distribution function is the same as Eqs. (2.48a)–(2.48f) except for the subscript h, that is,

$$\hat{\rho}_h = \int \hat{f}_h \mathbf{d}\boldsymbol{\zeta}, \tag{6.6a}$$

$$\hat{\rho}_h \hat{v}_{ih} = \int \zeta_i \hat{f}_h \mathbf{d}\boldsymbol{\zeta}, \tag{6.6b}$$

$$\frac{3}{2}\hat{\rho}_h \hat{T}_h = \int (\zeta_i - \hat{v}_{ih})^2 \hat{f}_h \mathbf{d}\boldsymbol{\zeta}, \tag{6.6c}$$

$$\hat{p}_h = \hat{\rho}_h \hat{T}_h, \tag{6.6d}$$

$$\hat{p}_{ijh} = 2\int (\zeta_i - \hat{v}_{ih})(\zeta_j - \hat{v}_{jh}) \hat{f}_h \mathbf{d}\boldsymbol{\zeta}, \tag{6.6e}$$

$$\hat{q}_{ih} = \int (\zeta_i - \hat{v}_{ih})(\zeta_j - \hat{v}_{jh})^2 \hat{f}_h \mathbf{d}\boldsymbol{\zeta}. \tag{6.6f}$$

Corresponding to the expansion (6.5), the macroscopic variable $\hat{h}_h$, where $\hat{h}$ represents $\hat{\rho}$, $\hat{v}_i$, $\hat{T}$, etc., is also expanded in ε:

$$\hat{h}_h = \hat{h}_{h0} + \hat{h}_{h1}\varepsilon + \hat{h}_{h2}\varepsilon^2 + \cdots. \tag{6.7}$$

The component function $\hat{h}_{hm}$ is related to the component function of the velocity distribution function as follows:

$$\hat{\rho}_{h0} = \int \hat{f}_{h0} \mathbf{d}\boldsymbol{\zeta}, \tag{6.8a}$$

$$\hat{\rho}_{h0} \hat{v}_{ih0} = \int \zeta_i \hat{f}_{h0} \mathbf{d}\boldsymbol{\zeta}, \tag{6.8b}$$

$$\frac{3}{2}\hat{\rho}_{h0} \hat{T}_{h0} = \int (\zeta_i - \hat{v}_{ih0})^2 \hat{f}_{h0} \mathbf{d}\boldsymbol{\zeta}, \tag{6.8c}$$

$$\hat{p}_{h0} = \hat{\rho}_{h0} \hat{T}_{h0}, \tag{6.8d}$$

$$\hat{p}_{ijh0} = 2\int (\zeta_i - \hat{v}_{ih0})(\zeta_j - \hat{v}_{jh0}) \hat{f}_{h0} \mathbf{d}\boldsymbol{\zeta}, \tag{6.8e}$$

$$\hat{q}_{ih0} = \int (\zeta_i - \hat{v}_{ih0})(\zeta_j - \hat{v}_{jh0})^2 \hat{f}_{h0} \mathbf{d}\boldsymbol{\zeta}, \tag{6.8f}$$

$$\hat{\rho}_{h1} = \int \hat{f}_{h1} \mathbf{d}\boldsymbol{\zeta}, \tag{6.9a}$$

$$\hat{\rho}_{h0}\hat{v}_{ih1} = \int (\zeta_i - \hat{v}_{ih0})\hat{f}_{h1} \mathbf{d}\boldsymbol{\zeta}, \tag{6.9b}$$

$$\frac{3}{2}\hat{\rho}_{h0}\hat{T}_{h1} = \int (\zeta_i - \hat{v}_{ih0})^2 \hat{f}_{h1} \mathbf{d}\boldsymbol{\zeta} - \frac{3}{2}\hat{\rho}_{h1}\hat{T}_{h0}, \tag{6.9c}$$

$$\hat{p}_{h1} = \hat{\rho}_{h0}\hat{T}_{h1} + \hat{\rho}_{h1}\hat{T}_{h0}, \tag{6.9d}$$

$$\hat{p}_{ijh1} = 2\int (\zeta_i - \hat{v}_{ih0})(\zeta_j - \hat{v}_{jh0})\hat{f}_{h1} \mathbf{d}\boldsymbol{\zeta}, \tag{6.9e}$$

$$\hat{q}_{ih1} = \int (\zeta_i - \hat{v}_{ih0})(\zeta_j - \hat{v}_{jh0})^2 \hat{f}_{h1} \mathbf{d}\boldsymbol{\zeta} - \frac{3}{2}\hat{p}_{h0}\hat{v}_{ih1} - \hat{p}_{ijh0}\hat{v}_{jh1}, \tag{6.9f}$$

$$\hat{\rho}_{h2} = \int \hat{f}_{h2} \mathbf{d}\boldsymbol{\zeta}, \tag{6.10a}$$

$$\hat{\rho}_{h0}\hat{v}_{ih2} = \int (\zeta_i - \hat{v}_{ih0})\hat{f}_{h2} \mathbf{d}\boldsymbol{\zeta} - \hat{\rho}_{h1}\hat{v}_{ih1}, \tag{6.10b}$$

$$\frac{3}{2}\hat{\rho}_{h0}\hat{T}_{h2} = \int (\zeta_i - \hat{v}_{ih0})^2 \hat{f}_{h2} \mathbf{d}\boldsymbol{\zeta} - \frac{3}{2}(\hat{\rho}_{h1}\hat{T}_{h1} + \hat{\rho}_{h2}\hat{T}_{h0}) - \hat{\rho}_{h0}\hat{v}_{ih1}^2, \tag{6.10c}$$

$$\hat{p}_{h2} = \hat{\rho}_{h2}\hat{T}_{h0} + \hat{\rho}_{h1}\hat{T}_{h1} + \hat{\rho}_{h0}\hat{T}_{h2}, \tag{6.10d}$$

$$\hat{p}_{ijh2} = 2\int (\zeta_i - \hat{v}_{ih0})(\zeta_j - \hat{v}_{jh0})\hat{f}_{h2} \mathbf{d}\boldsymbol{\zeta} - 2\hat{\rho}_{h0}\hat{v}_{ih1}\hat{v}_{jh1}, \tag{6.10e}$$

$$\begin{aligned}\hat{q}_{ih2} = &\int (\zeta_i - \hat{v}_{ih0})(\zeta_j - \hat{v}_{jh0})^2 \hat{f}_{h2} \mathbf{d}\boldsymbol{\zeta} - \frac{3}{2}(\hat{p}_{h1}\hat{v}_{ih1} + \hat{p}_{h0}\hat{v}_{ih2}) \\ &- \hat{p}_{ijh1}\hat{v}_{jh1} - \hat{p}_{ijh0}\hat{v}_{jh2},\end{aligned} \tag{6.10f}$$

. . . .

Now return to obtaining the solution $\hat{f}_h$. Substituting Eq. (6.5) into Eq. (6.1), we obtain the following integral equations for the component functions $\hat{f}_{hm}$ ($m = 0, 1, \ldots$) of the velocity distribution function $\hat{f}_h$:[2]

$$\hat{J}(\hat{f}_{h0}, \hat{f}_{h0}) = 0, \tag{6.11}$$

$$\hat{J}(\hat{f}_{h0}, \hat{f}_{h1}) = 0, \tag{6.12}$$

$$2\hat{J}(\hat{f}_{h0}, \hat{f}_{hm}) = \zeta_i \frac{\partial \hat{f}_{hm-2}}{\partial x_i} - \sum_{l=1}^{m-1} \hat{J}(\hat{f}_{hl}, \hat{f}_{hm-l}) \quad (m \geq 2). \tag{6.13}$$

The solution of Eq. (6.11) is Maxwellian, that is,

$$\hat{f}_{h0} = \frac{\hat{\rho}_{h0}}{(\pi \hat{T}_{h0})^{3/2}} \exp\left(-\frac{(\zeta_i - \hat{v}_{ih0})^2}{\hat{T}_{h0}}\right), \tag{6.14}$$

[2]For the BKW equation, the collision integrals on the left-hand sides in Eqs. (6.12) and (6.13) are taken as the collision integrals linearized around the Maxwellian $\hat{f}_{h0}$ (see Footnotes 12 in Section 2.9 and 26 in Appendix A.7). See Footnote 1 in Section 4.2 for the meaning of the terms under $\sum$ sign on the right-hand side.

where the relations (6.8a)–(6.8c) are taken into account. The homogeneous linear integral equation (6.12) has five independent solutions $\hat{f}_{h0}$, $\hat{f}_{h0}\zeta_i$, and $\hat{f}_{h0}\zeta_j^2$, which is easily seen from Eq. (A.110) and the relation $g' + g'_* = g + g_*$ ($g = 1,\ \zeta_i,\ \zeta_i^2$). Thus, in view of the relations (6.9a)–(6.9c), $\hat{f}_{h1}$ is expressed as

$$\hat{f}_{h1} = \hat{f}_{h0}\left\{\frac{\hat{\rho}_{h1}}{\hat{\rho}_{h0}} + \frac{2\hat{v}_{ih1}(\zeta_i - \hat{v}_{ih0})}{\hat{T}_{h0}} + \frac{\hat{T}_{h1}}{\hat{T}_{h0}}\left[\frac{(\zeta_j - \hat{v}_{jh0})^2}{\hat{T}_{h0}} - \frac{3}{2}\right]\right\}. \tag{6.15}$$

Then, the sum $\hat{f}_{h0} + \varepsilon\hat{f}_{h1}$ can be written in a Maxwellian form as

$$\hat{f}_{h0}+\varepsilon\hat{f}_{h1} = \frac{\hat{\rho}_{h0} + \varepsilon\hat{\rho}_{h1}}{\pi^{3/2}(\hat{T}_{h0} + \varepsilon\hat{T}_{h1})^{3/2}}\exp\left(-\frac{[\zeta_i - (\hat{v}_{ih0} + \varepsilon\hat{v}_{ih1})]^2}{\hat{T}_{h0} + \varepsilon\hat{T}_{h1}}\right) + O(\varepsilon^2). \tag{6.16}$$

The Hilbert solution of Eq. (6.1) is naturally Maxwellian up to the order of ε, since ε^2 is the natural parameter of the equation and the ε term is introduced in consideration of the boundary condition.

Equation (6.13) is an inhomogeneous linear integral equation for $\hat{f}_{hm}$ ($m \geq 2$). The homogeneous equation corresponding to Eq. (6.13) is the same form as Eq. (6.12) and has five nontrivial solutions $\hat{f}_{h0}$, $\hat{f}_{h0}\zeta_i$, and $\hat{f}_{h0}\zeta_j^2$. From the relation $\int g\hat{J}(\hat{f}_{h0}, \hat{f}_{hm})\mathbf{d}\boldsymbol{\zeta} = 0$ [Eq. (2.44)], the inhomogeneous term (say, Ih_m) of the integral equation (6.13) must satisfy the following relation (solvability condition) for Eq. (6.13) to have a solution:

$$\int(1, \zeta_i, \zeta_j^2)Ih_m\mathbf{d}\boldsymbol{\zeta} = 0 \quad (m \geq 2), \tag{6.17}$$

which is reduced to

$$\int(1, \zeta_i, \zeta_j^2)\zeta_k\frac{\partial\hat{f}_{hm-2}}{\partial x_k}\mathbf{d}\boldsymbol{\zeta} = 0 \quad (m \geq 2), \tag{6.18}$$

owing to the relation (2.44).

When the condition (6.18) is satisfied, the solution of Eq. (6.13) is expressed in the form

$$\hat{f}_{hm} = \hat{f}_{h0}(c_{m0} + c_{mi}\zeta_i + c_{m4}\zeta_j^2) + \hat{\Psi}_{hm} \qquad (m \geq 2), \tag{6.19}$$

where $\hat{\Psi}_{hm}$ is the particular solution of Eq. (6.13), and c_{m0}, c_{mi}, and c_{m4} are undetermined functions of x_i and related to the macroscopic variables $\hat{\rho}_{hm}$, $\hat{v}_{ihm}$, and $\hat{T}_{hm}$ [these variables are determined by c_{n0}, c_{ni}, and c_{n4} ($n \leq m$)]. In order to have the explicit expression of the relations without knowing $\hat{\Psi}_{hm}$, the following orthogonal relations:

$$\int(1, \zeta_i, \zeta_j^2)\hat{\Psi}_{hm}\mathbf{d}\boldsymbol{\zeta} = 0, \tag{6.20}$$

may be imposed.

From the solvability condition (6.18) with $m = 2$, we obtain the following Euler set of equations:

$$\frac{\partial \hat{\rho}_{h0}\hat{v}_{ih0}}{\partial x_i} = 0, \tag{6.21}$$

$$\hat{\rho}_{h0}\hat{v}_{jh0}\frac{\partial \hat{v}_{ih0}}{\partial x_j} + \frac{1}{2}\frac{\partial \hat{p}_{h0}}{\partial x_i} = 0, \tag{6.22}$$

$$\hat{\rho}_{h0}\hat{v}_{jh0}\frac{\partial}{\partial x_j}\left(\hat{v}_{ih0}^2 + \frac{5}{2}\hat{T}_{h0}\right) = 0. \tag{6.23}$$

From the solvability condition (6.18) with $m = 3$, we have

$$\frac{\partial(\hat{\rho}_{h0}\hat{v}_{ih1} + \hat{\rho}_{h1}\hat{v}_{ih0})}{\partial x_i} = 0, \tag{6.24}$$

$$\hat{\rho}_{h0}\hat{v}_{jh0}\frac{\partial \hat{v}_{ih1}}{\partial x_j} + (\hat{\rho}_{h0}\hat{v}_{jh1} + \hat{\rho}_{h1}\hat{v}_{jh0})\frac{\partial \hat{v}_{ih0}}{\partial x_j} + \frac{1}{2}\frac{\partial \hat{p}_{h1}}{\partial x_i} = 0, \tag{6.25}$$

$$\hat{\rho}_{h0}\hat{v}_{jh0}\frac{\partial}{\partial x_j}\left(2\hat{v}_{ih0}\hat{v}_{ih1} + \frac{5}{2}\hat{T}_{h1}\right) + (\hat{\rho}_{h0}\hat{v}_{jh1} + \hat{\rho}_{h1}\hat{v}_{jh0})\frac{\partial}{\partial x_j}\left(\hat{v}_{ih0}^2 + \frac{5}{2}\hat{T}_{h0}\right) = 0. \tag{6.26}$$

We can proceed with the analysis in a similar way with the information of $\hat{\Psi}_{hm}$. Naturally the two sets of equations, Eqs. (6.21)–(6.23) and Eqs. (6.24)–(6.26), are derived from the Euler set of equations obtained by the original Hilbert expansion by re-expanding in the variable ε. In solving Eqs. (6.21)–(6.23) or Eqs. (6.24)–(6.26), we have to use the additional condition (6.8d) or (6.9d). This note will not generally be repeated in the following discussions.

The stress tensor $\hat{p}_{ijhm}$ and the heat-flow vector $\hat{q}_{ihm}$ are expressed by the macroscopic variables $\hat{\rho}_{hn}$, $\hat{v}_{ihn}$, and $\hat{T}_{hn}$ (or $\hat{p}_{hn}$) with $n \leq m$ and their derivatives with $n \leq m - 1$, since the velocity distribution function is expressed by them. Corresponding to the fact that the velocity distribution function is Maxwellian up to the order ε, we have

$$\hat{p}_{ijh0} = \hat{p}_{h0}\delta_{ij}, \qquad \hat{p}_{ijh1} = \hat{p}_{h1}\delta_{ij}, \tag{6.27}$$

$$\hat{q}_{ih0} = 0, \qquad \hat{q}_{ih1} = 0 \tag{6.28}$$

The distribution function (up to the order ε) can be made to satisfy the diffuse reflection condition by taking

$$\hat{v}_{ih0} = \hat{v}_{wi}, \quad \hat{T}_{h0} = \hat{T}_w,$$
$$\hat{v}_{ih1} = 0, \quad \hat{T}_{h1} = 0.$$

However, these conditions, where $\hat{v}_{ih0}n_i = \hat{v}_{wi}n_i = 0$, are too strong for the two sets of equations, Eqs. (6.21)–(6.23) and Eqs. (6.24)–(6.26), to have solutions. Thus, we look for the solution of the boundary-value problem by relaxing the restriction $\partial\hat{f}/\partial x_i = O(\hat{f})$ on the Hilbert solution.

6.3 Viscous boundary-layer solution

In the system where the Knudsen number is small, a flow whose Mach number is of the order of unity is of a very high Reynolds number. In such a flow of a gas governed by the Navier–Stokes set of equations around a solid boundary, there appears a thin layer, called a viscous boundary layer, adjacent to the boundary. In the layer the variation of the state is strongly anisotropic, that is, the length scale of the variation in the direction parallel to the boundary is of the order of the characteristic length L, but in the direction normal to the boundary it shrinks by the factor of the inverse of the square root of the Reynolds number or the factor ε.

In view of this fact, assuming a similar situation, we will analyze the behavior of the gas in the neighborhood of the boundary, to obtain the correction to the Hilbert solution discussed in the preceding section. For the convenience of the analysis near the boundary, we introduce the viscous boundary-layer variables (χ_1, χ_2, y):

$$x_i = \varepsilon y n_i(\chi_1, \chi_2) + x_{wi}(\chi_1, \chi_2), \tag{6.29}$$

where $y = 0$ is the surface of the boundary, the y coordinate (normal coordinate) is stretched by $1/\varepsilon$, and χ_1 and χ_2 coordinates are taken to be orthogonal to each other on the boundary ($y = 0$). The unit vector in the direction of the χ_α coordinate on the boundary will be denoted by $t_i^{(\alpha)}$ ($\alpha = 1$ or 2), which is a function of the two variables χ_1 and χ_2. For definiteness, the set $(t_i^{(1)}, t_i^{(2)}, n_i)$ is taken to form a right-hand system. Thus,

$$t_i^{(\alpha)} t_i^{(\beta)} = \delta_{\alpha\beta}, \quad \left(\frac{\partial \chi_\alpha}{\partial x_i}\right)_0 \parallel t_i^{(\alpha)}, \quad \left(\frac{\partial \chi_\alpha}{\partial x_i}\right)_0 \left(\frac{\partial \chi_\beta}{\partial x_i}\right)_0 = 0 \quad (\alpha \neq \beta), \tag{6.30}$$

where, and in the following analysis, the quantities in the parentheses with subscript 0 are evaluated on the boundary (or at $x_i = x_{wi}$), and the summation convention is not applied to the subscript (or superscript) α or β related to the χ_α coordinate system. For simplicity, $\chi_{\alpha,\beta}$ will be used for $t_i^{(\beta)} (\partial \chi_\alpha / \partial x_i)_0$, i.e.,

$$\chi_{\alpha,\beta} = t_i^{(\beta)} \left(\frac{\partial \chi_\alpha}{\partial x_i}\right)_0, \tag{6.31}$$

where $\chi_{\alpha,\beta} = 0$ for $\alpha \neq \beta$. In this coordinate system, the Boltzmann equation (6.1) is rewritten as

$$\frac{1}{\varepsilon} \zeta_i n_i \frac{\partial \hat{f}}{\partial y} + \zeta_i \left(\frac{\partial \chi_1}{\partial x_i} \frac{\partial \hat{f}}{\partial \chi_1} + \frac{\partial \chi_2}{\partial x_i} \frac{\partial \hat{f}}{\partial \chi_2} \right) = \frac{1}{\varepsilon^2} \hat{J}(\hat{f}, \hat{f}). \tag{6.32}$$

The solution in the layer $[O(\varepsilon L)]$ whose length scale of variation is of the order of εL, i.e., $\partial \hat{f}/\partial y = O(\hat{f})$, is looked for in a power series of ε:

$$\hat{f}_V = \hat{f}_{V0} + \hat{f}_{V1}\varepsilon + \hat{f}_{V2}\varepsilon^2 + \cdots, \tag{6.33}$$

where the subscript V is attached to discriminate the type of solution (or viscous boundary-layer solution).

The relation between the macroscopic variables and the distribution function is the same as Eqs. (2.48a)–(2.48f) except for the subscript V, that is,

$$\hat{\rho}_V = \int \hat{f}_V \mathbf{d}\boldsymbol{\zeta}, \tag{6.34a}$$

$$\hat{\rho}_V \hat{v}_{iV} = \int \zeta_i \hat{f}_V \mathbf{d}\boldsymbol{\zeta}, \tag{6.34b}$$

$$\frac{3}{2}\hat{\rho}_V \hat{T}_V = \int (\zeta_i - \hat{v}_{iV})^2 \hat{f}_V \mathbf{d}\boldsymbol{\zeta}, \tag{6.34c}$$

$$\hat{p}_V = \hat{\rho}_V \hat{T}_V, \tag{6.34d}$$

$$\hat{p}_{ijV} = 2\int (\zeta_i - \hat{v}_{iV})(\zeta_j - \hat{v}_{jV}) \hat{f}_V \mathbf{d}\boldsymbol{\zeta}, \tag{6.34e}$$

$$\hat{q}_{iV} = \int (\zeta_i - \hat{v}_{iV})(\zeta_j - \hat{v}_{jV})^2 \hat{f}_V \mathbf{d}\boldsymbol{\zeta}. \tag{6.34f}$$

Corresponding to this expansion, we also expand the macroscopic variables $\hat{\rho}_V$, $\hat{v}_{iV}$, $\hat{T}_V$, $\hat{p}_V$, $\hat{p}_{ijV}$, and $\hat{q}_{iV}$ in power series of ε:

$$\hat{h}_V = \hat{h}_{V0} + \hat{h}_{V1}\varepsilon + \hat{h}_{V2}\varepsilon^2 + \cdots, \tag{6.35}$$

where $\hat{h}_V$ represents any of the macroscopic variables $\hat{\rho}_V$, $\hat{v}_{iV}$, $\hat{T}_V$, $\hat{p}_V$, $\hat{p}_{ijV}$, and $\hat{q}_{iV}$. By substitution of Eq. (6.33) into Eqs. (6.34a)–(6.34f), the component function $\hat{h}_{Vm}$ of $\hat{h}_V$ is expressed in terms of moments of $\hat{f}_{Vn}$ $(n \leq m)$ as follows [Note the nonlinearity of Eqs. (6.34b)–(6.34f)]:

$$\hat{\rho}_{V0} = \int \hat{f}_{V0} \mathbf{d}\boldsymbol{\zeta}, \tag{6.36a}$$

$$\hat{\rho}_{V0} \hat{v}_{iV0} = \int \zeta_i \hat{f}_{V0} \mathbf{d}\boldsymbol{\zeta}, \tag{6.36b}$$

$$\frac{3}{2}\hat{\rho}_{V0} \hat{T}_{V0} = \int (\zeta_i - \hat{v}_{iV0})^2 \hat{f}_{V0} \mathbf{d}\boldsymbol{\zeta}, \tag{6.36c}$$

$$\hat{p}_{V0} = \hat{\rho}_{V0} \hat{T}_{V0}, \tag{6.36d}$$

$$\hat{p}_{ijV0} = 2\int (\zeta_i - \hat{v}_{iV0})(\zeta_j - \hat{v}_{jV0}) \hat{f}_{V0} \mathbf{d}\boldsymbol{\zeta}, \tag{6.36e}$$

$$\hat{q}_{iV0} = \int (\zeta_i - \hat{v}_{iV0})(\zeta_j - \hat{v}_{jV0})^2 \hat{f}_{V0} \mathbf{d}\boldsymbol{\zeta}, \tag{6.36f}$$

$$\hat{\rho}_{V1} = \int \hat{f}_{V1} \mathbf{d}\boldsymbol{\zeta}, \tag{6.37a}$$

$$\hat{\rho}_{V0}\hat{v}_{iV1} = \int (\zeta_i - \hat{v}_{iV0}) \hat{f}_{V1} \mathbf{d}\boldsymbol{\zeta}, \tag{6.37b}$$

$$\frac{3}{2}\hat{\rho}_{V0}\hat{T}_{V1} = \int (\zeta_i - \hat{v}_{iV0})^2 \hat{f}_{V1} \mathbf{d}\boldsymbol{\zeta} - \frac{3}{2}\hat{\rho}_{V1}\hat{T}_{V0}, \tag{6.37c}$$

$$\hat{p}_{V1} = \hat{\rho}_{V0}\hat{T}_{V1} + \hat{\rho}_{V1}\hat{T}_{V0}, \tag{6.37d}$$

$$\hat{p}_{ijV1} = 2\int (\zeta_i - \hat{v}_{iV0})(\zeta_j - \hat{v}_{jV0}) \hat{f}_{V1} \mathbf{d}\boldsymbol{\zeta}, \tag{6.37e}$$

$$\hat{q}_{iV1} = \int (\zeta_i - \hat{v}_{iV0})(\zeta_j - \hat{v}_{jV0})^2 \hat{f}_{V1} \mathbf{d}\boldsymbol{\zeta} - \frac{3}{2}\hat{p}_{V0}\hat{v}_{iV1} - \hat{p}_{ijV0}\hat{v}_{jV1}, \tag{6.37f}$$

$$\hat{\rho}_{V2} = \int \hat{f}_{V2} \mathbf{d}\boldsymbol{\zeta}, \tag{6.38a}$$

$$\hat{\rho}_{V0}\hat{v}_{iV2} = \int (\zeta_i - \hat{v}_{iV0}) \hat{f}_{V2} \mathbf{d}\boldsymbol{\zeta} - \hat{\rho}_{V1}\hat{v}_{iV1}, \tag{6.38b}$$

$$\frac{3}{2}\hat{\rho}_{V0}\hat{T}_{V2} = \int (\zeta_i - \hat{v}_{iV0})^2 \hat{f}_{V2} \mathbf{d}\boldsymbol{\zeta} - \frac{3}{2}(\hat{\rho}_{V1}\hat{T}_{V1} + \hat{\rho}_{V2}\hat{T}_{V0}) - \hat{\rho}_{V0}\hat{v}_{iV1}^2, \tag{6.38c}$$

$$\hat{p}_{V2} = \hat{\rho}_{V2}\hat{T}_{V0} + \hat{\rho}_{V1}\hat{T}_{V1} + \hat{\rho}_{V0}\hat{T}_{V2}, \tag{6.38d}$$

$$\hat{p}_{ijV2} = 2\int (\zeta_i - \hat{v}_{iV0})(\zeta_j - \hat{v}_{jV0}) \hat{f}_{V2} \mathbf{d}\boldsymbol{\zeta} - 2\hat{\rho}_{V0}\hat{v}_{iV1}\hat{v}_{jV1}, \tag{6.38e}$$

$$\hat{q}_{iV2} = \int (\zeta_i - \hat{v}_{iV0})(\zeta_j - \hat{v}_{jV0})^2 \hat{f}_{V2} \mathbf{d}\boldsymbol{\zeta} - \frac{3}{2}(\hat{p}_{V1}\hat{v}_{iV1} + \hat{p}_{V0}\hat{v}_{iV2}) - \hat{p}_{ijV1}\hat{v}_{jV1} - \hat{p}_{ijV0}\hat{v}_{jV2}. \tag{6.38f}$$

We introduce some notation for economy of space in the following analysis. The Maxwellian distribution with the macroscopic variables $\hat{\rho}_V$, $\hat{v}_{iV}$, and $\hat{T}_V$ is indicated by $\hat{f}_{eV}$, that is,

$$\hat{f}_{eV} = \frac{\hat{\rho}_V}{(\pi\hat{T}_V)^{3/2}} \exp\left(-\frac{(\zeta_i - \hat{v}_{iV})^2}{\hat{T}_V}\right), \tag{6.39}$$

and the component function of its expansion in ε, obtained by the substitution of the expansion (6.35), is indicated by $\hat{f}_{eVm}$, that is,

$$\hat{f}_{eV} = \hat{f}_{eV0} + \varepsilon\hat{f}_{eV1} + \varepsilon^2\hat{f}_{eV2} + \cdots. \tag{6.40}$$

Some of the explicit forms of $\hat{f}_{eVm}$ $(m = 0, 1, \ldots)$ are

$$\hat{f}_{eV0} = \frac{\hat{\rho}_{V0}}{(\pi\hat{T}_{V0})^{3/2}} \exp\left(-\frac{(\zeta_i - \hat{v}_{iV0})^2}{\hat{T}_{V0}}\right) = \frac{\hat{\rho}_{V0}}{\hat{T}_{V0}{}^{3/2}} E(\tilde{\zeta}), \tag{6.41a}$$

$$\hat{f}_{eV1} = \hat{f}_{eV0}\left[\frac{\hat{\rho}_{V1}}{\hat{\rho}_{V0}} + \frac{2(\zeta_i - \hat{v}_{iV0})}{\hat{T}_{V0}^{1/2}}\frac{\hat{v}_{iV1}}{\hat{T}_{V0}^{1/2}} + \frac{\hat{T}_{V1}}{\hat{T}_{V0}}\left(\frac{(\zeta_i - \hat{v}_{iV0})^2}{\hat{T}_{V0}} - \frac{3}{2}\right)\right]$$
$$= \frac{\hat{\rho}_{V0}}{\hat{T}_{V0}{}^{3/2}}E(\tilde{\zeta})\left[\left(\frac{\hat{\rho}_{V1}}{\hat{\rho}_{V0}}\right) + 2\left(\frac{\hat{v}_{iV1}}{\hat{T}_{V0}^{1/2}}\right)\tilde{\zeta}_i + \left(\frac{\hat{T}_{V1}}{\hat{T}_{V0}}\right)\left(\tilde{\zeta}^2 - \frac{3}{2}\right)\right], \tag{6.41b}$$

where

$$E(\tilde{\zeta}) = \frac{1}{\pi^{3/2}}\exp\left(-\tilde{\zeta}^2\right), \quad \tilde{\zeta}_i = \frac{\zeta_i - \hat{v}_{iV0}}{\hat{T}_{V0}^{1/2}}, \quad \tilde{\zeta} = \sqrt{\tilde{\zeta}_i^2}.$$

It is noted that a moment of $\hat{f}_{eVm}$ with respect to ζ_i is faster calculated by taking the moment of $\hat{f}_{eV}$ and then expanding it in ε than by finding higher order $\hat{f}_{eVm}$ and then taking the moment. Thus a lengthy expression for the explicit form of $\hat{f}_{eV2}$ is not shown here.

Now return to obtaining the viscous boundary-layer solution $\hat{f}_V$. The analysis is similar to that in Section 6.2. The series (6.33) is substituted into Eq. (6.32), and the same order terms in ε are arranged. In this process, the expansions of $\partial\chi_1/\partial x_i$ and $\partial\chi_2/\partial x_i$ in power series of y, i.e.,

$$\frac{\partial\chi_\alpha}{\partial x_i} = \left(\frac{\partial\chi_\alpha}{\partial x_i}\right)_0 + \left(\frac{\partial^2\chi_\alpha}{\partial x_i\partial x_j}\right)_0 n_j\varepsilon y + \cdots, \tag{6.42}$$

are applied since they are slowly varying in the variable y and $y = O(1)$ in the viscous boundary layer. Then we obtain the following series of integral equations for $\hat{f}_{Vm}$:[3]

$$\hat{J}(\hat{f}_{V0}, \hat{f}_{V0}) = 0, \tag{6.43}$$

$$2\hat{J}(\hat{f}_{V0}, \hat{f}_{V1}) = \zeta_i n_i\frac{\partial\hat{f}_{V0}}{\partial y}, \tag{6.44}$$

$$2\hat{J}(\hat{f}_{V0}, \hat{f}_{V2}) = -\hat{J}(\hat{f}_{V1}, \hat{f}_{V1}) + \zeta_i n_i\frac{\partial\hat{f}_{V1}}{\partial y} + \zeta_i\left[\left(\frac{\partial\chi_1}{\partial x_i}\right)_0\frac{\partial\hat{f}_{V0}}{\partial\chi_1} + \left(\frac{\partial\chi_2}{\partial x_i}\right)_0\frac{\partial\hat{f}_{V0}}{\partial\chi_2}\right], \tag{6.45}$$

$$2\hat{J}(\hat{f}_{V0}, \hat{f}_{V3}) = -\sum_{l=1}^{2}\hat{J}(\hat{f}_{Vl}, \hat{f}_{V3-l}) + \zeta_i n_i\frac{\partial\hat{f}_{V2}}{\partial y} + \zeta_i\left[\left(\frac{\partial\chi_1}{\partial x_i}\right)_0\frac{\partial\hat{f}_{V1}}{\partial\chi_1} + \left(\frac{\partial\chi_2}{\partial x_i}\right)_0\frac{\partial\hat{f}_{V1}}{\partial\chi_2}\right] + y\zeta_i n_j\left[\left(\frac{\partial^2\chi_1}{\partial x_i\partial x_j}\right)_0\frac{\partial\hat{f}_{V0}}{\partial\chi_1} + \left(\frac{\partial^2\chi_2}{\partial x_i\partial x_j}\right)_0\frac{\partial\hat{f}_{V0}}{\partial\chi_2}\right]. \tag{6.46}$$

[3]For the BKW equation, the collision integrals on the left-hand sides in Eqs. (6.44)–(6.46) are taken as the collision integrals linearized around the Maxwellian $\hat{f}_{V0}$ (see Footnotes 12 in Section 2.9 and 26 in Appendix A.7). See Footnote 1 in Section 4.2 for the meaning of the $\hat{J}$ terms on the right-hand sides.

The solution of Eq. (6.43) is a Maxwellian, i.e.,

$$\hat{f}_{V0} = \hat{f}_{eV0}. \tag{6.47}$$

The equations for $\hat{f}_{Vm}$ $(m \geq 1)$ are inhomogeneous linear integral equations whose homogeneous parts are of a common form. Their common associated homogenous equation has five independent solutions: $\hat{f}_{V0}$, $\hat{f}_{V0}\zeta_i$, $\hat{f}_{V0}\zeta_j^2$. From the relation $\int g\hat{J}(\hat{f}_{V0}, \hat{f}_{Vm})\mathbf{d}\boldsymbol{\zeta} = 0$ [Eq. (2.44)], the equation for $\hat{f}_{Vm}$ $(m \geq 1)$ has a solution only when its inhomogeneous term, say Ih_m, satisfies the condition (solvability condition)

$$\int (1, \zeta_i, \zeta_j^2) Ih_m \mathbf{d}\boldsymbol{\zeta} = 0 \quad (m \geq 1). \tag{6.48}$$

Here, as in Section 6.2, there is no contribution from $\hat{J}$ terms in Ih_m [see Eq. (2.44)]. When the condition is satisfied, the solution $\hat{f}_{Vm}$ is expressed in the form

$$\hat{f}_{Vm} = \hat{f}_{V0}(c_{m0} + c_{mi}\zeta_i + c_{m4}\zeta_j^2) + \hat{\Psi}_{Vm} \qquad (m \geq 1), \tag{6.49}$$

where c_{m0}, c_{mi}, and c_{m4} are undetermined functions of χ_α and y, and $\hat{\Psi}_{Vm}$ is a particular solution of the equation.

Before proceeding to the analysis of $\hat{f}_{Vm}$ $(m \geq 1)$, we will make a short detour through some discussion of the boundary condition to save some manipulation. The leading term $\hat{f}_{V0}$ of the expansion (6.33) can be made to satisfy the boundary condition (6.2) by assigning the boundary values of $\hat{v}_{iV0}$ and $\hat{T}_{V0}$ as follows:

$$\hat{v}_{iV0} = \hat{v}_{wi}, \quad \hat{T}_{V0} = \hat{T}_w \quad \text{at } y = 0, \tag{6.50}$$

because the kinetic boundary condition on a simple boundary satisfies the condition (2.23c).

Now return to the discussion of $\hat{f}_{Vm}$. From the solvability condition (6.48) with $m = 1$, where

$$Ih_1 = \zeta_i n_i \frac{\partial \hat{f}_{V0}}{\partial y}, \tag{6.51}$$

we obtain

$$\frac{\partial}{\partial y}(\hat{\rho}_{V0}\hat{v}_{iV0}n_i) = 0, \tag{6.52}$$

$$\frac{\partial \hat{p}_{V0}}{\partial y} = 0, \tag{6.53}$$

and the other three relations are reduced to identities with the aid of the following Eq. (6.54). From Eqs. (6.50) and (6.52),

$$\hat{v}_{iV0}n_i = 0, \tag{6.54}$$

and from Eq. (6.53), $\hat{p}_{V0}$ is a function of χ_1 and χ_2 only, i.e.,

$$\hat{p}_{V0} = \hat{p}_{V0}(\chi_1, \chi_2). \tag{6.55}$$

When the solvability condition is satisfied, the solution $\hat{f}_{V1}$ of Eq. (6.44) is obtained in the following form (Appendixes A.2, A.5, and A.7):

$$\begin{aligned}\hat{f}_{V1} = \hat{f}_{eV1} &- \left(\frac{1}{\hat{T}_{V0}^{5/2}}\frac{\partial \hat{T}_{V0}}{\partial y}\right)\tilde{\zeta}_i n_i \mathcal{A}(\tilde{\zeta}, \hat{T}_{V0})E(\tilde{\zeta}) \\ &- \left(\frac{n_j}{\hat{T}_{V0}^2}\frac{\partial \hat{v}_{iV0}}{\partial y}\right)\left(\tilde{\zeta}_i\tilde{\zeta}_j - \frac{\tilde{\zeta}^2}{3}\delta_{ij}\right)\mathcal{B}(\tilde{\zeta}, \hat{T}_{V0})E(\tilde{\zeta}),\end{aligned} \tag{6.56}$$

where $\mathcal{A}(\tilde{\zeta}, \hat{T}_{V0})$ and $\mathcal{B}(\tilde{\zeta}, \hat{T}_{V0})$ are functions of $\tilde{\zeta}$ and $\hat{T}_{V0}$, defined in Appendix A.5 as solutions of integral equations related to the collision integral. There is no contribution of $\tilde{\zeta}^2\delta_{ij}/3$ of the third term in Eq. (6.56), since $\hat{v}_{iV0}n_i = 0$. This kind of term will be eliminated without notice in the following analysis. With this form of $\hat{f}_{V1}$ in the solvability condition (6.48) with $m = 2$, we obtain the following set of equations:

$$\frac{\partial}{\partial y}(\hat{p}_{V0}\hat{v}_{iV1}n_i) = -\sum_{\alpha=1}^{2}\left(\chi_{\alpha,\alpha}\frac{\partial \hat{p}_{V0}\hat{v}_{iV0}t_i^{(\alpha)}}{\partial \chi_\alpha} - (-1)^\alpha g_{3-\alpha}\hat{p}_{V0}\hat{v}_{iV0}t_i^{(\alpha)}\right), \tag{6.57}$$

$$\begin{aligned}&\hat{p}_{V0}\left[\hat{v}_{jV1}n_j\frac{\partial \hat{v}_{iV0}t_i^{(\alpha)}}{\partial y} + \sum_{\beta=1}^{2}\hat{v}_{kV0}t_k^{(\beta)}\left(\chi_{\beta,\beta}\frac{\partial \hat{v}_{iV0}t_i^{(\alpha)}}{\partial \chi_\beta} + (-1)^\alpha g_\beta \hat{v}_{iV0}t_i^{(3-\alpha)}\right)\right] \\ &= -\frac{1}{2}\chi_{\alpha,\alpha}\frac{\partial \hat{p}_{V0}}{\partial \chi_\alpha} + \frac{1}{2}\frac{\partial}{\partial y}\left(\hat{\gamma}_1(\hat{T}_{V0})\hat{T}_{V0}^{1/2}\frac{\partial \hat{v}_{iV0}t_i^{(\alpha)}}{\partial y}\right),\end{aligned} \tag{6.58}$$

$$\frac{1}{2}\frac{\partial \hat{p}_{V1}}{\partial y} = -\hat{p}_{V0}\sum_{\alpha=1}^{2}\left(\kappa_\alpha(\hat{v}_{iV0}t_i^{(\alpha)})^2 + \vartheta\hat{v}_{iV0}t_i^{(\alpha)}\hat{v}_{jV0}t_j^{(3-\alpha)}\right), \tag{6.59}$$

$$\begin{aligned}&\frac{3}{2}\hat{p}_{V0}\left(\hat{v}_{jV1}n_j\frac{\partial \hat{T}_{V0}}{\partial y} + \sum_{\alpha=1}^{2}\hat{v}_{kV0}t_k^{(\alpha)}\chi_{\alpha,\alpha}\frac{\partial \hat{T}_{V0}}{\partial \chi_\alpha}\right) \\ &= -\hat{p}_{V0}\left[\sum_{\alpha=1}^{2}\left(\chi_{\alpha,\alpha}\frac{\partial \hat{v}_{kV0}t_k^{(\alpha)}}{\partial \chi_\alpha} + (-1)^\alpha g_\alpha \hat{v}_{kV0}t_k^{(3-\alpha)}\right) + \frac{\partial \hat{v}_{kV1}n_k}{\partial y}\right] \\ &\quad + \hat{\gamma}_1(\hat{T}_{V0})\hat{T}_{V0}^{1/2}\sum_{\alpha=1}^{2}\left(\frac{\partial \hat{v}_{kV0}t_k^{(\alpha)}}{\partial y}\right)^2 + \frac{5}{4}\frac{\partial}{\partial y}\left(\hat{\gamma}_2(\hat{T}_{V0})\hat{T}_{V0}^{1/2}\frac{\partial \hat{T}_{V0}}{\partial y}\right).\end{aligned} \tag{6.60}$$

For these equations the following notes are required: First: Eqs. (6.58) and (6.60) are not the forms derived directly from the solvability condition; the former is deformed with the aid of Eq. (6.57), and the latter with the aid of Eqs. (6.57)–(6.59), to make them in forms more familiar in the classical fluid dynamics. The second: κ_α, g_α, and ϑ_α are determined by the geometric properties of the boundary surface and the χ_α-coordinate curve on it:

$$\chi_{\alpha,\alpha}\frac{\partial t_i^{(\alpha)}}{\partial \chi_\alpha} = \kappa_\alpha n_i - (-1)^\alpha g_\alpha t_i^{(3-\alpha)}, \quad \chi_{\alpha,\alpha}\frac{\partial t_i^{(3-\alpha)}}{\partial \chi_\alpha} = (-1)^\alpha(-\vartheta_\alpha n_i + g_\alpha t_i^{(\alpha)}),$$
$$\chi_{\alpha,\alpha}\frac{\partial n_i}{\partial \chi_\alpha} = -\kappa_\alpha t_i^{(\alpha)} + (-1)^\alpha \vartheta_\alpha t_i^{(3-\alpha)}, \tag{6.61}$$

where κ_α, g_α, and ϑ_α are, respectively, the normal curvature, geodesic curvature, and geodesic torsion, multiplied by L, on the χ_α-coordinate curve [Cartan (1977), Reinhardt & Soeder (1997)]. In the present case, where χ_1 and χ_2 curves (on the boundary) are taken orthogonal,

$$\bar{\kappa} = \frac{1}{2}(\kappa_1 + \kappa_2), \tag{6.62}$$

where $\bar{\kappa}$ is the mean curvature, multiplied by L, of the boundary, and

$$\vartheta_1 = -\vartheta_2 = \vartheta. \tag{6.63}$$

The last: $\hat{\gamma}_1(\hat{T}_{V0})$ and $\hat{\gamma}_2(\hat{T}_{V0})$, related to transport coefficients, are nondimensional functions of $\hat{T}_{V0}$ defined in Appendix A.5 as integrals of functions $\mathcal{A}(\tilde{\zeta}, \hat{T}_{V0})$ and $\mathcal{B}(\tilde{\zeta}, \hat{T}_{V0})$. They, multiplied by $\hat{T}_{V0}^{1/2}$, are related to viscosity and thermal conductivity of the gas at temperature $\hat{T}_{V0}$. For a hard-sphere gas,

$$\hat{\gamma}_1 = 1.270042427, \qquad \hat{\gamma}_2 = 1.922284066, \tag{6.64}$$

which are independent of $\hat{T}_{V0}$, and for the BKW model,

$$\hat{\gamma}_1/\hat{T}_{V0}^{1/2} = 1, \qquad \hat{\gamma}_2/\hat{T}_{V0}^{1/2} = 1. \tag{6.65}$$

Equations (6.57)–(6.60) being satisfied, then the solution $\hat{f}_{V2}$ of Eq. (6.45) is given in the following form (Appendixes A.2, A.5, and A.7):

$$\hat{f}_{V2} = \hat{f}_{eV2} + \hat{\psi}_{V2}E(\tilde{\zeta}), \tag{6.66}$$

$$\hat{\psi}_{V2} = -\left(\frac{n_i}{\hat{T}_{V0}^{5/2}}\frac{\partial \hat{T}_{V1}}{\partial y}\right)\tilde{\zeta}_i\mathcal{A} - \left(\frac{n_j}{\hat{T}_{V0}^2}\frac{\partial \hat{v}_{iV1}}{\partial y}\right)\left(\tilde{\zeta}_i\tilde{\zeta}_j - \frac{\tilde{\zeta}^2}{3}\delta_{ij}\right)\mathcal{B}$$
$$+\left(\frac{n_j\hat{v}_{iV1}}{\hat{T}_{V0}^3}\frac{\partial \hat{T}_{V0}}{\partial y}\right)\left(\frac{\tilde{\zeta}_i\tilde{\zeta}_j}{\tilde{\zeta}}\frac{\partial \mathcal{A}}{\partial \tilde{\zeta}} - (2\tilde{\zeta}_i\tilde{\zeta}_j - \delta_{ij})\mathcal{A}\right)$$
$$+\left(\frac{n_i\hat{T}_{V1}}{\hat{T}_{V0}^{7/2}}\frac{\partial \hat{T}_{V0}}{\partial y}\right)\tilde{\zeta}_i\left(\frac{\tilde{\zeta}}{2}\frac{\partial \mathcal{A}}{\partial \tilde{\zeta}} - (\tilde{\zeta}^2 - 3)\mathcal{A} - \hat{T}_{V0}\frac{\partial \mathcal{A}}{\partial \hat{T}_{V0}}\right) +$$

$$
\begin{aligned}
&+\left(\frac{n_j\hat{v}_{kV1}}{\hat{T}_{V0}^{5/2}}\frac{\partial\hat{v}_{iV0}}{\partial y}\right)\left(\frac{\tilde{\zeta}_i\tilde{\zeta}_j\tilde{\zeta}_k}{\tilde{\zeta}}\frac{\partial\mathcal{B}}{\partial\tilde{\zeta}}-(2\tilde{\zeta}_i\tilde{\zeta}_j\tilde{\zeta}_k-\tilde{\zeta}_i\delta_{jk}-\tilde{\zeta}_j\delta_{ki})\mathcal{B}\right)\\
&+\left(\frac{n_i\hat{T}_{V1}}{\hat{T}_{V0}^{3}}\frac{\partial\hat{v}_{jV0}}{\partial y}\right)\tilde{\zeta}_i\tilde{\zeta}_j\left(\frac{\tilde{\zeta}}{2}\frac{\partial\mathcal{B}}{\partial\tilde{\zeta}}-(\tilde{\zeta}^2-3)\mathcal{B}-\hat{T}_{V0}\frac{\partial\mathcal{B}}{\partial\hat{T}_{V0}}\right)\\
&-\frac{\mathcal{D}_j\hat{T}_{V0}}{\hat{T}_{V0}^{5/2}}\tilde{\zeta}_j\mathcal{A}-\frac{\mathcal{D}_j\hat{v}_{iV0}}{\hat{T}_{V0}^{2}}\left(\tilde{\zeta}_i\tilde{\zeta}_j-\frac{\tilde{\zeta}^2}{3}\delta_{ij}\right)\mathcal{B}\\
&+\frac{1}{\hat{p}_{V0}\hat{T}_{V0}^{3/2}}\frac{\partial^2\hat{T}_{V0}}{\partial y^2}\left[n_in_j\left(\tilde{\zeta}_i\tilde{\zeta}_j-\frac{\tilde{\zeta}^2}{3}\delta_{ij}\right)\mathcal{B}_1+\mathcal{N}^B\right]\\
&+\frac{1}{2\hat{p}_{V0}\hat{T}_{V0}^{5/2}}\left(\frac{\partial\hat{T}_{V0}}{\partial y}\right)^2\left[n_in_j\left(\tilde{\zeta}_i\tilde{\zeta}_j-\frac{\tilde{\zeta}^2}{3}\delta_{ij}\right)\mathcal{B}_2+\mathcal{N}^A\right]\\
&+\frac{n_in_j}{\hat{p}_{V0}\hat{T}_{V0}}\frac{\partial^2\hat{v}_{kV0}}{\partial y^2}\left(\tilde{\zeta}_i\tilde{\zeta}_j\tilde{\zeta}_k\mathcal{T}_2^A+\delta_{ij}\tilde{\zeta}_k\mathcal{T}_1^A\right)\\
&+\frac{n_in_j}{\hat{p}_{V0}\hat{T}_{V0}^2}\frac{\partial\hat{T}_{V0}}{\partial y}\frac{\partial\hat{v}_{kV0}}{\partial y}\left(\tilde{\zeta}_i\tilde{\zeta}_j\tilde{\zeta}_k\mathcal{T}_2^B+\delta_{ij}\tilde{\zeta}_k\mathcal{T}_1^B\right)\\
&+\frac{n_in_j}{\hat{p}_{V0}\hat{T}_{V0}^{3/2}}\frac{\partial\hat{v}_{kV0}}{\partial y}\frac{\partial\hat{v}_{lV0}}{\partial y}\left(\tilde{\zeta}_i\tilde{\zeta}_j\tilde{\zeta}_k\tilde{\zeta}_l\mathcal{Q}_4+\tilde{\zeta}_i\tilde{\zeta}_j\delta_{kl}\mathcal{Q}_3+\delta_{ij}\tilde{\zeta}_k\tilde{\zeta}_l\mathcal{Q}_2+\delta_{ij}\delta_{kl}\mathcal{Q}_1\right),
\end{aligned}
\tag{6.67}
$$

where

$$
\mathcal{D}_i=\left(\frac{\partial\chi_1}{\partial x_i}\right)_0\frac{\partial}{\partial\chi_1}+\left(\frac{\partial\chi_2}{\partial x_i}\right)_0\frac{\partial}{\partial\chi_2},
$$

and $\mathcal{A}$, $\mathcal{B}$, $\mathcal{B}_1$, $\mathcal{B}_2$, $\mathcal{N}^A$, $\mathcal{N}^B$, $\mathcal{T}_1^A$, $\mathcal{T}_2^A$, $\mathcal{T}_1^B$, $\mathcal{T}_2^B$, $\mathcal{Q}_1$, $\mathcal{Q}_2$, $\mathcal{Q}_3$, and $\mathcal{Q}_4$, the abbreviations of $\mathcal{A}(\tilde{\zeta},\hat{T}_{V0})$, $\mathcal{B}(\tilde{\zeta},\hat{T}_{V0})$, etc., are functions of $\tilde{\zeta}$ and $\hat{T}_{V0}$ defined by solutions of linear integral equations in Appendix A.5.

With this form of $\hat{f}_{V2}$ in the solvability condition (6.48) with $m=3$, we obtain the following set of equations:

$$
\begin{aligned}
&\frac{\partial}{\partial y}(\hat{p}_{V0}\hat{v}_{iV2}n_i+\hat{p}_{V1}\hat{v}_{iV1}n_i)+\sum_{\alpha=1}^{2}\chi_{\alpha,\alpha}\frac{\partial(\hat{p}_{V0}\hat{v}_{iV1}t_i^{(\alpha)}+\hat{p}_{V1}\hat{v}_{iV0}t_i^{(\alpha)})}{\partial\chi_\alpha}\\
&-2\bar{\kappa}\hat{p}_{V0}\hat{v}_{iV1}n_i-\sum_{\alpha=1}^{2}(-1)^\alpha g_{3-\alpha}(\hat{p}_{V0}\hat{v}_{iV1}t_i^{(\alpha)}+\hat{p}_{V1}\hat{v}_{iV0}t_i^{(\alpha)})\\
&+\sum_{\alpha=1}^{2}y\left(\kappa_\alpha\chi_{\alpha,\alpha}\frac{\partial\hat{p}_{V0}\hat{v}_{iV0}t_i^{(\alpha)}}{\partial\chi_\alpha}+\vartheta\chi_{\alpha,\alpha}\frac{\partial\hat{p}_{V0}\hat{v}_{iV0}t_i^{(3-\alpha)}}{\partial\chi_\alpha}\right)\\
&-\sum_{\alpha=1}^{2}(-1)^\alpha y(\kappa_{3-\alpha}g_{3-\alpha}+\vartheta g_\alpha)\hat{p}_{V0}\hat{v}_{iV0}t_i^{(\alpha)}=0,
\end{aligned}
\tag{6.68}
$$

$$\hat{\rho}_{V0}\hat{v}_{jV1}n_j\frac{\partial \hat{v}_{iV1}t_i^{(\alpha)}}{\partial y} + (\hat{\rho}_{V0}\hat{v}_{jV2}n_j + \hat{\rho}_{V1}\hat{v}_{jV1}n_j)\frac{\partial \hat{v}_{iV0}t_i^{(\alpha)}}{\partial y}$$
$$+\sum_{\beta=1}^{2}\hat{\rho}_{V0}\hat{v}_{jV0}t_j^{(\beta)}\chi_{\beta,\beta}\frac{\partial \hat{v}_{iV1}t_i^{(\alpha)}}{\partial \chi_\beta} - \hat{\rho}_{V0}\hat{v}_{iV1}n_i\left(\kappa_\alpha\hat{v}_{jV0}t_j^{(\alpha)} + \vartheta\hat{v}_{jV0}t_j^{(3-\alpha)}\right)$$
$$+(-1)^\alpha\hat{\rho}_{V0}\hat{v}_{jV1}t_j^{(3-\alpha)}\left(\sum_{\beta=1}^{2}g_\beta\hat{v}_{iV0}t_i^{(\beta)}\right)$$
$$+\sum_{\beta=1}^{2}(\hat{\rho}_{V0}\hat{v}_{jV1}t_j^{(\beta)} + \hat{\rho}_{V1}\hat{v}_{jV0}t_j^{(\beta)})\chi_{\beta,\beta}\frac{\partial \hat{v}_{iV0}t_i^{(\alpha)}}{\partial \chi_\beta}$$
$$+(-1)^\alpha\hat{v}_{iV0}t_i^{(3-\alpha)}\left(\sum_{\beta=1}^{2}g_\beta(\hat{\rho}_{V0}\hat{v}_{jV1}t_j^{(\beta)} + \hat{\rho}_{V1}\hat{v}_{jV0}t_j^{(\beta)})\right)$$
$$+\sum_{\beta=1}^{2}y\hat{\rho}_{V0}\left(\kappa_\beta\hat{v}_{jV0}t_j^{(\beta)} + \vartheta\hat{v}_{jV0}t_j^{(3-\beta)}\right)\chi_{\beta,\beta}\frac{\partial \hat{v}_{iV0}t_i^{(\alpha)}}{\partial \chi_\beta}$$
$$+(-1)^\alpha y\hat{\rho}_{V0}\left(\sum_{\beta=1}^{2}(\kappa_\beta g_\beta + \vartheta g_{3-\beta})\,\hat{v}_{iV0}t_i^{(\beta)}\right)\hat{v}_{jV0}t_j^{(3-\alpha)}$$
$$= -\frac{1}{2}\chi_{\alpha,\alpha}\frac{\partial \hat{p}_{V1}}{\partial \chi_\alpha} - \frac{1}{2}y\left(\kappa_\alpha\chi_{\alpha,\alpha}\frac{\partial \hat{p}_{V0}}{\partial \chi_\alpha} + \vartheta\chi_{3-\alpha,3-\alpha}\frac{\partial \hat{p}_{V0}}{\partial \chi_{3-\alpha}}\right)$$
$$+\frac{1}{2}\frac{\partial}{\partial y}\left[\hat{\gamma}_1\hat{T}_{V0}^{1/2}\left(\frac{\partial \hat{v}_{iV1}t_i^{(\alpha)}}{\partial y} + \kappa_\alpha\hat{v}_{iV0}t_i^{(\alpha)} + \vartheta\hat{v}_{iV0}t_i^{(3-\alpha)}\right)\right.$$
$$\left. +\hat{T}_{V1}\frac{\mathrm{d}\hat{\gamma}_1\hat{T}_{V0}^{1/2}}{\mathrm{d}\hat{T}_{V0}}\frac{\partial \hat{v}_{iV0}t_i^{(\alpha)}}{\partial y}\right]$$
$$-\frac{1}{2}\hat{\gamma}_1\hat{T}_{V0}^{1/2}\left[(2\bar{\kappa} + \kappa_\alpha)\frac{\partial \hat{v}_{iV0}t_i^{(\alpha)}}{\partial y} + \vartheta\frac{\partial \hat{v}_{iV0}t_i^{(3-\alpha)}}{\partial y}\right], \tag{6.69}$$

$$\sum_{\alpha=1}^{2}\hat{\rho}_{V0}\hat{v}_{jV0}t_j^{(\alpha)}\left(\chi_{\alpha,\alpha}\frac{\partial \hat{v}_{iV1}n_i}{\partial \chi_\alpha} + \kappa_\alpha\hat{v}_{iV1}t_i^{(\alpha)} + \vartheta\hat{v}_{iV1}t_i^{(3-\alpha)}\right)$$
$$+\hat{\rho}_{V0}\hat{v}_{jV1}n_j\frac{\partial \hat{v}_{iV1}n_i}{\partial y}$$
$$+\sum_{\alpha=1}^{2}(\hat{\rho}_{V0}\hat{v}_{jV1}t_j^{(\alpha)} + \hat{\rho}_{V1}\hat{v}_{jV0}t_j^{(\alpha)})\left(\kappa_\alpha\hat{v}_{iV0}t_i^{(\alpha)} + \vartheta\hat{v}_{iV0}t_i^{(3-\alpha)}\right)$$
$$+y\hat{\rho}_{V0}\sum_{\alpha=1}^{2}\left(\kappa_\alpha\hat{v}_{iV0}t_i^{(\alpha)} + \vartheta\hat{v}_{iV0}t_i^{(3-\alpha)}\right)^2 =$$

$$
\begin{aligned}
&= -\frac{1}{2}\frac{\partial \hat{p}_{V2}}{\partial y} + \frac{2}{3}\frac{\partial}{\partial y}\left(\hat{\gamma}_1 \hat{T}_{V0}^{1/2}\frac{\partial \hat{v}_{iV1}n_i}{\partial y}\right) \\
&+ \frac{1}{2}\sum_{\alpha=1}^{2}\left[\chi_{\alpha,\alpha}\frac{\partial}{\partial \chi_\alpha}\left(\hat{\gamma}_1\hat{T}_{V0}^{1/2}\frac{\partial \hat{v}_{iV0}t_i^{(\alpha)}}{\partial y}\right) + (-1)^\alpha g_\alpha \hat{\gamma}_1 \hat{T}_{V0}^{1/2}\frac{\partial \hat{v}_{iV0}t_i^{(3-\alpha)}}{\partial y}\right] \\
&- \frac{1}{3}\frac{\partial}{\partial y}\left[\hat{\gamma}_1\hat{T}_{V0}^{1/2}\sum_{\alpha=1}^{2}\left(\chi_{\alpha,\alpha}\frac{\partial \hat{v}_{iV0}t_i^{(\alpha)}}{\partial \chi_\alpha} + (-1)^\alpha g_\alpha \hat{v}_{iV0}t_i^{(3-\alpha)}\right)\right] \\
&- \frac{1}{3\hat{\rho}_{V0}}\frac{\partial}{\partial y}\left[\hat{\gamma}_3\hat{T}_{V0}\frac{\partial^2 \hat{T}_{V0}}{\partial y^2} + \hat{\gamma}_7\left(\frac{\partial \hat{T}_{V0}}{\partial y}\right)^2\right] \\
&- \frac{\partial}{\partial y}\left[\frac{\hat{\gamma}_8 - 2\hat{\gamma}_9}{3\hat{\rho}_{V0}}\sum_{\alpha=1}^{2}\left(\frac{\partial \hat{v}_{jV0}t_j^{(\alpha)}}{\partial y}\right)^2\right],
\end{aligned}
\tag{6.70}
$$

$$
\begin{aligned}
&\frac{3}{2}\hat{\rho}_{V0}\hat{v}_{jV1}n_j\frac{\partial \hat{T}_{V1}}{\partial y} + \frac{3}{2}(\hat{\rho}_{V0}\hat{v}_{jV2}n_j + \hat{\rho}_{V1}\hat{v}_{jV1}n_j)\frac{\partial \hat{T}_{V0}}{\partial y} \\
&+ \sum_{\alpha=1}^{2}\frac{3}{2}\chi_{\alpha,\alpha}\left(\hat{\rho}_{V0}\hat{v}_{jV0}t_j^{(\alpha)}\frac{\partial \hat{T}_{V1}}{\partial \chi_\alpha} + (\hat{\rho}_{V0}\hat{v}_{jV1}t_j^{(\alpha)} + \hat{\rho}_{V1}\hat{v}_{jV0}t_j^{(\alpha)})\frac{\partial \hat{T}_{V0}}{\partial \chi_\alpha}\right) \\
&+ \sum_{\alpha=1}^{2}\frac{3}{2}y\hat{\rho}_{V0}(\kappa_\alpha \hat{v}_{jV0}t_j^{(\alpha)} + \vartheta \hat{v}_{jV0}t_j^{(3-\alpha)})\chi_{\alpha,\alpha}\frac{\partial \hat{T}_{V0}}{\partial \chi_\alpha} \\
&= -\hat{p}_{V0}\left[\frac{\partial \hat{v}_{jV2}n_j}{\partial y} + \sum_{\alpha=1}^{2}\left(\chi_{\alpha,\alpha}\frac{\partial \hat{v}_{jV1}t_j^{(\alpha)}}{\partial \chi_\alpha} - (-1)^\alpha g_{3-\alpha}\hat{v}_{jV1}t_j^{(\alpha)}\right) - 2\bar{\kappa}\hat{v}_{jV1}n_j\right. \\
&\qquad + \sum_{\alpha=1}^{2}y\left(\kappa_\alpha\chi_{\alpha,\alpha}\frac{\partial \hat{v}_{jV0}t_j^{(\alpha)}}{\partial \chi_\alpha} + \vartheta\chi_{\alpha,\alpha}\frac{\partial \hat{v}_{jV0}t_j^{(3-\alpha)}}{\partial \chi_\alpha}\right. \\
&\qquad\qquad \left.\left. - (-1)^\alpha(\kappa_{3-\alpha}g_{3-\alpha} + \vartheta g_\alpha)\hat{v}_{jV0}t_j^{(\alpha)}\right)\right] \\
&- \hat{p}_{V1}\left[\frac{\partial \hat{v}_{jV1}n_j}{\partial y} + \sum_{\alpha=1}^{2}\left(\chi_{\alpha,\alpha}\frac{\partial \hat{v}_{jV0}t_j^{(\alpha)}}{\partial \chi_\alpha} - (-1)^\alpha g_{3-\alpha}\hat{v}_{jV0}t_j^{(\alpha)}\right)\right] \\
&+ \sum_{\alpha=1}^{2}\left[2\hat{\gamma}_1\hat{T}_{V0}^{1/2}\left(\frac{\partial \hat{v}_{jV1}t_j^{(\alpha)}}{\partial y} + \kappa_\alpha \hat{v}_{jV0}t_j^{(\alpha)} + \vartheta \hat{v}_{jV0}t_j^{(3-\alpha)}\right)\right. \\
&\qquad \left. + \hat{T}_{V1}\frac{\mathrm{d}\hat{\gamma}_1\hat{T}_{V0}^{1/2}}{\mathrm{d}\hat{T}_{V0}}\frac{\partial \hat{v}_{jV0}t_j^{(\alpha)}}{\partial y}\right]\frac{\partial \hat{v}_{iV0}t_i^{(\alpha)}}{\partial y} \\
&+ \frac{5}{4}\frac{\partial}{\partial y}\left(\hat{\gamma}_2\hat{T}_{V0}^{1/2}\frac{\partial \hat{T}_{V1}}{\partial y} + \hat{T}_{V1}\frac{\mathrm{d}\hat{\gamma}_2\hat{T}_{V0}^{1/2}}{\mathrm{d}\hat{T}_{V0}}\frac{\partial \hat{T}_{V0}}{\partial y}\right) - \frac{5}{2}\bar{\kappa}\hat{\gamma}_2\hat{T}_{V0}^{1/2}\frac{\partial \hat{T}_{V0}}{\partial y},
\end{aligned}
\tag{6.71}
$$

where the functions $\hat{\gamma}_1(\hat{T}_{V0})$, $\hat{\gamma}_2(\hat{T}_{V0})$, $\hat{\gamma}_3(\hat{T}_{V0})$, $\hat{\gamma}_7(\hat{T}_{V0})$, $\hat{\gamma}_8(\hat{T}_{V0})$, and $\hat{\gamma}_9(\hat{T}_{V0})$, defined in Appendix A.5, are denoted by $\hat{\gamma}_1$, $\hat{\gamma}_2$, $\hat{\gamma}_3$, $\hat{\gamma}_7$, $\hat{\gamma}_8$, and $\hat{\gamma}_9$ by omitting their argument $\hat{T}_{V0}$ for brevity.

The series of equations obtained from the solvability condition (6.48) constitutes systems of equations that determine the component functions $\hat{\rho}_{Vm}$, $\hat{v}_{iVm}$, $\hat{T}_{Vm}$, and $\hat{p}_{Vm}$ of the macroscopic variables $\hat{\rho}_V$, $\hat{v}_{iV}$, $\hat{T}_V$, and $\hat{p}_V$. That is, Eqs. (6.52), (6.53), (6.58), and (6.60) with Eq. (6.57) determine $\hat{\rho}_{V0}$, $\hat{v}_{iV0}$, $\hat{T}_{V0}$, and $\hat{p}_{V0}$; Eqs. (6.57), (6.59), (6.69), and (6.71) with Eq. (6.68) determine $\hat{\rho}_{V1}$, $\hat{v}_{iV1}$, $\hat{T}_{V1}$, and $\hat{p}_{V1}$.[4] We can obtain the higher-order equations in a similar way. Corresponding to the degeneration of three relations in the solvability condition (6.48) with $m = 1$ to identities, the arrangement of equations from the solvability conditions (6.48) for $m = 1, 2, \ldots$ is staggered, but it is consistent. In the leading system $(\hat{\rho}_{V0}, \hat{v}_{iV0}, \hat{T}_{V0}, \hat{p}_{V0})$, we already have the boundary condition (6.50), so called nonslip condition, on the boundary $y = 0$, with the aid of which Eq. (6.54) is obtained. In the next section, we will discuss the boundary condition of the next system $(\hat{\rho}_{V1}, \hat{v}_{iV1}, \hat{T}_{V1}, \hat{p}_{V1})$.

The stress tensor $\hat{p}_{ijVm}$ and the heat-flow vector $\hat{q}_{iVm}$ are expressed by the macroscopic variables $\hat{\rho}_{Vn}$, $\hat{v}_{iVn}$, and $\hat{T}_{Vn}$ (or $\hat{p}_{Vn}$) with $n \leq m$, since the velocity distribution function is expressed by them. They are

$$\hat{p}_{ijV0} = \hat{p}_{V0}\delta_{ij}, \tag{6.72}$$

$$\hat{p}_{ijV1}n_i n_j = \hat{p}_{ijV1}t_i^{(\alpha)}t_j^{(\alpha)} = \hat{p}_{V1}, \tag{6.73a}$$

$$\hat{p}_{ijV1}n_i t_j^{(\alpha)} = -\hat{\gamma}_1\hat{T}_{V0}^{1/2}\frac{\partial \hat{v}_{jV0}t_j^{(\alpha)}}{\partial y}, \tag{6.73b}$$

$$\hat{p}_{ijV1}t_i^{(\alpha)}t_j^{(3-\alpha)} = 0, \tag{6.73c}$$

$$\begin{aligned}\hat{p}_{ijV2}n_i n_j &= \hat{p}_{V2}\\ &- \frac{4}{3}\hat{\gamma}_1\hat{T}_{V0}^{1/2}\left[\frac{\partial \hat{v}_{jV1}n_j}{\partial y} - \frac{1}{2}\sum_{\alpha=1}^{2}\left(\chi_{\alpha,\alpha}\frac{\partial \hat{v}_{jV0}t_j^{(\alpha)}}{\partial \chi_\alpha} + (-1)^\alpha g_\alpha \hat{v}_{jV0}t_j^{(3-\alpha)}\right)\right]\\ &+ \frac{2}{3\hat{p}_{V0}}\left[\hat{\gamma}_3\hat{T}_{V0}\frac{\partial^2 \hat{T}_{V0}}{\partial y^2} + \hat{\gamma}_7\left(\frac{\partial \hat{T}_{V0}}{\partial y}\right)^2\right] + \frac{2(\hat{\gamma}_8 - 2\hat{\gamma}_9)}{3\hat{p}_{V0}}\sum_{\alpha=1}^{2}\left(\frac{\partial \hat{v}_{jV0}t_j^{(\alpha)}}{\partial y}\right)^2,\end{aligned} \tag{6.74a}$$

[4] In the discussion, Eq. (6.57) appears twice. Its role may be better explained briefly here, though the detailed discussion taking the boundary condition into account will be given in Sections 6.6 and 6.7. With the aid of Eq. (6.57), the first-order $\hat{v}_{iV1}n_i$ is expressed in terms of the leading-order quantities. With this expression of $\hat{v}_{iV1}n_i$, Eqs. (6.58) and (6.60) are expressed with the leading-order quantities. Equation (6.68) plays the same role in the determination of the first-order quantities.

$$\begin{aligned}\hat{p}_{ijV2}t_i^{(\alpha)}t_j^{(\alpha)} &= \hat{p}_{V2} + \frac{2}{3}\hat{\gamma}_1\hat{T}_{V0}^{1/2}\frac{\partial \hat{v}_{jV1}n_j}{\partial y} \\ &- 2\hat{\gamma}_1\hat{T}_{V0}^{1/2}\left(\chi_{\alpha,\alpha}\frac{\partial \hat{v}_{jV0}t_j^{(\alpha)}}{\partial \chi_\alpha} + (-1)^\alpha g_\alpha \hat{v}_{jV0}t_j^{(3-\alpha)}\right) \\ &+ \frac{2}{3}\hat{\gamma}_1\hat{T}_{V0}^{1/2}\sum_{\beta=1}^{2}\left(\chi_{\beta,\beta}\frac{\partial \hat{v}_{jV0}t_j^{(\beta)}}{\partial \chi_\beta} + (-1)^\beta g_\beta \hat{v}_{jV0}t_j^{(3-\beta)}\right) \\ &- \frac{1}{3\hat{p}_{V0}}\left[\hat{\gamma}_3\hat{T}_{V0}\frac{\partial^2 \hat{T}_{V0}}{\partial y^2} + \hat{\gamma}_7\left(\frac{\partial \hat{T}_{V0}}{\partial y}\right)^2\right] + \frac{2\hat{\gamma}_8}{\hat{p}_{V0}}\left(\frac{\partial \hat{v}_{jV0}t_j^{(\alpha)}}{\partial y}\right)^2 \\ &+ \frac{2(\hat{\gamma}_9 - 2\hat{\gamma}_8)}{3\hat{p}_{V0}}\sum_{\beta=1}^{2}\left(\frac{\partial \hat{v}_{jV0}t_j^{(\beta)}}{\partial y}\right)^2, \end{aligned} \tag{6.74b}$$

$$\begin{aligned}\hat{p}_{ijV2}n_i t_j^{(\alpha)} &= -\hat{\gamma}_1\hat{T}_{V0}^{1/2}\left(\frac{\partial \hat{v}_{jV1}t_j^{(\alpha)}}{\partial y} + \kappa_\alpha \hat{v}_{jV0}t_j^{(\alpha)} + \vartheta \hat{v}_{jV0}t_j^{(3-\alpha)}\right) \\ &- \hat{T}_{V1}\frac{\mathrm{d}\hat{\gamma}_1\hat{T}_{V0}^{1/2}}{\mathrm{d}\hat{T}_{V0}}\frac{\partial \hat{v}_{jV0}t_j^{(\alpha)}}{\partial y}, \end{aligned} \tag{6.74c}$$

$$\begin{aligned}\hat{p}_{ijV2}t_i^{(\alpha)}t_j^{(3-\alpha)} &= -\hat{\gamma}_1\hat{T}_{V0}^{1/2}\sum_{\beta=1}^{2}\left(\chi_{\beta,\beta}\frac{\partial \hat{v}_{jV0}t_j^{(3-\beta)}}{\partial \chi_\beta} - (-1)^\beta g_\beta \hat{v}_{jV0}t_j^{(\beta)}\right) \\ &+ \frac{2\hat{\gamma}_8}{\hat{p}_{V0}}\frac{\partial \hat{v}_{iV0}t_i^{(\alpha)}}{\partial y}\frac{\partial \hat{v}_{jV0}t_j^{(3-\alpha)}}{\partial y}, \end{aligned} \tag{6.74d}$$

$$\hat{q}_{iV0} = 0, \tag{6.75}$$

$$\hat{q}_{iV1}n_i = -\frac{5}{4}\hat{\gamma}_2\hat{T}_{V0}^{1/2}\frac{\partial \hat{T}_{V0}}{\partial y}, \tag{6.76a}$$

$$\hat{q}_{iV1}t_i^{(\alpha)} = 0, \tag{6.76b}$$

$$\hat{q}_{iV2}n_i = -\frac{5}{4}\left(\hat{\gamma}_2\hat{T}_{V0}^{1/2}\frac{\partial \hat{T}_{V1}}{\partial y} + \hat{T}_{V1}\frac{\mathrm{d}\hat{\gamma}_2\hat{T}_{V0}^{1/2}}{\mathrm{d}\hat{T}_{V0}}\frac{\partial \hat{T}_{V0}}{\partial y}\right), \tag{6.77a}$$

$$\begin{aligned}\hat{q}_{iV2}t_i^{(\alpha)} &= -\frac{5}{4}\hat{\gamma}_2\hat{T}_{V0}^{1/2}\chi_{\alpha,\alpha}\frac{\partial \hat{T}_{V0}}{\partial \chi_\alpha} + \frac{\hat{\gamma}_3\hat{T}_{V0}}{2\hat{p}_{V0}}\frac{\partial^2 \hat{v}_{iV0}t_i^{(\alpha)}}{\partial y^2} \\ &+ \frac{4\hat{\gamma}_{10}}{\hat{p}_{V0}}\frac{\partial \hat{T}_{V0}}{\partial y}\frac{\partial \hat{v}_{iV0}t_i^{(\alpha)}}{\partial y}, \end{aligned} \tag{6.77b}$$

where the function $\hat{\gamma}_{10}(\hat{T}_{V0})$ defined in Appendix A.5, as well as $\hat{\gamma}_1(\hat{T}_{V0})$, $\hat{\gamma}_2(\hat{T}_{V0})$, etc., is denoted by $\hat{\gamma}_{10}$ by omitting its argument $\hat{T}_{V0}$ for brevity.

The terms with the factor $\hat{\gamma}_1$ in $\hat{p}_{ijVm}$ are the viscous stress given by Newton's law, and the terms with the factor $\hat{\gamma}_2$ in $\hat{q}_{iVm}$ are the heat flow by Fourier's law. The $\hat{\gamma}_1\hat{T}_{V0}^{1/2}$ and $\hat{\gamma}_2\hat{T}_{V0}^{1/2}$ are, respectively, the (nondimensional) viscosity and thermal conductivity of the gas, and $\hat{T}_{V1}\mathrm{d}\hat{\gamma}_1\hat{T}_{V0}^{1/2}/\mathrm{d}\hat{T}_{V0}$ in $\hat{p}_{ijV2}$ and $\hat{T}_{V1}\mathrm{d}\hat{\gamma}_2\hat{T}_{V0}^{1/2}/\mathrm{d}\hat{T}_{V0}$ in $\hat{q}_{iV2}$ are due to their temperature dependence [see Appendix A.6]. The complicated forms of the $\hat{\gamma}_1$ and $\hat{\gamma}_2$ terms are due to the curved coordinates and the dependence of the viscosity and thermal conductivity on the temperature of the gas. The other terms are due to non-Navier–Stokes stress and heat flow. Among them, the terms with the factors $\hat{\gamma}_3$ and $\hat{\gamma}_7$ in the stress formulas are called thermal stress. The stress tensor and heat-flow vector up to the second order, i.e., $\hat{p}_{ijV2}$ and $\hat{q}_{iV2}$, contribute to Eqs. (6.69)–(6.71), that is, the information of $\hat{f}_{V2}$ is required to derive Eqs. (6.69)–(6.71). However, only the $\hat{\gamma}_1$ and $\hat{\gamma}_2$ terms contribute to Eqs. (6.69) and (6.71). As we have discussed, these equations, Eqs. (6.69) and (6.71), determine $\hat{\rho}_{V1}$, $\hat{v}_{iV1}$, $\hat{T}_{V1}$, and $\hat{p}_{V1}$ with the aid of Eq. (6.68). That is, these variables [thus, $\hat{f}_{V1}$ with the aid of Eq. (6.56)] are determined by the equations without non-Navier–Stokes stress and heat flow, which disagrees with the statement of Darrozes (1969). Incidentally, Eq. (6.70), one of the equations that determine the next-order macroscopic variables, contains non-Navier–Stokes terms.

The nondimensional transport coefficients $\hat{\gamma}_8$, $\hat{\gamma}_9$, and $\hat{\gamma}_{10}$, as well as $\hat{\gamma}_1$, $\hat{\gamma}_2$, $\hat{\gamma}_3$, and $\hat{\gamma}_7$, generally depend on the reference temperature T_0 besides $\hat{T}_{V0}$ (see Footnote 13 in Appendix A.2). Their dimensional forms are naturally independent of T_0, but it is not clear from the nondimensional forms, since their dependences on T_0 and $\hat{T}_{V0}$ are not obtained explicitly. This is clarified in Appendix A.6.

6.4 Knudsen-layer solution and slip condition

The boundary condition (6.2) is matched with the first term $\hat{f}_{V0}$ by imposing the nonslip condition (6.50). For the discussion at the higher orders, we consider the case of the diffuse reflection (6.3a) to make the discussion concrete. Then, the higher order term $\hat{f}_{Vm}$ $(m \geq 1)$ should satisfy the same form of condition as Eq. (6.3a), with $\hat{f}$ in $\hat{\sigma}_w$ replaced by $\hat{f}_{Vm}$, for the solution $\hat{f}_V$ to satisfy the boundary condition.[5] However, the solution $\hat{f}_V$ does not have this freedom. In fact, the functional form of $\hat{f}_{V1}$, as a function of ζ_i, is different from that

[5]Explicitly,

$$\hat{f}_{Vm} = \hat{\sigma}_{wm}(\pi\hat{T}_w)^{-3/2}\exp\left(-(\zeta_i - \hat{v}_{wi})^2/\hat{T}_w\right) \quad (\zeta_j n_j > 0),$$

$$\hat{\sigma}_{wm} = -2(\pi/\hat{T}_w)^{1/2}\int_{\zeta_j n_j<0} \zeta_j n_j \hat{f}_{Vm}\mathbf{d}\boldsymbol{\zeta}.$$

of the boundary condition owing to the second and third terms of Eq. (6.56). These two terms are already determined by the macroscopic variables $\hat{v}_{iV0}$, $\hat{T}_{V0}$, $\partial\hat{T}_{V0}/\partial y$, and $\partial\hat{v}_{iV0}/\partial y$ obtained from Eqs. (6.36d), (6.55), (6.57), (6.58), and (6.60) under the boundary condition (6.50) at $y = 0$ and that at $y = \infty$ to be discussed later, and they generally are not zero. Thus, we cannot construct the solution satisfying the boundary condition at $y = 0$ with $\hat{f}_V$. This can also be seen by the fact that Eq. (6.32) for $\hat{f}_V$ is of a singular type, where the differential term $\partial/\partial y$ is, in practice, multiplied by the small parameter ε. In view of this character of Eq. (6.32), we assume that the state of the gas varies appreciably in the direction normal to the boundary over the distance of $y = O(\varepsilon)$ in the neighborhood of the boundary, and try to find the solution of the boundary-value problem by modifying $\hat{f}_V$ only in the neighborhood of the boundary. That is, we put the solution in the form

$$\hat{f} = \hat{f}_V + \hat{f}_K, \tag{6.78}$$

where the correction term $\hat{f}_K$ is assumed to vary appreciably in the direction n_i over the distance of the order of the mean free path l_0 or $y = O(\varepsilon)$ and to be appreciable only in the neighborhood [or in $y = O(\varepsilon)$] of the boundary, that is, $\partial\hat{f}_K/\partial y = O(\hat{f}_K/\varepsilon)$, and $\hat{f}_K$ vanishes very rapidly (or faster than any inverse power of y/ε) as y/ε tends to infinity (Knudsen-layer correction). For the convenience of analysis, we further stretch the coordinate y by $1/\varepsilon$:[6]

$$x_i = \varepsilon^2\eta n_i(\chi_1, \chi_2) + x_{wi}(\chi_1, \chi_2). \tag{6.79}$$

Then, the solution is conveniently expressed by the two sets (χ_1, χ_2, y) and (χ_1, χ_2, η) of variables:

$$\hat{f} = \hat{f}_V(\chi_1, \chi_2, y) + \hat{f}_K(\chi_1, \chi_2, \eta). \tag{6.80}$$

The Knudsen-layer correction $\hat{f}_K(\chi_1, \chi_2, \eta)$ is also expanded in a power series of ε:

$$\hat{f}_K = \hat{f}_{K1}\varepsilon + \hat{f}_{K2}\varepsilon^2 + \cdots, \tag{6.81}$$

where the series starts from the term of order ε, since $\hat{f}_{V0}$ satisfies the boundary condition at $y = 0$ by imposing Eq. (6.50).

In the new coordinate system (χ_1, χ_2, η), the Boltzmann equation (6.1) or (6.32) is rewritten as

$$\zeta_i n_i \frac{\partial \hat{f}}{\partial \eta} + \varepsilon^2 \zeta_i \left(\frac{\partial \chi_1}{\partial x_i}\frac{\partial \hat{f}}{\partial \chi_1} + \frac{\partial \chi_2}{\partial x_i}\frac{\partial \hat{f}}{\partial \chi_2} \right) = \hat{J}(\hat{f}, \hat{f}). \tag{6.82}$$

Equation (6.80) with the expansions (6.33) and (6.81) is substituted into the Boltzmann equation (6.82), and the same order terms in ε are arranged. In this

[6] The variables (η, χ_1, χ_2) are the same as the Knudsen-layer variables defined in Eq. (3.41).

process, the fact that $\hat{f}_V$ satisfies the Boltzmann equation is used, and the series expansion of $\hat{f}_V$ in a power series of y, i.e.,

$$\hat{f}_V = (\hat{f}_{V0})_0 + \varepsilon[(\hat{f}_{V1})_0 + \eta(\partial \hat{f}_{V0}/\partial y)_0] + \cdots, \tag{6.83}$$

is applied to the product of $\hat{f}_V$ and $\hat{f}_K$ in the collision term, since $\hat{f}_K$ is rapidly decaying. In Eq. (6.83), the quantities in the parentheses with subscript 0 are evaluated at $y = 0$. As a result, a series of equations for $\hat{f}_{Km}$ $(m \geq 1)$ is obtained as follows:[7]

$$\zeta_i n_i \frac{\partial \hat{f}_{K1}}{\partial \eta} = 2\hat{J}((\hat{f}_{V0})_0, \hat{f}_{K1}), \tag{6.84}$$

$$\zeta_i n_i \frac{\partial \hat{f}_{K2}}{\partial \eta} = 2\hat{J}((\hat{f}_{V0})_0, \hat{f}_{K2}) + 2\hat{J}((\hat{f}_{V1})_0 + \eta(\partial \hat{f}_{V0}/\partial y)_0, \hat{f}_{K1}) + \hat{J}(\hat{f}_{K1}, \hat{f}_{K1}), \tag{6.85}$$

$$\cdots\cdots\cdots.$$

Equation (6.84) is the one-dimensional Boltzmann equation linearized around the Maxwellian $(\hat{f}_{V0})_0$, where $\hat{v}_{iV0} = \hat{v}_{wi}$ and $\hat{T}_{V0} = \hat{T}_w$.

The boundary condition for $\hat{f}_{Km}$ is given as follows. The Knudsen-layer correction $\hat{f}_K$ is introduced as the correction to $\hat{f}_V$ in the neighborhood of the boundary of $\eta = O(1)$ and is assumed to vanish faster than any inverse power of η. Thus,

$$\hat{f}_{Km} \to 0 \quad \text{as } \eta \to \infty. \tag{6.86}$$

The rapid decay is required for the application of the expansion (6.83) in Eq. (6.82) [see also the comment below Eq. (3.52)]. This decay is verified for a hard-sphere gas by Bardos's theorem in Appendix A.9. From Eqs. (6.3a), (6.3b), and (6.78), $\hat{f}$ (or $\hat{f}_V + \hat{f}_K$) should satisfy the following condition: at $\eta = 0$,

$$\hat{f}_K = \frac{\hat{\sigma}_w}{(\pi \hat{T}_w)^{3/2}} \exp\left(-\frac{(\zeta_i - \hat{v}_{wi})^2}{\hat{T}_w}\right) - \hat{f}_V \quad (\zeta_i n_i > 0), \tag{6.87}$$

$$\hat{\sigma}_w = -2\left(\frac{\pi}{\hat{T}_w}\right)^{1/2} \int_{\zeta_j n_j < 0} \zeta_j n_j (\hat{f}_V + \hat{f}_K) \mathbf{d}\boldsymbol{\zeta}. \tag{6.88}$$

On the other hand, by Eq. (6.50), at $y = 0$,

$$\begin{aligned} \hat{f}_{V0} &= (\hat{f}_{V0})_0 \\ &= \frac{(\hat{\rho}_{V0})_0}{(\pi \hat{T}_w)^{3/2}} \exp\left(-\frac{(\zeta_i - \hat{v}_{wi})^2}{\hat{T}_w}\right). \end{aligned} \tag{6.89}$$

Therefore, at $\eta = 0$,

$$\hat{f}_{Km} = \frac{\hat{\sigma}_{wm}}{(\pi \hat{T}_w)^{3/2}} \exp\left(-\frac{(\zeta_i - \hat{v}_{wi})^2}{\hat{T}_w}\right) - \hat{f}_{Vm} \quad (\zeta_i n_i > 0) \quad \text{for } m \geq 1, \tag{6.90}$$

[7]For the first terms on the right-hand sides of Eqs. (6.84) and (6.85), see Footnotes 26 and 28 in Appendix A. For the other terms, see Footnotes 12 in Section 2.9 and 1 in Section 4.2.

where

$$\hat{\sigma}_{wm} = -2\left(\frac{\pi}{\hat{T}_w}\right)^{1/2}\int_{\zeta_i n_i<0}\zeta_j n_j(\hat{f}_{Vm}+\hat{f}_{Km})\mathbf{d}\boldsymbol{\zeta}. \tag{6.91}$$

Especially, with Eq. (6.56) for $\hat{f}_{V1}$, the boundary condition for $\hat{f}_{K1}$ at $\eta = 0$ is given as follows:

$$\begin{aligned}\hat{f}_{K1} = -\hat{f}_{V0}\Bigg[&\frac{\hat{\rho}_{V1}-\hat{\sigma}_{w1}}{\hat{\rho}_{V0}}+\frac{2(\zeta_i-\hat{v}_{wi})\hat{v}_{iV1}}{\hat{T}_w}+\frac{\hat{T}_{V1}}{\hat{T}_w}\left(\frac{(\zeta_i-\hat{v}_{wi})^2}{\hat{T}_w}-\frac{3}{2}\right)\\&-\frac{\zeta_i n_i\mathcal{A}(|\zeta_j-\hat{v}_{wj}|/\hat{T}_w^{1/2},\hat{T}_w)}{\hat{\rho}_{V0}\hat{T}_w^{3/2}}\frac{\partial\hat{T}_{V0}}{\partial y}\\&-\frac{(\zeta_i-\hat{v}_{wi})\zeta_j n_j\mathcal{B}(|\zeta_k-\hat{v}_{wk}|/\hat{T}_w^{1/2},\hat{T}_w)}{\hat{\rho}_{V0}\hat{T}_w^{3/2}}\frac{\partial\hat{v}_{iV0}}{\partial y}\Bigg]\quad(\zeta_i n_i>0),\end{aligned} \tag{6.92}$$

$$\hat{\sigma}_{w1} = \hat{\rho}_{V1}+\frac{\hat{\rho}_{V0}\hat{T}_{V1}}{2\hat{T}_w}-\left(\frac{\pi}{\hat{T}_w}\right)^{1/2}\left(\hat{\rho}_{V0}\hat{v}_{iV1}n_i+2\int_{\zeta_i n_i<0}\zeta_j n_j\hat{f}_{K1}\mathbf{d}\boldsymbol{\zeta}\right), \tag{6.93}$$

where the relations (6.50), (6.54), and $\hat{v}_{wi}n_i = 0$ and the subsidiary condition to Eq. (A.84) are used.

By a transformation similar to Eq. (5.67), the equation and boundary conditions for f_{Km} [(6.84), (6.85), etc. and (6.86), (6.90)] are reduced to the same type of a half-space problem of the linearized Boltzmann equation as discussed in the foregoing chapters. In place of the Knudsen-layer variable η, the molecular velocity ζ_i, and the distribution function $\hat{f}_{Km}$, the following $\tilde{\eta}$, $\tilde{\zeta}_i$, and $\phi_{Km}(\tilde{\eta},\tilde{\zeta}_i)$, locally normalized along the boundary, are introduced:

$$\tilde{\eta} = \frac{(\hat{\rho}_{V0})_0}{g_M(\hat{T}_w)}\eta,\quad \tilde{\zeta}_i = \frac{\zeta_i-\hat{v}_{wi}}{\hat{T}_w^{1/2}},\quad \phi_{Km}(\tilde{\eta},\tilde{\zeta}_i)E(\tilde{\zeta}) = \frac{\hat{T}_w^{3/2}}{(\hat{\rho}_{V0})_0}\hat{f}_{Km}, \tag{6.94}$$

where the quantities in the parentheses with subscript 0, i.e., $(**)_0$, are evaluated on the boundary, and a function $g_M(\hat{T}_w)$ of $\hat{T}_w$ is introduced, as in Chapter 5 for simplicity of the results for some molecular models, in the following way: if the linearized collision operator $\mathcal{L}_{\hat{T}_w}(*)$ [or $\hat{B}_{\hat{T}_w}$ in $\mathcal{L}_{\hat{T}_w}(*)$], defined in Eq. (A.23), can be reduced to be independent of $\hat{T}_w$ by multiplication of some function of $\hat{T}_w$, then $g_M(\hat{T}_w)$ is chosen in such a way that $g_M(\hat{T}_w)\mathcal{L}_{\hat{T}_w}(*) = \mathcal{L}_1(*)$ [$= \mathcal{L}(*)$], and if not, $g_M(\hat{T}_w) = 1$. For hard-sphere gas, $\mathcal{L}_{\hat{T}_w}(*)$ itself is independent of $\hat{T}_w$ and for the BKW model, $\hat{T}_w^{1/2}\mathcal{L}_{\hat{T}_w}(*)$ is independent of $\hat{T}_w$. Then, according to Appendix A.9, the equation (6.84) for $\hat{f}_{K1}$, for example, is reduced to

$$\tilde{\zeta}_i n_i\frac{\partial\phi_{K1}(\tilde{\eta},\tilde{\zeta}_i)}{\partial\tilde{\eta}} = g_M(\hat{T}_w)\mathcal{L}_{\hat{T}_w}(\phi_{K1}(\tilde{\eta},\tilde{\zeta}_i))\quad(\tilde{\eta}>0). \tag{6.95}$$

The boundary conditions (6.86) and (6.92) are, respectively, reduced to

$$\phi_{K1} \to 0 \qquad \text{as } \tilde{\eta} \to \infty, \tag{6.96}$$

and

$$\phi_{K1} = -\varpi_1 - \frac{2\tilde{\zeta}_i \hat{v}_{iV1}}{\hat{T}_w^{1/2}} - \frac{\hat{T}_{V1}}{\hat{T}_w}\left(\tilde{\zeta}_i^{\,2} - \frac{3}{2}\right) + \frac{\tilde{\zeta}_i n_i \mathcal{A}(\tilde{\zeta}, \hat{T}_w)}{\hat{\rho}_{V0}\hat{T}_w}\frac{\partial \hat{T}_{V0}}{\partial y}$$
$$+ \frac{\tilde{\zeta}_j n_j \tilde{\zeta}_i \mathcal{B}(\tilde{\zeta}, \hat{T}_w)}{\hat{\rho}_{V0}\hat{T}_w^{1/2}}\frac{\partial \hat{v}_{iV0}}{\partial y} \qquad (\tilde{\zeta}_i n_i > 0) \qquad \text{at } \tilde{\eta} = 0, \tag{6.97}$$

where

$$\varpi_1 = -\frac{\hat{T}_{V1}}{2\hat{T}_w} + \sqrt{\pi}\left(\frac{\hat{v}_{iV1} n_i}{\hat{T}_w^{1/2}} + 2\int_{\tilde{\zeta}_i n_i < 0} \tilde{\zeta}_j n_j \phi_{K1}(\tilde{\boldsymbol{\zeta}}) E(\tilde{\zeta}) \mathbf{d}\tilde{\boldsymbol{\zeta}}\right). \tag{6.98}$$

By Bardos's theorem (Appendix A.9), the above boundary-value problem has a solution if and only if the undetermined boundary data $\hat{v}_{iV1}$ and $\hat{T}_{V1}$ take special values, and the solution as well as the set of the values is unique. Their theorem also assures that the solution decays with an exponential speed as $\tilde{\eta} \to \infty$. By a similar discussion to that in Section 3.4 based on Bardos's theorem, the boundary data $\hat{v}_{iV1}$ and $\hat{T}_{V1}$ are specified in the following forms: at $x_i = x_{wi}$,

$$\hat{v}_{iV1} t_i^{(\alpha)} = -\frac{\hat{k}_0}{\hat{\rho}_{V0}}\frac{\partial \hat{v}_{iV0} t_i^{(\alpha)}}{\partial y}, \tag{6.99a}$$

$$\hat{v}_{iV1} n_i = 0, \tag{6.99b}$$

$$\hat{T}_{V1} = \frac{\hat{d}_1}{\hat{\rho}_{V0}}\frac{\partial \hat{T}_{V0}}{\partial y}, \tag{6.99c}$$

where the slip coefficients $\hat{k}_0$ and $\hat{d}_1$ are functions of $\hat{T}_w$ determined by molecular model. For a hard-sphere gas and the BKW model,

$$\hat{k}_0 = k_0 = -1.2540, \qquad \hat{d}_1 = d_1 = 2.4001 \quad \text{(hard-sphere)}, \tag{6.100a}$$

$$\hat{k}_0/\hat{T}_w^{1/2} = k_0 = -1.01619, \quad \hat{d}_1/\hat{T}_w^{1/2} = d_1 = 1.30272 \quad \text{(BKW)}. \tag{6.100b}$$

The Knudsen-layer part ($\hat{\rho}_K$, $\hat{v}_{iK}$, $\hat{T}_K$, $\hat{p}_K$, $\hat{p}_{ijK}$, $\hat{q}_{iK}$) of the macroscopic variables ($\hat{\rho}$, $\hat{v}_i$, $\hat{T}$, $\hat{p}$, $\hat{p}_{ij}$, $\hat{q}_i$) are also defined as the remainders ($\hat{\rho} - \hat{\rho}_V$, $\hat{v}_i - \hat{v}_{iV}$, $\hat{T} - \hat{T}_V$, etc.). Then, they depend on $\hat{f}_V$ as well as $\hat{f}_K$, since the relations between the macroscopic variables and the velocity distribution functions, i.e., Eqs. (2.48a)-(2.48f), are nonlinear. For example,

$$(\hat{\rho}_V + \hat{\rho}_K)\hat{v}_{iK} = \int \zeta_i \hat{f}_K \mathbf{d}\boldsymbol{\zeta} - \hat{v}_{iV}\hat{\rho}_K. \tag{6.101}$$

Corresponding to the expansion (6.81), the Knudsen-layer parts of the macroscopic variables are also expanded in power series of ε:

$$\hat{h}_K = \hat{h}_{K1}\varepsilon + \hat{h}_{K2}\varepsilon^2 + \cdots, \tag{6.102}$$

where h represents $\hat{\rho}$, $\hat{v}_i$, $\hat{T}$, $\hat{p}$, $\hat{p}_{ij}$, and $\hat{q}_i$. The relations of the component functions of the expansions to $\hat{f}_{Km}$ are given as follows:

$$\hat{\rho}_{K1} = \int \hat{f}_{K1}\mathbf{d}\boldsymbol{\zeta}, \tag{6.103a}$$

$$(\hat{\rho}_{V0})_0\hat{v}_{iK1} = \int \zeta_i \hat{f}_{K1}\mathbf{d}\boldsymbol{\zeta} - \hat{\rho}_{K1}(\hat{v}_{iV0})_0, \tag{6.103b}$$

$$\frac{3}{2}(\hat{\rho}_{V0})_0\hat{T}_{K1} = \int [\zeta_j - (\hat{v}_{jV0})_0]^2 \hat{f}_{K1}\mathbf{d}\boldsymbol{\zeta} - \frac{3}{2}\hat{\rho}_{K1}(\hat{T}_{V0})_0, \tag{6.103c}$$

$$\hat{p}_{K1} = (\hat{\rho}_{V0})_0\hat{T}_{K1} + \hat{\rho}_{K1}(\hat{T}_{V0})_0, \tag{6.103d}$$

$$\hat{p}_{ijK1} = 2\int [\zeta_i - (\hat{v}_{iV0})_0][\zeta_j - (\hat{v}_{jV0})_0]\hat{f}_{K1}\mathbf{d}\boldsymbol{\zeta}, \tag{6.103e}$$

$$\hat{q}_{iK1} = \int [\zeta_i - (\hat{v}_{iV0})_0][\zeta_j - (\hat{v}_{jV0})_0]^2 \hat{f}_{K1}\mathbf{d}\boldsymbol{\zeta} - \frac{3}{2}(\hat{\rho}_{V0})_0\hat{v}_{iK1} - (\hat{p}_{ijV0})_0\hat{v}_{jK1}, \tag{6.103f}$$

....

The macroscopic variables $\hat{\rho}_K$, $\hat{v}_{iK}$, $\hat{T}_K$, $\hat{p}_K$, $\hat{p}_{ijK}$, and $\hat{q}_{iK}$ are expressed in the following forms:

$$\hat{\rho}_{K1} = \frac{1}{\hat{T}_w}\left(\frac{\partial \hat{T}_{V0}}{\partial y}\right)_0 \hat{\Omega}_1(\tilde{\eta}), \tag{6.104a}$$

$$\hat{v}_{iK1}t_i^{(\alpha)} = -\frac{1}{(\hat{\rho}_{V0})_0}\left(\frac{\partial \hat{v}_{iV0}t_i^{(\alpha)}}{\partial y}\right)_0 \hat{Y}_0(\tilde{\eta}), \quad \hat{v}_{iK1}n_i = 0, \tag{6.104b}$$

$$\hat{T}_{K1} = \frac{1}{(\hat{\rho}_{V0})_0}\left(\frac{\partial \hat{T}_{V0}}{\partial y}\right)_0 \hat{\Theta}_1(\tilde{\eta}), \tag{6.104c}$$

$$\left.\begin{aligned} &\hat{p}_{ijK1}n_i n_j = \hat{p}_{ijK1}n_i t_j^{(\alpha)} = \hat{p}_{ijK1}t_i^{(\alpha)}t_j^{(3-\alpha)} = 0,\\ &\hat{p}_{ijK1}t_i^{(\alpha)}t_j^{(\alpha)} = \frac{3}{2}\left(\frac{\partial \hat{T}_{V0}}{\partial y}\right)_0\left[\hat{\Omega}_1(\tilde{\eta}) + \hat{\Theta}_1(\tilde{\eta})\right], \end{aligned}\right\} \tag{6.104d}$$

$$\hat{q}_{iK1}n_i = 0, \quad \hat{q}_{iK1}t_i^{(\alpha)} = \left(\hat{T}_{V0}\frac{\partial \hat{v}_{iV0}t_i^{(\alpha)}}{\partial y}\right)_0 \hat{H}_A(\tilde{\eta}), \tag{6.104e}$$

where the Knudsen-layer functions $\hat{\Omega}_1$, $\hat{Y}_0$, $\hat{\Theta}_1$, and $\hat{H}_A$ are functions of $\tilde{\eta}$ and $\hat{T}_w$, and their functional forms depend on molecular models. It may be noted that the Knudsen-layer functions $\hat{\Omega}_1$, $\hat{Y}_0$, $\hat{\Theta}_1$, and $\hat{H}_A$ as well as the nondimensional slip coefficients $\hat{k}_0$ and $\hat{d}_1$ generally depend on the reference temperature T_0, since the operator $\mathcal{L}_{\hat{T}_w}$ depends on T_0 besides the parameter $\hat{T}_w$ (see Footnotes 11 in Section 2.9 and 13 in Appendix A.2). If the linearized collision operator $\mathcal{L}_{\hat{T}_w}(*)$ multiplied by $g_M(\hat{T}_w)$ in Eq. (6.95) is independent of $\hat{T}_w$, as for a hard-sphere gas $[g_M(\hat{T}_w) = 1]$ and for the BKW model $[g_M(\hat{T}_w) = \hat{T}_w^{1/2}]$, then the functions $\hat{\Omega}_1(\tilde{\eta})/g_M(\hat{T}_w)$, $\hat{Y}_0(\tilde{\eta})/g_M(\hat{T}_w)$, $\hat{\Theta}_1(\tilde{\eta})/g_M(\hat{T}_w)$, and $\hat{H}_A(\tilde{\eta})/g_M(\hat{T}_w)$ as well as $\hat{k}_0/g_M(\hat{T}_w)$ and $\hat{d}_1/g_M(\hat{T}_w)$ are independent of $\hat{T}_w$ (see Footnote 8 in Section 5.4). These functions, denoted by $\Omega_1(\eta)$, $Y_0(\eta)$, $\Theta_1(\eta)$, and $H_A(\eta)$ with η (there) $= \tilde{\eta}$ (here), are tabulated for a hard-sphere gas and the BKW model in Tables 3.2 and 3.3 [note $H_A(\eta) = Y_0(\eta)/2$ for BKW].

The slip in the boundary condition for the viscous boundary-layer equations and the Knudsen-layer correction appear at the order of ε (or $\sqrt{\mathrm{Kn}}$). This is because the length scale of variation in the direction normal to the boundary is of the order of εL, e.g., $n_i \partial\hat{T}/\partial x_i = O(\hat{T}/\varepsilon)$. Accordingly, the thermal creep flow (see Section 3.11.1), which appeared in the slip conditions in the foregoing chapters, does not appear at this order, since it is proportional to the mean free path and the tangential derivative of the boundary temperature.

In the following analysis, the boundary condition for the normal velocity $\hat{v}_{iV2}n_i$ at $y = 0$ is required, for which the information of $\hat{v}_{iK2}n_i$ is also necessary. From the expansion of Eq. (6.101), this is expressed by the velocity distribution function $\hat{f}_{K2}$ as

$$(\hat{\rho}_{V0})_0\hat{v}_{iK2}n_i = \int \zeta_i n_i \hat{f}_{K2}\mathbf{d}\boldsymbol{\zeta}, \tag{6.105}$$

which is simplified thanks to Eq. (6.99b) and the second relation in Eq. (6.104b). Then, from Eq. (6.85) integrated over the whole space of $\boldsymbol{\zeta}$,

$$(\hat{\rho}_{V0})_0\hat{v}_{iK2}n_i = \int_\infty^\eta \left(\int \Big[2\hat{J}((\hat{f}_{V0})_0, \hat{f}_{K2}) + 2\hat{J}((\hat{f}_{V1})_0 + \eta(\partial \hat{f}_{V0}/\partial y)_0, \hat{f}_{K1}) \right.$$
$$\left. + \hat{J}(\hat{f}_{K1}, \hat{f}_{K1})\Big]\, \mathbf{d}\boldsymbol{\zeta}\right)\mathrm{d}\eta = 0. \tag{6.106}$$

The last relation is obtained with the aid of the general property (2.44) of the collision integral. On the other hand, on a simple boundary, $\hat{v}_i n_i = 0$. Thus, at $y = 0$,

$$\hat{v}_{iV2}n_i = -\hat{v}_{iK2}n_i = 0. \tag{6.107}$$

That is, the displacement effect (terminology introduced in viscous boundary-layer theory; see, e.g., Schlichting [1979]) of the Knudsen layer vanishes (or is of higher order). Then from Eq. (6.68), $\hat{\rho}_{V0}\hat{v}_{iV2}n_i$ can be expressed by the lower-order quantities as

$$\begin{aligned}
\hat{\rho}_{V0}\hat{v}_{iV2}n_i = {} & -\hat{\rho}_{V1}\hat{v}_{iV1}n_i - \sum_{\alpha=1}^{2}\chi_{\alpha,\alpha}\int_0^y \frac{\partial(\hat{\rho}_{V0}\hat{v}_{iV1} + \hat{\rho}_{V1}\hat{v}_{iV0})t_i^{(\alpha)}}{\partial\chi_\alpha}\mathrm{d}y \\
& + 2\bar{\kappa}\int_0^y \hat{\rho}_{V0}\hat{v}_{iV1}n_i\mathrm{d}y + \sum_{\alpha=1}^{2}(-1)^\alpha g_{3-\alpha}\int_0^y (\hat{\rho}_{V0}\hat{v}_{iV1}t_i^{(\alpha)} + \hat{\rho}_{V1}\hat{v}_{iV0}t_i^{(\alpha)})\mathrm{d}y \\
& - \sum_{\alpha=1}^{2}\int_0^y y\left(\kappa_\alpha\chi_{\alpha,\alpha}\frac{\partial\hat{\rho}_{V0}\hat{v}_{iV0}t_i^{(\alpha)}}{\partial\chi_\alpha} + \vartheta\chi_{\alpha,\alpha}\frac{\partial\hat{\rho}_{V0}\hat{v}_{iV0}t_i^{(3-\alpha)}}{\partial\chi_\alpha}\right)\mathrm{d}y \\
& + \sum_{\alpha=1}^{2}(-1)^\alpha(\kappa_{3-\alpha}g_{3-\alpha} + \vartheta g_\alpha)\int_0^y y\hat{\rho}_{V0}\hat{v}_{iV0}t_i^{(\alpha)}\mathrm{d}y.
\end{aligned} \tag{6.108}$$

To summarize, at the leading order (or in the continuum limit), under a very general kinetic boundary condition, there is no Knudsen layer and no slip (or no jump) of the viscous boundary part of velocity and temperature over a simple boundary. The Knudsen layer, which appears at higher orders, is governed by a spatially one-dimensional (homogeneous or inhomogeneous) linearized Boltzmann equation as that for the Knudsen layer in the foregoing chapters. The first-order Knudsen layer and slip coefficients are practically the same as in the linearized problem; the apparent difference is due to the choice of the reference state. At higher orders, different terms appear (see, e.g., Sone & Yamamoto [1970], Sone & Onishi [1977]). Thus, the generalization of the present analysis under the diffuse reflection to other kinetic boundary conditions is straightforward as in the linearized problem (see Sections 3.4). The slip conditions (6.99a)–(6.99c) and the Knudsen layer (6.104a)–(6.104c) remain in the same forms for a general locally isotropic simple boundary where the reflection law is given by Eq. (2.22) with a finite diffuse-reflection part. The differences are the slip coefficients $\hat{k}_0$ and $\hat{d}_1$ and the Knudsen-layer functions $\hat{Y}_0(\bar{\eta})$, $\hat{\Omega}_1(\bar{\eta})$, $\hat{\Theta}_1(\bar{\eta})$, and $\hat{H}_A(\bar{\eta})$. They depend on the reflection law on the boundary as well as on the molecular model of the gas.

6.5 Connection of Hilbert and viscous boundary-layer solutions

In the study of a flow with its Mach number of the order of unity around a simple boundary in the foregoing sections, we found possible types of solutions: the solution (Hilbert solution) in the general domain and that (viscous boundary-layer solution + Knudsen-layer solution) in the neighborhood (within the distance of the order of the square root of the Knudsen number) of the boundary, but have not finished the connection of these solutions. Here, we will do that. For this purpose, we will review the Hilbert and viscous boundary-layer solutions.

The velocity distribution function of the Hilbert solution is expressed as a given function of the molecular velocity with the macroscopic variables, density, flow velocity, and temperature, as the parametric functions. Up to the order of ε, it is Maxwellian, and the variation of the parametric macroscopic variables of the Maxwellian is determined by the Euler set of equations. In the neighborhood εL of a boundary, the parametric functions are expressed in the form

$$\hat{h}_h = \left(\hat{h}_{h0}\right)_0 + \varepsilon\left[\left(\hat{h}_{h1}\right)_0 + y n_i \left(\frac{\partial \hat{h}_{h0}}{\partial x_i}\right)_0\right] + \cdots, \tag{6.109}$$

where the order with respect to ε is reshuffled.

In Section 6.4, it is shown that the solution $\hat{f}$ in this neighborhood is given as

$$\hat{f} = \hat{f}_V + \hat{f}_K. \tag{6.110}$$

If $\hat{f}_V$ is expressed as the sum

$$\hat{f}_V = \hat{f}_h + \hat{f}_{VC}, \tag{6.111}$$

where $\hat{f}_{VC}$ is the correction function decaying rapidly (or faster than any inverse power of y) as $y \to \infty$, then the connection of this solution and the Hilbert solution can be done. We will derive the connection conditions of the two solutions.

The viscous solution $\hat{f}_V$ is also expressed by a given function of molecular velocity with the macroscopic variables as the parametric functions. The first three terms of the expansion in ε are given by Eqs. (6.47), (6.56), and (6.66) with (6.67), the first two of which are repeated here for easier understanding of the process of analysis:

$$\hat{f}_{V0} = \frac{\hat{\rho}_{V0}}{(\pi \hat{T}_{V0})^{3/2}} \exp(-\tilde{\zeta}_i^2), \tag{6.112}$$

$$\begin{aligned} \hat{f}_{V1} = \hat{f}_{V0} \Bigg[& \frac{\hat{\rho}_{V1}}{\hat{\rho}_{V0}} + 2\tilde{\zeta}_i \frac{\hat{v}_{iV1}}{\hat{T}_{V0}^{1/2}} + \frac{\hat{T}_{V1}}{\hat{T}_{V0}} \left(\tilde{\zeta}_i^2 - \frac{3}{2} \right) \\ & - \frac{\tilde{\zeta}_i n_i \mathcal{A}(\tilde{\zeta}, \hat{T}_{V0})}{\hat{\rho}_{V0} \hat{T}_{V0}} \frac{\partial \hat{T}_{V0}}{\partial y} - \frac{\tilde{\zeta}_i \tilde{\zeta}_j n_j \mathcal{B}(\tilde{\zeta}, \hat{T}_{V0})}{\hat{\rho}_{V0} \hat{T}_{V0}^{1/2}} \frac{\partial \hat{v}_{iV0}}{\partial y} \Bigg], \end{aligned} \tag{6.113}$$

$$\tilde{\zeta}_i = \frac{\zeta_i - \hat{v}_{iV0}}{\hat{T}_{V0}^{1/2}}, \quad \tilde{\zeta} = \sqrt{\tilde{\zeta}_i^2}.$$

The function $\hat{f}_{V0}$ is Maxwellian. This is a favorable feature since $\hat{f}_{h0}$ is also Maxwellian. If we consider a point where y is large but εy is small, then the correction $\hat{f}_{VC}$ is negligible there, since its exponential decay is assumed. If the $\hat{f}_{V0}$ agrees with $\hat{f}_{h0}$ there, $\hat{f}_{V0}$ is connected with $\hat{f}_{h0}$ [The reshuffled form of $\hat{f}_{hm}$, or Eq. (6.109), in the thin layer should be used here]. This agreement is satisfied if the following conditions are fulfilled:

$$\hat{\rho}_{V0} \sim (\hat{\rho}_{h0})_0 \qquad \text{as } y \to \infty, \tag{6.114a}$$

$$\hat{v}_{iV0} \sim (\hat{v}_{ih0})_0 \qquad \text{as } y \to \infty, \tag{6.114b}$$

$$\hat{T}_{V0} \sim (\hat{T}_{h0})_0 \qquad \text{as } y \to \infty. \tag{6.114c}$$

The next-order viscous boundary-layer solution $\hat{f}_{V1}$ is not Maxwellian, but it reduces to it in the region where y is large but εy is small. In fact, $\partial \hat{T}_{V0}/\partial y$ and $\partial \hat{v}_{iV0}/\partial y$ in the non-Maxwellian terms of $\hat{f}_{V1}$ are small quantities of the order of ε, because the rapidly varying part [or the terms of $\partial/\partial y = O(1)$] is negligibly small there owing to its exponential decay and therefore only the moderately varying part [or the terms of $\partial/\partial y = O(\varepsilon)$] remains there. Thus, $\hat{f}_{V1}$ as well as $\hat{f}_{h1}$ is Maxwellian there. Therefore, by the same reason as in the leading-order case, the connection of $\hat{f}_{V1}$ with $\hat{f}_{h1}$ is made if the macroscopic variables satisfy the following conditions:

$$\hat{\rho}_{V1} \sim (\hat{\rho}_{h1})_0 + yn_j \left(\frac{\partial \hat{\rho}_{h0}}{\partial x_j}\right)_0 \quad \text{as } y \to \infty, \tag{6.115a}$$

$$\hat{v}_{iV1} \sim (\hat{v}_{ih1})_0 + yn_j \left(\frac{\partial \hat{v}_{ih0}}{\partial x_j}\right)_0 \quad \text{as } y \to \infty, \tag{6.115b}$$

$$\hat{T}_{V1} \sim \left(\hat{T}_{h1}\right)_0 + yn_j \left(\frac{\partial \hat{T}_{h0}}{\partial x_j}\right)_0 \quad \text{as } y \to \infty. \tag{6.115c}$$

Thus, if we can find the solutions $\hat{\rho}_{hm}$, $\hat{v}_{ihm}$, $\hat{T}_{hm}$ $(m = 0, 1)$ of the Euler-type sets of equations and those $\hat{\rho}_{Vm}$, $\hat{v}_{iVm}$, $\hat{T}_{Vm}$ $(m = 0, 1)$ of the viscous boundary-layer-type sets that satisfy the connection conditions (6.114a)–(6.114c) and (6.115a)–(6.115c), and confirm that the viscous boundary-layer solution approaches the state at infinity compatible with the connection conditions (6.114a)–(6.114c) and (6.115a)–(6.115c) with an exponential speed, then the connection process is finished up to the order of ε. In the next section, we will show how the solution satisfying the connection conditions is constructed.

6.6 Recipe for construction of solution

We have seen in Section 6.3 that from Eq. (6.52) and the boundary condition (6.50), $\hat{v}_{iV0}n_i$ vanishes, i.e., Eq. (6.54) holds, and that from Eq. (6.53), $\hat{p}_{V0}$ is independent of y, i.e., Eq. (6.55) holds. Then, from Eq. (6.114b),

$$(\hat{v}_{ih0})_0\, n_i = 0. \tag{6.116}$$

This is used as the boundary condition for the Euler set (6.21)–(6.23) on a simple boundary. This is the standard boundary condition for the Euler set of equations in the classical gas dynamics. Solving the Euler set (6.21)–(6.23) under the boundary condition (6.116) [and the other necessary condition, e.g., the condition at infinity], we obtain the leading macroscopic variables $\hat{\rho}_{h0}$, $\hat{v}_{ih0}$, $\hat{T}_{h0}$, and $\hat{p}_{h0}$, thus $(\hat{\rho}_{h0})_0$, $(\hat{v}_{ih0})_0\, t_i^{(\alpha)}$, $(\hat{T}_{h0})_0$, and $(\hat{p}_{h0})_0$. Then, from the connection conditions (6.114a)–(6.114c) with the aid of Eq. (6.36d), the behavior of $\hat{\rho}_{V0}$, $\hat{v}_{iV0}t_i^{(\alpha)}$, $\hat{T}_{V0}$, and $\hat{p}_{V0}$ as $y \to \infty$ is determined. From Eq. (6.55) and the condition on $\hat{p}_{V0}$ as $y \to \infty$ just obtained, $\hat{p}_{V0}$ is determined:

$$\hat{p}_{V0} = (\hat{p}_{h0})_0\,. \tag{6.117}$$

The conditions on $\hat{v}_{iV0}t_i^{(\alpha)}$ and $\hat{T}_{V0}$ as $y \to \infty$ just obtained, together with the conditions on $\hat{v}_{iV0}t_i^{(\alpha)}$ and $\hat{T}_{V0}$ at $y = 0$ given in Eq. (6.50), form the boundary conditions for viscous boundary-layer equations (6.58) and (6.60). In Eqs. (6.58) and (6.60), $\hat{p}_{V0}$ is given by Eq. (6.117), $\hat{\rho}_{V0} = \hat{p}_{V0}/\hat{T}_{V0}$ by Eq. (6.36d), and a higher order quantity $\hat{\rho}_{V0}\hat{v}_{jV1}n_j$ is expressed by lower order quantities by Eq. (6.57) and the boundary condition (6.99b) as

$$\hat{\rho}_{V0}\hat{v}_{iV1}n_i = -\sum_{\alpha=1}^{2}\int_0^y \left(\chi_{\alpha,\alpha}\frac{\partial \hat{\rho}_{V0}\hat{v}_{iV0}t_i^{(\alpha)}}{\partial \chi_\alpha} - (-1)^\alpha g_{3-\alpha}\hat{\rho}_{V0}\hat{v}_{iV0}t_i^{(\alpha)}\right)\mathrm{d}y. \tag{6.118}$$

Thus, Eqs. (6.58) and (6.60) with the supplementary equations (6.36d), (6.117), and (6.118) are reduced to the equations for $\hat{v}_{iV0}t_i^{(\alpha)}$ and $\hat{T}_{V0}$. These equations with the above-mentioned boundary conditions at infinity and at $y = 0$ determine $\hat{v}_{iV0}t_i^{(\alpha)}$ and $\hat{T}_{V0}$, thus all the macroscopic variables at the leading order of the expansion.

Now we proceed to determination of the quantities of the order ε. From the connection conditions (6.114a) and (6.115b), including the condition of rapid approach, and the expression (6.118) of $\hat{v}_{iV1}n_i$, the value of $\hat{v}_{ih1}n_i$ on the boundary is given as

$$(\hat{v}_{ih1})_0\, n_i = -\int_0^\infty \left[\frac{1}{(\hat{\rho}_{h0})_0}\sum_{\alpha=1}^{2}\left(\chi_{\alpha,\alpha}\frac{\partial \hat{\rho}_{V0}\hat{v}_{iV0}t_i^{(\alpha)}}{\partial \chi_\alpha} - (-1)^\alpha g_{3-\alpha}\hat{\rho}_{V0}\hat{v}_{iV0}t_i^{(\alpha)}\right) + \left(\frac{\partial \hat{v}_{ih0}}{\partial x_j}\right)_0 n_i n_j\right]\mathrm{d}y. \tag{6.119}$$

This is the boundary condition for the linearized Euler set (6.24)–(6.26). In Eqs. (6.25) and (6.26) of the set, the coefficients of the normal derivatives of $\hat{v}_{ih1}$ and $\hat{T}_{h1}$ on the boundary vanish owing to Eq. (6.116). Equation (6.119) is an appropriate boundary condition for this set.[8] From the linearized Euler set (6.24)–(6.26) and the boundary condition (6.119) [and the other necessary condition, e.g., the condition at infinity], the first-order quantities $\hat{\rho}_{h1}$, $\hat{v}_{ih1}$, $\hat{T}_{h1}$, and $\hat{p}_{h1}$ [thus $(\hat{\rho}_{h1})_0$, $(\hat{v}_{ih1})_0\, t_i^{(\alpha)}$, $(\hat{T}_{h1})_0$, and $(\hat{p}_{h1})_0$] are obtained. Then, from the connection conditions (6.115a)–(6.115c), with the aid of Eq. (6.37d), the behavior of $\hat{\rho}_{V1}$, $\hat{v}_{iV1}t_i^{(\alpha)}$, $\hat{T}_{V1}$, and $\hat{p}_{V1}$ as $y \to \infty$ is determined. These conditions give the boundary conditions for the viscous boundary-layer set of equations (6.59), (6.69), and (6.71) as $y \to \infty$. Their boundary conditions at $y = 0$ are given by Eqs. (6.99a) and (6.99c). The higher-order normal velocity $\hat{v}_{jV2}n_j$ in Eqs. (6.69) and (6.71) is replaced by Eq. (6.108). Equations (6.59), (6.69), and (6.71) with the supplementary equations (6.37d) and (6.108) and the boundary conditions at $y = 0$ and at infinity determine $\hat{\rho}_{V1}$, $\hat{v}_{iV1}t_i^{(\alpha)}$, $\hat{p}_{V1}$, and $\hat{T}_{V1}$, thus all the macroscopic variables at the order of ε.

[8](a) A simpler example of this type of boundary-value problem for the linearized Euler equations is found in the analysis of a uniform flow of an inviscid gas past a thin body (see, e.g., Liepmann & Roshko [1957]). (b) If $(\hat{v}_{ih0})_0\, n_i > 0$, specification of the normal component $\hat{v}_{ih1}n_i$ only is not sufficient. We will give a simple explanatory example of this difference in Appendix A.10.

6.7 Discussions

We have obtained a system of equations and boundary conditions that describes a flow of a slightly rarefied gas with a finite Mach number up to the order of the square root of the Knudsen number and described the features of the system and the procedure of solution of the system. The viscous boundary-layer equations obtained for the three dimensional case, however, are complicated and their character is not transparent. Thus, we here list the equations and boundary conditions in the two-dimensional case and give supplementary explanation. In the following discussion, the quantities are assumed to be uniform along the χ_2 coordinate, and χ_1, which is orthogonal to χ_2 and is taken in such a way that $\chi_{1,1} = 1$, is simply denoted by χ. For simplicity, $\hat{v}_{iVm}t_i^{(1)}$ and $\hat{v}_{iVm}n_i$ are denoted, respectively, by u_m and v_m. Generally, the order of expansion is denoted by the subscript of each quantity, and the subscripts V, α, and β and the superscripts (α) and (β) are omitted (e.g., $t_i^{(1)} \to t_i$, $\kappa_1 \to \kappa$, $\kappa_2 \to 0$, $g_\alpha \to 0$, $\vartheta_\alpha \to 0$). The viscous boundary-layer equations and their boundary conditions obtained in Sections 6.3–6.5 are simplified as follows. The leading-order equations are

$$v_0 = 0, \qquad \hat{p}_0 = \hat{p}_0(\chi) = (\hat{p}_{h0})_0, \tag{6.120}$$

$$\hat{\rho}_0\left(u_0\frac{\partial u_0}{\partial \chi} + v_1\frac{\partial u_0}{\partial y}\right) = -\frac{1}{2}\frac{\mathrm{d}\hat{p}_0}{\mathrm{d}\chi} + \frac{1}{2}\frac{\partial}{\partial y}\left(\hat{\gamma}_1\hat{T}_0^{1/2}\frac{\partial u_0}{\partial y}\right), \tag{6.121}$$

$$\frac{3}{2}\hat{\rho}_0\left(u_0\frac{\partial \hat{T}_0}{\partial \chi} + v_1\frac{\partial \hat{T}_0}{\partial y}\right) = -\hat{p}_0\left(\frac{\partial u_0}{\partial \chi} + \frac{\partial v_1}{\partial y}\right) + \hat{\gamma}_1\hat{T}_0^{1/2}\left(\frac{\partial u_0}{\partial y}\right)^2 + \frac{5}{4}\frac{\partial}{\partial y}\left(\hat{\gamma}_2\hat{T}_0^{1/2}\frac{\partial \hat{T}_0}{\partial y}\right), \tag{6.122}$$

where

$$\hat{\rho}_0 = \hat{p}_0/\hat{T}_0, \tag{6.123a}$$

$$v_1 = -\frac{1}{\hat{\rho}_0}\int_0^y \frac{\partial \hat{\rho}_0 u_0}{\partial \chi}\mathrm{d}y. \tag{6.123b}$$

Their boundary conditions are

$$u_0 = \hat{v}_{wi}t_i, \qquad \hat{T}_0 = \hat{T}_w \qquad \text{at } y = 0, \tag{6.124}$$

$$u_0 \sim (\hat{v}_{ih0})_0 t_i, \qquad \hat{T}_0 \sim (\hat{T}_{h0})_0 \qquad \text{as } y \to \infty. \tag{6.125}$$

The $(\hat{p}_{h0})_0$ in Eq. (6.120) and $(\hat{v}_{ih0})_0 t_i$ and $(\hat{T}_{h0})_0$ in Eq. (6.125) are given by the solution of the Euler set of equations (6.21)–(6.23) under the boundary condition $(\hat{v}_{ih0})_0 n_i = v_0(y \to \infty) = 0$.

The next-order equations are

$$\hat{\rho}_0 u_0 \frac{\partial u_1}{\partial \chi} + (\hat{\rho}_0 u_1 + \hat{\rho}_1 u_0)\frac{\partial u_0}{\partial \chi} + \hat{\rho}_0 v_1 \frac{\partial u_1}{\partial y}$$
$$+ (\hat{\rho}_0 v_2 + \hat{\rho}_1 v_1)\frac{\partial u_0}{\partial y} - \kappa \hat{\rho}_0 u_0 v_1 + \kappa y \hat{\rho}_0 u_0 \frac{\partial u_0}{\partial \chi}$$
$$= -\frac{1}{2}\left(\frac{\partial \hat{p}_1}{\partial \chi} + \kappa y \frac{\partial \hat{p}_0}{\partial \chi}\right)$$
$$+ \frac{1}{2}\frac{\partial}{\partial y}\left[\hat{\gamma}_1 \hat{T}_0^{1/2}\left(\frac{\partial u_1}{\partial y} + \kappa u_0\right) + \hat{T}_1 \frac{\mathrm{d}\hat{\gamma}_1 \hat{T}_0^{1/2}}{\mathrm{d}\hat{T}_0}\frac{\partial u_0}{\partial y}\right]$$
$$- \kappa \hat{\gamma}_1 \hat{T}_0^{1/2}\frac{\partial u_0}{\partial y}, \tag{6.126}$$

$$\frac{\partial \hat{p}_1}{\partial y} = -2\kappa \hat{\rho}_0 {u_0}^2, \tag{6.127}$$

$$\frac{3}{2}\hat{\rho}_0 v_1 \frac{\partial \hat{T}_1}{\partial y} + \frac{3}{2}(\hat{\rho}_0 v_2 + \hat{\rho}_1 v_1)\frac{\partial \hat{T}_0}{\partial y}$$
$$+ \frac{3}{2}\hat{\rho}_0 u_0 \frac{\partial \hat{T}_1}{\partial \chi} + \frac{3}{2}(\hat{\rho}_0 u_1 + \hat{\rho}_1 u_0)\frac{\partial \hat{T}_0}{\partial \chi} + \frac{3}{2}\kappa y \hat{\rho}_0 u_0 \frac{\partial \hat{T}_0}{\partial \chi}$$
$$= -\hat{p}_0\left[\frac{\partial v_2}{\partial y} + \frac{\partial u_1}{\partial \chi} + \kappa\left(y\frac{\partial u_0}{\partial \chi} - v_1\right)\right] - \hat{p}_1\left(\frac{\partial v_1}{\partial y} + \frac{\partial u_0}{\partial \chi}\right)$$
$$+ \left[2\hat{\gamma}_1 \hat{T}_0^{1/2}\left(\frac{\partial u_1}{\partial y} + \kappa u_0\right) + \hat{T}_1 \frac{\mathrm{d}\hat{\gamma}_1 \hat{T}_0^{1/2}}{\mathrm{d}\hat{T}_0}\frac{\partial u_0}{\partial y}\right]\frac{\partial u_0}{\partial y}$$
$$+ \frac{5}{4}\frac{\partial}{\partial y}\left[\hat{\gamma}_2 \hat{T}_0^{1/2}\frac{\partial \hat{T}_1}{\partial y} + \hat{T}_1 \frac{\mathrm{d}\hat{\gamma}_2 \hat{T}_0^{1/2}}{\mathrm{d}\hat{T}_0}\frac{\partial \hat{T}_0}{\partial y}\right] - \frac{5}{4}\kappa \hat{\gamma}_2 \hat{T}_0^{1/2}\frac{\partial \hat{T}_0}{\partial y}, \tag{6.128}$$

where $\hat{\rho}_1$ and v_2 are given by

$$\hat{\rho}_1 \hat{T}_0 = \hat{p}_1 - \hat{\rho}_0 \hat{T}_1, \tag{6.129a}$$

$$v_2 = -\frac{\hat{\rho}_1 v_1}{\hat{\rho}_0} - \frac{1}{\hat{\rho}_0}\int_0^y \left(\frac{\partial(\hat{\rho}_0 u_1 + \hat{\rho}_1 u_0)}{\partial \chi} - \kappa \hat{\rho}_0 v_1 + \kappa y \frac{\partial \hat{\rho}_0 u_0}{\partial \chi}\right) \mathrm{d}y. \tag{6.129b}$$

The boundary conditions at $y = 0$ are

$$u_1 = -\frac{\hat{k}_0}{\hat{\rho}_0}\frac{\partial u_0}{\partial y}, \quad \hat{T}_1 = \frac{\hat{d}_1}{\hat{\rho}_0}\frac{\partial \hat{T}_0}{\partial y}. \tag{6.130}$$

The conditions at infinity are

$$\hat{\rho}_1 \sim (\hat{\rho}_{h1})_0 + yn_j \left(\frac{\partial \hat{\rho}_{h0}}{\partial x_j}\right)_0 \qquad \text{as } y \to \infty, \tag{6.131a}$$

$$u_1 \sim (\hat{v}_{ih1})_0 t_i + yn_j \left(\frac{\partial \hat{v}_{ih0}}{\partial x_j}\right)_0 t_i \quad \text{as } y \to \infty, \tag{6.131b}$$

$$\hat{T}_1 \sim \left(\hat{T}_{h1}\right)_0 + yn_j \left(\frac{\partial \hat{T}_{h0}}{\partial x_j}\right)_0 \qquad \text{as } y \to \infty. \tag{6.131c}$$

The data on the right-hand sides are given by the solution of the Euler set of equations (6.24)–(6.26) under the boundary condition

$$(\hat{v}_{ih1})_0 n_i = -\int_0^\infty \left[\left(\frac{\partial \hat{v}_{ih0}}{\partial x_j}\right)_0 n_i n_j + \frac{1}{(\hat{\rho}_{h0})_0}\frac{\partial \hat{\rho}_0 u_0}{\partial \chi}\right] \mathrm{d}y. \tag{6.132}$$

The viscous boundary-layer equations (6.121)–(6.123b) and boundary conditions (6.124) and (6.125) at the leading order are the same as those for the Navier–Stokes equations for a compressible gas (the compressible gas version of the Prandtl boundary-layer equations and their boundary conditions). Equations (6.126)–(6.129b) for the next-order variables also do not contain the non-Navier–Stokes stress and heat-flow terms. The term containing $\hat{T}_1 \mathrm{d}\hat{\gamma}_1 \hat{T}_0^{1/2}/\mathrm{d}\hat{T}_0$ or $\hat{T}_1 \mathrm{d}\hat{\gamma}_2 \hat{T}_0^{1/2}/\mathrm{d}\hat{T}_0$ is due to the fact that the viscosity or thermal conductivity of the gas depends on its temperature. These equations are derived from the Navier–Stokes set of equations for a compressible gas where the coordinate normal to the boundary is stretched by the factor of the square root $\sqrt{\mathrm{Re}}$ of the Reynolds number, by a power series expansion in $\mathrm{Re}^{-1/2}$ [Note the relation (3.95), which depends on the stress formula (6.73b) in converting the mean free path to the viscosity, for the transfer from the $\mathrm{Re}^{-1/2}$-expansion to ε-expansion]. The result does not support the claim by Darrozes [1969] that the boundary-layer equations describing the leading effect of gas rarefaction (or $\hat{\rho}_1$, u_1, v_1, $\hat{p}_1$, and $\hat{T}_1$) should contain a non-Navier–Stokes stress term. (In his paper no explicit equations are given.) Owing to special anisotropic character of the viscous boundary-layer equations, a higher-order quantity, or v_2, which is expressed by lower-order quantities, enters the equations that determine the behavior of $\hat{\rho}_1$, u_1, $\hat{p}_1$, and $\hat{T}_1$. However, in view of the fact that its boundary value $(v_2)_0$ vanishes owing to the displacement effect of the Knudsen layer being of higher order, the contributions up to the order of $\sqrt{\mathrm{Kn}}$ are included in the system of the Navier–Stokes equations (in the nonexpanded original form) and the (correspondingly rearranged) slip conditions consisting of tangential velocity slip due to the shear of flow and temperature jump due to the temperature gradient normal to the boundary.

In the analysis of the viscous boundary layer, we did not follow the method of our foregoing analyses of this type of problem, that is, we did not split the solution into two parts: the Hilbert solution and its correction. This is because the viscous boundary-layer equation is usually obtained in a form without splitting and we want to keep the resemblance here. We can, of course, carry out the analysis in the split form: $\hat{f}_V = \hat{f}_h + \hat{f}_{VC}$, where $\hat{f}_{VC}$ is the viscous boundary-layer correction to the Hilbert solution and is assumed to decay rapidly away from the boundary. Correspondingly, a macroscopic variable $\hat{h}_V$ is split as

$$\hat{h}_{V0} = (\hat{h}_{h0})_0 + \hat{h}_{VC0}, \quad \hat{h}_{V1} = (\hat{h}_{h1})_0 + yn_i(\partial\hat{h}_{h0}/\partial x_i)_0 + \hat{h}_{VC1}, \quad \cdots,$$

where $\hat{h}$ represents $\hat{\rho}$, $\hat{v}_i$, etc. The analysis of $\hat{f}_{VC}$ is carried out in a similar way to that of $\hat{f}_V$. The equations that determine the behavior of $\hat{h}_{VC0}$, $\hat{h}_{VC1}$, etc. are naturally those obtained by substituting the above relations into the equations for $\hat{h}_{V0}$, $\hat{h}_{V1}$, etc. An equation for $\hat{h}_{VCm}$ is denoted by the superscript * on the number of its corresponding equation. From Eq. (6.54)* and the condition of rapid decay of $\hat{h}_{VCm}$, it is found that $(\hat{v}_{ih0})_0 n_i = 0$ and $\hat{v}_{iVC0}n_i = 0$, the former of which is the boundary condition (6.116) for the Euler set (6.21)–(6.23). From the solution of the Euler set, $(\hat{\rho}_{h0})_0$, $(\hat{v}_{ih0})_0 t_i^{(\alpha)}$, $(\hat{T}_{h0})_0$, and $(\hat{p}_{h0})_0$ are determined. From Eq. (6.55)* and the condition of rapid decay of $\hat{p}_{VC0}$, it is found that $\hat{p}_{VC0} = 0$. Equations (6.58)* and (6.60)* with the supplementary equations (6.118)* and (6.36d)* are the equations for the correction functions $\hat{\rho}_{VC0}$, $\hat{v}_{iVC0}t_i^{(\alpha)}$, $\hat{v}_{iVC1}n_i$, and $\hat{T}_{VC0}$; Eq. (6.50)* and the condition of rapid decay of $\hat{h}_{VC0}$ as $y \to \infty$ are their boundary conditions. This system of equations and boundary conditions contains an undetermined boundary value $(\hat{v}_{ih1})_0 n_i$, as well as the determined data $(\hat{\rho}_{h0})_0$, $(\hat{v}_{ih0})_0 t_i^{(\alpha)}$, $(\partial\hat{v}_{ih0}/\partial x_j)_0 n_i n_j$, $(\hat{T}_{h0})_0$, and $(\hat{p}_{h0})_0$ (and their derivatives with respect to χ_1 or χ_2). The undetermined data $(\hat{v}_{ih1})_0 n_i$ should be a special value for the system to have a solution of rapid decay. That is, taking the limit as $y \to \infty$ of Eq. (6.118)* with $y(\hat{\rho}_{h0}\partial\hat{v}_{ih0}/\partial x_j)_0 n_i n_j$ term shifted to the right-hand side and taking into account the condition of rapid decay, then we obtain Eq. (6.119)*. (The condition does not generally assure the existence of a solution.) This is the boundary condition for the linearized Euler set (6.24)–(6.26). The discussion for further steps can be carried out in a similar way. In this procedure, the connection conditions are replaced by the condition of rapid decay, and the discussion may be more comprehensible. The two ways of analysis are equivalent.

We have not mentioned the shock layer (Grad [1969]). A thin layer with sharp variation across it may appear inside the gas region when a local Mach number in the flow is larger than unity. It is not difficult to incorporate the layer in the present solution if the layer is outside the viscous boundary layer (or in the Euler region). Up to the first order of ε, in the same way as in the classical gas dynamics, the layer may be taken as a discontinuity, and the regions across the discontinuity have only to be connected by the shock wave condition (Rankine–Hugoniot relation). The internal structure of the layer is expressed by that of the plane shock wave (Grad [1969], Caflisch & Nicolaenko [1982]).[9] The connection condition being given by the shock wave condition, the analysis of the layer is unnecessary for the overall behavior of the gas in

[9]The structure of a plane shock wave to which the flow is normal (normal shock wave) is obtained as the solution of the one-dimensional Boltzmann equation in an infinite domain, that approaches uniform states (Maxwellians) at upstream and downstream infinities. The upstream and downstream states (density, velocity, and temperature) are related for the so-

contrast to the Knudsen-layer analysis, from which the slip conditions (6.99a)–(6.99c) are derived. (Incidentally, the boundary condition on an interface of a gas and its condensed phase for the fluid-dynamic-type equations derived by the Knudsen-layer analysis in the foregoing and following chapters affects the overall behavior of the gas even in the continuum limit.) A higher-order correction (the effect of the curvature of the layer) of the connection condition is discussed by Cercignani & Lancellotti [1991].

In the present study, we do not consider the case where the boundary is the condensed phase of a gas and evaporation or condensation takes place there. The case where the Mach number of the gas flow evaporating from or condensing onto the boundary is of the order of unity will be studied in the next chapter. Then, the fluid-dynamic region governed by the Euler set of equations is directly connected to a Knudsen layer. This Knudsen layer, however, is governed by the one-dimensional original (or nonlinear) Boltzmann equation. When the Mach number of evaporating or condensing flow is of the order of the square root of the Knudsen number but the Mach number of the flow along the boundary is of the order of unity, the analysis can be carried out in a similar way to that in the present case. The difference comes in the slip condition and Knudsen layer of the ε-order term [thus, also in Eqs. (6.108) and (6.118)]. In a flow with its Mach number of the order of unity, however, the pressure variation (divided by pressure) along the boundary is of the order of unity except in special cases, and thus the difference between the gas pressure and the saturated gas pressure at the temperature of the condensed phase (divided by pressure) is generally of the order of unity, which induces an evaporating or condensing flow with its Mach number of the order of unity. Thus, the situation just considered does not generally occur except in cases with symmetry such as in the plane or cylindrical Couette flow.

In the present analysis, we considered a flow of a finite Mach number and obtained the Euler or viscous boundary-layer equations for a compressible gas as the fluid-dynamic-type equations. The analysis for a flow of small Mach number and temperature variation but of a large Reynolds number can be carried out with small modification, and the Euler set of equations and "the set of viscous boundary-layer equations" for an incompressible gas[10] are derived as the leading-order system. The situation where the Mach number Ma is of the order of $\mathrm{Kn}^{n/m}$ [m and n $(< m)$: positive integers] and the variation of tem-

lution to exist (shock condition). The shock condition is obtained, without analyzing the Boltzmann equation, simply by integration of the conservation equations (2.11)–(2.13) with $\partial/\partial t = 0$ and $F_i = 0$ [or (2.51)–(2.53) with $\partial/\partial \hat{t} = 0$ and $\hat{F}_i = 0$] or more simply by application of the conservation of mass, momentum, and energy fluxes between the two uniform (Maxwellian) states. The thickness of the shock wave is of the order of the mean free path. The solution of a shock wave to which the flow is oblique (oblique shock wave) is obtained by simply translating the solution of a normal shock wave along the wave surface with an appropriate velocity. The effect of the curvature (or nonuniformity of the strength) of the shock layer in the solution in the Euler region appears at the order of ε^2 (or at the order of the Knudsen number, more precisely, based on the curvature as the reference length).

[10]There is a slight difference of the energy equation of the set of viscous boundary-layer equations from that for an incompressible fluid. See Footnote 6 in Section 4.3.

perature of the boundary is the same order as the Mach number is considered. Then, the leading Maxwellian distribution $\hat{f}_{h0}$ or $\hat{f}_{V0}$ is a uniform state at rest (or $\hat{\rho}_{h0}$ and $\hat{T}_{h0}$ are constant and $\hat{v}_{ih0} = 0$, or $\hat{\rho}_{V0}$ and $\hat{T}_{V0}$ are constant and $\hat{v}_{iV0} = 0$), and a nonuniform term starts at the order of $\mathrm{Kn}^{n/m}$. The Reynolds number Re ($\propto$ Ma/Kn) is of the order of $\mathrm{Kn}^{-(m-n)/m}$. Thus, the thickness of the viscous boundary layer may be considered to be of the order of $\mathrm{Kn}^{(m-n)/2m}L$; correspondingly, the stretching factor of the viscous boundary-layer coordinate is $\mathrm{Kn}^{-(m-n)/2m}$. With these differences in mind and taking $\mathrm{Kn}^{1/2m}$ as the expansion parameter instead of ε (or $\mathrm{Kn}^{1/2}$), the analysis (the Euler region, the viscous boundary layer, the Knudsen layer, and their connection) can be carried out in a similar way to that in the present chapter.

In the two-dimensional Couette flow, where the physical variables depend only on one spatial coordinate, the length scale of variation of variables is different in directions: the direction normal to the boundary and the directions parallel to the boundary. Let L_C be the width of the channel (or the distance between the two boundaries). With X_i the space coordinates, an appropriate nondimensional coordinate system $(\tilde{y}, \tilde{\chi}_1, \tilde{\chi}_2)$ and Knudsen number Kn_C (or the corresponding k, say k_C) are

$$X_i = \tilde{y}L_C n_i + X_{wi}(\tilde{\chi}_1, \tilde{\chi}_2), \qquad \mathrm{Kn}_C = \frac{\ell_0}{L_C} \quad \left(\text{or } k_C = \frac{\sqrt{\pi}}{2}\frac{\ell_0}{L_C}\right),$$

and the physical variables are independent of $\tilde{\chi}_1$ and $\tilde{\chi}_2$. The analysis of the viscous boundary layer, where the length scale of variation of variables depends on directions, can be applied to the asymptotic analysis for small Knudsen numbers Kn_C of this system. That is, L_C corresponds to εL in the boundary-layer analysis, and therefore, k_C corresponds to ε, since $k_C = \sqrt{\pi}\ell_0/2L_C \to \sqrt{\pi}\ell_0/2\varepsilon L = k/\varepsilon = \varepsilon$. Thus, with these correspondences, the viscous boundary-layer equations with the slip boundary conditions and Knudsen-layer corrections, where the derivatives with respect to χ_1 and χ_2 are eliminated, determine the behavior of the system. The result shows that the viscous effect spreads over the whole field (or the Euler region is absent) in the Couette flow for large Reynolds numbers. This is understood in the following way: in this degenerate situation, the convection terms of the conservation equations degenerate and the viscous effect spreads over the whole flow field. Similarly, in the cylindrical Couette flow between coaxial circular cylinders, the convection terms degenerate, except the term due to centripetal acceleration in the momentum equation in the radial direction, and the viscous effect spreads over the whole field. For the BKW equation and diffuse reflection, the higher-order systems of the fluid-dynamic equations, slip conditions, and Knudsen-layer corrections are discussed in Sone & Yamamoto [1970] and Sone & Onishi [1977].

The result discussed above is for a simple boundary. When condensation or evaporation is taking place on a boundary, the situation is different. The (weak or strong) convecting effect of evaporating or condensing flow will be discussed in the next two chapters.[11]

[11] Another interesting example is given in Sone, Takata & Sugimoto [1996].

Chapter 7

Nonlinear Theory III – Finite Speed of Evaporation and Condensation

In the preceding chapter, we have discussed a flow with a finite Mach number around a simple boundary, across which there is no mass flux. In the present chapter we consider a flow of a gas around its condensed phase where evaporation or condensation with speed of a finite Mach number is taking place. Owing to the convection effect of condensing or evaporating flow of the order of the Mach number, the viscous boundary layer, which appears in a flow with a finite Mach number around a simple boundary (Chapter 6), shrinks to merge into the Knudsen layer over a condensing boundary, or the layer spreads into the whole flow field (or the Euler region) over an evaporating boundary and the viscous effect is reduced to a secondary one there. The Euler region is connected directly to the Knudsen layer. Instead, the Knudsen layer is governed by a nonlinear equation in contrast to that in the foregoing chapters.

7.1 Problem

Consider a system composed of a gas and its condensed phase of smooth shape, on the surface of which evaporation or condensation is taking place. No external force is acting on the gas [or F_i in Eq. (2.5) is zero], and the state of the gas is steady (or time-independent). The Mach number of the speed of evaporation and condensation of the gas on the interface of the gas and its condensed phase is of the order of unity. We will investigate the asymptotic behavior of the gas for small Knudsen numbers of the system. First, we discuss the problem under the complete condensation boundary condition and then extend it to a more general boundary condition (Section 7.6).

With the notation in Section 2.9, the basic equation [Eq. (2.40a)] is

$$\zeta_i \frac{\partial \hat{f}}{\partial x_i} = \frac{1}{k} \hat{J}(\hat{f}, \hat{f}), \tag{7.1}$$

and the complete condensation condition [Eq. (2.56a) with $\alpha_c = 1$] is

$$\hat{f}(x_i, \zeta_i) = \frac{\hat{p}_w}{\pi^{3/2} \hat{T}_w^{5/2}} \exp\left(-\frac{(\zeta_i - \hat{v}_{wi})^2}{\hat{T}_w}\right) \quad (\zeta_j n_j > 0), \tag{7.2}$$

where the boundary parameters $\hat{v}_{wi}$ ($\hat{v}_{wi} n_i = 0$), $\hat{T}_w$, and $\hat{p}_w$ are, for simplicity, assumed to be independent of k in this chapter.

The asymptotic behavior of the solution $\hat{f}$ of the boundary-value problem (7.1) with (7.2) for small k (or $k \ll 1$) is studied without restriction on the size of the variables on the basis of Aoki & Sone [1991].

7.2 Hilbert solution

First we look for a moderately varying solution $\hat{f}_H$ (Hilbert solution or expansion), whose length scale of variation is of the order of the characteristic length L [i.e., $\partial \hat{f}_H / \partial x_i = O(\hat{f}_H)$], in a power series of k without restriction on the size of the variables:

$$\hat{f}_H = \hat{f}_{H0} + \hat{f}_{H1} k + \cdots. \tag{7.3}$$

This is the pioneering work by Hilbert on the solution of the Boltzmann equation (Hilbert [1912]). We have seen that various sets of fluid-dynamic-type equations describing the behavior of a gas in the continuum limit are derived by this principle of expansion, but with various restrictions on the size of the variables.

The relation between the macroscopic variables and the velocity distribution function is the same as Eqs. (2.48a)–(2.48f) except for the subscript H, that is,

$$\hat{\rho}_H = \int \hat{f}_H \mathbf{d}\boldsymbol{\zeta}, \tag{7.4a}$$

$$\hat{\rho}_H \hat{v}_{iH} = \int \zeta_i \hat{f}_H \mathbf{d}\boldsymbol{\zeta}, \tag{7.4b}$$

$$\frac{3}{2} \hat{\rho}_H \hat{T}_H = \int (\zeta_i - \hat{v}_{iH})^2 \hat{f}_H \mathbf{d}\boldsymbol{\zeta}, \tag{7.4c}$$

$$\hat{p}_H = \hat{\rho}_H \hat{T}_H, \tag{7.4d}$$

$$\hat{p}_{ijH} = 2 \int (\zeta_i - \hat{v}_{iH})(\zeta_j - \hat{v}_{jH}) \hat{f}_H \mathbf{d}\boldsymbol{\zeta}, \tag{7.4e}$$

$$\hat{q}_{iH} = \int (\zeta_i - \hat{v}_{iH})(\zeta_j - \hat{v}_{jH})^2 \hat{f}_H \mathbf{d}\boldsymbol{\zeta}. \tag{7.4f}$$

Corresponding to the expansion (7.3), the macroscopic variable $\hat{h}_H$, where $\hat{h}$ represents $\hat{\rho}$, $\hat{v}_i$, $\hat{T}$, etc., is also expanded in k:

$$\hat{h}_H = \hat{h}_{H0} + \hat{h}_{H1} k + \cdots. \tag{7.5}$$

The component function $\hat{h}_{Hm}$ is related to the component function $\hat{f}_{Hn}$ $(n \leq m)$ of the velocity distribution function $\hat{f}_H$ as follows:

$$\hat{\rho}_{H0} = \int \hat{f}_{H0} \mathbf{d}\boldsymbol{\zeta}, \tag{7.6a}$$

$$\hat{\rho}_{H0}\hat{v}_{iH0} = \int \zeta_i \hat{f}_{H0} \mathbf{d}\boldsymbol{\zeta}, \tag{7.6b}$$

$$\frac{3}{2}\hat{\rho}_{H0}\hat{T}_{H0} = \int (\zeta_i - \hat{v}_{iH0})^2 \hat{f}_{H0} \mathbf{d}\boldsymbol{\zeta}, \tag{7.6c}$$

$$\hat{p}_{H0} = \hat{\rho}_{H0}\hat{T}_{H0}, \tag{7.6d}$$

$$\hat{p}_{ijH0} = 2\int (\zeta_i - \hat{v}_{iH0})(\zeta_j - \hat{v}_{jH0}) \hat{f}_{H0} \mathbf{d}\boldsymbol{\zeta}, \tag{7.6e}$$

$$\hat{q}_{iH0} = \int (\zeta_i - \hat{v}_{iH0})(\zeta_j - \hat{v}_{jH0})^2 \hat{f}_{H0} \mathbf{d}\boldsymbol{\zeta}, \tag{7.6f}$$

$$\hat{\rho}_{H1} = \int \hat{f}_{H1} \mathbf{d}\boldsymbol{\zeta}, \tag{7.7a}$$

$$\hat{\rho}_{H0}\hat{v}_{iH1} = \int (\zeta_i - \hat{v}_{iH0}) \hat{f}_{H1} \mathbf{d}\boldsymbol{\zeta}, \tag{7.7b}$$

$$\frac{3}{2}\hat{\rho}_{H0}\hat{T}_{H1} = \int (\zeta_i - \hat{v}_{iH0})^2 \hat{f}_{H1} \mathbf{d}\boldsymbol{\zeta} - \frac{3}{2}\hat{\rho}_{H1}\hat{T}_{H0}, \tag{7.7c}$$

$$\hat{p}_{H1} = \hat{\rho}_{H0}\hat{T}_{H1} + \hat{\rho}_{H1}\hat{T}_{H0}, \tag{7.7d}$$

$$\hat{p}_{ijH1} = 2\int (\zeta_i - \hat{v}_{iH0})(\zeta_j - \hat{v}_{jH0}) \hat{f}_{H1} \mathbf{d}\boldsymbol{\zeta}, \tag{7.7e}$$

$$\hat{q}_{iH1} = \int (\zeta_i - \hat{v}_{iH0})(\zeta_j - \hat{v}_{jH0})^2 \hat{f}_{H1} \mathbf{d}\boldsymbol{\zeta} - \frac{3}{2}\hat{p}_{H0}\hat{v}_{iH1} - \hat{p}_{ijH0}\hat{v}_{jH1}, \tag{7.7f}$$

$$\cdots\cdots\cdots\cdots .$$

Now, return to obtaining the Hilbert solution. Substituting Eq. (7.3) into the Boltzmann equation (7.1) and arranging the same order terms in k, we obtain the following series of integral equations for the component function $\hat{f}_{Hm}$:[1]

$$\hat{J}(\hat{f}_{H0}, \hat{f}_{H0}) = 0, \tag{7.8}$$

$$2\hat{J}(\hat{f}_{H0}, \hat{f}_{Hm}) = \zeta_i \frac{\partial \hat{f}_{Hm-1}}{\partial x_i} - \sum_{r=1}^{m-1} \hat{J}(\hat{f}_{Hr}, \hat{f}_{Hm-r}) \quad (m \geq 1), \tag{7.9}$$

where the $\sum$ term is absent when $m = 1$.

The solution $\hat{f}_{H0}$ of the integral equation (7.8) is given by

$$\hat{f}_{H0} = \frac{\hat{\rho}_{H0}}{\pi^{3/2}\hat{T}_{H0}^{5/2}} \exp\left(-\frac{(\zeta_i - \hat{v}_{iH0})^2}{\hat{T}_{H0}}\right), \tag{7.10}$$

[1] For the BKW equation, the collision integral on the left-hand side in Eq. (7.9) is taken as the collision integral linearized around the Maxwellian $\hat{f}_{H0}$ (see Footnotes 12 in Section 2.9 and 26 in Appendix A.7). See Footnote 1 in Section 4.2 for the meaning of the terms under $\sum$ sign on the right-hand side.

where the relations (7.6a)–(7.6d) are used. With this $\hat{f}_{H0}$, the equation (7.9) is an inhomogeneous linear integral equation for $\hat{f}_{Hm}$ ($m \geq 1$). The homogeneous equation corresponding to Eq. (7.9), i.e.,

$$\hat{J}(\hat{f}_{H0}, \hat{g}) = 0,$$

has five independent solutions:

$$\hat{g} = \hat{f}_{H0} g, \qquad g = 1, \;\; \zeta_i, \;\; \zeta_i^2,$$

which is easily seen with the aid of the relations $g' + g'_* = g + g_*$ and $\hat{f}'_{H0}\hat{f}'_{H0*} = \hat{f}_{H0}\hat{f}_{H0*}$. From the relation $\int g\hat{J}(\hat{f}_{H0}, \hat{f}_{Hm})\mathrm{d}\boldsymbol{\zeta} = 0$ [Eq. (2.44)], the inhomogeneous term of the integral equation (7.9) must satisfy the following relation (solvability condition) for (7.9) to have a solution:

$$\int g\left(\zeta_i \frac{\partial \hat{f}_{Hm-1}}{\partial x_i} - \sum_{r=1}^{m-1} \hat{J}(\hat{f}_{Hr}, \hat{f}_{Hm-r})\right) \mathrm{d}\boldsymbol{\zeta} = 0. \tag{7.11}$$

With the general relation (2.44) of the collision integral, the above solvability condition is reduced to

$$\int g\zeta_i \frac{\partial \hat{f}_{Hm-1}}{\partial x_i} \mathrm{d}\boldsymbol{\zeta} = 0. \tag{7.12}$$

For $m = 1$, $\zeta_i \partial \hat{f}_{Hm-1}/\partial x_i$ in Eq. (7.12) is given by

$$\zeta_i \frac{\partial \hat{f}_{H0}}{\partial x_i} = \zeta_i \hat{f}_{H0} \times \left[\frac{1}{\hat{\rho}_{H0}} \frac{\partial \hat{\rho}_{H0}}{\partial x_i} + \frac{2(\zeta_j - \hat{v}_{jH0})}{\hat{T}_{H0}} \frac{\partial \hat{v}_{jH0}}{\partial x_i} + \left(\frac{(\zeta_j - \hat{v}_{jH0})^2}{\hat{T}_{H0}} - \frac{3}{2}\right) \frac{1}{\hat{T}_{H0}} \frac{\partial \hat{T}_{H0}}{\partial x_i}\right].$$

Thus, the solvability condition (7.12) for $m = 1$ is reduced to the set of equations

$$\frac{\partial \hat{\rho}_{H0}\hat{v}_{jH0}}{\partial x_j} = 0, \tag{7.13}$$

$$\frac{\partial \hat{\rho}_{H0}\hat{v}_{iH0}\hat{v}_{jH0}}{\partial x_j} + \frac{1}{2}\frac{\partial \hat{p}_{H0}}{\partial x_i} = 0, \tag{7.14}$$

$$\frac{\partial}{\partial x_j}\left[\hat{\rho}_{H0}\hat{v}_{jH0}\left(\frac{5}{2}\hat{T}_{H0} + \hat{v}_{iH0}^2\right)\right] = 0, \tag{7.15}$$

which is equivalent to the Euler set of equations for an ideal gas. We can proceed with the analysis in a similar way.

7.3 Knudsen layer

The Hilbert solution cannot be made to satisfy the complete condensation condition even at the leading order when the gas is evaporating or condensing at a finite speed [$\hat{v}_i n_i = O(1)$ or $\hat{v}_{iH0} n_i \neq 0$], because the Maxwellian $\hat{f}_{H0}$ with nonzero velocity ($\hat{v}_{iH0} n_i \neq 0$) cannot agree with the boundary condition (7.2) with $\hat{v}_{wi} n_i = 0$. As in Chapters 3–5, we try to construct the solution by introducing a Knudsen-layer correction directly to the Hilbert solution.[2] That is, the solution of the boundary-value problem is put in the form

$$\hat{f} = \hat{f}_H + \hat{f}_K, \tag{7.16}$$

where the correction $\hat{f}_K$ is a rapidly varying function with its length scale of variation of the order of the mean free path in the direction normal to the surface of the condensed phase [i.e., $k n_i \partial \hat{f}_K / \partial x_i = O(\hat{f}_K)$] and is appreciable only in a thin layer, with thickness of the order of the mean free path, adjacent to the condensed phase (Knudsen-layer part). Substituting Eq. (7.16) into the Boltzmann equation (7.1) and taking into account that the Hilbert solution $\hat{f}_H$ itself satisfies the Boltzmann equation (7.1), then we find that the equation for $\hat{f}_K$ is given as[3]

$$\zeta_i \frac{\partial \hat{f}_K}{\partial x_i} = \frac{1}{k}[2\hat{J}(\hat{f}_H, \hat{f}_K) + \hat{J}(\hat{f}_K, \hat{f}_K)]. \tag{7.17}$$

Here the same Knudsen-layer variables (η, χ_1, χ_2) as in Section 3.4 are introduced, i.e.,

$$x_i = k\eta n_i(\chi_1, \chi_2) + x_{wi}(\chi_1, \chi_2), \tag{7.18}$$

where x_{wi} is the boundary surface, η is a stretched coordinate normal to the boundary, χ_1 and χ_2 are (unstretched) coordinates within a parallel surface $\eta = \text{const}$, and the normal vector n_i is a function of χ_1 and χ_2. Then, Eq. (7.17) is rewritten as

$$\zeta_i n_i \frac{\partial \hat{f}_K}{\partial \eta} = 2\hat{J}(\hat{f}_H, \hat{f}_K) + \hat{J}(\hat{f}_K, \hat{f}_K) - k\zeta_i \left(\frac{\partial \chi_1}{\partial x_i} \frac{\partial \hat{f}_K}{\partial \chi_1} + \frac{\partial \chi_2}{\partial x_i} \frac{\partial \hat{f}_K}{\partial \chi_2} \right). \tag{7.19}$$

The Knudsen-layer correction $\hat{f}_K$ is expanded in a power series of k, i.e.,

$$\hat{f}_K = \hat{f}_{K0} + \hat{f}_{K1} k + \cdots, \tag{7.20}$$

---

[2]In contrast to the analysis of the preceding chapter, owing to the convection effect of condensing or evaporating flow with a finite Mach number, the viscous boundary layer with thickness of the order of the square root of the Knudsen number, i.e., $O(k^{1/2}L)$ shrinks to merge into the Knudsen layer over a condensing boundary or spreads into the whole flow field over an evaporating boundary to reduce the viscous effect to a secondary one in the Euler region. If the viscous boundary-layer analysis in Section 6.3 is carried out under the assumption of a finite normal velocity $\hat{v}_{iV0} n_i$, it is found that the flow variables $\hat{v}_{iV0}, \hat{T}_{V0}$, etc. are independent of the viscous boundary-layer variable y. That is, their length scale of variation is much larger (and is found to be of the order of L, with more precise discussion, which means that the layer is expressed by the Euler solution). A solution with length scale of variation of the order of $k^{1/2}L$ is absent.

[3]For the BKW equation, $2\hat{J}(\hat{f}_H, \hat{f}_K) + \hat{J}(\hat{f}_K, \hat{f}_K)$ on the right-hand side should be understood as

$$\hat{J}(\hat{f}_H + \hat{f}_K, \hat{f}_H + \hat{f}_K) - \hat{J}(\hat{f}_H, \hat{f}_H),$$

(see Footnote 12 in Section 2.9).

where the series starts from the order of unity, because the correction is required at the leading order as explained at the beginning of this section.[4] Substituting the series (7.20) and the reshuffled expansion of $\hat{f}_H$, i.e.,

$$\hat{f}_H = (\hat{f}_{H0})_0 + \left[(\hat{f}_{H1})_0 + \left(\frac{\partial \hat{f}_{H0}}{\partial x_i}\right)_0 n_i\eta\right] k + \cdots, \tag{7.21}$$

where the quantities in the parentheses with subscript 0, i.e., $(\cdots)_0$, are evaluated on the interface, into Eq. (7.19), we obtain a series of equations for $\hat{f}_{Km}$. The equation for the leading term $(\hat{f}_{H0})_0 + \hat{f}_{K0}$ of the sum of the reshuffled Hilbert expansion (7.21) and the expansion (7.20) of the Knudsen-layer correction is given by the one-dimensional Boltzmann equation

$$\zeta_i n_i \frac{\partial \hat{f}_{HK0}}{\partial \eta} = \hat{J}(\hat{f}_{HK0}, \hat{f}_{HK0}), \tag{7.22}$$

where

$$\hat{f}_{HK0} = (\hat{f}_{H0})_0 + \hat{f}_{K0}. \tag{7.23}$$

The boundary condition for $\hat{f}_{Km}$ at $\eta = 0$ is, for $\zeta_i n_i > 0$,

$$\hat{f}_{K0} = \frac{\hat{p}_w}{\pi^{3/2}\hat{T}_w^{5/2}} \exp\left(-\frac{(\zeta_i - \hat{v}_{wi})^2}{\hat{T}_w}\right) - \frac{(\hat{p}_{H0})_0}{\pi^{3/2}(\hat{T}_{H0})_0^{5/2}} \exp\left(-\frac{[\zeta_i - (\hat{v}_{iH0})_0]^2}{(\hat{T}_{H0})_0}\right), \tag{7.24a}$$

$$\hat{f}_{Km} = -\hat{f}_{Hm} \quad (m \geq 1), \tag{7.24b}$$

and the condition at infinity is

$$\hat{f}_{Km} \to 0 \quad \text{as } \eta \to \infty. \tag{7.25}$$

Thus, the boundary condition for $\hat{f}_{HK0}$ is

$$\hat{f}_{HK0} = \frac{\hat{p}_w}{\pi^{3/2}\hat{T}_w{}^{5/2}} \exp\left(-\frac{(\zeta_i - \hat{v}_{wi})^2}{\hat{T}_w}\right) \quad (\zeta_i n_i > 0) \quad \text{at } \eta = 0, \tag{7.26a}$$

$$\hat{f}_{HK0} \to \frac{(\hat{p}_{H0})_0}{\pi^{3/2}(\hat{T}_{H0})_0^{5/2}} \exp\left(-\frac{[\zeta_i - (\hat{v}_{iH0})_0]^2}{(\hat{T}_{H0})_0}\right) \quad \text{as } \eta \to \infty. \tag{7.26b}$$

The Knudsen layer in the present problem is governed by a nonlinear equation (the original one-dimensional Boltzmann equation) in contrast to the Knudsen layers in the foregoing chapters. Any mathematical theory corresponding to Bardos's theorem for the linearized Boltzmann equation is not available to the nonlinear equation. However, in view of Bardos's theorem, it is expected

[4]We have already seen in Chapters 3 and 4 that the Knudsen-layer correction is required at the leading order when evaporation or condensation is taking place.

in the nonlinear problem that some relations among the parameters $\hat{p}_w$, $\hat{T}_w$, $\hat{v}_{wi}$, $(\hat{p}_{H0})_0$, $(\hat{T}_{H0})_0$, and $(\hat{v}_{iH0})_0$ must be satisfied for the half-space problem to have a solution. These relations serve as the boundary condition on an interface of the gas and its condensed phase for the Euler set of equations [i.e., Eqs. (7.13)–(7.15)]. In the continuum limit (or the limit $k \to 0$), the Knudsen layer is flattened on the condensed phase, but the Knudsen layer must be analyzed to obtain the relations as the boundary condition for the Euler set on the interface. We will discuss the solution of this half-space problem in the next section.

7.4 Half-space problem of evaporation and condensation

The solution of the half-space problem in the preceding section [i.e., Eqs. (7.22), (7.26a), and (7.26b)] is obtained, by a simple transformation and an appropriate choice of reference quantities, from the solution of the following half-space problem of evaporation and condensation: A semi-infinite expanse ($X_1 > 0$) of a gas is bounded by its plane condensed phase at $X_1 = 0$; the plane condensed phase is at rest and is kept at a uniform temperature T_w; the gas far away from the condensed phase is in the uniform state with pressure p_∞, temperature T_∞, and velocity $(v_{1\infty}, v_{2\infty}, 0)$; and the steady behavior of the gas evaporating from or condensing onto the condensed phase (depending on $v_{1\infty} > 0$ or $v_{1\infty} < 0$) is to be studied. With the choice of the reference quantities as $T_0 = T_w$ and $p_0 = p_w$ in the list (2.36), the corresponding mathematical problem is given by the following half-space boundary-value problem: The governing equation is

$$\zeta_1 \frac{\partial \hat{f}}{\partial x} = \hat{J}(\hat{f}, \hat{f}), \tag{7.27}$$

and the boundary conditions for $\hat{f}$ are

$$\hat{f} = \frac{1}{\pi^{3/2}} \exp\left(-\zeta_i^2\right) \quad (\zeta_1 > 0) \quad \text{at } x = 0, \tag{7.28a}$$

$$\hat{f} \to \frac{\hat{p}_\infty}{\pi^{3/2}\hat{T}_\infty^{5/2}} \exp\left(-\frac{(\zeta_1 - \hat{v}_{1\infty})^2 + (\zeta_2 - \hat{v}_{2\infty})^2 + \zeta_3^2}{\hat{T}_\infty}\right) \quad \text{as } x \to \infty, \tag{7.28b}$$

where

$$\hat{p}_\infty = \frac{p_\infty}{p_w}, \quad \hat{T}_\infty = \frac{T_\infty}{T_w}, \quad \hat{v}_{i\infty} = \frac{v_{i\infty}}{(2RT_w)^{1/2}}, \quad x = \frac{2}{\sqrt{\pi}} \frac{X_1}{\ell_w},$$

where ℓ_w is the mean free path in the equilibrium state at rest with pressure p_w and temperature T_w [see, e.g., Eqs. (2.20c) and (2.34)]. We introduce the parameters $M_{\infty 1}$ and $M_{\infty 2}$ defined by

$$M_{\infty 1} = \sqrt{\frac{6T_w}{5T_\infty}} \hat{v}_{1\infty}, \quad M_{\infty 2} = \sqrt{\frac{6T_w}{5T_\infty}} |\hat{v}_{2\infty}|,$$

and use the set of four parameters p_∞/p_w, T_∞/T_w, $M_{\infty 1}$, and $M_{\infty 2}$ instead of p_∞/p_w, T_∞/T_w, $\hat{v}_{1\infty}$, and $\hat{v}_{2\infty}$ to characterize the problem in the following discussions.

To find the overall feature of the sets of parameters that allow the solution of the half-space problem by numerical analysis without a clue like Bardos's theorem requires an extensive study. This was done for the BKW equation in Sone, Aoki & Yamashita [1986], Sone, Aoki, Sugimoto & Yamada [1988], Sone & Sugimoto [1990], Aoki, Sone & Yamada [1990], Aoki, Nishino, Sone & Sugimoto [1991]. The problem is attacked as a time-evolution problem under extensive cases of initial and boundary data. The time-evolution process, especially the long-time behavior of the solution, is examined in detail. From the long-time behavior of the solution together with supplementary time-independent computations, a large number of solutions of the (time-independent) half-space problem are determined together with the data of the parameters (p_∞/p_w, T_∞/T_w, $M_{\infty 1}$, $M_{\infty 2}$) that allow the solutions. By careful variation of the initial and boundary values of the time-evolution analysis, the range of parameters that allow the solution of the (time-independent) half-space problem is determined. The problem is studied analytically for a weak evaporation or condensation ($|M_{\infty 1}| \ll 1$) (Sone [1978]) and for a transonic condensation ($|1 + M_{\infty 1}| \ll 1$) (Sone, Golse, Ohwada & Doi [1998]). Once the feature of the range of existence of solution is clarified, it becomes easier to obtain its approximate solution. The half-space problem was also studied for other molecular models by Kogan & Abramov [1991] and Kryukov [1991], and similar results were obtained.[5]

Summarizing the numerical studies, the half-space problem has a solution only when the following relation or relations among the four parameters are satisfied:

(I) $M_{\infty 1} \geq 0$ (evaporating flow)

(a) $M_{\infty 1} \leq 1$ (subsonic evaporation)

$$p_\infty/p_w = h_1(M_{\infty 1}), \tag{7.29a}$$

$$T_\infty/T_w = h_2(M_{\infty 1}), \tag{7.29b}$$

$$M_{\infty 2} = 0, \tag{7.29c}$$

where the functions $h_1(M_{\infty 1})$ and $h_2(M_{\infty 1})$ depend on molecular models.

(b) $M_{\infty 1} > 1$ (supersonic evaporation)

There is no solution.

(II) $M_{\infty 1} < 0$ (condensing flow)

(a) $M_{\infty 1} > -1$ (subsonic condensation)

$$p_\infty/p_w = F_s(M_{\infty 1}, M_{\infty 2}, T_\infty/T_w), \tag{7.30}$$

where the function $F_s(M_{\infty 1}, M_{\infty 2}, T_\infty/T_w)$ depends on molecular models.

[5] In some works on the half-space problem, the form of the solution is assumed from the first before analysis, which restricts a possible solution. Thus, this method of solution does not serve to determine the range of the parameters where the solution exists; in fact, an incorrect result was derived from this type of analysis.

(b) $M_{\infty 1} < -1$ (supersonic condensation)

$$p_\infty/p_w > F_b(M_{\infty 1}, M_{\infty 2}, T_\infty/T_w), \tag{7.31}$$

where the function $F_b(M_{\infty 1}, M_{\infty 2}, T_\infty/T_w)$ depends on molecular models.

(c) $M_{\infty 1} = -1$ (sonic condensation)

The limiting values of the two functions F_s and F_b agree as $M_{\infty 1} \to -1_\pm$, and the sonic solution exists in the range

$$p_\infty/p_w \geq F_b(-1_-, M_{\infty 2}, T_\infty/T_w) = F_s(-1_+, M_{\infty 2}, T_\infty/T_w). \tag{7.32}$$

To summarize, an evaporation solution exists along a hypercurve in the parameter space $(M_{\infty 1}, M_{\infty 2}, T_\infty/T_w, p_\infty/p_w)$, a subsonic condensation solution exists on a hypersurface in the space, and a supersonic condensation solution exists in a domain of the space. The schematic figure of the range of existence of a solution is shown in Fig. 7.1. The data of the functions h_1, h_2, F_s, and F_b for the BKW equation will be given together with the boundary conditions for the Euler set of equations in the next section. Further discussion of these functions will be made in Section 7.7. The dimension of the range of existence of a solution in the parameter space $(M_{\infty 1}, M_{\infty 2}, T_\infty/T_w, p_\infty/p_w)$ changes from evaporation to condensation and from subsonic condensation to supersonic condensation. The structure of the transition from evaporation to subsonic condensation, studied analytically in Sone [1978], is explained in Section 4.10.1. The structure of the transition from subsonic to supersonic condensation, studied analytically, with detailed exemplifying numerical computation, for the case $M_{\infty 2} = 0$ in Sone, Golse, Ohwada & Doi [1998], is briefly explained in the next paragraph. Incidentally, the studies of the half-space problem with $M_{\infty 2} = 0$ are reviewed in Sone [2000b].

The structure of the transition from subsonic condensation to supersonic is similar to that of the transition from evaporation to condensation. As explained in Section 4.10.1, the solution of the half-space problem of weak evaporation and condensation is expressed by the combination of the S solution and its Knudsen-layer correction. The Knudsen-layer correction varies appreciably over the length of the mean free path, but the S solution is slowly varying on the scale of the mean free path. The S solution expressing an evaporating flow is limited to a uniform flow, but that for a condensing flow has a structure, the size of which can be chosen at our disposal under the limitation of the weakly nonlinear theory. This introduces the slowly varying structure in the temperature and tangential velocity. As the speed of condensation increases, this structure shrinks to merge into the Knudsen layer owing to the convection effect of the condensing flow, keeping the freedom of arbitrary choice of $M_{\infty 2}$ and T_∞/T_w. This type of solution (or a Knudsen-layer-type solution with more freedom than an evaporating flow) continues to exist further in the supersonic region. In a transonic region, there are two types of supersonic condensing flows slowly varying on the scale of the mean free path: one is a weak shock wave and the other is an accelerating supersonic flow. The combination of a subsonic Knudsen-layer-type solution and a weak-shock-wave solution or the

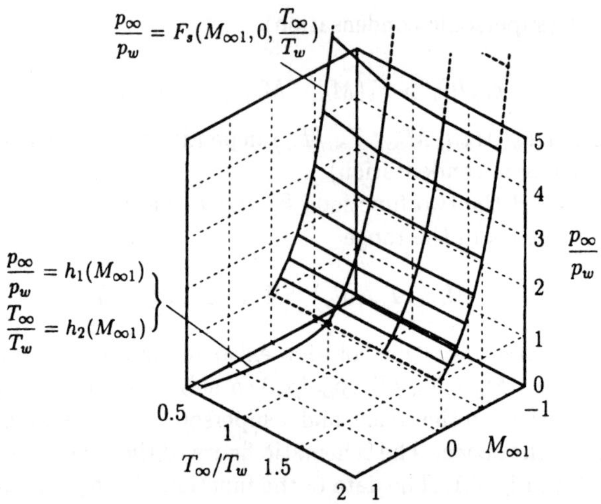

(a)

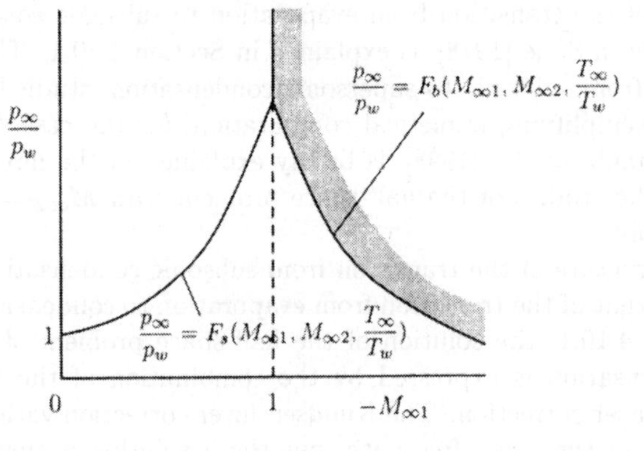

(b)

Figure 7.1. Schematic figures of the range of existence of a solution in the parameter space $(M_{\infty 1}, M_{\infty 2}, T_\infty/T_w, p_\infty/p_w)$. (a) Existence range of an evaporation solution and a subsonic condensation solution on the section $M_{\infty 2} = 0$ in the parameter space; (b) existence range of a condensation solution on a section ($M_{\infty 2}$ = const, T_∞/T_w = const) in the parameter space.

combination of a supersonic Knudsen-layer-type solution and a weak-shock-wave solution or an accelerating supersonic solution gives another supersonic condensation solution. The leading term, showing the deviation from the sonic state, of each slowly varying solution is Maxwellian, as well as the upstream state of the Knudsen-layer-type solution, as in the case of the S solution expressing weak condensation. Therefore, the connection of the two solutions is simple. A subsonic Knudsen-layer-type solution can be joined to a point on the subsonic part of weak-shock-wave solutions of various strengths. A supersonic Knudsen-layer-type solution can be joined to a point on the supersonic part of various weak-shock-wave solutions and to a point on various accelerating supersonic solutions. Thus, the dimension of the range where a solution exists in the parameter space $(M_{\infty 1}, M_{\infty 2}, T_\infty/T_w, p_\infty/p_w)$ increases by one (the freedom to choose p_∞/p_w) on transition from subsonic to supersonic condensation. For further increase of the speed of the condensing flow, the slowly varying solution shrinks to merge into the Knudsen-layer-type solution owing to the convection effect of the condensing flow, keeping the new freedom.

7.5 System of equations and boundary conditions in the continuum limit

From the results of the half-space problem of evaporation and condensation, we can derive the boundary conditions on an interface of a gas and its condensed phase for the Euler set of equations in the continuum limit ($k \to 0$) by a simple transformation. Here we list the system of equations and boundary conditions. They are presented in dimensional variables, since the reference quantities are redundant in the present nonlinear and continuum-limit case. Furthermore, the subscripts H and 0 are omitted for simplicity, since only the Hilbert part and the continuum limit are considered here.

The Euler set of equations describing the behavior of the gas is

$$\frac{\partial \rho v_i}{\partial X_i} = 0, \tag{7.33a}$$

$$\rho v_j \frac{\partial v_i}{\partial X_j} + \frac{\partial p}{\partial X_i} = 0, \tag{7.33b}$$

$$\rho v_j \frac{\partial}{\partial X_j}\left(\frac{5}{2}RT + \frac{1}{2}v_i^2\right) = 0, \tag{7.33c}$$

where

$$p = R\rho T. \tag{7.34}$$

The two half-space problems: the system (7.22), (7.26a), and (7.26b) and the system (7.27), (7.28a), and (7.28b) being compared, it is seen that the two problems are equivalent with the difference that the coordinate systems are moving relatively with velocity v_{wi} and that the reference quantities T_0 and p_0 are taken to be special values T_w and p_w in the latter system. Thus, the solution of the first problem including the relations among the parameters (thus, the boundary conditions for the Euler set of equations) is derived from that of the second problem. The notation

$$M_n = \frac{v_i n_i}{(5RT/3)^{1/2}} \quad \text{and} \quad \overline{M}_t = \frac{|v_i - v_j n_j n_i - v_{wi}|}{(5RT/3)^{1/2}}, \tag{7.35}$$

Table 7.1. The functions $h_1(M_n)$ and $h_2(M_n)$ (BKW & complete condensation; Sone & Sugimoto [1990, 1993]).

M_n	h_1	h_2	M_n	h_1	h_2	M_n	h_1	h_2
0.0000	1.0000	1.0000	0.4000	0.4900	0.8470	0.8000	0.2695	0.7088
0.04999	0.9083	0.9798	0.4400	0.4593	0.8326	0.8400	0.2553	0.6956
0.07998	0.8582	0.9679	0.4800	0.4310	0.8184	0.8800	0.2420	0.6824
0.1200	0.7966	0.9521	0.5200	0.4050	0.8043	0.9200	0.2297	0.6693
0.1600	0.7404	0.9365	0.5600	0.3809	0.7904	0.9600	0.2182	0.6563
0.2000	0.6891	0.9212	0.6000	0.3586	0.7765	0.9700	0.2155	0.6530
0.2400	0.6421	0.9060	0.6400	0.3380	0.7628	0.9800	0.2128	0.6498
0.2800	0.5991	0.8910	0.6800	0.3189	0.7492	0.9900	0.2101	0.6466
0.3200	0.5596	0.8761	0.7200	0.3012	0.7356	1.0000	0.2075	0.6434
0.3600	0.5233	0.8615	0.7600	0.2848	0.7222			

Table 7.2. $F_s(M_n, 0, T/T_w)$ (BKW & complete condensation; Aoki & Sone [1991]). The symbol * indicates the limiting value as M_n tends to zero. The solution at $M_n = 0$ exists only for $T/T_w = 1$.

	F_s						
$-M_n$	$T/T_w = 0.5$	0.75	1.0	1.5	2.0	3.0	4.0
0.00	1.000*	1.000*	1.000	1.000*	1.000*	1.000*	1.000*
0.05	1.114	1.104	1.104	1.106	1.112	1.119	1.130
0.10	1.232	1.221	1.220	1.225	1.233	1.250	1.267
0.15	1.367	1.354	1.352	1.359	1.370	1.396	1.421
0.20	1.525	1.506	1.502	1.511	1.526	1.559	1.592
0.25	1.707	1.679	1.673	1.683	1.701	1.742	1.785
0.30	1.914	1.878	1.869	1.879	1.900	1.951	2.002
0.35	2.146	2.106	2.092	2.103	2.130	2.186	2.254
0.40	2.427	2.369	2.350	2.359	2.385	2.454	2.521
0.45	2.757	2.675	2.649	2.654	2.685	2.761	2.838
0.50	3.122	3.031	2.998	2.995	3.026	3.111	3.212
0.55	3.583	3.449	3.396	3.389	3.423	3.513	3.612
0.60	4.092	3.942	3.869	3.849	3.870	3.975	4.107
0.65	4.734	4.525	4.424	4.385	4.411	4.509	4.639
0.70	5.527	5.225	5.077	5.014	5.029	5.124	5.284
0.75	6.411	6.074	5.873	5.758	5.747	5.838	5.993
0.80	7.626	7.105	6.826	6.640	6.597	6.666	6.829
0.85	9.092	8.385	8.040	7.695	7.602	7.630	7.758
0.90	11.11	9.993	9.443	8.968	8.790	8.754	8.902

being introduced, the boundary conditions on the interface of a gas and its condensed phase for the Euler set of equations (7.33a)–(7.33c) are given as follows:

(I) $M_n \geq 0$ (evaporating flow)

(a) $M_n \leq 1$ (subsonic evaporation)

$$p/p_w = h_1(M_n), \quad T/T_w = h_2(M_n), \quad \overline{M}_t = 0. \tag{7.36}$$

The functions $h_1(M_n)$ and $h_2(M_n)$ for the BKW equation with the complete condensation condition are tabulated in Table 7.1.

Table 7.3. $F_b(M_n, 0, T/T_w)$ (BKW & complete condensation; Aoki & Sone [1991]).

	F_b				
$-M_n$	$T/T_w = 0.5$	0.75	1.0	1.5	2.0
1.1	9.009	8.130	7.703	7.331	7.210
1.2	5.586	5.185	5.002	4.864	4.850
1.3	3.825	3.614	3.526	3.477	3.498
1.4	2.793	2.673	2.629	2.619	2.650
1.5	2.137	2.064	2.042	2.051	2.085
1.6	1.692	1.647	1.638	1.654	1.686
1.7	1.376	1.348	1.346	1.367	1.396
1.8	1.143	1.126	1.129	1.151	1.180
1.9	0.9666	0.9573	0.9623	0.9852	1.011
2.0	0.8296	0.8252	0.8318	0.8545	0.8797
2.1	0.7209	0.7199	0.7274	0.7496	0.7733
2.2	0.6331	0.6346	0.6428	0.6640	0.6861
2.3	0.5611	0.5644	0.5730	0.5936	0.6141
2.4	0.5014	0.5059	0.5145	0.5342	0.5534
2.5	0.4513	0.4566	0.4653	0.4841	0.5023
2.6	0.4086	0.4146	0.4231	0.4411	0.4583
2.7	0.3722	0.3799	0.3872	0.4044	0.4208
2.8	0.3407	0.3474	0.3557	0.3723	0.3879
2.9	0.3133	0.3202	0.3283	0.3443	0.3591
3.0	0.2893	0.2963	0.3043	0.3196	0.3338

(b) $M_n > 1$ (supersonic evaporation)
There is no solution.

(II) $M_n < 0$ (condensing flow)
(a) $M_n > -1$ (subsonic condensation)

$$p/p_w = F_s(M_n, \overline{M}_t, T/T_w). \tag{7.37}$$

The function $F_s(M_n, \overline{M}_t, T/T_w)$ for the BKW equation with the complete condensation condition is shown in Fig. 7.2 and Table 7.2.[6]
(b) $M_n < -1$ (supersonic condensation)

$$p/p_w > F_b(M_n, \overline{M}_t, T/T_w). \tag{7.38}$$

The function $F_b(M_n, \overline{M}_t, T/T_w)$ for the BKW equation with the complete condensation condition is shown in Fig. 7.3 and Table 7.3.
(c) $M_n = -1$ (sonic condensation)
The limiting values of the two functions F_s and F_b agree as $M_n \to -1_\pm$, and the sonic solution exists in the range

$$p/p_w \geq F_b(-1_-, \overline{M}_t, T/T_w) = F_s(-1_+, \overline{M}_t, T/T_w). \tag{7.39}$$

[6]Direct simulation Monte-Carlo computation for hard-sphere molecules for the cases $\overline{M}_t = 0$ and $T/T_w = 0.5$, 1, and 2 (Sone & Sasaki [unpublished]) shows that the relative differences from the results p/p_w for the BKW model are less than 1% except in the range $0.9 \leq M_n \leq 1$ at $T/T_w = 0.5$, where the differences are bounded by 5%.

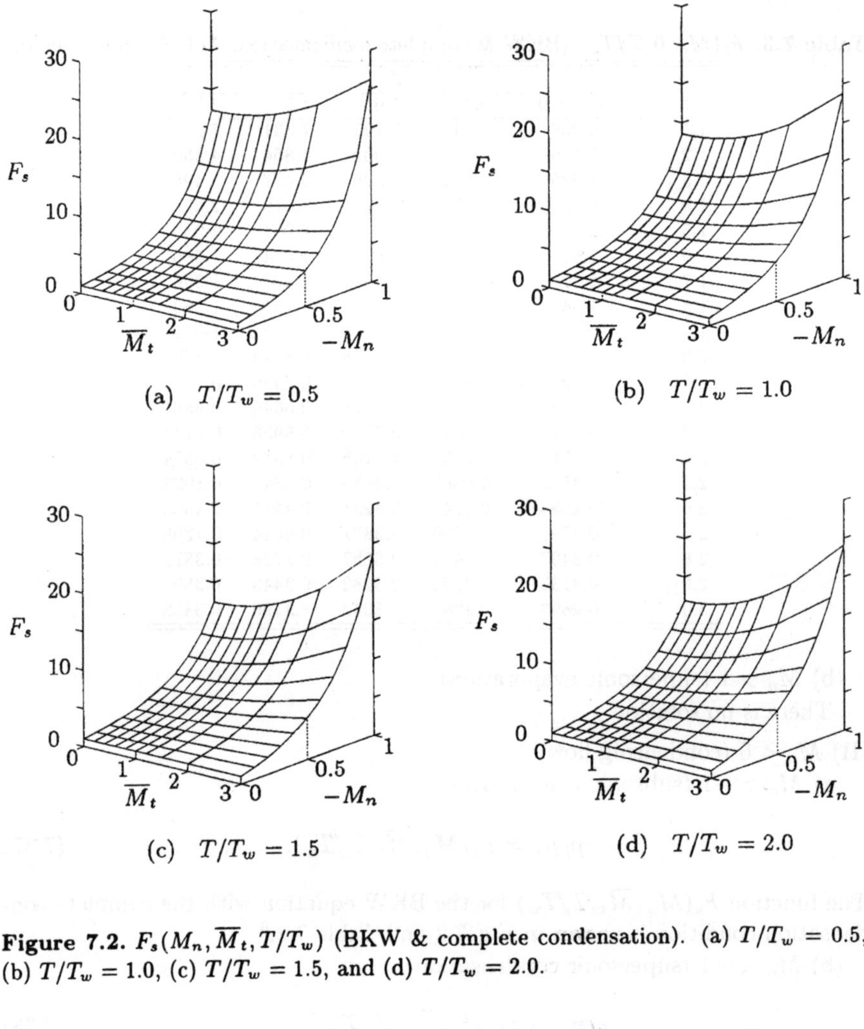

(a) $T/T_w = 0.5$ (b) $T/T_w = 1.0$

(c) $T/T_w = 1.5$ (d) $T/T_w = 2.0$

Figure 7.2. $F_s(M_n, \overline{M}_t, T/T_w)$ **(BKW & complete condensation). (a)** $T/T_w = 0.5$, **(b)** $T/T_w = 1.0$, **(c)** $T/T_w = 1.5$, **and (d)** $T/T_w = 2.0$.

7.6 Generalized kinetic boundary condition

Up to this point, we have proceeded with the discussion under the complete condensation condition (7.2). In this section, we extend the results to a more general boundary condition. The condition we take here is the condition (2.56a) with $\alpha = 1$, that is,

$$\hat{f}(x_i, \zeta_i) = \frac{\alpha_c \hat{\rho}_w + (1 - \alpha_c)\hat{\sigma}_w}{(\pi \hat{T}_w)^{3/2}} \exp\left(-\frac{(\zeta_i - \hat{v}_{wi})^2}{\hat{T}_w}\right) \quad (\zeta_j n_j > 0), \tag{7.40a}$$

$$\hat{\sigma}_w = -2\left(\frac{\pi}{\hat{T}_w}\right)^{1/2} \int_{\zeta_j n_j < 0} \zeta_j n_j \hat{f}(x_i, \zeta_i) \mathbf{d}\boldsymbol{\zeta}, \tag{7.40b}$$

where α_c $(0 \leq \alpha_c \leq 1)$ is determined by the condition of the surface of the condensed phase and is called the condensation coefficient of the boundary. In

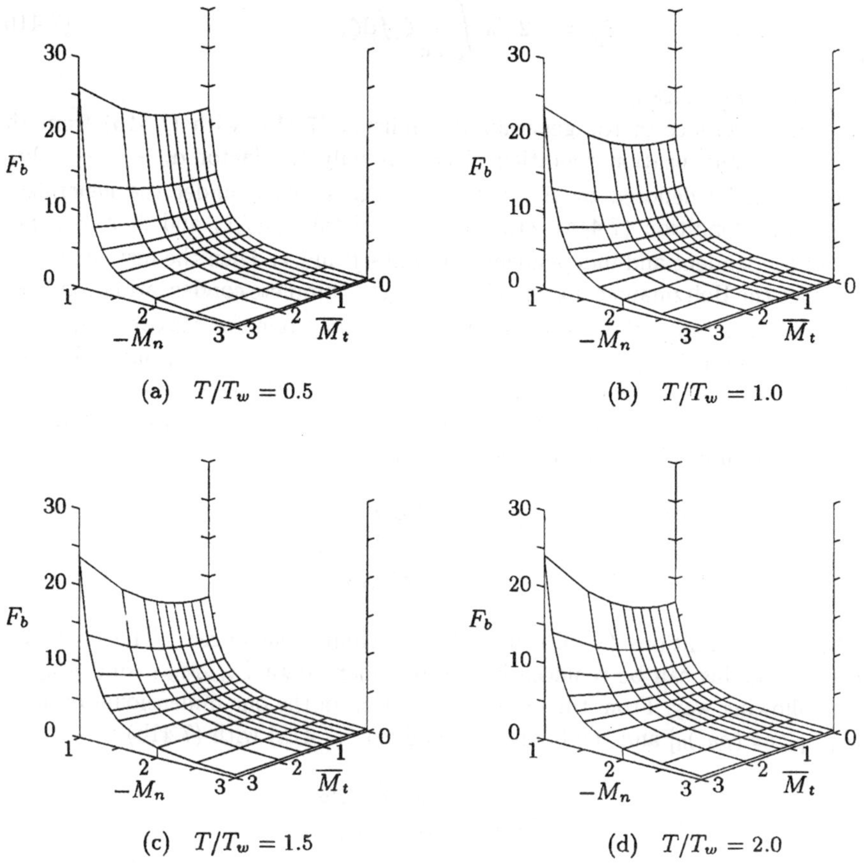

Figure 7.3. $F_b(M_n, \overline{M}_t, T/T_w)$ (BKW & complete condensation). (a) $T/T_w = 0.5$, (b) $T/T_w = 1.0$, (c) $T/T_w = 1.5$, and (d) $T/T_w = 2.0$.

this model of the boundary condition, the $1-\alpha_c$ part of the impinging molecules makes diffuse reflection. The model is widely mentioned as a generalization of the complete condensation condition (e.g., Kogan & Makashev [1971], Mager, Adomeit & Wortberg [1989], Sone & Sugimoto [1990], Kogan & Abramov [1991], Aoki & Sone [1991], Sone, Ohwada & Aoki [1991], Sugimoto & Sone [1992], Sone & Sugimoto [1993], Aoki & Masukawa [1994], Ytrehus [1997], Sone, Takata & Golse [2001]).

In the analysis in the foregoing sections in this chapter, the boundary condition affects only the Knudsen-layer analysis. Thus, we consider the half-space problem in Section 7.4 under the generalized condition, that is, Eqs. (7.27), (7.28b), and the boundary condition

$$\hat{f} = \frac{\alpha_c + (1-\alpha_c)\hat{\sigma}_w}{\pi^{3/2}} \exp\left(-\zeta_i^2\right) \quad (\zeta_1 > 0), \tag{7.41a}$$

$$\hat{\sigma}_w = -2\sqrt{\pi}\int_{\zeta_1<0}\zeta_1\hat{f}\mathrm{d}\boldsymbol{\zeta}, \tag{7.41b}$$

instead of Eq. (7.28a).

The difference of the generalized condition (7.41a) with (7.41b) from the complete condensation condition (7.28a) is only the factor $\alpha_c + (1-\alpha_c)\hat{\sigma}_w$, which will be denoted by $\bar{p}_w$. Let $\hat{f}_G$ be the solution under the generalized boundary condition (7.41a) with (7.41b) and take $\hat{f}_G/\bar{p}_w$ and $\bar{p}_w x$ as a new set of (dependent and independent) variables $\hat{f}$ and x. Then the new function $\hat{f}$ satisfies the Boltzmann equation (7.27) and the complete condensation condition (7.28a). At infinity, the new distribution function $\hat{f}$ remains Maxwellian, but its parameters are $\hat{p}_\infty/\bar{p}_w$, $\hat{T}_\infty$, $\hat{v}_{1\infty}$ (or $M_{\infty 1}$), and $\hat{v}_{2\infty}$ (or $M_{\infty 2}$), instead of $\hat{p}_\infty$, $\hat{T}_\infty$, $\hat{v}_{1\infty}$ (or $M_{\infty 1}$), and $\hat{v}_{2\infty}$ (or $M_{\infty 2}$). Therefore, the new set of parameters $\hat{p}_\infty/\bar{p}_w$, $\hat{T}_\infty$, $M_{\infty 1}$, and $M_{\infty 2}$ satisfy the relations (7.29a)–(7.32) for the complete condensation condition under the transcription

$$\frac{p_\infty}{p_w}\,(=\hat{p}_\infty) \longleftrightarrow \frac{\hat{p}_\infty}{\bar{p}_w}, \quad \frac{T_\infty}{T_w}\left(=\hat{T}_\infty\right) \longleftrightarrow \hat{T}_\infty,$$
$$M_{\infty 1} \longleftrightarrow M_{\infty 1}, \quad M_{\infty 2} \longleftrightarrow M_{\infty 2}.$$

Thus, the conditions (7.29b) and (7.29c) remain unchanged. The $\bar{p}_w$ in the other formulas can be replaced by the boundary data $\hat{T}_\infty$, $M_{\infty 1}$, and $M_{\infty 2}$ in the following way. From the mass conservation in the one-dimensional problem $[\hat{\rho}_\infty\hat{v}_{1\infty} = (\hat{\rho}\hat{v}_1)_0]$ and the boundary condition (7.41a) with (7.41b),[7]

$$\hat{\sigma}_w = 1 - \frac{2\sqrt{\pi}}{\alpha_c}\frac{\hat{v}_{1\infty}\hat{p}_\infty}{\hat{T}_\infty}.$$

Thus,

$$\begin{aligned}\bar{p}_w &= \alpha_c + (1-\alpha_c)\hat{\sigma}_w \\ &= 1 - \left(\frac{10\pi}{3}\right)^{1/2}\frac{(1-\alpha_c)}{\alpha_c}\frac{M_{\infty 1}\hat{p}_\infty}{\hat{T}_\infty^{1/2}}.\end{aligned} \tag{7.42}$$

Then, the following relation holds between $\hat{p}_\infty$ and $(\hat{p}_\infty/\bar{p}_w)$:

$$\hat{p}_\infty = \frac{(\hat{p}_\infty/\bar{p}_w)}{1+\left(\dfrac{10\pi}{3}\right)^{1/2}\dfrac{(1-\alpha_c)}{\alpha_c}\dfrac{M_{\infty 1}}{\hat{T}_\infty^{1/2}}\left(\dfrac{\hat{p}_\infty}{\bar{p}_w}\right)}, \tag{7.43}$$

which is independent of molecular models. The conditions on p_∞/p_w (and also on T_∞/T_w for the case of evaporation) for the complete condensation condition given by Eqs. (7.29a)–(7.32) being substituted into $(\hat{p}_\infty/\bar{p}_w)$ (and also into T_∞/T_w for the case of evaporation) on the right-hand side of the relation (7.43)

[7] Note the relation $\hat{\rho}_\infty\hat{v}_{1\infty} = \left(\int\zeta_1\hat{f}\mathrm{d}\boldsymbol{\zeta}\right)_0 = \left(\int_{\zeta_1<0}\zeta_1\hat{f}\mathrm{d}\boldsymbol{\zeta}\right)_0 + \left(\int_{\zeta_1>0}\zeta_1\hat{f}\mathrm{d}\boldsymbol{\zeta}\right)_0.$

[with p_∞/p_w in the former formulas as $(\hat{p}_\infty/\bar{p}_w)$ in Eq. (7.43)], the corresponding formulas for the generalized conditions are obtained. In this process, we have to confirm that both $\hat{p}_\infty$ and $\bar{p}_w$ are positive.

After the transformation corresponding to that from the formulas (7.29a)–(7.32) to the formulas (7.36)–(7.39), the boundary conditions for the Euler set of equations (7.33a)–(7.33c) under the generalized kinetic boundary condition (7.40a) with (7.40b) are obtained as follows:

(I) $M_n \geq 0$ (evaporating flow)

(a) $M_n \leq 1$ (subsonic evaporation)

$$\frac{p}{p_w} = \frac{h_1(M_n)}{1 + \left(\dfrac{10\pi}{3}\right)^{1/2} \dfrac{(1-\alpha_c)}{\alpha_c} \dfrac{M_n h_1(M_n)}{[h_2(M_n)]^{1/2}}}, \tag{7.44a}$$

$$T/T_w = h_2(M_n), \tag{7.44b}$$

$$\overline{M}_t = 0. \tag{7.44c}$$

The pressure ratio p/p_w increases with α_c and takes its maximum value for $\alpha_c = 1$.

(b) $M_n > 1$ (supersonic evaporation)

There is no solution.

(II) $M_n < 0$ (condensing flow)

(a) $M_n > -1$ (subsonic condensation)

$$\frac{p}{p_w} = \frac{F_s(M_n, \overline{M}_t, T/T_w)}{1 + \left(\dfrac{10\pi}{3}\right)^{1/2} \dfrac{(1-\alpha_c)}{\alpha_c} \left(\dfrac{T_w}{T}\right)^{1/2} M_n F_s(M_n, \overline{M}_t, T/T_w)}. \tag{7.45}$$

The denominator of the fraction on the right-hand side can be negative, since M_n is negative. In this range of parameters, the pressure ratio p/p_w is negative; then, the solution is physically meaningless, that is, there is no solution. The pressure ratio p/p_w increases as α_c decreases from $\alpha_c = 1$ (the complete condensation).

(b) $M_n < -1$ (supersonic condensation)

$$\frac{p}{p_w} > \frac{F_b(M_n, \overline{M}_t, T/T_w)}{1 + \left(\dfrac{10\pi}{3}\right)^{1/2} \dfrac{(1-\alpha_c)}{\alpha_c} \left(\dfrac{T_w}{T}\right)^{1/2} M_n F_b(M_n, \overline{M}_t, T/T_w)}, \tag{7.46}$$

for the range where the denominator of the fraction on the right-hand side is positive. For the range where the denominator is negative, the direction of the inequality sign in Eq. (7.46) should be opposite, and thus the pressure ratio p/p_w is negative. Then the solution is physically meaningless, or in other words, there is no solution. The boundary for the pressure ratio p/p_w shifts upwards as α_c decreases from $\alpha_c = 1$.

When the condensation coefficient α_c decreases from $\alpha_c = 1$, the denominator of the fraction in Eq. (7.45) becomes zero at some value of α_c, say α_c^{cr}. For $\alpha_c \leq \alpha_c^{cr}$, there is no solution. Similarly, the denominator in the inequality (7.46) vanishes at some α_c, say α_c^{Cr}. For $\alpha_c \leq \alpha_c^{Cr}$, there is no solution. Consider the case where F_s is a decreasing function of M_n, F_b is an increasing function of M_n, and $F_s(M_n = -1_+) = F_b(M_n = -1_-)$ as in the BKW model. When α_c decreases from $\alpha_c = 1$, the two denominators on the right-hand sides of Eqs. (7.45) and (7.46) first vanish at $M_n = -1$ at some common values of α_c, i.e., $\alpha_c = \alpha_c^{cr}(M_n = -1) = \alpha_c^{Cr}(M_n = -1)$. For smaller values of α_c, there is no solution for a band region of M_n around $M_n = -1$; the width of the band widens as α_c decreases.

7.7 Boundary-condition functions $h_1(M_n)$, $h_2(M_n)$, $F_s(M_n, \overline{M}_t, T/T_w)$, and $F_b(M_n, \overline{M}_t, T/T_w)$

The boundary conditions for the Euler set of equations on an interface of a gas and its condensed phase are determined by the functions $h_1(M_n)$, $h_2(M_n)$, $F_s(M_n, \overline{M}_t, T/T_w)$, and $F_b(M_n, \overline{M}_t, T/T_w)$ not only in the case of the complete condensation condition but also in the case of the generalized condition. Their information is essential to application of the theory. For these functions, however, solution of the nonlinear Boltzmann equation is required, and only limited numerical solutions are available. Here we summarize their information common to molecular models.

First we make some conjectures on the behavior of the functions of h_1, h_2, F_s, and F_b, which can be derived from various numerical computations (e.g., Kogan & Makashev [1971], Ytrehus [1977], besides the results mentioned in Section 7.4), supplemented with unpublished data by Sone & Sasaki for a hard-sphere gas and by Andries & Aoki for a modified BKW equation (Andries, Le Tellac, Perlat & Perthame [2000]).

Conjecture. *The functions $h_1(M_n)$ and $h_2(M_n)$ are monotonically decreasing functions of M_n in $0 \leq M_n \leq 1$, and they take the value unity at $M_n = 0$, i.e., $h_1(0) = 1$ and $h_2(0) = 1$.*

Conjecture. *The function $F_s(M_n, \overline{M}_t, T/T_w)$ is a decreasing function of M_n in $-1 < M_n < 0$, and takes the limiting value unity as $M_n \to 0_-$, i.e., $\lim_{M_n \to 0_-} F_s(M_n, \overline{M}_t, T/T_w) = 1$.*

Conjecture. *The function $F_b(M_n, \overline{M}_t, T/T_w)$ is an increasing function of M_n in $-\infty < M_n < -1$, and takes a common limiting value at $M_n = -1$ to $F_s(M_n, \overline{M}_t, T/T_w)$, i.e.,*

$$\lim_{M_n \to -1_-} F_b(M_n, \overline{M}_t, T/T_w) = \lim_{M_n \to -1_+} F_s(M_n, \overline{M}_t, T/T_w). \tag{7.47}$$

Conjecture. *A point $(M_n, \overline{M}_t, T/T_w, p/p_w)$ on the hypersurface $p/p_w =$*

$F_b(M_n, \overline{M}_t, T/T_w)$ is related to a point on the hypersurface $p^*/p_w = F_s(M_n^*, \overline{M}_t^*, T^*/T_w)$ by the shock wave relation (Liepmann & Roshko [1957]) between the upstream and downstream conditions. That is, if the set of the parameters $(M_n, \overline{M}_t, T/T_w, p/p_w)$ lies on the hypersurface $p/p_w = F_b(M_n, \overline{M}_t, T/T_w)$, then the set of parameters $(M_n^*, \overline{M}_t^*, T^*/T_w, p^*/p_w)$ given by

$$M_n^* = -\sqrt{\frac{M_n^2+3}{5M_n^2-1}}, \tag{7.48a}$$

$$\overline{M}_t^* = \frac{4|M_n|\overline{M}_t}{\sqrt{(5M_n^2-1)(M_n^2+3)}}, \tag{7.48b}$$

$$\frac{T^*}{T_w} = \frac{(5M_n^2-1)(M_n^2+3)}{16M_n^2}\frac{T}{T_w}, \tag{7.48c}$$

$$\frac{p^*}{p_w} = \frac{5M_n^2-1}{4}\frac{p}{p_w}, \tag{7.48d}$$

lies on the hypersurface $p^*/p_w = F_s(M_n^*, \overline{M}_t^*, T^*/T_w)$.

According to Sone, Takata & Sugimoto [1996] and Sone, Takata & Golse [2001], the following upper and lower bounds of the functions h_1, h_2, F_s, and F_b are derived with the aid of the mass, momentum, and energy laws (or equations) and the H theorem:[8]

(I) In an evaporating flow ($M_n > 0$), the following inequalities hold.

(a)

$$\frac{10\pi}{3}[M_n h_1(M_n)]^2 \le h_2(M_n) \le \frac{1}{\pi}\left(\frac{6}{5}\right)^3\left[\frac{1}{M_n\,(M_n^2+3)\,h_1(M_n)}\right]^2. \tag{7.49}$$

(b)

$$h_1(M_n) \le [h_2(M_n)]^{5/2}\exp\left\{\frac{5}{2}\left[1-\left(1+\frac{1}{3}M_n^2\right)h_2(M_n)\right]\right\}. \tag{7.50}$$

(c)

$$\frac{1}{2(1+5M_n^2/3)} \le h_1(M_n) \le \frac{1}{(1+M_n^2/3)^{5/2}}, \tag{7.51}$$

where the second inequality is derived from the condition (7.50), but it is listed because of its simplicity without $h_2(M_n)$.

(d) From the condition (7.49) and the first relation in the condition (7.51),

$$\frac{5\pi}{6}\left(\frac{M_n}{1+5M_n^2/3}\right)^2 \le h_2(M_n) \le \frac{4}{\pi}\left(\frac{6}{5}\right)^3\left[\frac{1+5M_n^2/3}{M_n\,(M_n^2+3)}\right]^2. \tag{7.52}$$

[8] Recently Bobylev, Grzhibovskis & Heintz [2001] have also discussed the bounds on a similar basis to that of Sone, Takata, & Sugimoto [1996]. The relation between the two works is explained in Footnote 35 of Sone, Takata & Golse [2001]. They also computed the functions h_1 and h_2 by their maximal entropy production hypothesis. The results practically agree with those by Sone & Sugimoto [1990].

(e) From the condition (7.50) and the first relation in the condition (7.51),

$$\frac{(1+M_n^2/3)^{5/2}}{2(1+5M_n^2/3)} \leq [(1+M_n^2/3)h_2(M_n)]^{5/2} \exp\left\{\frac{5}{2}\left[1-\left(1+\frac{1}{3}M_n^2\right)h_2(M_n)\right]\right\}. \tag{7.53}$$

This relation gives the upper and lower bounds of $h_2(M_n)$, because the function on the right-hand side has its maximum value unity at $(1+M_n^2/3)h_2(M_n)=1$ and is monotonic as a function of $(1+M_n^2/3)h_2(M_n)$ on both sides of the maximum point.

The bounds on $h_1(M_n)$ given by Eq. (7.51) and those on $h_2(M_n)$ by Eqs. (7.52) and (7.53) are shown in Fig. 7.4, where the curves of $h_1(M_n)$ and $h_2(M_n)$ (Sone & Sugimoto [1990]) for the BKW equation together with the data in Kogan & Makashev [1971], Murakami & Oshima [1974], and Ytrehus [1977] are shown for reference. The curve of $h_1(M_n)$ is close to the upper bound for small M_n, but it is close to the lower bound near $M_n = 1$.

(II) In a condensing flow, both functions $F_s(M_n, \overline{M}_t, T/T_w)$ and $F_b(M_n, \overline{M}_t, T/T_w)$ are bounded from below by the function $B_{\max}(M_n, \overline{M}_t, T/T_w)$ constructed by the larger value of the following two functions B_1 and B_2:

$$B_1 = \frac{1}{2(1+5M_n^2/3)}, \tag{7.54a}$$

$$B_2 = \left(\frac{T}{T_w}\right)^{5/2} \exp\left(\frac{5}{2}\left\{1-\left[1+\frac{1}{3}\left(M_n^2+\overline{M}_t^2\right)\right]\left(\frac{T}{T_w}\right)\right\}\right). \tag{7.54b}$$

That is,

$$F_s(M_n, \overline{M}_t, T/T_w) \geq B_{\max}(M_n, \overline{M}_t, T/T_w), \tag{7.55a}$$

$$F_b(M_n, \overline{M}_t, T/T_w) \geq B_{\max}(M_n, \overline{M}_t, T/T_w), \tag{7.55b}$$

where

$$B_{\max}(M_n, \overline{M}_t, T/T_w) = \max(B_1, B_2). \tag{7.56}$$

The functions B_1 and B_2 are plotted in Fig. 7.5, where $\overline{M}_t = 0$ and $T/T_w = 0.5, 1,$ and 1.5 are chosen for B_2. [The B_2 for other values of $\overline{M}_t$ can be read from the figure by taking $-(M_n^2+\overline{M}_t^2)^{1/2}$ as M_n.] It should be noted that the function B_1 is independent of $\overline{M}_t$ and T/T_w.

The condition (7.55a) is not strong enough to give the modest conclusion $F_s(M_n, \overline{M}_t, T/T_w) \geq 1$ from the numerical results.[9] The condition (7.55b), though very weak quantitatively compared with the results of numerical computation near the sonic point, gives an important result. That is, it shows that supersonic solutions exist only for a limited range of the parameters M_n, $\overline{M}_t$, T/T_w, and p/p_w and that the range is uniformly bounded with respect to $\overline{M}_t$ and T/T_w. The limitation of the range of existence of supersonic solutions was proved mathematically not to exist for the solution of the Boltzmann equation

[9] The numerical results for a hard-sphere gas and the BKW model show $F_s(M_n=-1, \overline{M}_t=0, T/T_w=1) > 10$.

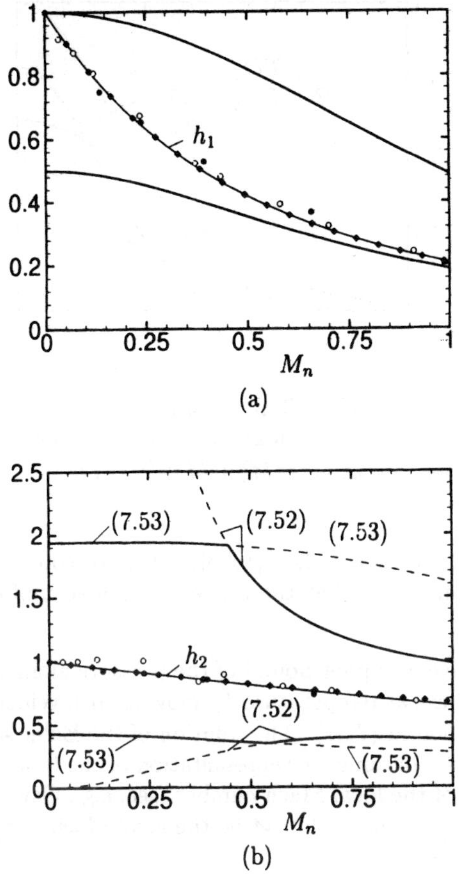

Figure 7.4. The upper and lower bounds of $h_1(M_n)$ and $h_2(M_n)$. (a) The bounds of $h_1(M_n)$ [Eq. (7.51)]; (b) two kinds of the bounds of $h_2(M_n)$ [Eqs. (7.52) and (7.53)]. In panel (a), thick solid lines indicate the bounds (7.51). In panel (b), the thick solid lines indicate the stronger bounds of the two: Eqs. (7.52) and (7.53), and the dashed lines the weaker. The data of the functions $h_1(M_n)$ and $h_2(M_n)$ by several authors are also shown for reference. The thin solid line indicates the numerical result (finite difference method) for the BKW equation in Sone & Sugimoto [1990]; the symbol ●: Kogan & Makashev [1971] (BKW); ○: Murakami & Oshima [1974] (Maxwell molecules, DSMC method); and ⧫: Ytrehus [1977] (moment method).

linearized around the uniform state at infinity (Coron, Golse & Sulem [1988]).[10] The present bound supports results (Sone, Aoki & Yamashita [1986], Sone, Aoki, Sugimoto & Yamada [1988], Aoki, Sone & Yamada [1990], Aoki, Nishino, Sone & Sugimoto [1991], Kogan & Abramov [1991], and Kryukov [1991]) of numerical analysis theoretically and supplements the theoretical work near the transonic

[10]This difference comes from the property of the linearized problem that the speed of a wave propagation is independent of its amplitude.

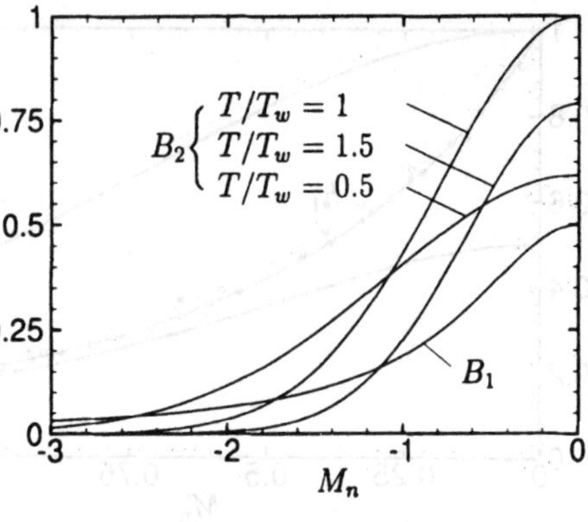

Figure 7.5. B_1 and B_2 versus M_n [Eqs. (7.54a) and (7.54b)]. The other arguments $\overline{M}_t$ and T/T_w of B_2 are set as $\overline{M}_t = 0$ and $T/T_w = 0.5$, 1, and 1.5. The B_2 for other values of $\overline{M}_t$ can be read by taking $-(M_n^2 + \overline{M}_t^2)^{1/2}$ as M_n.

region (Sone, Golse, Ohwada & Doi [1998]). It is, of course, not derived from the numerical computations that the range is bounded uniformly with respect to $\overline{M}_t$ and T/T_w.

Now we will derive an upper bound of the pressure along a flow evaporating from a condensed phase at temperature T_w moving with velocity v_{wi} ($v_{wi}n_i = 0$) with the aid of the bound (7.51). The solution of the Euler set of equations for a steady flow admits a parametric representation of the flow variables along the streamline in terms of the local Mach number (see, e.g., Oswatitsch [1956], Landau & Lifshitz [1987]). That is, let $\mathcal{M}$ be the local Mach number $|v_i|/\sqrt{5RT/3}$ in the usual sense, then

$$\frac{p}{p_0} = \left(\frac{1+\mathcal{M}_0^2/3}{1+\mathcal{M}^2/3}\right)^{5/2}, \tag{7.57}$$

where the subscript 0 indicates the values at some reference point. The reference state being taken on the condensed phase, then from the boundary condition (7.44a)–(7.44c) [the generalized form of Eq. (7.36)], which is derived under the generalized kinetic boundary condition (7.40a) with (7.40b), the parameters p_0 and $\mathcal{M}_0^2$ are given by

$$p_0 = (p)_0 = \frac{p_w h_1(M_{n0})}{1+\left(\dfrac{10\pi}{3}\right)^{1/2}\dfrac{(1-\alpha_c)}{\alpha_c}\dfrac{M_{n0}h_1(M_{n0})}{[h_2(M_{n0})]^{1/2}}} \le p_w h_1(M_{n0}),$$

$$\mathcal{M}_0^2 = \left(v_i^2/(5RT/3)\right)_0 = M_{n0}^2 + M_w^2,$$

$$M_{n0} = \left(v_i n_i/\sqrt{5RT/3}\right)_0, \quad M_w^2 = v_{wi}^2/(5RT/3)_0,$$

where the quantities in the parentheses with subscript 0, i.e., $(\cdots)_0$, are evaluated on the condensed phase and the condition $v_{wi}n_i = 0$ is used. With these relations in Eq. (7.57) and the second relation (7.51) being applied, then

$$\begin{aligned} p\left(1+\mathcal{M}^2/3\right)^{5/2} &\le p_w h_1(M_{n0})\left(1+\mathcal{M}_0^2/3\right)^{5/2} \\ &= p_w h_1(M_{n0})\left(1+M_{n0}^2/3\right)^{5/2}\left(\frac{1+\mathcal{M}_0^2/3}{1+M_{n0}^2/3}\right)^{5/2} \\ &\le p_w\left(1+\frac{M_w^2/3}{1+M_{n0}^2/3}\right)^{5/2} \quad \text{[see Eq. (7.51)]} \\ &\le p_w\left(1+M_w^2/3\right)^{5/2}. \end{aligned} \tag{7.58}$$

It is noted that the temperature T in M_w is not T_w but $(T)_0$. Comparing Eq. (7.58) with Eq. (7.57), we find that the pressure p along the flow is bounded by the pressure at the point with the same $\mathcal{M}$ of the isentropic flow starting from the state at pressure p_w, temperature $(T)_0$, and flow speed $|v_{wi}|$. The maximum pressure occurs at $\mathcal{M}=0$. In a flow evaporating from a condensed phase at rest ($v_{wi}=0$), the pressure is bounded as $p \le p_w$.

7.8 Applications

7.8.1 Two-surface problem of evaporation and condensation

Consider a gas between its two parallel condensed phases at rest and with different uniform temperatures T_{w1} and T_{w2} ($T_{w1} > T_{w2}$). The saturated gas pressure at temperature T_{w1} is denoted by p_{w1}, and that at T_{w2} by p_{w2} ($p_{w1} > p_{w2}$). Evaporation will take place on the hotter condensed phase and condensation on the colder, and a mass flow will occur from the hotter side to the colder. The behavior of this process in the limit as the Knudsen number $\to 0$ can be obtained as the solution of the Euler set of equations (7.33a)–(7.33c) under the boundary conditions (7.36)–(7.39).

In the present one-dimensional problem, the variables M_1 $(= v_1/\sqrt{5RT/3})$, p, and T are all constant from the Euler set, where the X_1 axis is normal to the condensed phase and is directed from the hotter to the colder. A shock discontinuity is excluded because the flow cannot be supersonic from the discussion below. On the hotter condensed phase, the condition (7.36) for evaporation is applied, i.e.,

$$p/p_{w1} = h_1(M_1), \quad T/T_{w1} = h_2(M_1) \quad (M_1 \le 1). \tag{7.59}$$

On the colder condensed phase, the condition (7.37), (7.38), or (7.39) for condensation holds, i.e.,

$$p/p_{w2} = F_s(-M_1, 0, T/T_{w2}) \quad (0 < M_1 < 1), \tag{7.60}$$

or

$$p/p_{w2} > F_b(-M_1, 0, T/T_{w2}) \quad (M_1 > 1), \tag{7.61}$$

$$p/p_{w2} \ge F_b(-1, 0, T/T_{w2}) \quad (M_1 = 1). \tag{7.62}$$

Table 7.4. Two-surface problem of evaporation and condensation (BKW & complete condensation; Aoki & Sone [1991]).

T_{w1}/T_{w2}	p_{w1}/p_{w2}	M_1	T/T_{w2}	p/p_{w2}	$v_1/\sqrt{2RT_{w2}}$
1.2	2	0.178	1.12	1.43	0.171
1.2	4	0.354	1.04	2.11	0.329
1.2	6	0.456	0.992	2.69	0.415
1.2	10	0.581	0.940	3.69	0.514
2	2	0.175	1.86	1.44	0.218
2	6	0.455	1.66	2.69	0.534
2	10	0.582	1.57	3.68	0.665
2	20	0.751	1.45	5.77	0.825
4	6	0.443	3.33	2.74	0.738
4	10	0.573	3.14	3.73	0.928
4	20	0.748	2.91	5.79	1.16
4	30	0.848	2.77	7.58	1.29

From Eq. (7.59), a flow with $M_1 > 1$ never occurs. When $M_1 < 1$, eliminating p and T from Eq. (7.59) and (7.60), we have

$$(p_{w1}/p_{w2})h_1(M_1) = F_s(-M_1, 0, (T_{w1}/T_{w2})h_2(M_1)), \tag{7.63}$$

from which M_1 is determined. Then, p and T are obtained from Eq. (7.59). When $M_1 = 1$, Eqs. (7.59) and (7.62) lead to

$$(p_{w1}/p_{w2})h_1(1) \geq F_b(-1, 0, (T_{w1}/T_{w2})h_2(1)).$$

From the numerical data for the BKW equation (Tables 7.1 and 7.2 and Fig. 7.2), the function $F_s(-M_1,\, 0,\, (T_{w1}/T_{w2})h_2(M_1))/h_1(M_1)$ increases monotonically from 1 to $F_b(-1, 0, (T_{w1}/T_{w2})h_2(1))/h_1(1)$ as M_1 increases from 0 to 1. Thus, from Eq. (7.63), we find that M_1 increases monotonically from 0 to 1 as p_{w1}/p_{w2} increases from 1 to $F_b(-1, 0, (T_{w1}/T_{w2})h_2(1))/h_1(1)$. For larger values of p_{w1}/p_{w2}, $M_1 = 1$ (the flow remains sonic). Some of the solutions of Eq. (7.63), obtained in Aoki & Sone [1991], are shown in Table 7.4. The solution of M_1 of Eq. (7.63) is insensitive to T_{w1}/T_{w2} and mostly determined by p_{w1}/p_{w2}, reflecting the weak dependence of $F_s(-M_1, 0, T/T_{w2})$ on T/T_{w2}.

The problem is studied mathematically by Arkeryd & Nouri [2000].

7.8.2 Evaporating flow from a spherical condensed phase into a vacuum

Consider a steady evaporating flow from a spherical condensed phase (radius: L, temperature: T_w, the saturated gas pressure at temperature T_w: p_w) into an infinite expanse of a vacuum. Let ℓ_w be the mean free path in the equilibrium state at rest with pressure p_w and temperature T_w, and let Kn_w be the Knudsen number ℓ_w/L. We try to obtain the limiting solution as $\mathrm{Kn}_w \to 0$ of the problem on the basis of the Euler set of equations (7.33a)–(7.33c) and the boundary condition (7.36). The solution requires examination of its validity, because the density of the gas decreases to vanish at downstream infinity.

The solution of the Euler set of equations (isentropic flow) admits the parametric representation (see, e.g., Oswatitsch [1956], Landau & Lifshitz [1987]): Along a streamline, for example,

$$\frac{p}{p_0} = \left(\frac{3+\mathcal{M}_0^2}{3+\mathcal{M}^2}\right)^{5/2}, \quad \frac{T}{T_0} = \left(\frac{3+\mathcal{M}_0^2}{3+\mathcal{M}^2}\right),$$

where $\mathcal{M}$ is the Mach number, i.e., the local flow speed $|v_i|$ divided by the local speed of sound $\sqrt{5RT/3}$ and the subscript 0 indicates the values at a reference point, which is on the sphere here. Further for a radial flow, from the continuity equation (7.33a),

$$\frac{r}{L} = \left(\frac{\mathcal{M}_0}{\mathcal{M}}\right)^{1/2} \left(\frac{3+\mathcal{M}^2}{3+\mathcal{M}_0^2}\right),$$

where r is the distance from the center of the sphere.

In an evaporating flow into a vacuum, the Mach number $\mathcal{M}$ ranges from $\mathcal{M} = 1$ to infinite. In view of no supersonic evaporation on the condensed phase [(I).(b) in Section 7.5], the starting Mach number $\mathcal{M}_0$ is unity, i.e.,

$$\mathcal{M}_0 = 1.$$

Identifying M_n in the boundary condition (7.36) with $\mathcal{M}_0\,(=1)$, then we obtain the flow field as follows:[11]

$$\left.\begin{aligned} &\frac{\rho}{\rho_0} = \frac{8}{(3+\mathcal{M}^2)^{3/2}}, \quad \frac{v_r}{(5RT_0/3)^{1/2}} = \frac{2\mathcal{M}}{(3+\mathcal{M}^2)^{1/2}}, \quad \frac{T}{T_0} = \frac{4}{3+\mathcal{M}^2}, \\ &\frac{p}{p_0} = \left(\frac{4}{3+\mathcal{M}^2}\right)^{5/2}, \quad \frac{r}{L} = \frac{(3+\mathcal{M}^2)}{4\mathcal{M}^{1/2}}, \end{aligned}\right\} \tag{7.64}$$

where v_r is the radial component of the flow velocity (only nonvanishing one here) and

$$\rho_0 = \frac{h_1(1)p_w}{h_2(1)RT_w}, \quad T_0 = h_2(1)T_w, \quad p_0 = h_1(1)p_w.$$

For the BKW model,

$$h_1(1) = 0.2075, \quad h_2(1) = 0.6434, \quad h_1(1)/h_2(1) = 0.3225.$$

For large r/L,

$$\left.\begin{aligned} &\frac{\rho}{\rho_0} = 8\left(\frac{L}{4r}\right)^2, \quad \frac{v_r}{(5RT_0/3)^{1/2}} = 2 - 3\left(\frac{L}{4r}\right)^{4/3}, \\ &\frac{T}{T_0} = 4\left(\frac{L}{4r}\right)^{4/3}, \quad \mathcal{M} = \left(\frac{4r}{L}\right)^{2/3} \end{aligned}\right\} \tag{7.65}$$

[11] In the isentropic flow, $d\rho/dr, dv_r/dr$, etc. are infinite at $\mathcal{M} = 1$ or $r = L$. This violates the assumption imposed in derivation of the Euler system from the Boltzmann equation. Rigorously, a discussion on this point is required.

In the far field from the sphere, the variations of the variables become small. The length scale of variation [e.g., $\rho/(d\rho/dr)$] of the solution is of the order of r and increases indefinitely with the distance from the sphere. On the other hand, as the density decreases along the flow, the (local) mean free path l_r, defined by the flow speed divided by collision frequency,[12] increases along the flow. Rewriting the definition of the mean free path l_r as

$$l_r = \frac{v_r}{\bar{\nu}_c} = \frac{v_r}{\sqrt{RT_w}} \frac{(\bar{\nu}_c)_w}{\bar{\nu}_c} \frac{\sqrt{RT_w}}{(\bar{\nu}_c)_w} \propto \frac{v_r}{\sqrt{RT_w}} \frac{(\bar{\nu}_c)_w}{\bar{\nu}_c} \ell_w, \tag{7.66}$$

and taking into account the behavior of the far field, i.e., Eq. (7.65), then we have

$$l_r \propto (r/L)^{8/3} \ell_w \quad \text{(a molecule with a finite } d_m),$$
$$l_r \propto (r/L)^2 \ell_w \quad \text{(Maxwell molecule, BKW)},$$

where the difference comes from the collision frequency [see Eq. (2.20a) and the explanation in Section 2.8]. Thus the local Knudsen number $Kn_r = l_r/r$, which characterizes the flow, is given as

$$Kn_r \propto (r/L)^{5/3} \mathrm{Kn}_w \quad \text{(a molecule with a finite } d_m), \tag{7.67a}$$
$$Kn_r \propto (r/L) \mathrm{Kn}_w \quad \text{(Maxwell molecule, BKW)}. \tag{7.67b}$$

The local Knudsen number Kn_r ranges from $O(\mathrm{Kn}_w)$ to ∞ and does not remain small for both cases. Thus, the flow field deviates from the solution of the Euler set of equations given above in the far field, however small Kn_w may be. An example showing the deviation will be given at the last paragraph in this subsection.

When Kn_w is very small, there is a region where Kn_r is small but $\mathcal{M}$ is large.[13] The isentropic solution is valid up to this region. The behavior downstream of this region can be studied by the hypersonic approximation (Hamel & Willis [1966], Edwards & Cheng [1966], Freeman [1967], Grundy [1969]), where a simplification is made under the assumption that the width of the velocity distribution function is much smaller than the flow speed. According to Hamel & Willis [1966] and Edwards & Cheng [1966], the solution of this approximation of the BKW equation for a spherically expanding flow is expressed by the

[12] In the far field, where $\mathcal{M}$ is large (hypersonic flow), the mean free path based on the flow speed, instead of the thermal speed, characterizes the variation of flow in its direction (note the operator $\xi_i \partial/\partial X_i$ in the Boltzmann equation).

[13] The relation among Kn_r, $\mathcal{M}$, and Kn_w is $Kn_r \propto \mathcal{M}^{5/2}\mathrm{Kn}_w$ (a molecule with a finite d_m) and $Kn_r \propto \mathcal{M}^{3/2}\mathrm{Kn}_w$ (Maxwell molecule, BKW). Thus, Kn_r can be small for a large $\mathcal{M}$ if Kn_w is very small.

confluent hypergeometric functions as follows:

$$\rho/\rho_w = c_0\hat{r}^{-2}, \tag{7.68a}$$

$$v_r/\sqrt{2RT_w} = c_1, \tag{7.68b}$$

$$\frac{T}{T_w} = \frac{1}{\hat{r}^2}\left[c_2 U\left(\frac{2}{3}, 3, \frac{\alpha}{\mathrm{Kn}_w\hat{r}}\right) + c_3 M\left(\frac{2}{3}, 3, \frac{\alpha}{\mathrm{Kn}_w\hat{r}}\right)\right], \tag{7.68c}$$

$$\alpha = 2c_0/\sqrt{\pi}c_1, \quad \hat{r} = r/L,$$

where c_0, c_1, c_2, and c_3 are undetermined constants, and $U(a,b,x)$ and $M(a,b,x)$ are Kummer's functions (Abramowitz & Stegun [1972]; the same notation is used as in this reference).

Making use of the above solution in the hypersonic range, we can construct the solution of the BKW equation for the evaporating flow from a spherical condensed phase in the whole range of r. We connect the isentropic solution and that of the hypersonic approximation in the region where both solutions are valid, i.e., in the region where $\mathcal{M}$ is large and Kn_r is small. In this region, $\mathrm{Kn}_w\hat{r}$ is small from Eq. (7.67b). Thus, the solution (7.68a)–(7.68c) is reduced there to the following form (inner expansion):

$$\rho/\rho_w = c_0\hat{r}^{-2}, \tag{7.69a}$$

$$v_r/\sqrt{2RT_w} = c_1, \tag{7.69b}$$

$$\frac{T}{T_w} = c_2\left(\frac{\mathrm{Kn}_w}{\alpha}\right)^{2/3}\frac{1}{\hat{r}^{4/3}} + \frac{2c_3}{\Gamma(2/3)}\left(\frac{\mathrm{Kn}_w}{\alpha}\right)^{7/3}\hat{r}^{1/3}\exp\left(\frac{\alpha}{\mathrm{Kn}_w\hat{r}}\right), \tag{7.69c}$$

where $\Gamma(x)$ is the gamma function (Abramowitz & Stegun [1972]). Matching the leading terms of the two solutions: Eq. (7.65) and Eqs. (7.69a)–(7.69c), we have

$$c_0 = \frac{h_1(1)}{2h_2(1)}, \quad c_1 = \left(\frac{10}{3}h_2(1)\right)^{1/2}, \quad c_2 = \left(\frac{3[h_1(1)]^2}{40\pi}\right)^{1/3}\mathrm{Kn}_w^{-2/3}, \quad c_3 = 0. \tag{7.70}$$

With these coefficients, the limiting value of T/T_w as $r/L \to \infty$ is given as

$$\frac{T}{T_w} \to A_0\mathrm{Kn}_w^{4/3},$$

$$A_0 = \left(\frac{5\pi}{3}\right)^{2/3}\frac{[h_2(1)]^3}{\Gamma(2/3)[h_1(1)]^{4/3}} = 4.828.$$

The temperature at infinity remains at a finite value, which is called the frozen temperature. The transverse temperature $T_\perp$,[14] which is often used to indicate the anisotropic character of the velocity distribution function by comparison

[14]The transverse and parallel temperatures $T_\perp$ and $T_\parallel$ are defined as follows: Let the local

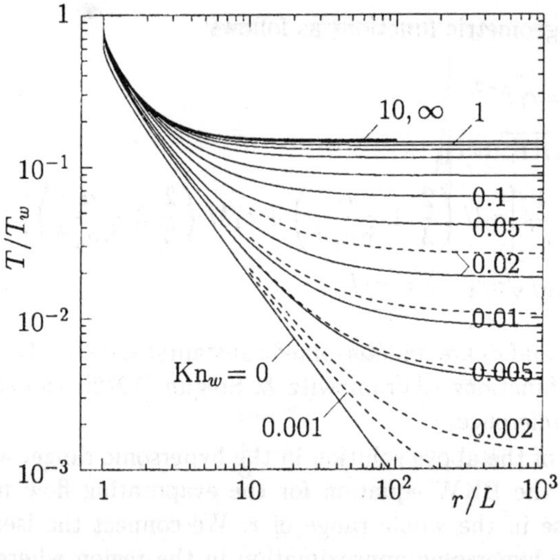

Figure 7.6. The profiles of the temperature field T/T_w versus r/L for various Knudsen numbers ($\mathrm{Kn}_w = 0$, 0.005, 0.01, 0.02, 0.05, 0.1, 0.2, 0.5, 1, 2, 5, 10, and ∞) (BKW & complete condensation; Sone & Sugimoto [1993]). The curve $\mathrm{Kn}_w = 0$ is the solution (7.64), the solid line for a finite Kn_w is the numerical solution, and the dashed line is the solution by the hypersonic approximation, i.e., Eq. (7.68c) with Eq. (7.70).

with the parallel temperature $T_{\shortparallel}$ or the temperature T, is shown to decay as

$$\frac{T_\perp}{T_w} \to A_1 \mathrm{Kn}_w^{1/3} \frac{L}{r},$$

$$A_1 = \left(\frac{5\pi}{24}\right)^{1/6} \frac{[h_2(1)]^{3/2}}{\Gamma(2/3)[h_1(1)]^{1/3}} = 0.5999.$$

In a far field the velocity distribution function is no longer Maxwellian, or the flow is not isentropic.

The profiles of temperature versus r/L of the three solutions of the BKW equations: the asymptotic solution (7.64) at $\mathrm{Kn}_w = 0$, the numerical solution for various Knudsen numbers, and the solution by the hypersonic approximation, i.e., Eq. (7.68c) with Eq. (7.70), are shown in Fig. 7.6. The convergence of the numerical solution to the solution at $\mathrm{Kn}_w = 0$ as $\mathrm{Kn}_w \to 0$ becomes slower with

---

flow direction be in X_1 direction, for example, then

$$2RT_\perp = \frac{1}{\rho}\int [(\xi_2 - v_2)^2 + (\xi_3 - v_3)^2] f(\boldsymbol{X}, \boldsymbol{\xi}, t) d\boldsymbol{\xi},$$

$$RT_{\shortparallel} = \frac{1}{\rho}\int (\xi_1 - v_1)^2 f(\boldsymbol{X}, \boldsymbol{\xi}, t) d\boldsymbol{\xi}.$$

Thus,

$$T = (T_{\shortparallel} + 2T_\perp)/3.$$

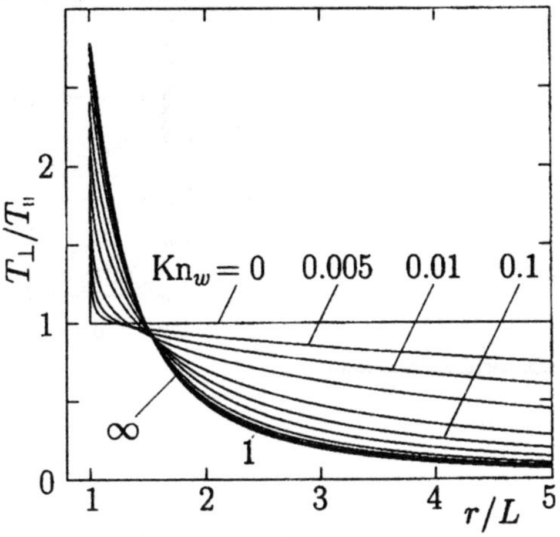

Figure 7.7. The profiles of the ratio $T_\perp/T_\shortparallel$ versus r/L for various Knudsen numbers (Kn_w = 0, 0.005, 0.01, 0.02, 0.05, 0.1, 0.2, 0.5, 1, 2, 5, 10, and ∞) (BKW & complete condensation; Sone & Sugimoto [1993]).

the distance from the sphere. The solution is well approximated uniformly by the hypersonic solution. In Fig. 7.7, the profiles of the ratio $T_\perp/T_\shortparallel$ versus r/L are shown for various Knudsen numbers. Here the convergence is not uniform. That is, the deviation of the ratio $T_\perp/T_\shortparallel$ for a small but finite value of Kn_w from unity increases with the distance from the sphere. These data are taken from Sone & Sugimoto [1993], where the detailed numerical solutions of the problem for the whole range of the Knudsen number Kn_w are presented.

7.8.3 Evaporating flow from a cylindrical condensed phase into a vacuum

Consider the same problem as in the preceding subsection with the difference that the condensed phase is a circular cylinder instead of a sphere. The solution for a cylindrical condensed phase is given, with obvious reinterpretation of the notations corresponding to the cylindrical condensed phase, as follows:

$$\frac{\rho}{\rho_0} = \frac{8}{(3+\mathcal{M}^2)^{3/2}}, \quad \frac{v_r}{(5RT_0/3)^{1/2}} = \frac{2\mathcal{M}}{(3+\mathcal{M}^2)^{1/2}}, \quad \frac{T}{T_0} = \frac{4}{3+\mathcal{M}^2},$$

$$\frac{p}{p_0} = \left(\frac{4}{3+\mathcal{M}^2}\right)^{5/2}, \quad \frac{r}{L} = \frac{(3+\mathcal{M}^2)^2}{16\mathcal{M}},$$

where

$$\rho_0 = \frac{h_1(1)p_w}{h_2(1)RT_w}, \quad T_0 = h_2(1)T_w, \quad p_0 = h_1(1)p_w.$$

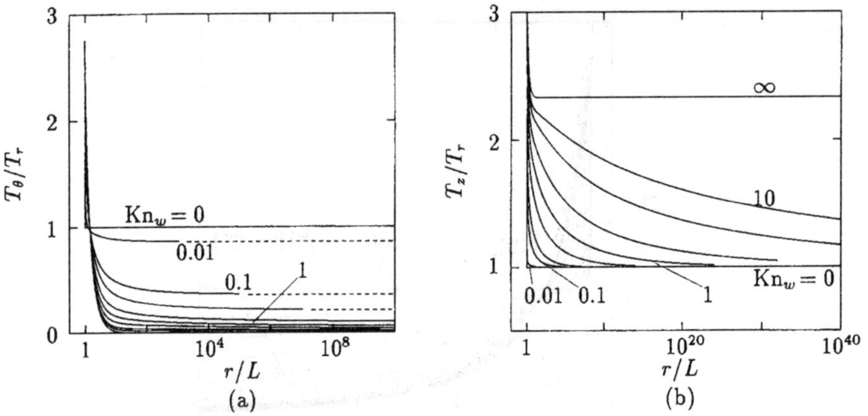

Figure 7.8. The profiles of the ratios T_θ/T_r and T_z/T_r versus r/L for various Knudsen numbers (Kn_w = 0, 0.01, 0.1, 0.2, 0.5, 1, 2, 5, 10, and ∞) (BKW & complete condensation; Sone & Sugimoto [1995]). (a) T_θ/T_r versus r/L and (b) T_z/T_r versus r/L. The dashed lines in panel (a) indicate estimated values at infinity for $\mathrm{Kn}_w = 0.01$, 0.1, and 0.2.

The difference from the spherical case is only the relation between r and $\mathcal{M}$, as easily understood. For large r/L, these relations are reduced to

$$\left.\begin{aligned} &\frac{\rho}{\rho_0} = \frac{L}{2r}, \quad \frac{v_r}{(5RT_0/3)^{1/2}} = 2 - \frac{3}{4}\left(\frac{L}{2r}\right)^{2/3}, \\ &\frac{T}{T_0} = \left(\frac{L}{2r}\right)^{2/3}, \quad \mathcal{M} = 2\left(\frac{2r}{L}\right)^{1/3}. \end{aligned}\right\} \tag{7.71}$$

The local characteristic length in the far field is r as in the spherical case, and the local mean free path there is

$$l_r \propto (r/L)^{4/3}\ell_w \quad \text{(a molecule with a finite } d_m\text{)},$$

$$l_r \propto (r/L)\ell_w \quad \text{(Maxwell molecule, BKW)}.$$

Thus the local Knudsen number Kn_r is

$$Kn_r \propto (r/L)^{1/3}\mathrm{Kn}_w \quad \text{(a molecule with a finite } d_m\text{)},$$

$$Kn_r \propto \mathrm{Kn}_w \quad \text{(Maxwell molecule, BKW)}.$$

For a gas consisting of molecules with a finite d_m, including a hard-sphere gas, the local Knudsen number increases indefinitely with the distance from the cylinder. Therefore, the present solution becomes invalid there. On the other hand, for a gas of Maxwell molecules and the BKW model, the local Knudsen number is invariant along the flow. Thus, the present solution is a good approximation for small Kn_w up to downstream infinity, where the density of the gas vanishes. This is shown by comparison with the numerical solution in Sone & Sugimoto [1995], where detailed numerical computations of the flow are carried out for the whole range of the Knudsen number Kn_w on the basis of the BKW equation. For example, the ratios T_θ/T_r and T_z/T_r, measures of anisotropy of the velocity distribution function, of the three temperatures T_r, T_θ, and T_z[15] for the BKW model both converge to unity uniformly except in a finite region around the cylinder as $\mathrm{Kn}_w \to 0$, as shown in Figs. 7.8 (a) and (b). This is a good contrast to the case of evaporation from a sphere in the preceding subsection.

[15]The three temperatures T_r, T_θ, and T_z are the temperatures in the three directions r, θ, and z in the cylindrical coordinate system (r, θ, z) with the axis of the cylinder taken as the z axis, i.e.,

$$RT_r = \frac{1}{\rho}\int(\xi_r - v_r)^2 f d\boldsymbol{\xi}, \quad RT_\theta = \frac{1}{\rho}\int(\xi_\theta - v_\theta)^2 f d\boldsymbol{\xi}, \quad RT_z = \frac{1}{\rho}\int(\xi_z - v_z)^2 f d\boldsymbol{\xi}.$$

Chapter 8

Bifurcation of Cylindrical Couette Flow with Evaporation and Condensation

In Chapters 3–7, various important situations in the continuum limit were discussed in terms of the Hilbert expansion by paying attention to the order of the quantities in the situation under consideration. There are some marginal cases for which only short notes are given as supplementary discussions, since the process of analysis and their general results can be understood or inferred rather easily from the main discussions. In this last chapter, we choose a special problem as an example of analysis in the marginal cases. The problem, a cylindrical Couette flow with evaporation and condensation, is chosen, since it shows a type of bifurcation of flow in a very simple situation that cannot be seen in the case without evaporation and condensation. For the convenience of description and ease of reading, the explanation of notation is repeated since some new notation is inevitably to be introduced in the special example, although it is consistent with that defined in Section 2.10.

8.1 Problem

Consider a gas between two rotating coaxial cylinders made of the condensed phase of the gas. Let the radius, temperature, and circumferential velocity of the inner cylinder be, respectively, L_A, T_A, and $V_{\theta A}$, and let the corresponding quantities of the outer cylinder be L_B, T_B, and $V_{\theta B}$. The saturated gas pressure at temperature T_A (T_B) is denoted by p_A (p_B). The following situation is concerned here: (i) the behavior of the gas is time-independent; (ii) the gas molecules leaving the cylinders obey the complete condensation condition; (iii) the flow field is axially symmetric and uniform (or axially and circumferentially uniform); and (iv) the speeds of rotation of the cylinders and the Knudsen number of the system are small.

The variables in the system being reduced to nondimensional variables appropriately with the aid of boundary data, the system is found to be determined by the following six parameters:

$$\hat{r}_B = \frac{L_B}{L_A}, \quad \frac{V_{\theta A}}{\sqrt{2RT_A}}, \quad \frac{V_{\theta B}}{\sqrt{2RT_A}}, \quad \frac{T_B}{T_A}, \quad \frac{p_B}{p_A}, \quad k, \tag{8.1}$$

where R is the specific gas constant and k is defined by

$$k = \frac{\sqrt{\pi}}{2}\frac{\ell_A}{L_A}, \tag{8.2}$$

where ℓ_A is the mean free path of the gas molecules at the saturated equilibrium state at rest at temperature T_A. In Sections 8.2–8.4, taking a small quantity ε, we study the asymptotic solution of the problem described above when the above six parameters are limited to the following conditions:

$$\left.\begin{aligned} &\hat{r}_B - 1 = O(1), \quad \frac{V_{\theta A}}{\sqrt{2RT_A}} = \varepsilon u_{\theta A1}, \quad \frac{V_{\theta B}}{\sqrt{2RT_A}} = \varepsilon u_{\theta B1}, \\ &\frac{T_B}{T_A} - 1 = \varepsilon^2 \tau_{B2}, \quad \frac{p_B}{p_A} - 1 = \varepsilon^2 P_{B2}, \quad k = \varepsilon^m, \end{aligned}\right\} \tag{8.3}$$

where $u_{\theta A1}$, $u_{\theta B1}$, τ_{B2}, and P_{B2} are quantities of the order of unity and $m \geq 3$. The other cases, where $p_B/p_A - 1$ is of the order of ε instead of ε^2, or $m = 1$ or 2, are discussed briefly in Section 8.5. The content of this chapter is based on Sone & Doi [2000]. The analysis in Section 8.2 is a variation of that of Chapter 6 and the analysis in Section 8.3 is that of Chapter 4. Thus, only the outline of analysis is given here. The detailed description is found in Sone & Doi [2000].

For analysis in the above range of the parameters, the nondimensional Boltzmann equation and the boundary condition for a perturbed velocity distribution function in the cylindrical coordinate system introduced in Sections 2.10 and 2.12 are convenient. The Boltzmann equation is given by

$$\zeta_r \frac{\partial \phi}{\partial \hat{r}} + \frac{\zeta_\theta^2}{\hat{r}}\frac{\partial \phi}{\partial \zeta_r} - \frac{\zeta_r \zeta_\theta}{\hat{r}}\frac{\partial \phi}{\partial \zeta_\theta} = \frac{1}{\varepsilon^m}[\mathcal{L}(\phi) + \mathcal{J}(\phi,\phi)], \tag{8.4}$$

where the radial distance r from the axis of the cylinders, the molecular velocity $(\xi_r, \xi_\theta, \xi_z)$ in the cylindrical coordinate system (r, θ, z) with the axis of the cylinders being taken as z axis, and the velocity distribution function $f(r, \xi_r, \xi_\theta, \xi_z)$ are expressed, respectively, by

$$\left.\begin{aligned} &r = L_A \hat{r}, \quad (\xi_r, \xi_\theta, \xi_z) = \sqrt{2RT_A}(\zeta_r, \zeta_\theta, \zeta_z), \\ &f = 2p_A(2RT_A)^{-5/2}E(\zeta)[1 + \phi(\hat{r}, \zeta_r, \zeta_\theta, \zeta_z)], \\ &E(\zeta) = \pi^{-3/2}\exp(-\zeta^2), \quad \zeta = (\zeta_r^2 + \zeta_\theta^2 + \zeta_z^2)^{1/2}, \end{aligned}\right\} \tag{8.5}$$

and $\mathcal{L}(\phi)$ and $\mathcal{J}(\phi,\phi)$ are, respectively, nondimensional linearized and original collision integrals defined by Eqs. (2.58b) and (2.58c) in Section 2.10. The boundary conditions on the two cylinders are

$$\phi = \phi_e(0,0,0,\varepsilon u_{\theta A1}) \quad (\zeta_r > 0) \ \text{ at } \hat{r} = 1, \tag{8.6a}$$

$$\phi = \phi_e(\varepsilon^2 P_{B2}, \varepsilon^2 \tau_{B2}, 0, \varepsilon u_{\theta B1}) \quad (\zeta_r < 0) \ \text{ at } \hat{r} = \hat{r}_B, \tag{8.6b}$$

where ϕ_e is the perturbed part of a local Maxwellian from the Maxwellian at rest with temperature T_A and pressure p_A, that is,

$$E(\zeta)[1+\phi_e(a,b,c,d)] = \frac{1+a}{\pi^{3/2}(1+b)^{5/2}} \exp\left(-\frac{(\zeta_r - c)^2 + (\zeta_\theta - d)^2 + \zeta_z^2}{1+b}\right).$$

The density $(p_A/RT_A)(1+\omega)$, flow velocity $\sqrt{2RT_A}(u_r, u_\theta, 0)$, temperature $T_A(1+\tau)$, and pressure $p_A(1+P)$ of the gas are given by the following moments of ϕ:

$$\omega = \int \phi E \mathbf{d}\boldsymbol{\zeta}, \tag{8.7a}$$

$$(u_r, u_\theta) = \frac{1}{1+\omega}\int (\zeta_r, \zeta_\theta)\phi E \mathbf{d}\boldsymbol{\zeta}, \tag{8.7b}$$

$$\frac{3}{2}\tau = \frac{1}{1+\omega}\int \left(\zeta^2 - \frac{3}{2}\right)\phi E \mathbf{d}\boldsymbol{\zeta} - u_r^2 - u_\theta^2, \tag{8.7c}$$

$$1 + P = (1+\omega)(1+\tau), \tag{8.7d}$$

$$\mathbf{d}\boldsymbol{\zeta} = d\zeta_r d\zeta_\theta d\zeta_z.$$

8.2 Solution type I

8.2.1 Analysis

The solution that describes the overall behavior of the gas is obtained in a power series of ε under the assumption that $\partial\phi/\partial\hat{r} = O(\phi)$:

$$\phi_H = \phi_{H1}\varepsilon + \phi_{H2}\varepsilon^2 + \cdots, \tag{8.8}$$

where the subscript H is attached to discriminate the solution of this class (Hilbert solution)[1]. Corresponding to this expansion, the moments of ϕ_H, the macroscopic variables: the density, flow velocity, pressure, and temperature, are also expanded in power series of ε:

$$h_H = h_{H1}\varepsilon + h_{H2}\varepsilon^2 + \cdots,$$

[1]This expansion is a variation of the S expansion. Here, we do not make such detailed classification. We have only to make distinctions among solutions with different length scales of variation.

where h represents ω, u_r, etc. Substitution of the series (8.8) into the Boltzmann equation (8.4) leads to a series of integral equations for ϕ_{Hs} $(s = 1, 2, \ldots)$ of the form

$$\mathcal{L}(\phi_{Hs}) = Ih_s \quad (s = 1, 2, \ldots), \tag{8.9}$$

where Ih_s $(Ih_1 = 0)$ is the inhomogeneous term determined by ϕ_{Hn} for $n < s$. The corresponding homogeneous equation $\mathcal{L}(\phi_{Hs}) = 0$ has five independent solutions 1, ζ_r, ζ_θ, ζ_z, ζ^2, but ζ_z is excluded in the present problem from the symmetry request. Up to the order $s = m$, the inhomogeneous term Ih_s consists only of $\mathcal{J}(\phi_{Hs-n}, \phi_{Hn})$ terms $(n = 1, \ldots, s-1)$, and thus, ϕ_{Hs} is the corresponding term of the expansion of the local Maxwellian $\phi_e(P_H, \tau_H, u_{rH}, u_{\theta H})$. For $s \geq m+1$, additional inhomogeneous terms, which consist of the derivative terms of Eq. (8.4), enter. The newly entered derivative terms must satisfy the solvability condition

$$\int (1, \zeta_r, \zeta_\theta, \zeta^2) \left(\zeta_r \frac{\partial \phi_{Hs-m}}{\partial \hat{r}} + \frac{\zeta_\theta^2}{\hat{r}} \frac{\partial \phi_{Hs-m}}{\partial \zeta_r} - \frac{\zeta_r \zeta_\theta}{\hat{r}} \frac{\partial \phi_{Hs-m}}{\partial \zeta_\theta} \right) E \mathbf{d}\boldsymbol{\zeta} = 0. \tag{8.10}$$

The solution ϕ_{Hs} of Eq. (8.9) is expressed in the form

$$\phi_{Hs} = a_{0s} + a_{rs}\zeta_r + a_{\theta s}\zeta_\theta + a_{4s}\zeta^2 + \hat{\phi}_s,$$

where the coefficients a_{0s}, a_{rs}, $a_{\theta s}$, and a_{4s} are undetermined constants and $\hat{\phi}_s$ is the particular solution of Eq. (8.9). If $\hat{\phi}_s$ is chosen to be orthogonal to $(1, \zeta_r, \zeta_\theta, \zeta^2)$, i.e., $\int (1, \zeta_r, \zeta_\theta, \zeta^2)\hat{\phi}_s E(\zeta)\mathbf{d}\boldsymbol{\zeta} = 0$, then a_{0s}, a_{rs}, $a_{\theta s}$, and a_{4s} are determined by the macroscopic variables ω_{Hn} (or P_{Hn}), u_{rHn}, $u_{\theta Hn}$, and τ_{Hn} $(n \leq s)$. From the structure of the integral equation (8.9), the component function ϕ_{Hs} also is expressed with ω_{Hn} (or P_{Hn}), u_{rHn}, $u_{\theta Hn}$, and τ_{Hn} $(n \leq s)$.

The solvability condition (8.10) at each s $(\geq m+1)$ gives four differential equations containing a_{0n-m}, a_{rn-m}, $a_{\theta n-m}$, and a_{4n-m} [or ω_{Hn-m} (or P_{Hn-m}), u_{rHn-m}, $u_{\theta Hn-m}$, and τ_{Hn-m}] $(n \leq s)$. The expansion can be carried out consistently with a special assumption $u_{rH1} = P_{H1} = 0$, which is consistent with the boundary conditions (8.6a) and (8.6b) to be studied later. (This is derived by considering the boundary condition if we start from $u_{rH1} \neq 0$ and $P_{H1} \neq 0$, but the process is omitted for brevity of description.) Owing to the degeneracy of u_{rH1} and P_{H1}, the series of differential equations obtained from the solvability condition (8.10) $(s \geq m+1)$ is arranged in the staggered combination $(u_{\theta Hn}, u_{rHn+1}, \tau_{Hn}, P_{Hn+1};\ n = 1, 2, \ldots)$ as a set. The differential equations for the leading order are

$$\frac{\mathrm{d}\, u_{rH2}\hat{r}}{\mathrm{d}\hat{r}} = 0, \tag{8.11a}$$

$$\frac{u_{\theta H1}^2}{\hat{r}} = \frac{1}{2}\frac{\mathrm{d}\, P_{H2}}{\mathrm{d}\hat{r}}, \tag{8.11b}$$

$$u_{rH2}\frac{\mathrm{d}\, u_{\theta H1}\hat{r}}{\mathrm{d}\hat{r}} = 0, \tag{8.11c}$$

$$u_{rH2}\frac{\mathrm{d}\, \tau_{H1}}{\mathrm{d}\hat{r}} = 0. \tag{8.11d}$$

The Hilbert solution ϕ_H, however, does not have freedom enough to be matched to the kinetic boundary conditions (8.6a) and (8.6b). For example, ϕ_{H1} is Maxwellian with $u_{rH1} = 0$, and thus, if the following conditions

$$u_{\theta H1} = u_{\theta A1}, \ \tau_{H1} = 0 \ \text{ at } \hat{r} = 1 \ \text{ and } \ u_{\theta H1} = u_{\theta B1}, \ \tau_{H1} = 0 \ \text{ at } \hat{r} = \hat{r}_B,$$

are satisfied, ϕ_{H1} is matched to the boundary conditions (8.6a) and (8.6b) at the order of ε; these four conditions are, however, too strong to find $u_{\theta H1}$ and τ_{H1} that satisfy Eqs. (8.11c) and (8.11d).[2] Thus, we introduce the correction to ϕ_H in two steps.

First, we consider a solution that varies sharply in the neighborhood of the cylinders with the length scale of variation of the order of $\varepsilon^{m-2}L_A$, that is, $\partial\phi/\partial\hat{r} = O(\varepsilon^{-m+2}\phi)$ in the neighborhood of the cylinders and that is continuously transformed into the Hilbert solution as going away from the cylinders.[3] With these assumptions, the solution of Eq. (8.4) is obtained in a power series of ε. The process of analysis is similar to that of the Hilbert expansion except that the derivative term in the integral equation (8.9) is upgraded by $m - 2$. The solution is discriminated by the subscript V.[4] We start under the assumption $u_{rV1} = P_{V1} = 0$ by the same reason as in the analysis of the Hilbert solution. The differential equations for the macroscopic variables for the leading order are

$$\frac{\mathrm{d}\,u_{rV2}}{\mathrm{d}\,y} = 0, \tag{8.12a}$$

$$\frac{\mathrm{d}\,P_{V2}}{\mathrm{d}\,y} = 0, \tag{8.12b}$$

$$u_{rV2}\frac{\mathrm{d}\,u_{\theta V1}}{\mathrm{d}\,y} = \frac{\gamma_1}{2}\frac{\mathrm{d}^2 u_{\theta V1}}{\mathrm{d}\,y^2}, \tag{8.12c}$$

$$u_{rV2}\frac{\mathrm{d}\,\tau_{V1}}{\mathrm{d}\,y} = \frac{\gamma_2}{2}\frac{\mathrm{d}^2 \tau_{V1}}{\mathrm{d}\,y^2}, \tag{8.12d}$$

$$y = \frac{r - r_L}{\varepsilon^{m-2}L_A} = \frac{\hat{r} - \hat{r}_L}{\varepsilon^{m-2}},$$

where r_L and $\hat{r}_L$ are, respectively, L_A and 1 for the solution V on the inner cylinder, or L_B and $\hat{r}_B$ for the solution V on the outer cylinder, and γ_1 and γ_2 are constants defined in Eqs. (3.33) and (3.34), which depend on molecular

---

[2]This situation was more generally explained, e.g., in the first paragraph of Sections 3.4, 6.4, or 7.3.

[3]This length scale of variation of the solution normal to the boundary is suggested by the following consideration: In the Navier–Stokes system, a boundary layer with thickness of the order of kinematic viscosity (viscosity divided by density) divided by the suction speed is developed in a uniform flow along an infinite plane wall with suction (Schlichting [1979]). This implies a boundary layer with thickness of the order of $\ell_A/\varepsilon^2|u_{rH2}|$, which is ε^{m-2} in the $\hat{r}$ variable, on the cylinder of the condensing side, since the kinematic viscosity is proportional to the product of the sound speed, which is the same order as $\sqrt{2RT_A}$, and the mean free path ℓ_A [see Eq. (3.90)].

[4]This is a variation of the viscous-boundary-layer solution introduced in Section 6.3.

models, e.g.,

$$\gamma_1 = 1.270042427, \quad \gamma_2 = 1.922284066 \quad \text{(a hard-sphere gas)},$$
$$\gamma_1 = \gamma_2 = 1 \quad \text{(BKW model)}.$$

Equations (8.12a)–(8.12d) are simplified by the condition that the equations are applied in the region $y = O(1)$ in addition to the above assumptions.

By a similar discussion to that in Section 6.5, the condition of the connection of the Hilbert solution and the solution V is given as follows: As $y \to \infty$ on the inner cylinder or as $y \to -\infty$ on the outer cylinder,

$$h_{V1} \sim (h_{H1})_L, \quad h_{V2} \sim (h_{H2})_L + y\left(\frac{\mathrm{d}\, h_{H1}}{\mathrm{d}\hat{r}}\right)_L \delta_3(m), \tag{8.13}$$

where h represents P, u_r, u_θ, and τ; $\delta_3(m) = 1$ $(m = 3)$ and $\delta_3(m) = 0$ $(m \neq 3)$; and the quantities in the parentheses with subscript L are evaluated on the inner or outer cylinder depending on $L = A$ or $L = B$.

On an evaporating cylinder [on the inner cylinder for $u_{rV2} > 0$ (or $u_{rH2} > 0$ by Eq. (8.13)) or on the outer cylinder for $u_{rV2} < 0$ (or $u_{rH2} < 0$)], the solutions $u_{\theta V1}$ and τ_{V1} in Eqs. (8.12c) and (8.12d) diverge exponentially as $y \to \infty$ on the inner cylinder or as $y \to -\infty$ on the outer cylinder. The solution V that can be connected continuously to the Hilbert solution is possible only on the condensing cylinder. Thus, we call the solution V the suction-boundary-layer solution.

The suction-boundary-layer solution ϕ_V has more freedom to be adjusted to the boundary conditions than the Hilbert solution ϕ_H because Eqs. (8.12c) and (8.12d) are of the second degree in contrast to Eqs. (8.11c) and (8.11d). At the level of order ε, the ϕ_{H1} and ϕ_{V1} being the corresponding components of the Maxwellians with $u_{rH1} = u_{rV1} = P_{H1} = P_{V1} = 0$, the combination of the two solutions ϕ_H and ϕ_V on the condensing cylinder is made to satisfy the boundary conditions (8.6a) and (8.6b) by matching the boundary values of $u_{\theta H1}$ or $u_{\theta V1}$ and τ_{H1} or τ_{V1} as follows:

$$\left.\begin{aligned} u_{\theta G1} &= u_{\theta A1}, \quad \tau_{G1} = 0 \quad \text{at } \hat{r} = 1, \\ u_{\theta G1} &= u_{\theta B1}, \quad \tau_{G1} = 0 \quad \text{at } \hat{r} = \hat{r}_B, \end{aligned}\right\} \tag{8.14}$$

where G is H or V depending on whether the gas is evaporating or condensing there.[5] At the higher orders, the Hilbert and suction-boundary-layer solutions ϕ_H and ϕ_V do not have freedom enough to be matched to the kinetic boundary conditions (8.6a) and (8.6b). For example, both ϕ_H and ϕ_V are Maxwellian up to the order of ε^2, but u_{rH2} and u_{rV2} are not zero; they do not match with the boundary conditions (8.6a) and (8.6b) at the corresponding order (see Footnote 2). Thus, as the second step, we introduce a Knudsen-layer correction to obtain the solution of the boundary-value problem.

[5]The combined system of Eqs. (8.11a)–(8.11d) and Eqs. (8.12a)–(8.12d) with the connection condition (8.13) has more freedom than the simple system (8.11a)–(8.11d).

The solution of the problem is put in the form

$$\phi = \phi_G + \phi_K, \tag{8.15}$$

where $\phi_G = \phi_H$ or ϕ_V depending on the correction on the evaporating or condensing cylinder. The Knudsen-layer correction ϕ_K is assumed to vary appreciably over the distance of the order of the mean free path $[\partial\phi_K/\partial\hat{r} = O(\varepsilon^{-m}\phi_K)]$ and to decay rapidly away from the cylinders. The correction ϕ_K is also expanded in a power series of ε:

$$\phi_K = \phi_{K2}\varepsilon^2 + \phi_{K3}\varepsilon^3 + \cdots, \tag{8.16}$$

where the series starts from the second order of ε because the boundary conditions are satisfied only by ϕ_H and ϕ_V at the first order.

The equation for ϕ_{K2} is given by

$$\zeta_r \frac{\partial\phi_{K2}}{\partial\eta} = \mathcal{L}(\phi_{K2}), \tag{8.17}$$

where $\eta = (\hat{r}-1)/\varepsilon^m$ on the inner cylinder and $\eta = (\hat{r}-\hat{r}_B)/\varepsilon^m$ on the outer cylinder. The boundary condition for ϕ_{K2} is given as follows: For the Knudsen layer on the inner cylinder ($\hat{r}=1$),

$$\phi_{K2} = -P_{G2} - 2\zeta_\theta u_{\theta G2} - \left(\zeta^2 - \frac{5}{2}\right)\tau_{G2} - 2\zeta_r u_{rG2} \quad (\zeta_r > 0) \quad \text{at } \eta = 0,$$

$$\phi_{K2} \to 0 \quad \text{as } \eta \to \infty,$$

and for the Knudsen layer on the outer cylinder ($\hat{r} = \hat{r}_B$),

$$\phi_{K2} = P_{B2} - P_{G2} - 2\zeta_\theta u_{\theta G2} + \left(\zeta^2 - \frac{5}{2}\right)(\tau_{B2} - \tau_{G2}) - 2\zeta_r u_{rG2}$$

$$(\zeta_r < 0) \quad \text{at } \eta = 0,$$

$$\phi_{K2} \to 0 \quad \text{as } \eta \to -\infty,$$

where the subscript G represents H on the evaporating cylinder and V on the condensing cylinder.

These Knudsen-layer problems are special cases of the Knudsen-layer problem given by Eqs. (3.44), (3.51), and (3.52) with $m = 0$ in Section 3.4. Thus, the solution ϕ_{K2} exists uniquely when and only when the boundary values of P_{G2}, $u_{\theta G2}$, τ_{G2}, and u_{rG2} satisfy the following conditions: for the Knudsen layer on the inner cylinder,

$$P_{G2} = C_4^* u_{rG2}, \quad \tau_{G2} = d_4^* u_{rG2}, \quad u_{\theta G2} = 0 \quad \text{at } \hat{r} = 1, \tag{8.18a}$$

and for the Knudsen layer on the outer cylinder,

$$P_{G2} = P_{B2} - C_4^* u_{rG2}, \quad \tau_{G2} = \tau_{B2} - d_4^* u_{rG2}, \quad u_{\theta G2} = 0 \quad \text{at } \hat{r} = \hat{r}_B, \tag{8.18b}$$

where C_4^* and d_4^* are the slip coefficients encountered in Section 3.5.2. That is,

$$C_4^* = -2.1412, \quad d_4^* = -0.4557 \quad \text{(hard-sphere)},$$
$$C_4^* = -2.13204, \quad d_4^* = -0.44675 \quad \text{(BKW)}.$$

We can proceed with the analysis in a similar way.

Corresponding to the staggered combination $(u_{\theta Hs}, \tau_{Hs}, P_{Hs+1}, u_{rHs+1})$ or $(u_{\theta Vs}, \tau_{Vs}, P_{Vs+1}, u_{rVs+1})$ for the set of equations, we choose a staggered combination of the conditions on their boundary data as the boundary condition for the set of equations. The condition (8.14) on the boundary values of $u_{\theta H1}$ or $u_{\theta V1}$ and τ_{H1} or τ_{V1} and the first relations between the boundary values of P_{G2} and u_{rG2} in Eqs. (8.18a) and (8.18b) form the boundary conditions for the equations (8.11a)–(8.11d) or (8.12a)–(8.12d) on the cylinders. That is,

$$u_{\theta G1} = u_{\theta A1}, \quad \tau_{G1} = 0, \quad P_{G2} = C_4^* u_{rG2} \qquad \text{at } \hat{r} = 1, \tag{8.19a}$$

$$u_{\theta G1} = u_{\theta B1}, \quad \tau_{G1} = 0, \quad P_{G2} = P_{B2} - C_4^* u_{rG2} \quad \text{at } \hat{r} = \hat{r}_B. \tag{8.19b}$$

Now enough material has been prepared to obtain the asymptotic solution at the leading order as $\varepsilon \to 0$. We only have to solve Eqs. (8.11a)–(8.11d) and Eqs. (8.12a)–(8.12d) under the connection condition (8.13) and the boundary conditions (8.19a) and (8.19b).

In the above analysis, the correction to the Hilbert solution is carried out in two steps: the suction-boundary-layer solution and the Knudsen-layer correction. Thanks to this procedure, the equation of the Knudsen layer, for which the discussion at the level of the velocity distribution function is required, is reduced to a linear equation. If we try to find the correction in one step, we have to discuss the problem at the level of the velocity distribution function with two different length scales of variation. It is described by a nonlinear equation.

8.2.2 Solution

At this stage, we do not know on which cylinder condensation is taking place. Thus, we will obtain the solutions of the fore-mentioned system of equations and boundary conditions for both cases and examine their consistency with the assumption. The two solutions are given as follows:

(i) Case $u_{rH2} > 0$ (condensation is taking place on the outer cylinder)

$$u_{\theta H1} = \frac{u_{\theta A1}}{\hat{r}}, \quad \tau_{H1} = 0, \quad u_{rH2} = \frac{c_{r2}}{\hat{r}}, \quad P_{H2} = c_{p2} - \left(\frac{u_{\theta A1}}{\hat{r}}\right)^2, \tag{8.20}$$

$$\left.\begin{aligned} u_{\theta V1} &= \frac{u_{\theta A1}}{\hat{r}_B} + \left(u_{\theta B1} - \frac{u_{\theta A1}}{\hat{r}_B}\right)\exp\left(\frac{2c_{r2}y}{\gamma_1 \hat{r}_B}\right), \quad \tau_{V1} = 0, \\ u_{rV2} &= \frac{c_{r2}}{\hat{r}_B}, \quad P_{V2} = c_{p2} - \left(\frac{u_{\theta A1}}{\hat{r}_B}\right)^2, \end{aligned}\right\} \tag{8.21}$$

where

$$\left.\begin{aligned} c_{r2} &= \frac{-1}{C_4^*}\left[-\frac{\hat{r}_B}{\hat{r}_B+1}P_{B2} + \frac{\hat{r}_B - 1}{\hat{r}_B}u_{\theta A1}^2\right], \\ c_{p2} &= \frac{1}{\hat{r}_B}u_{\theta A1}^2 + \frac{\hat{r}_B}{\hat{r}_B+1}P_{B2}. \end{aligned}\right\} \tag{8.22}$$

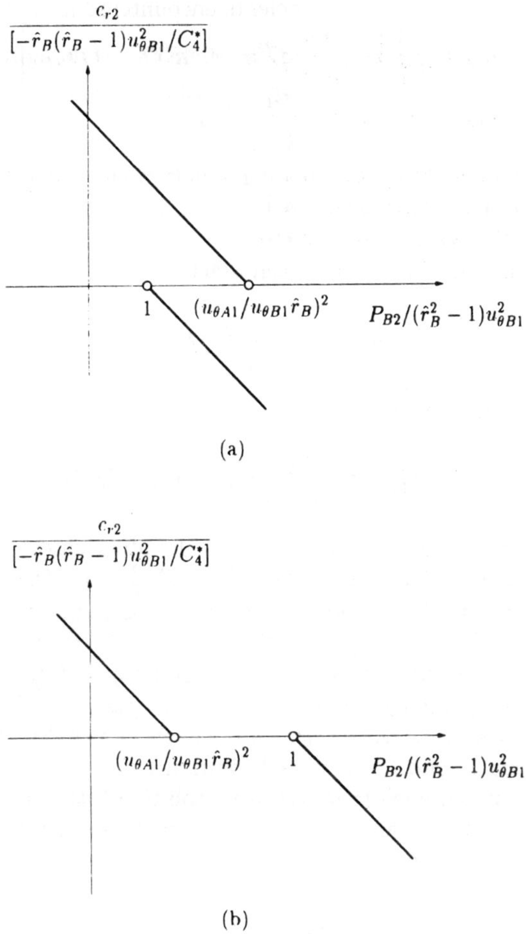

Figure 8.1. Bifurcation diagram of the solution obtained in Section 8.2 (c_{r2} versus P_{B2}). (a) $u_{\theta A1}^2 > (u_{\theta B1}\hat{r}_B)^2$ and (b) $u_{\theta A1}^2 < (u_{\theta B1}\hat{r}_B)^2$. The slope is -1 irrespective of the parameters.

The solution satisfies the condition $u_{rH2} > 0$, or exists when $P_{B2} < (\hat{r}_B^2 - 1)(u_{\theta A1}/\hat{r}_B)^2$ because $C_4^* < 0$. This solution will be called solution I_+.

(ii) Case $u_{rH2} < 0$ (condensation is taking place on the inner cylinder)

$$u_{\theta H1} = \frac{u_{\theta B1}\hat{r}_B}{\hat{r}}, \ \tau_{H1} = 0, \ u_{rH2} = \frac{c_{r2}}{\hat{r}}, \ P_{H2} = c_{p2} - \left(\frac{u_{\theta B1}\hat{r}_B}{\hat{r}}\right)^2, \tag{8.23}$$

$$\left.\begin{aligned} &u_{\theta V1} = u_{\theta B1}\hat{r}_B + (u_{\theta A1} - u_{\theta B1}\hat{r}_B)\exp\left(\frac{2c_{r2}y}{\gamma_1}\right), \quad \tau_{V1} = 0, \\ &u_{rV2} = c_{r2}, \quad P_{V2} = c_{p2} - (u_{\theta B1}\hat{r}_B)^2, \end{aligned}\right\} \tag{8.24}$$

where

$$\left.\begin{aligned} c_{r2} &= \frac{-1}{C_4^*}\left[-\frac{\hat{r}_B}{\hat{r}_B+1}P_{B2}+\hat{r}_B(\hat{r}_B-1)u_{\theta B1}^2\right], \\ c_{p2} &= \hat{r}_B u_{\theta B1}^2 + \frac{\hat{r}_B}{\hat{r}_B+1}P_{B2}. \end{aligned}\right\} \tag{8.25}$$

The solution satisfies the condition $u_{rH2} < 0$, or exists when $P_{B2} > (\hat{r}_B^2-1)u_{\theta B1}^2$. This solution will be called solution I$_-$.

Combining the results, we find that
(A) the two solutions, I$_+$ and I$_-$, exist when

$$(u_{\theta B1}\hat{r}_B)^2 < \frac{\hat{r}_B^2}{\hat{r}_B^2-1}P_{B2} < u_{\theta A1}^2; \tag{8.26}$$

(B) no solution exists when

$$u_{\theta A1}^2 < \frac{\hat{r}_B^2}{\hat{r}_B^2-1}P_{B2} < (u_{\theta B1}\hat{r}_B)^2; \tag{8.27}$$

(C) one of the solutions, I$_+$ or I$_-$, exists in the other cases.

As the parameter P_{B2} approaches $(\hat{r}_B^2-1)u_{\theta A1}^2/\hat{r}_B^2$ from below, c_{r2} (>0) tends to zero and finally solution I$_+$ ceases to exist, and on the other hand, as P_{B2} approaches $(\hat{r}_B^2-1)u_{\theta B1}^2$ from above, c_{r2} (<0) tends to zero and finally solution I$_-$ ceases to exist. We will call the point $P_{B2} = (\hat{r}_B^2-1)u_{\theta A1}^2/\hat{r}_B^2$ critical point C_+ and the point $P_{B2} = (\hat{r}_B^2-1)u_{\theta B1}^2$ critical point C_-. The bifurcation diagrams in terms of c_{r2} versus P_{B2} are shown in Fig. 8.1. Hereafter, we call the two solutions I$_+$ and I$_-$ combined solution I.

In the solution obtained in this section, solution I, the effect of viscosity and thermal conductivity (γ_1 and γ_2) is confined in the suction boundary layer and is a secondary one in most of the flow field.

8.3 Solution type II

The two solutions I$_+$ and I$_-$ are disconnected unless the two critical points C_+ and C_- coincide, or unless $u_{\theta A1}^2 = u_{\theta B1}^2\hat{r}_B^2$. In the present section, we present another type of solution (say, solution II) connecting the two solutions. The Hilbert expansion (8.8) of the Boltzmann equation (8.4) can be carried out consistently under the assumption that $u_{rHs} = 0$ ($s = 1, \ldots, m-1$) as well as $P_{H1} = 0$, and a series of differential equations that determine the variables $(u_{\theta Hs}, \tau_{Hs}, P_{Hs+1}, u_{rHs+m-1})$ as a set is obtained from the solvability conditions of the linear integral equations for the component functions ϕ_{Hs} of the expansion of ϕ_H in ε. The set of equations for $(u_{\theta H1}, \tau_{H1}, P_{H2}, u_{rHm})$ of the leading order is given as follows:[6]

$$\frac{\mathrm{d}\, u_{rHm}\hat{r}}{\mathrm{d}\,\hat{r}} = 0, \tag{8.28a}$$

[6]The radial velocity u_{rHm} does not appear in the highest differential terms in Eqs. (8.28c) and (8.28d). Thus, the case $u_{rHm} = 0$ has not to be treated separately.

$$\frac{u_{\theta H1}^2}{\hat{r}} = \frac{1}{2}\frac{\mathrm{d}P_{H2}}{\mathrm{d}\hat{r}}, \tag{8.28b}$$

$$\frac{u_{rHm}}{\hat{r}}\frac{\mathrm{d}u_{\theta H1}\hat{r}}{\mathrm{d}\hat{r}} = \frac{\gamma_1}{2}\left[\frac{1}{\hat{r}}\frac{\mathrm{d}}{\mathrm{d}\hat{r}}\left(\hat{r}\frac{\mathrm{d}u_{\theta H1}}{\mathrm{d}\hat{r}}\right) - \frac{u_{\theta H1}}{\hat{r}^2}\right], \tag{8.28c}$$

$$u_{rHm}\frac{\mathrm{d}\tau_{H1}}{\mathrm{d}\hat{r}} = \frac{\gamma_2}{2}\frac{1}{\hat{r}}\frac{\mathrm{d}}{\mathrm{d}\hat{r}}\left(\hat{r}\frac{\mathrm{d}\tau_{H1}}{\mathrm{d}\hat{r}}\right). \tag{8.28d}$$

Equations (8.28c) and (8.28d) for $u_{\theta H1}$ and τ_{H1} are of the second degree in contrast to Eqs. (8.11c) and (8.11d). The effect of viscosity and thermal conductivity (γ_1 and γ_2) is no longer confined in a suction boundary layer but spreads over the whole field. Equations (8.28a)–(8.28d) play the combined role of Eqs. (8.11a)–(8.11d) and Eqs. (8.12a)–(8.12d). The sets of equations at the higher orders also have the same character.

The component function ϕ_{Hs} from $s = 1$ to $s = m-1$ is found to be the corresponding term of the expansion of the local Maxwellian $\phi_e(P_H, \tau_H, u_{rH}, u_{\theta H})$ with $u_{rH} = 0$ from the discussion in Section 8.2.1 and the assumption introduced at the beginning of this section. Thus, ϕ_{Hs} ($s = 1, \ldots, m-1$) can be made to satisfy the kinetic boundary conditions (8.6a) and (8.6b) only by adjusting the boundary values of $u_{\theta Hs}, \tau_{Hs}$, and P_{Hs} (or ω_{Hs}) at $\hat{r} = 1$ and $\hat{r} = \hat{r}_B$. The six conditions with staggered combination $(u_{\theta Hs}, \tau_{Hs}, P_{Hs+1})$ chosen from the condition obtained for $u_{\theta Hs}$, τ_{Hs}, and P_{Hs} ($s = 1, \ldots, m-1$) are appropriate conditions to determine the solution of the set of equations for $(u_{\theta Hs}, \tau_{Hs}, P_{Hs+1}, u_{rHs+m-1})$ with $s = 1, \ldots, m-2$. For the higher orders ($s \geq m$), Knudsen-layer corrections ϕ_{Ks} are required to obtain the solution that satisfies the kinetic boundary conditions (8.6a) and (8.6b) of the corresponding order. From the Knudsen-layer analysis, the relations among the boundary values of $(u_{\theta Hs}, u_{rHs}, \tau_{Hs}, P_{Hs}; s \geq m)$ are derived. From an appropriate staggered combination of the relations obtained up to this point, the boundary conditions for the set of equations for $(u_{\theta Hs}, \tau_{Hs}, P_{Hs+1}; s \geq m-1)$ are obtained. The boundary conditions for the set of equations (8.28a)–(8.28d) are

$$u_{\theta H1} = u_{\theta A1}, \quad \tau_{H1} = 0, \quad P_{H2} = 0 \qquad \text{at } \hat{r} = 1, \tag{8.29a}$$

$$u_{\theta H1} = u_{\theta B1}, \quad \tau_{H1} = 0, \quad P_{H2} = P_{B2} \quad \text{at } \hat{r} = \hat{r}_B. \tag{8.29b}$$

The solution of this boundary-value problem, Eqs. (8.28a)–(8.28d) with (8.29a) and (8.29b), is easily obtained in the following form:

$$\left.\begin{aligned} &u_{\theta H1} = \frac{a}{\hat{r}} + b\hat{r}^{1+\alpha}, \quad \tau_{H1} = 0, \quad u_{rHm} = \frac{\gamma_1\alpha}{2\hat{r}}, \\ &P_{H2} = P_{B2} + a^2\left(\frac{1}{\hat{r}_B^2} - \frac{1}{\hat{r}^2}\right) + \frac{b^2}{\alpha+1}(\hat{r}^{2(1+\alpha)} - \hat{r}_B^{2(1+\alpha)}) \\ &\quad + \frac{4ab}{\alpha}(\hat{r}^{\alpha} - \hat{r}_B^{\alpha}), \end{aligned}\right\} \tag{8.30}$$

where

$$a = \frac{\hat{r}_B(u_{\theta A1}\hat{r}_B^{1+\alpha} - u_{\theta B1})}{\hat{r}_B^{2+\alpha} - 1}, \quad b = \frac{u_{\theta B1}\hat{r}_B - u_{\theta A1}}{\hat{r}_B^{2+\alpha} - 1},$$

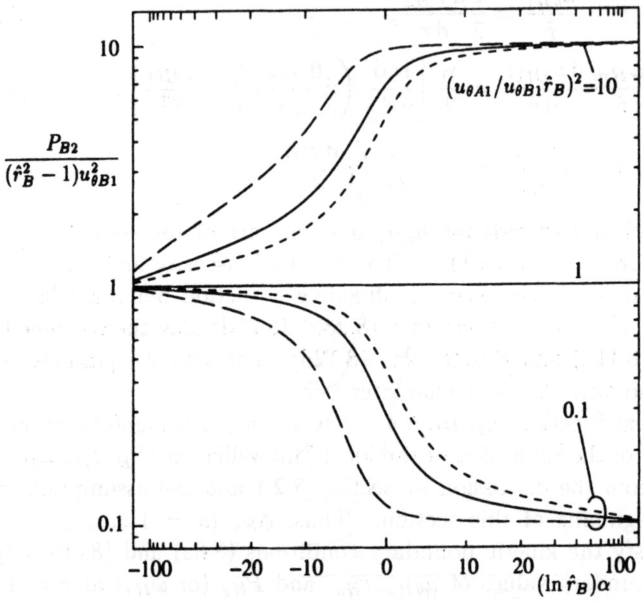

Figure 8.2. $P_{B2}/(\hat{r}_B^2-1)u_{\theta B1}^2$ **versus** α **[Eq. (8.31)] I: The two cylinders are rotating in the same direction ($u_{\theta A1}u_{\theta B1} > 0$). The solid lines —— indicate the case $\hat{r}_B = 2$, the long dashed lines — — —: $\hat{r}_B = 10$, and the short dashed lines - - -: $\hat{r}_B = 1.1$.**

and α is the solution of the transcendental equation[7]

$$\frac{P_{B2}}{(\hat{r}_B^2-1)u_{\theta B1}^2} = \frac{(\hat{r}_B^{2+\alpha}M_u-1)^2}{(\hat{r}_B^{2+\alpha}-1)^2} + \frac{\hat{r}_B^2(\hat{r}_B^{2(1+\alpha)}-1)(M_u-1)^2}{(\alpha+1)(\hat{r}_B^2-1)(\hat{r}_B^{2+\alpha}-1)^2} - \frac{4\hat{r}_B^2(\hat{r}_B^\alpha-1)(M_u-1)(\hat{r}_B^{2+\alpha}M_u-1)}{\alpha(\hat{r}_B^2-1)(\hat{r}_B^{2+\alpha}-1)^2}, \tag{8.31}$$

$$M_u = \frac{u_{\theta A1}}{u_{\theta B1}\hat{r}_B}.$$

The $P_{B2}/(\hat{r}_B^2-1)u_{\theta B1}^2$ is a continuous function of α, and it approaches 1 as $\alpha \to -\infty$ and $(u_{\theta A1}/u_{\theta B1}\hat{r}_B)^2$ as $\alpha \to \infty$. Thus, Eq. (8.31) for α has, at least, one solution when $P_{B2}/(\hat{r}_B^2-1)u_{\theta B1}^2$ lies in the range

$$\left[\frac{P_{B2}}{(\hat{r}_B^2-1)u_{\theta B1}^2}-1\right]\left[\frac{P_{B2}}{(\hat{r}_B^2-1)u_{\theta B1}^2}-\left(\frac{u_{\theta A1}}{u_{\theta B1}\hat{r}_B}\right)^2\right] < 0. \tag{8.32}$$

The point $\alpha = \infty$ corresponds to critical point C_+, and the point $\alpha = -\infty$ corresponds to critical point C_-. Therefore, there is another solution that connects

[7]In the solution obtained and the following expression, there are apparent singularities, i.e., $\alpha = 0, -1$, and -2. The correct solutions for these cases are obtained by the corresponding limiting processes.

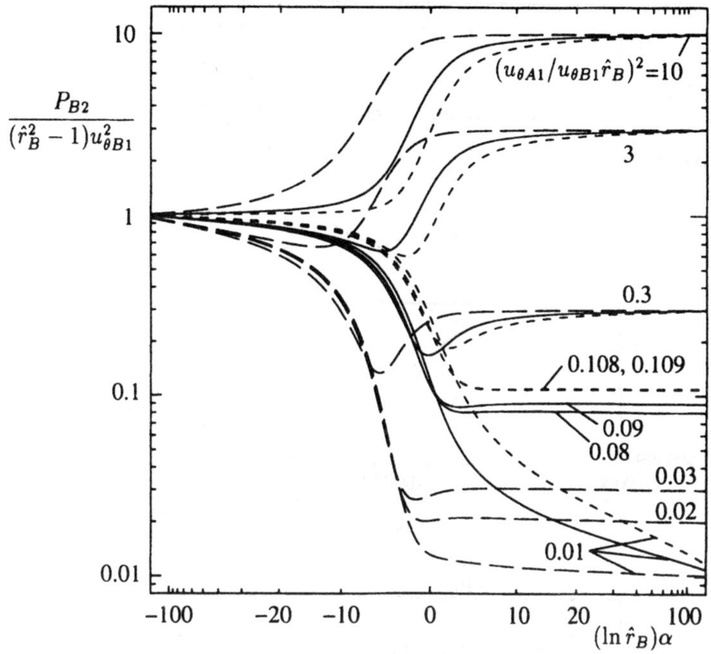

Figure 8.3. $P_{B2}/(\hat{r}_B^2-1)u_{\theta B1}^2$ versus α [Eq. (8.31)] II: The two cylinders are rotating in opposite directions ($u_{\theta A1}u_{\theta B1} < 0$). The solid lines —— indicate the case $\hat{r}_B = 2$, the long dashed lines — — —: $\hat{r}_B = 10$, and the short dashed lines - - -: $\hat{r}_B = 1.1$. The profile for $[(u_{\theta A1}/u_{\theta B1}\hat{r}_B)^2 = 0.109, \hat{r}_B = 1.1]$ and that for $[(u_{\theta A1}/u_{\theta B1}\hat{r}_B)^2 = 0.03, \hat{r}_B = 10]$ show a similar oscillatory behavior, with the minimum at a finite α, to that for $[(u_{\theta A1}/u_{\theta B1}\hat{r}_B)^2 = 0.09, \hat{r}_B = 2]$. The profile for $[(u_{\theta A1}/u_{\theta B1}\hat{r}_B)^2 = 0.108, \hat{r}_B = 1.1]$ and that for $[(u_{\theta A1}/u_{\theta B1}\hat{r}_B)^2 = 0.02, \hat{r}_B = 10]$ show a similar oscillatory behavior to that for $[(u_{\theta A1}/u_{\theta B1}\hat{r}_B)^2 = 0.08, \hat{r}_B = 2]$. See Fig. 8.4 for the detailed profiles of the two kinds of oscillatory behavior.

the two critical points C_+ and C_-.

The behavior of Eq. (8.31) is examined numerically with the aid of its asymptotic behavior as $\alpha \to \pm\infty$. When the two cylinders are rotating in the same direction (or $u_{\theta A1}u_{\theta B1} > 0$), $P_{B2}/(\hat{r}_B^2-1)u_{\theta B1}^2$ is a monotonic function of α, monotonic increasing for $(u_{\theta A1}/u_{\theta B1}\hat{r}_B)^2 > 1$ or decreasing for $(u_{\theta A1}/u_{\theta B1}\hat{r}_B)^2 < 1$ (Fig. 8.2). The solution α of Eq. (8.31) (or the solution II) uniquely exists in the range (8.32). On the other hand, when they are rotating in opposite directions (or $u_{\theta A1}u_{\theta B1} < 0$), $P_{B2}/(\hat{r}_B^2-1)u_{\theta B1}^2$ is not necessarily a monotonic function of α, and Eq. (8.31) for α (or solution II) may have three solutions in some range of Eq. (8.32) and furthermore may have two solutions in some range outside Eq. (8.32). Several examples of the profile $P_{B2}/(\hat{r}_B^2-1)u_{\theta B1}^2$ versus α are shown in Figs. 8.3 and 8.4. The bifurcation diagram of solution II for $u_{\theta A1}u_{\theta B1} < 0$ derived from these figures is classified into six types, which are shown in Fig. 8.5.

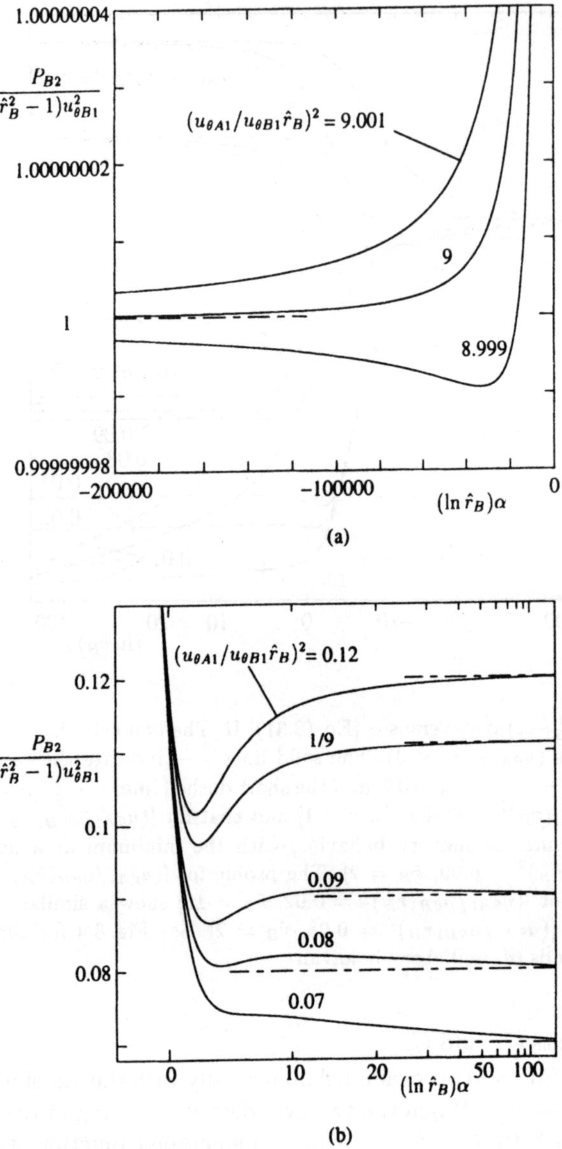

Figure 8.4. $P_{B2}/(\hat{r}_B^2 - 1)u_{\theta B1}^2$ versus α [Eq. (8.31)] III: Magnified local profiles ($u_{\theta A1}u_{\theta B1} < 0$). (a) Profiles ($\hat{r}_B = 2$) around $u_{\theta A1} = -3u_{\theta B1}\hat{r}_B$, and (b) profiles ($\hat{r}_B = 2$) around $u_{\theta A1} = -u_{\theta B1}\hat{r}_B/3$. In panel (a), the chain line —-— indicates the line $P_{B2}/(\hat{r}_B^2 - 1)u_{\theta B1}^2 = 1$. In panel (b), the chain lines —-— indicate the asymptotes as $\alpha \to \infty$ [or $P_{B2}/(\hat{r}_B^2 - 1)u_{\theta B1}^2 = (u_{\theta A1}/u_{\theta B1}\hat{r}_B)^2$].

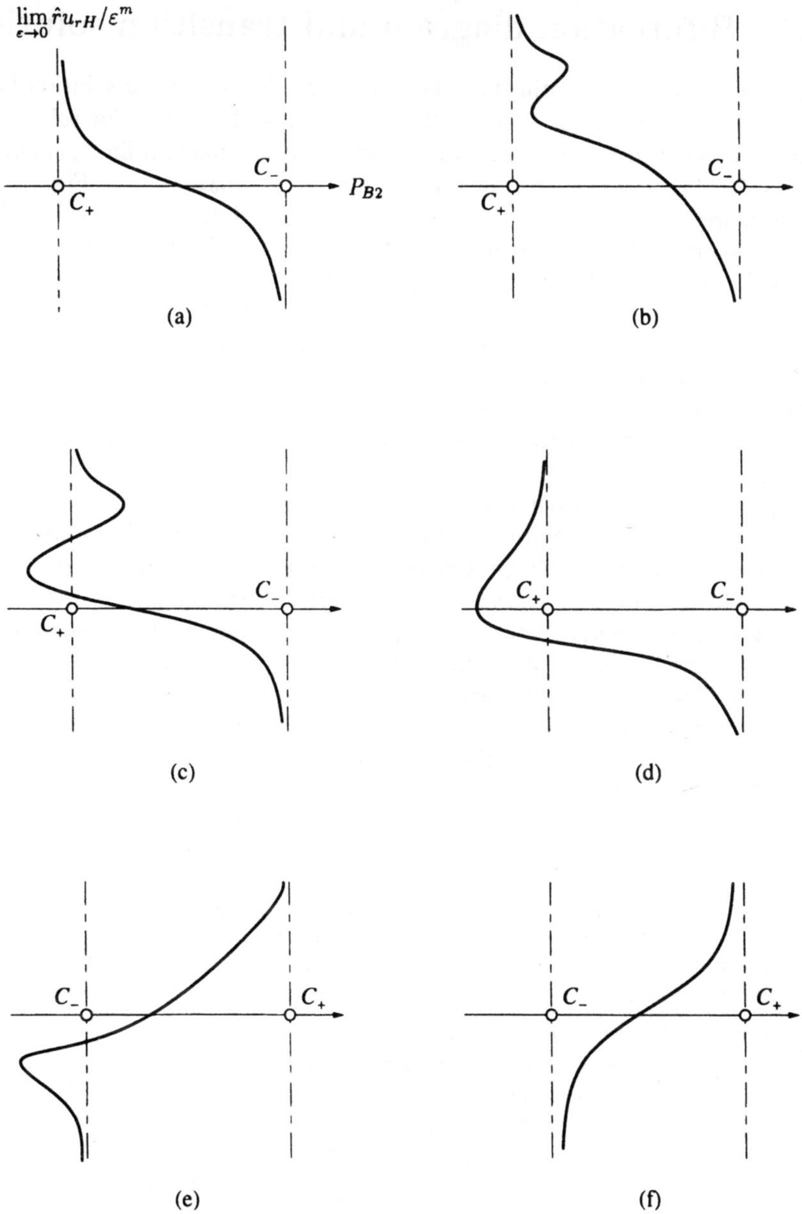

Figure 8.5. Bifurcation diagram of solution II (or schematic profiles $\lim_{\varepsilon\to0}\hat{r}u_{rH}/\varepsilon^m$ versus P_{B2}), when the two cylinders are rotating in opposite directions ($u_{\theta A1}u_{\theta B1} < 0$). (a) $(u_{\theta A1}/u_{\theta B1}\hat{r}_B)^2 \leq 1/9c_1$ ($c_1 > 1$), (b) $1/9c_1 < (u_{\theta A1}/u_{\theta B1}\hat{r}_B)^2 < 1/9c_2$ ($c_1 > c_2 > 1$), (c) $1/9c_2 < (u_{\theta A1}/u_{\theta B1}\hat{r}_B)^2 < 1/9$, (d) $1/9 \leq (u_{\theta A1}/u_{\theta B1}\hat{r}_B)^2 < 1$, (e) $1 < (u_{\theta A1}/u_{\theta B1}\hat{r}_B)^2 < 9$, and (f) $(u_{\theta A1}/u_{\theta B1}\hat{r}_B)^2 \geq 9$, where c_1 and c_2 depend on $\hat{r}_B$.

8.4 Bifurcation diagram and transition solution

Each solution can be identified by the value of c_{r2} (or $u_{rH2}\hat{r}$) for solution I and that of α (or $u_{rHm}\hat{r}$) for solution II. Summarizing the results for solutions I and II obtained in Sections 8.2 and 8.3, we show the diagram $\lim_{\varepsilon\to 0} u_{rH}\hat{r}/\varepsilon^2$ versus P_{B2} for $u_{\theta A1}u_{\theta B1} > 0$ in Fig. 8.6 and for $u_{\theta A1}u_{\theta B1} < 0$ in Fig. 8.7. In these figures, $\lim_{\varepsilon\to 0} u_{rH}\hat{r}/\varepsilon^2$ vanishes for all the solutions of type II. Thus, the microstructure is shown in these figures. When both cylinders are rotating in the same direction ($u_{\theta A1}u_{\theta B1} > 0$), the situation is simple: (a) If $(u_{\theta A1}/u_{\theta B1}\hat{r}_B)^2 > 1$, three solutions, I_+, I_-, and II, exist in the range $1 < P_{B2}/(\hat{r}_B^2 - 1)u_{\theta B1}^2 < (u_{\theta A1}/u_{\theta B1}\hat{r}_B)^2$, and the solution is unique in the other range; (b) if $(u_{\theta A1}/u_{\theta B1}\hat{r}_B)^2 < 1$, the solution is unique in the whole range. When the two cylinders are rotating in opposite directions ($u_{\theta A1}u_{\theta B1} < 0$), the situation is classified into the six cases shown in Fig. 8.7, depending on the values of $(u_{\theta A1}/u_{\theta B1}\hat{r}_B)^2$ and $\hat{r}_B$.

Up to now, we have excluded the discussion at the critical points C_+ and C_-. The two solutions I_+ and II or I_- and II are not confirmed to be transformed into each other continuously because of the difference of the order of the radial velocity. In the foregoing analyses, the asymptotic solution as $\varepsilon \to 0$ is sought for a fixed set of the parameters $\hat{r}_B$, $u_{\theta A1}$, $u_{\theta B1}$, τ_{B2}, and P_{B2}. The parameters may be arbitrarily close to a critical point as far as they are fixed in the process as $\varepsilon \to 0$. Therefore the bifurcation diagram as $\varepsilon \to 0$ is not affected. Thus, in the present monograph, we examine the transition solution between the two types of solutions, i.e., the solutions I and II, only for the case $m = 4$.

The analysis of the transition solution is carried out in the neighborhood of the critical point C_+ or C_- variable with ε. That is, putting

$$P_{B2} - (\hat{r}_B^2 - 1)(u_{\theta A1}/\hat{r}_B)^2 = \Delta_{(C+)}\varepsilon \quad \text{(around } C_+\text{)},$$

$$\text{or} \quad P_{B2} - (\hat{r}_B^2 - 1)u_{\theta B1}^2 = \Delta_{(C-)}\varepsilon \quad \text{(around } C_-\text{)},$$

we carry out an analysis similar to that in Section 8.2, which can be done consistently to higher orders. Brief summary of the results is as follows:

(i) In the neighborhood of critical point C_+

(a) When $(3u_{\theta A1} + u_{\theta B1}\hat{r}_B)(u_{\theta A1} - u_{\theta B1}\hat{r}_B) < 0$, the solution with positive u_{rH} exists uniquely for all values of $\Delta_{(C+)}$. This solution is continuously transformed into solution I_+ in the intermediate region $O(\varepsilon) \ll -\Delta_{(C+)}\varepsilon \ll O(1)$ and into solution II in a region $O(\varepsilon) \ll \Delta_{(C+)}\varepsilon \ll O(1)$.

(b) When $(3u_{\theta A1} + u_{\theta B1}\hat{r}_B)(u_{\theta A1} - u_{\theta B1}\hat{r}_B) > 0$, two solutions with positive u_{rH} exist for $\Delta_{(C+)}$ that is smaller than some negative value, and they are continuously joined at the threshold value of $\Delta_{(C+)}$. No solution exists for the other values of $\Delta_{(C+)}$. One of the solution with larger u_{rH} is transformed into solution I_+ and the other solution into II in an intermediate region $O(\varepsilon) \ll -\Delta_{(C+)}\varepsilon \ll O(1)$.

(ii) In the neighborhood of critical point C_-

(a) When $(u_{\theta A1} + 3u_{\theta B1}\hat{r}_B)(u_{\theta A1} - u_{\theta B1}\hat{r}_B) < 0$, the solution with negative u_{rH} exists uniquely for all values of $\Delta_{(C-)}$. This solution is continuously transformed

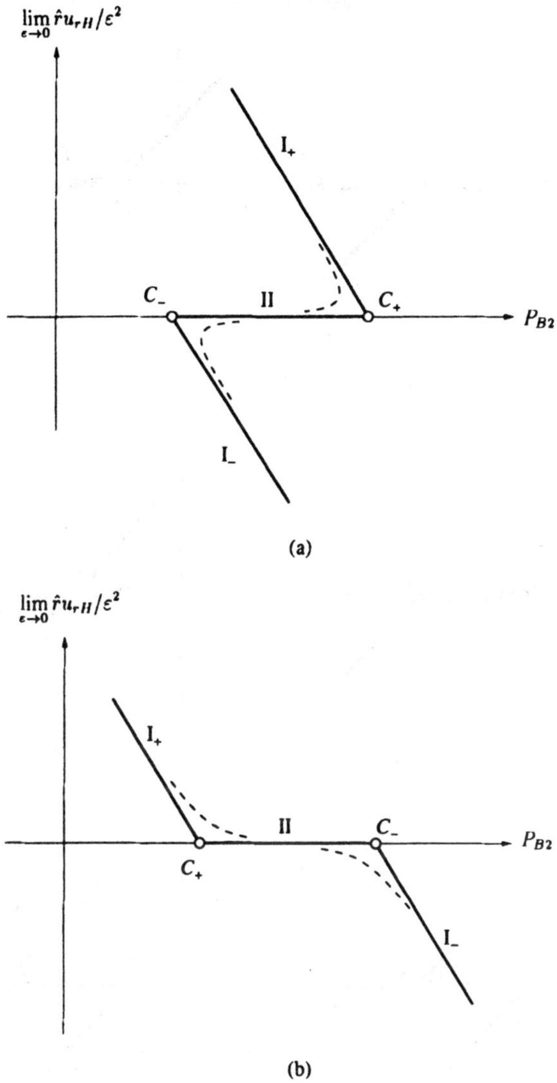

Figure 8.6. Bifurcation diagram (or schematic profiles $\lim_{\varepsilon\to 0} \hat{r}u_{rH}/\varepsilon^2$ versus P_{B2}) I: The two cylinders are rotating in the same direction ($u_{\theta A1}u_{\theta B1} > 0$). (a) $u_{\theta A1}^2 > (u_{\theta B1}\hat{r}_B)^2$ and (b) $u_{\theta A1}^2 < (u_{\theta B1}\hat{r}_B)^2$. The solution is indicated by thick solid lines. The connection of the two solutions I and II by a transition solution is shown by dashed lines.

into solution I$_-$ in the intermediate region $O(\varepsilon) \ll \Delta_{(C-)}\varepsilon \ll O(1)$ and into solution II in a region $O(\varepsilon) \ll -\Delta_{(C-)}\varepsilon \ll O(1)$.

(b) When $(u_{\theta A1} + 3u_{\theta B1}\hat{r}_B)(u_{\theta A1} - u_{\theta B1}\hat{r}_B) > 0$, two solutions with negative u_{rH} exist for $\Delta_{(C-)}$ that is larger than some positive value, and they are continuously joined at the threshold value of $\Delta_{(C-)}$. No solution exists for the other

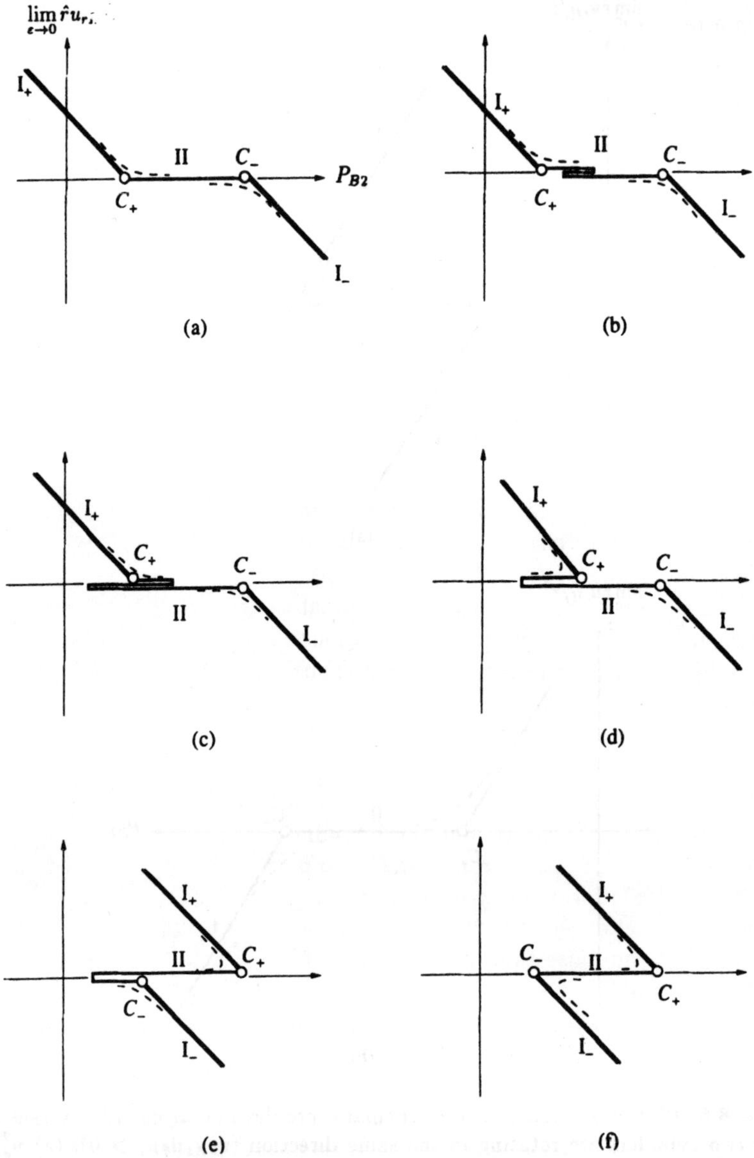

Figure 8.7. Bifurcation diagram (or schematic profiles $\lim_{\varepsilon\to 0} \hat{r}u_{rH}/\varepsilon^2$ versus P_{B2}) II: The two cylinders are rotating in opposite directions ($u_{\theta A1}u_{\theta B1} < 0$). (a) $(u_{\theta A1}/u_{\theta B1}\hat{r}_B)^2 \le 1/9c_1$, (b) $1/9c_1 < (u_{\theta A1}/u_{\theta B1}\hat{r}_B)^2 < 1/9c_2$, (c) $1/9c_2 < (u_{\theta A1}/u_{\theta B1}\hat{r}_B)^2 < 1/9$, (d) $1/9 \le (u_{\theta A1}/u_{\theta B1}\hat{r}_B)^2 < 1$, (e) $1 < (u_{\theta A1}/u_{\theta B1}\hat{r}_B)^2 < 9$, and (f) $(u_{\theta A1}/u_{\theta B1}\hat{r}_B)^2 \ge 9$, where c_1 and c_2 depend on $\hat{r}_B$. The solution is indicated by thick or medium solid lines. The connection of the two solutions I and II by a transition solution is shown by dashed lines.

values of $\Delta_{(C-)}$. These solutions are continuously transformed into solution I_- or II in a region $O(\varepsilon) \ll \Delta_{(C-)}\varepsilon \ll O(1)$.

The behavior of the transition solutions in the bifurcation diagrams in Figs. 8.6 and 8.7 is shown by dashed lines.

8.5 Discussions for the other parameter range

In this section, we briefly discuss the cases where $p_B/p_A - 1$ is of the order of ε, or $m = 1$ or 2, which are excluded in the foregoing analysis.[8]

When $p_B/p_A - 1$ is of the order of ε, according to the general theory in Chapter 4 or the extension of the present study, the radial velocity of the order of ε is uniquely determined by $p_B/p_A - 1$, and the tangential velocity is also determined uniquely and independently of the higher-order radial velocity. The general asymptotic theory for the case $m = 1$ is explained in Chapter 4. When $m = 1$ and $p_B/p_A - 1 = O(\varepsilon^2)$, the radial velocity is of the order of ε^2, and the leading tangential velocity is uniquely determined by the Stokes set of equations without convection terms (no radial velocity enters in this set).

When $m = 2$ and $p_B/p_A - 1 = P_{B2}\varepsilon^2$ $[P_{B2} = O(1)]$, the asymptotic analysis for small ε leads to the sets of equations similar to those obtained in Section 8.3. That is, $u_{rH1} = P_{H1} = 0$, and the equations governing $u_{\theta H1}$, τ_{H1}, u_{rH2}, and P_{H2} are Eqs. (8.28a)–(8.28d) with $m = 2$. The boundary conditions are

$$u_{\theta H1} = u_{\theta A1}, \quad \tau_{H1} = 0, \quad P_{H2} = C_4^* u_{rH2} \qquad \text{at } \hat{r} = 1,$$

$$u_{\theta H1} = u_{\theta B1}, \quad \tau_{H1} = 0, \quad P_{H2} = P_{B2} - C_4^* u_{rH2} \qquad \text{at } \hat{r} = \hat{r}_B.$$

The equations are of type II, but the boundary conditions are of type I. The solution exists for the whole range of P_{B2} and has mixed features of solutions I and II: The solution is of type II in the sense that the viscous effect is not confined in a suction boundary layer but extends over the whole domain, but is of type I in the sense that the solution exists for arbitrarily large P_{B2} and that $\lim_{\varepsilon\to 0} \hat{r}u_{rH}/\varepsilon^2$ diverges to ∞ or $-\infty$ as P_{B2} tends to $-\infty$ or ∞. The bifurcation diagram ($\lim_{\varepsilon\to 0} \hat{r}u_{rH}/\varepsilon^2$ versus P_{B2}) has two types as shown in Figs. 8.8 (a) and 8.8 (b) in the latter of which three solutions are possible.

8.6 Concluding remark and supplementary comment

We considered a gas between two coaxial circular cylinders made of the condensed phase of the gas, where evaporation or condensation takes place. The time-independent axially symmetric and uniform behavior of the gas was investigated analytically with special interest in bifurcation of the flow, when the

[8]The saturated gas pressure of a gas is a rapidly increasing function of its temperature. Thus, we consider only the case $T_B/T_A - 1 = O(p_B/p_A - 1)$.

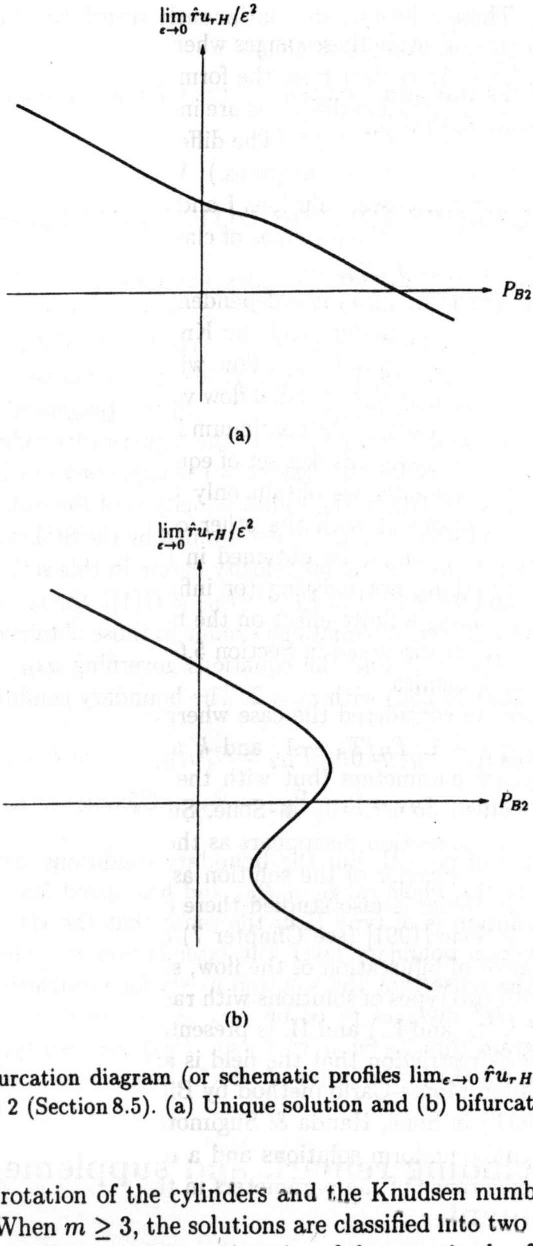

Figure 8.8. Bifurcation diagram (or schematic profiles $\lim_{\varepsilon\to 0}\hat{r}u_{rH}/\varepsilon^2$ versus P_{B2}) for the case $m = 2$ (Section 8.5). (a) Unique solution and (b) bifurcation.

speeds $O(\varepsilon)$ of rotation of the cylinders and the Knudsen number $O(\varepsilon^m)$ were small ($\varepsilon \ll 1$). When $m \geq 3$, the solutions are classified into two types, I and II, (Sections 8.2 and 8.3) according to the order of the magnitude of the radial flow velocity (or whether the effect of viscosity is confined in a thin layer or extends over the whole field). The type-I solutions are subclassified into two classes, I_+ and I_-, of positive or negative radial velocity. In some range of the parameters explicitly given, the two subclasses of solutions coexist and in some other range explicitly given, no solution I (neither I_+ nor I_-) exists. In these ranges, there exists at least one solution of type II. Two or three solutions of type II also exist in some range outside or inside these ranges when the two cylinders are rotating in opposite directions. As is clear from the formulas, the formulas of classification and range in the bifurcation diagrams are independent of molecular models (or γ_1, γ_2, and C_4^*) as far as $C_4^* < 0$.[9] (The difference appears only in the scale of the ordinate of the bifurcation diagrams.) When $m = 2$, the solution has mixed features of the two types, solutions I and II, and bifurcation with three solutions occurs. In this case the formulas of classification and range depend on molecular models (or γ_1 and C_4^*).

The feature of the bifurcation is independent of m (the relative size of the speeds of rotation of the cylinders and the Knudsen number) if $m \geq 3$. The limiting case $m \to \infty$ corresponds to a flow with small Mach numbers in the continuum limit. In solution II, the radial flow velocity u_r vanishes in this limit. That is, if we live in the world of the continuum limit, there is no radial velocity. However, if we solve the Navier–Stokes set of equations for small Mach numbers neglecting the radial velocity, we obtain only the solution II with $\alpha = 0$ in Section 8.3. The solutions II with the other α, which are also the solutions in the continuum limit, cannot be obtained in the framework of the classical gas dynamics. Something not existing (or infinitesimal) in the world of the continuum limit produces a finite effect on the behavior of a gas in that world, which is the ghost effect discussed in Section 5.6 and shows the incompleteness of the classical gas dynamics.

In this chapter, we considered the case where the parameters $V_{\theta A}/\sqrt{2RT_A}$, $V_{\theta B}/\sqrt{2RT_A}$, $p_B/p_A - 1$, $T_B/T_A - 1$, and k are all vanishingly small. For finite values of these parameters (but with the outer cylinder being at rest), the problem is studied numerically in Sone, Sugimoto & Aoki [1999], where how the bifurcation of solution disappears as the Knudsen number increases is shown. The limiting behavior of the solution as $k \to 0$ with other parameters being kept at finite values is also studied there on the basis of the asymptotic theories in Aoki & Sone [1991] (see Chapter 7) and Sone, Takata & Sugimoto [1996]. The behavior of bifurcation of the flow, similar to that for $m \geq 3$ in the present study, with two types of solutions with radial velocity of different orders, corresponding to I (I_+ and I_-) and II, is presented. Recently, the problem was studied without the restriction that the field is axially uniform numerically by the direct-simulation Monte-Carlo method by Bird [1963] (see also Bird [1994], Sone & Aoki [1994]) in Sone, Handa & Sugimoto [2002], where three types of solutions: two axially uniform solutions and a roll-type solution are shown to exist stably in some range of the parameters in the problem.

[9] $C_4^* < 0$ for a hard-sphere gas and the BKW model.

Appendix A

Supplementary Explanations and Formulas

A.1 Formal derivation of the Boltzmann equation from the Liouville equation

We will give a formal derivation of the Boltzmann equation from the Liouville equation on the basis of Grad [1958]. Consider a system consisting of N identical particles each of which has mass m. Take the $6N$-dimensional Γ space consisting of the position $\boldsymbol{X}^{(i)}$ and velocity $\boldsymbol{\xi}^{(i)}$ of the particle i ($i = 1, 2, \ldots, N$). Let f^N, a function of the $6N+1$ variables $\boldsymbol{X}^{(1)}, \ldots, \boldsymbol{X}^{(N)}, \boldsymbol{\xi}^{(1)}, \ldots, \boldsymbol{\xi}^{(N)}$, and t, be the N-particle probability density function finding the particle i ($i = 1, 2, \ldots, N$) at the position $\boldsymbol{X}^{(i)}$ and at velocity $\boldsymbol{\xi}^{(i)}$ in the Γ space with $6N$ dimensions at time t. The behavior of the density function f^N is determined by the Liouville equation[1] given by

$$\frac{\partial f^N}{\partial t} + \sum_{i=1}^{N} \left(\boldsymbol{\xi}^{(i)} \frac{\partial f^N}{\partial \boldsymbol{X}^{(i)}} + \boldsymbol{F}^{(i)} \frac{\partial f^N}{\partial \boldsymbol{\xi}^{(i)}} \right) = 0, \tag{A.1}$$

where $m\boldsymbol{F}^{(i)}$ is the force acting on the particle i. The meaning of the simplified notations of the gradients $\partial f^N/\partial \boldsymbol{X}^{(i)}$ and $\partial f^N/\partial \boldsymbol{\xi}^{(i)}$ and the scalar products $\boldsymbol{\xi}^{(i)}\partial f^N/\partial \boldsymbol{X}^{(i)}$ and $\boldsymbol{F}^{(i)}\partial f^N/\partial \boldsymbol{\xi}^{(i)}$, without dot $\cdot$, may be obvious. Here, the force $m\boldsymbol{F}^{(i)}$ acting on the particle i consists only of the inter-particle force, that is,

$$\boldsymbol{F}^{(i)} = \sum_{j=1\ (j\neq i)}^{N} \boldsymbol{F}_{i,j},$$

[1] The Liouville equation is the exact equation describing the behavior of the probability density function f^N in Γ space of the N-particle system obeying Newton's law of motion. Its derivation is found in a textbook of statistical mechanics (e.g., Tolman [1979], Hill [1987], Reif [1965], Diu, Guthmann, Lederer & Roulet [1989]).

where $m\boldsymbol{F}_{i,j}$ is the force by the particle j (thus, $\boldsymbol{F}_{i,j} = -\boldsymbol{F}_{j,i}$ by the law of action and its reaction). The extension to a system with an external force is straightforward. We further assume that the force $\boldsymbol{F}_{i,j}$ is given by a potential determined by the inter-particle distance $|\boldsymbol{X}^{(j)} - \boldsymbol{X}^{(i)}|$ and extends to only a finite distance d_m (or $\boldsymbol{F}_{i,j} = 0$ for $|\boldsymbol{X}^{(j)} - \boldsymbol{X}^{(i)}| \geq d_m$)[2] and that the density function is symmetric with respect to the N sets $(\boldsymbol{X}^{(i)}, \boldsymbol{\xi}^{(i)})$ $(i = 1, \ldots, N)$ of variables.

Now, we introduce the following truncated probability density functions $\tilde{f}_1$, $\tilde{f}_2$, $\tilde{f}_3, \ldots$:

$$\tilde{f}_1 = \int_{\Omega_1,\, \text{all } \boldsymbol{\xi}^{(2)}, \ldots, \text{all } \boldsymbol{\xi}^{(N)}} f^N \mathrm{d}\boldsymbol{X}^{(2)} \cdots \mathrm{d}\boldsymbol{X}^{(N)} \mathrm{d}\boldsymbol{\xi}^{(2)} \cdots \mathrm{d}\boldsymbol{\xi}^{(N)}, \tag{A.2a}$$

$$\tilde{f}_2 = \int_{\Omega_{12},\, \text{all } \boldsymbol{\xi}^{(3)}, \ldots, \text{all } \boldsymbol{\xi}^{(N)}} f^N \mathrm{d}\boldsymbol{X}^{(3)} \cdots \mathrm{d}\boldsymbol{X}^{(N)} \mathrm{d}\boldsymbol{\xi}^{(3)} \cdots \mathrm{d}\boldsymbol{\xi}^{(N)}, \tag{A.2b}$$

$$\tilde{f}_3 = \int_{\Omega_{123},\, \text{all } \boldsymbol{\xi}^{(4)}, \ldots, \text{all } \boldsymbol{\xi}^{(N)}} f^N \mathrm{d}\boldsymbol{X}^{(4)} \cdots \mathrm{d}\boldsymbol{X}^{(N)} \mathrm{d}\boldsymbol{\xi}^{(4)} \cdots \mathrm{d}\boldsymbol{\xi}^{(N)}, \tag{A.2c}$$

$$\cdots\cdots\cdots,$$

where the ranges Ω_1, Ω_{12}, and Ω_{123} are defined as follows: Let $\boldsymbol{X}^{(1)}$ be fixed and let the domain D_1^r $(r = 2, \ldots, N)$ of the space $\boldsymbol{X}^{(r)}$ be defined by

$$D_1^r = \{|\boldsymbol{X}^{(r)} - \boldsymbol{X}^{(1)}| \geq d_m\}.$$

That is, the particle r in D_1^r is not subject to the inter-particle force $\boldsymbol{F}_{r,1}$ due to the particle 1. Then, $\Omega_1, \Omega_{12}, \Omega_{123}, \ldots$ are

$$\Omega_1 = D_1^2 \times D_1^3 \times \cdots \times D_1^N,$$
$$\Omega_{12} = D_1^3 \times D_1^4 \times \cdots \times D_1^N,$$
$$\Omega_{123} = D_1^4 \times D_1^5 \times \cdots \times D_1^N,$$
$$\cdots\cdots\cdots.$$

The functions $\tilde{f}_1$, $\tilde{f}_2$, etc. are, respectively, functions of $(\boldsymbol{X}^{(1)}, \boldsymbol{\xi}^{(1)}, t)$, $(\boldsymbol{X}^{(1)}, \boldsymbol{\xi}^{(1)}, \boldsymbol{X}^{(2)}, \boldsymbol{\xi}^{(2)}, t)$, etc.; $\tilde{f}_2$ is not symmetric with respect to the two sets $(\boldsymbol{X}^{(1)}, \boldsymbol{\xi}^{(1)})$ and $(\boldsymbol{X}^{(2)}, \boldsymbol{\xi}^{(2)})$ of variables. Incidentally, we introduce the notation S_i^r for the sphere in the space $\boldsymbol{X}^{(r)}$ with $\boldsymbol{X}^{(i)}$ as its center:

$$S_i^r = \{|\boldsymbol{X}^{(r)} - \boldsymbol{X}^{(i)}| < d_m\}.$$

Thus, D_1^r is the complement of S_1^r.

Integrating the Liouville equation (A.1) with respect to $\boldsymbol{X}^{(2)}, \ldots, \boldsymbol{X}^{(N)}$, $\boldsymbol{\xi}^{(2)}, \ldots, \boldsymbol{\xi}^{(N)}$ over the domain Ω_1 and the whole spaces of $\boldsymbol{\xi}^{(2)}, \ldots, \boldsymbol{\xi}^{(N)}$ or

[2] For a hard-sphere particle, d_m is the diameter of the particle.

with respect to $\boldsymbol{X}^{(3)}, \ldots, \boldsymbol{X}^{(N)}, \boldsymbol{\xi}^{(3)}, \ldots, \boldsymbol{\xi}^{(N)}$ over the domain Ω_{12} and the whole spaces of $\boldsymbol{\xi}^{(3)}, \ldots, \boldsymbol{\xi}^{(N)}$ or ..., we obtain the following series of equations for $\tilde{f}_1, \tilde{f}_2, \ldots$:

$$\frac{\partial \tilde{f}_1}{\partial t} + \xi^{(1)} \frac{\partial \tilde{f}_1}{\partial \boldsymbol{X}^{(1)}} = (N-1) \int_{\partial S_1^2,\, \text{all}\, \boldsymbol{\xi}^{(2)}} (\boldsymbol{\xi}^{(2)} - \boldsymbol{\xi}^{(1)}) \cdot \boldsymbol{n}_1^2 \tilde{f}_2 \mathrm{d}^2 \boldsymbol{X}^{(2)} \mathbf{d}\boldsymbol{\xi}^{(2)}, \quad \text{(A.3a)}$$

$$\begin{aligned} \frac{\partial \tilde{f}_2}{\partial t} &+ \xi^{(1)} \frac{\partial \tilde{f}_2}{\partial \boldsymbol{X}^{(1)}} + \xi^{(2)} \frac{\partial \tilde{f}_2}{\partial \boldsymbol{X}^{(2)}} + \boldsymbol{F}_{1,2} \frac{\partial \tilde{f}_2}{\partial \boldsymbol{\xi}^{(1)}} + \boldsymbol{F}_{2,1} \frac{\partial \tilde{f}_2}{\partial \boldsymbol{\xi}^{(2)}} \\ &= (N-2) \int_{\partial S_1^3,\, \text{all}\, \boldsymbol{\xi}^{(3)}} (\boldsymbol{\xi}^{(3)} - \boldsymbol{\xi}^{(1)}) \cdot \boldsymbol{n}_1^3 \tilde{f}_3 \mathrm{d}^2 \boldsymbol{X}^{(3)} \mathbf{d}\boldsymbol{\xi}^{(3)} \\ &\quad - (N-2) \int_{D_1^3,\, \text{all}\, \boldsymbol{\xi}^{(3)}} \boldsymbol{F}_{2,3} \frac{\partial \tilde{f}_3}{\partial \boldsymbol{\xi}^{(2)}} \mathbf{d}\boldsymbol{X}^{(3)} \mathbf{d}\boldsymbol{\xi}^{(3)}, \qquad \text{(A.3b)} \end{aligned}$$

$$\ldots\ldots\ldots\ldots\ldots ,$$

where $\boldsymbol{n}_i^j = (\boldsymbol{X}^{(j)} - \boldsymbol{X}^{(i)})/|\boldsymbol{X}^{(j)} - \boldsymbol{X}^{(i)}|$; the range ∂S_1^2 of integration $\mathrm{d}^2\boldsymbol{X}^{(2)}$ with respect to $\boldsymbol{X}^{(2)}$ in Eq. (A.3a) is the surface ∂S_1^2 of the sphere S_1^2; the range ∂S_1^3 of integration $\mathrm{d}^2\boldsymbol{X}^{(3)}$ with respect to $\boldsymbol{X}^{(3)}$ in Eq. (A.3b) is the surface ∂S_1^3 of the sphere S_1^3; and the range D_1^3 of integration $\mathbf{d}\boldsymbol{X}^{(3)}$ with respect to $\boldsymbol{X}^{(3)}$ in the last term is practically $D_1^3 \cap S_2^3$, since the force $\boldsymbol{F}_{2,3}$ vanishes outside S_2^3. The above equations are considerably simplified by the assumptions of the symmetry of f^N with respect to the N sets $(\boldsymbol{X}^{(i)}, \boldsymbol{\xi}^{(i)})$ $(i = 1, \ldots, N)$ of variables and a finite range d_m of the inter-particle force $\boldsymbol{F}_{i,j}$. Owing to the choice of the ranges of integration Ω_1, Ω_{12}, etc., instead of the whole spaces of $\boldsymbol{X}^{(2)}, \ldots, \boldsymbol{X}^{(N)}$, those of $\boldsymbol{X}^{(3)}, \ldots, \boldsymbol{X}^{(N)}$, etc., the interaction forces $\boldsymbol{F}_{i,j}$ do not appear in Eq. (A.3a), the interaction forces other than $\boldsymbol{F}_{1,2}$, $\boldsymbol{F}_{2,1}$, and $\boldsymbol{F}_{2,3}$ do not in Eq. (A.3b), etc. In place of these simplifications, the surface integrals over ∂S_1^2 and ∂S_1^3 on the right-hand sides of Eqs. (A.3a) and (A.3b) arise from the boundaries of the domains Ω_1 and Ω_{12}. Some supplementary explanations may be in order on the derivation of the right-hand sides.

With the aid of Gauss's divergence theorem, the integral of $\boldsymbol{\xi}^{(k)} \partial f^N / \partial \boldsymbol{X}^{(k)}$ $(k \geq 2)$ over Ω_1 or that of $\boldsymbol{\xi}^{(k)} \partial f^N / \partial \boldsymbol{X}^{(k)}$ $(k \geq 3)$ over Ω_{12} can be transformed into an integral over the boundary of Ω_1 (or ∂S_1^k) or Ω_{12} (or ∂S_1^k). For example,

$$\begin{aligned} &\int_{\Omega_1,\, \text{all}\, \boldsymbol{\xi}^{(2)}, \ldots, \text{all}\, \boldsymbol{\xi}^{(N)}} \boldsymbol{\xi}^{(2)} \frac{\partial f^N}{\partial \boldsymbol{X}^{(2)}} \mathbf{d}\boldsymbol{X}^{(2)} \cdots \mathbf{d}\boldsymbol{X}^{(N)} \mathbf{d}\boldsymbol{\xi}^{(2)} \cdots \mathbf{d}\boldsymbol{\xi}^{(N)} \\ &\qquad = - \int_{\partial S_1^2,\, \text{all}\, \boldsymbol{\xi}^{(2)}} \boldsymbol{\xi}^{(2)} \cdot \boldsymbol{n}_1^2 \tilde{f}_2 \mathrm{d}^2 \boldsymbol{X}^{(2)} \mathbf{d}\boldsymbol{\xi}^{(2)}, \end{aligned}$$

$$\begin{aligned} &\int_{\Omega_{12},\, \text{all}\, \boldsymbol{\xi}^{(3)}, \ldots, \text{all}\, \boldsymbol{\xi}^{(N)}} \boldsymbol{\xi}^{(3)} \frac{\partial f^N}{\partial \boldsymbol{X}^{(3)}} \mathbf{d}\boldsymbol{X}^{(3)} \cdots \mathbf{d}\boldsymbol{X}^{(N)} \mathbf{d}\boldsymbol{\xi}^{(3)} \cdots \mathbf{d}\boldsymbol{\xi}^{(N)} \\ &\qquad = - \int_{\partial S_1^3,\, \text{all}\, \boldsymbol{\xi}^{(3)}} \boldsymbol{\xi}^{(3)} \cdot \boldsymbol{n}_1^3 \tilde{f}_3 \mathrm{d}^2 \boldsymbol{X}^{(3)} \mathbf{d}\boldsymbol{\xi}^{(3)}. \end{aligned}$$

In the integral of $\boldsymbol{\xi}^{(1)}\partial f^N/\partial \boldsymbol{X}^{(1)}$ over Ω_1 or Ω_{12}, a boundary term comes in through the process of interchanging the order of integration and differentiation, since the domain (Ω_1 or Ω_{12}) of integration varies with $\boldsymbol{X}^{(1)}$. The derivative, with respect to α, of the integral $\int_{\mathcal{D}(\alpha)} g(\boldsymbol{X},\alpha)\mathbf{d}\boldsymbol{X}$ over the domain $\mathcal{D}(\alpha)$ deformable with α is given in the following form:

$$\frac{\mathrm{d}}{\mathrm{d}\alpha}\int_{\mathcal{D}(\alpha)} g(\boldsymbol{X},\alpha)\mathbf{d}\boldsymbol{X} = \int_{\mathcal{D}(\alpha)} \frac{\partial g(\boldsymbol{X},\alpha)}{\partial \alpha}\mathbf{d}\boldsymbol{X} + \int_{\partial\mathcal{D}(\alpha)} g(\boldsymbol{X},\alpha)\frac{\partial \boldsymbol{X}_w}{\partial \alpha}\cdot\boldsymbol{n}_w \mathrm{d}^2\boldsymbol{X}, \tag{A.4}$$

where $\boldsymbol{X}_w$ is a point on the boundary $\partial\mathcal{D}(\alpha)$ of $\mathcal{D}(\alpha)$, $\boldsymbol{n}_w$ is the outward unit normal vector to the boundary and $\mathrm{d}^2\boldsymbol{X}$ is the surface element on $\partial\mathcal{D}(\alpha)$; $\partial\boldsymbol{X}_w/\partial\alpha$ is not unique but $\boldsymbol{n}_w\cdot\partial\boldsymbol{X}_w/\partial\alpha$ is unique. The derivation of the formula is given at the end of this section as a lemma. With the aid of this formula, the above-mentioned integrals are transformed into the following forms:[3]

$$\int_{\Omega_1,\,\mathrm{all}\,\boldsymbol{\xi}^{(2)},\dots,\,\mathrm{all}\,\boldsymbol{\xi}^{(N)}} \boldsymbol{\xi}^{(1)}\frac{\partial f^N}{\partial \boldsymbol{X}^{(1)}}\mathbf{d}\boldsymbol{X}^{(2)}\cdots\mathbf{d}\boldsymbol{X}^{(N)}\mathbf{d}\boldsymbol{\xi}^{(2)}\cdots\mathbf{d}\boldsymbol{\xi}^{(N)}$$

$$= \boldsymbol{\xi}^{(1)}\frac{\partial \tilde{f}_1}{\partial \boldsymbol{X}^{(1)}} + (N-1)\int_{\partial S_1^2,\,\mathrm{all}\,\boldsymbol{\xi}^{(2)}} \boldsymbol{\xi}^{(1)}\cdot\boldsymbol{n}_1^2\tilde{f}_2\mathrm{d}^2\boldsymbol{X}^{(2)}\mathbf{d}\boldsymbol{\xi}^{(2)},$$

$$\int_{\Omega_{12},\,\mathrm{all}\,\boldsymbol{\xi}^{(3)},\dots,\,\mathrm{all}\,\boldsymbol{\xi}^{(N)}} \boldsymbol{\xi}^{(1)}\frac{\partial f^N}{\partial \boldsymbol{X}^{(1)}}\mathbf{d}\boldsymbol{X}^{(3)}\cdots\mathbf{d}\boldsymbol{X}^{(N)}\mathbf{d}\boldsymbol{\xi}^{(3)}\cdots\mathbf{d}\boldsymbol{\xi}^{(N)}$$

$$= \boldsymbol{\xi}^{(1)}\frac{\partial \tilde{f}_2}{\partial \boldsymbol{X}^{(1)}} + (N-2)\int_{\partial S_1^3,\,\mathrm{all}\,\boldsymbol{\xi}^{(3)}} \boldsymbol{\xi}^{(1)}\cdot\boldsymbol{n}_1^3\tilde{f}_3\mathrm{d}^2\boldsymbol{X}^{(3)}\mathbf{d}\boldsymbol{\xi}^{(3)},$$

where the factors $N-1$ and $N-2$ are due to the symmetry of the density function.

Let us return to the mainstream of the discussion. The series of equations (A.3a), (A.3b), ... is not a closed system, that is, Eq. (A.3a) contains $\tilde{f}_2$ as well as $\tilde{f}_1$, Eq. (A.3b) contains $\tilde{f}_3$ as well as $\tilde{f}_2$, and so on. Now we will consider the simplification of Eqs. (A.3a) and (A.3b) under the situation of the Grad–Boltzmann limit. The **Grad–Boltzmann limit** is the limiting process where the number of particles tends to infinite ($N \to \infty$) and the range of the inter-particle force vanishes ($d_m \to 0$) keeping Nd_m^2 at a fixed value, and thus, inevitably, $Nd_m^3 \to 0$. The right-hand side of Eq. (A.3a) and the first term on the right-hand side of Eq. (A.3b) are integrals over the surface ∂S_1^2 or ∂S_1^3, and therefore they are of the order of Nd_m^2, which is of the order of unity in the Grad–Boltzmann limit, if none of $\tilde{f}_2$ and $\tilde{f}_3$ is singularly large. The second term on the right-hand side of Eq. (A.3b) is an integral over the small region $D_1^3 \cap S_2^3$ of the order of d_m^3, but its magnitude is not Nd_m^3 and is roughly estimated to be Nd_m^2, since the interaction force $\boldsymbol{F}_{2,3}$ can be infinitely large as $|\boldsymbol{X}^{(2)} - \boldsymbol{X}^{(3)}| \to 0$.[4]

---

[3]The integration over Ω_1 ($= D_1^2 \times D_1^3 \times \cdots \times D_1^N$) is carried out successively, for example, over D_1^N, D_1^{N-1}, ..., and D_1^2, and the formula (A.4) is applied at each step of integration.

[4]The inter-particle force $\boldsymbol{F}_{2,3}$ in the integral over $D_1^3 \cap S_2^3$ can be very large (or diverge)

The range ∂S_1^2 of integration in the integral in Eq. (A.3a) is conveniently split into two parts, one is the part ∂S_{1+}^2 of the sphere where $(\boldsymbol{\xi}^{(2)} - \boldsymbol{\xi}^{(1)})\cdot\boldsymbol{n}_1^2 > 0$ and the other is ∂S_{1-}^2 where $(\boldsymbol{\xi}^{(2)} - \boldsymbol{\xi}^{(1)})\cdot\boldsymbol{n}_1^2 < 0$ for a given set of $\boldsymbol{\xi}^{(2)}$ and $\boldsymbol{\xi}^{(1)}$. On ∂S_{1+}^2 the two particles 1 and 2 are going to separate after their interaction by inter-particle force, and on ∂S_{1-}^2 the two particles are going to enter the range of influence of their inter-particle force. Let the position of the particle 2 on ∂S_{1+}^2 be $\boldsymbol{X}_+^{(2)}$ and that on ∂S_{1-}^2 be $\boldsymbol{X}_-^{(2)}$. Here we introduce an important assumption, called **molecular chaos**, that the two-particle truncated density function $\tilde{f}_2$ at $(\boldsymbol{X}^{(1)}, \boldsymbol{\xi}^{(1)}, \boldsymbol{X}_-^{(2)}, \boldsymbol{\xi}^{(2)}, t)$ on ∂S_{1-}^2 can be expressed by the product of the two one-particle truncated density functions,[5] that is,

$$\tilde{f}_2(\boldsymbol{X}^{(1)}, \boldsymbol{\xi}^{(1)}, \boldsymbol{X}_-^{(2)}, \boldsymbol{\xi}^{(2)}, t) = \tilde{f}_1(\boldsymbol{X}^{(1)}, \boldsymbol{\xi}^{(1)}, t)\tilde{f}_1(\boldsymbol{X}_-^{(2)}, \boldsymbol{\xi}^{(2)}, t). \tag{A.5}$$

We will try to express the two-particle truncated density function $\tilde{f}_2$ on the part ∂S_{1+}^2 in terms of the one-particle truncated density function $\tilde{f}_1$. That is, first, $\tilde{f}_2$ on the part ∂S_{1+}^2 is expressed in terms of $\tilde{f}_2$ on the part ∂S_{1-}^2, and then the assumption of molecular chaos is applied to this $\tilde{f}_2$. Equation (A.3b) describes the variation of $\tilde{f}_2$ along the characteristic of Eq. (A.3b) or the trajectory in the space $(\boldsymbol{X}^{(1)}, \boldsymbol{\xi}^{(1)}, \boldsymbol{X}^{(2)}, \boldsymbol{\xi}^{(2)}, t)$ of the system of the two particles $\boldsymbol{X}^{(1)}$ and $\boldsymbol{X}^{(2)}$ interacting with each other by the inter-particle forces $\boldsymbol{F}_{1,2}$ and $\boldsymbol{F}_{2,1}$. The variation is determined by the right-hand side of Eq. (A.3b) or $\tilde{f}_3$. The trajectory passes the region $|\boldsymbol{X}^{(2)} - \boldsymbol{X}^{(1)}| < d_m$ in a short time of the order of $d_m/|\boldsymbol{\xi}^{(2)} - \boldsymbol{\xi}^{(1)}|$. In other words, in this short time, the two particles $\boldsymbol{X}^{(1)}$ and $\boldsymbol{X}^{(2)}$ move from ∂S_{1-}^2 to ∂S_{1+}^2. During this short period (or collision process), the two particles are subject to the strong inter-particle forces $\boldsymbol{F}_{1,2}$ and $\boldsymbol{F}_{2,1}$ and their velocities $\boldsymbol{\xi}^{(1)}$ and $\boldsymbol{\xi}^{(2)}$ vary considerably. On the other hand, the contribution of the right-hand side of Eq. (A.3b) during this short period is of the order of d_m, since the size of the right-hand side is of the order of unity under the Grad–Boltzmann limit (see the paragraph before the last and Footnote 4). Therefore, the two-particle truncated density function $\tilde{f}_2$ takes the same value at the two ends of the trajectory of this short interaction period within an error of the order of d_m.

Let the two particles 1 and 2 be at $(\boldsymbol{X}^{(1)}, \boldsymbol{\xi}^{(1)})$ and $(\boldsymbol{X}_+^{(2)}, \boldsymbol{\xi}^{(2)})$, which is in the situation ∂S_{1+}^2. In view of their situation that they are on ∂S_{1+}^2, they were

---

as $\boldsymbol{X}^{(3)}$ approaches $\boldsymbol{X}^{(2)}$. For a repulsive force, however, $\tilde{f}_3$ is going to vanish there, since the particles $\boldsymbol{X}^{(2)}$ and $\boldsymbol{X}^{(3)}$ cannot approach in a large repulsive force field. That is, the two particles cannot be closer than the distance d_{sp} limited by $|\boldsymbol{F}_{2,3}|(d_m - d_{sp}) < O(|\boldsymbol{\xi}^{(2)}|^2 + |\boldsymbol{\xi}^{(3)}|^2)$. Thus, the integral is roughly estimated to be of the order of d_m^2 (note that it is not d_m^3). [This can also be seen in the following way: The velocity $\boldsymbol{\xi}^{(3)}$ of the particle $\boldsymbol{X}^{(3)}$ is subject to a change of the order of $|\boldsymbol{F}_{2,3}|d_m/|\boldsymbol{\xi}^{(3)} - \boldsymbol{\xi}^{(2)}|$ during a short interaction period $d_m/|\boldsymbol{\xi}^{(3)} - \boldsymbol{\xi}^{(2)}|$, which is, on the other hand, of the order of $|\boldsymbol{\xi}^{(3)}|$. Thus, the average size of $\boldsymbol{F}_{2,3}$ during the interaction is of the order of $|\boldsymbol{\xi}^{(3)}|^2/d_m$.] Thus, the second term is of the same order of magnitude $[O(Nd_m^2)]$ as the first one.

[5]The assumption of molecular chaos is discussed in Grad [1958], on which the framework of this section is based.

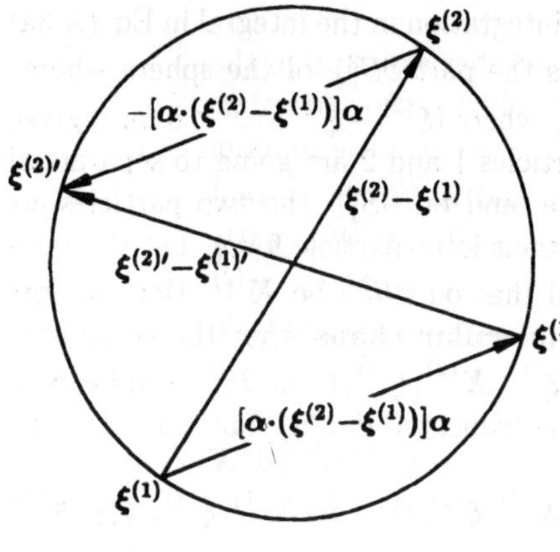

Figure A.1. Relation between $(\boldsymbol{\xi}^{(1)}, \boldsymbol{\xi}^{(2)})$ and $(\boldsymbol{\xi}^{(1)\prime}, \boldsymbol{\xi}^{(2)\prime})$. The velocities $\boldsymbol{\xi}^{(1)\prime}$ and $\boldsymbol{\xi}^{(2)\prime}$ are antipodal points on the sphere with $\boldsymbol{\xi}^{(1)}$ and $\boldsymbol{\xi}^{(2)}$ as its antipodal points.

at the points $(\boldsymbol{X}^{(1)\prime}, \boldsymbol{\xi}^{(1)\prime})$ and $(\boldsymbol{X}_-^{(2)\prime}, \boldsymbol{\xi}^{(2)\prime})$ in the situation ∂S_{1-}^2 at a close past time $t - \Delta t$, where Δt is of the order of $d_m/|\boldsymbol{\xi}^{(2)} - \boldsymbol{\xi}^{(1)}|$. Then, the particle velocities $\boldsymbol{\xi}^{(1)\prime}$ and $\boldsymbol{\xi}^{(2)\prime}$ at $t - \Delta t$ are expressed by the original ones $\boldsymbol{\xi}^{(1)}$ and $\boldsymbol{\xi}^{(2)}$ with a unit vector $\boldsymbol{\alpha}$ as a parameter as follows (see Fig. A.1):

$$\boldsymbol{\xi}^{(1)\prime} = \boldsymbol{\xi}^{(1)} + [\boldsymbol{\alpha} \cdot (\boldsymbol{\xi}^{(2)} - \boldsymbol{\xi}^{(1)})]\boldsymbol{\alpha}, \quad \boldsymbol{\xi}^{(2)\prime} = \boldsymbol{\xi}^{(2)} - [\boldsymbol{\alpha} \cdot (\boldsymbol{\xi}^{(2)} - \boldsymbol{\xi}^{(1)})]\boldsymbol{\alpha}, \quad \text{(A.6)}$$

where the dot $\cdot$ for the scalar product serves to avoid confusion and the unit vectors $\boldsymbol{\alpha}$ and $-\boldsymbol{\alpha}$ give the same result. This relation is the result of conservation of momentum and energy before and after the interaction. The dynamics of the system (in the past and future) is uniquely determined by $(\boldsymbol{X}^{(1)}, \boldsymbol{\xi}^{(1)}, \boldsymbol{X}_+^{(2)}, \boldsymbol{\xi}^{(2)})$ on ∂S_{1+}^2, and the parameter $\boldsymbol{\alpha}$ is expressed with $\boldsymbol{n}_{1+}^2$ $[= (\boldsymbol{X}_+^{(2)} - \boldsymbol{X}^{(1)})/d_m]$ and $\boldsymbol{\xi}^{(2)} - \boldsymbol{\xi}^{(1)}$, i.e., $\boldsymbol{\alpha} = \boldsymbol{\alpha}(\boldsymbol{n}_{1+}^2, \boldsymbol{\xi}^{(2)} - \boldsymbol{\xi}^{(1)})$, the functional form of which is determined by the inter-particle potential, containing d_m (see, for example, Goldstein [1950], Landau & Lifshitz [1960], Sone & Aoki [1994]).

According to the discussion in the paragraph before the last,

$$\tilde{f}_2(\boldsymbol{X}^{(1)}, \boldsymbol{\xi}^{(1)}, \boldsymbol{X}_+^{(2)}, \boldsymbol{\xi}^{(2)}, t) = \tilde{f}_2(\boldsymbol{X}^{(1)\prime}, \boldsymbol{\xi}^{(1)\prime}, \boldsymbol{X}_-^{(2)\prime}, \boldsymbol{\xi}^{(2)\prime}, t - \Delta t).$$

For $\tilde{f}_2$ on ∂S_{1-}^2, i.e., $\tilde{f}_2(\boldsymbol{X}^{(1)\prime}, \boldsymbol{\xi}^{(1)\prime}, \boldsymbol{X}_-^{(2)\prime}, \boldsymbol{\xi}^{(2)\prime}, t - \Delta t)$, the assumption of molecular chaos can be applied. Thus, the two-particle truncated density function $\tilde{f}_2$ at $(\boldsymbol{X}^{(1)}, \boldsymbol{\xi}^{(1)}, \boldsymbol{X}_+^{(2)}, \boldsymbol{\xi}^{(2)}, t)$ is expressed by the product of the one-particle truncated density functions in the form

$$\begin{aligned}\tilde{f}_2(\boldsymbol{X}^{(1)}, \boldsymbol{\xi}^{(1)}, \boldsymbol{X}_+^{(2)}, \boldsymbol{\xi}^{(2)}, t) &= \tilde{f}_2(\boldsymbol{X}^{(1)\prime}, \boldsymbol{\xi}^{(1)\prime}, \boldsymbol{X}_-^{(2)\prime}, \boldsymbol{\xi}^{(2)\prime}, t - \Delta t) \\ &= \tilde{f}_1(\boldsymbol{X}^{(1)\prime}, \boldsymbol{\xi}^{(1)\prime}, t - \Delta t)\tilde{f}_1(\boldsymbol{X}_-^{(2)\prime}, \boldsymbol{\xi}^{(2)\prime}, t - \Delta t).\end{aligned} \quad \text{(A.7)}$$

Now, we introduce another assumption that the one-particle truncated density function $\tilde{f}_1$ is a **slowly varying** function in time t and space $\boldsymbol{X}^{(1)}$ on the scale of d_m. Then, in the one-particle function $\tilde{f}_1$ of Eqs. (A.5) and (A.7), the arguments $\boldsymbol{X}_-^{(2)}$, $\boldsymbol{X}^{(1)\prime}$, and $\boldsymbol{X}_-^{(2)\prime}$ can be replaced by $\boldsymbol{X}^{(1)}$, and the argument $t-\Delta t$ by t, since $|\boldsymbol{X}_-^{(2)}-\boldsymbol{X}^{(1)}|$, $|\boldsymbol{X}^{(1)\prime}-\boldsymbol{X}^{(1)}|$, $|\boldsymbol{X}_-^{(2)\prime}-\boldsymbol{X}^{(1)}|$, and Δt are all of the order of d_m. With Eqs. (A.5) and (A.7) thus simplified, the integral of the right-hand side of Eq. (A.3a) is expressed by the one-particle truncated density function $\tilde{f}_1$ as

$$\int_{\partial S_1^2,\,\text{all}\,\boldsymbol{\xi}^{(2)}}(\boldsymbol{\xi}^{(2)}-\boldsymbol{\xi}^{(1)})\cdot\boldsymbol{n}_1^2\tilde{f}_2\mathrm{d}^2\boldsymbol{X}^{(2)}\mathrm{d}\boldsymbol{\xi}^{(2)}=I_-+I_+, \tag{A.8}$$

where

$$I_-=\int_{\substack{\partial S_{1-}^2\\ \text{all}\,\boldsymbol{\xi}^{(2)}}}(\boldsymbol{\xi}^{(2)}-\boldsymbol{\xi}^{(1)})\cdot\boldsymbol{n}_1^2\tilde{f}_1(\boldsymbol{X}^{(1)},\boldsymbol{\xi}^{(1)},t)\tilde{f}_1(\boldsymbol{X}^{(1)},\boldsymbol{\xi}^{(2)},t)\mathrm{d}^2\boldsymbol{X}^{(2)}\mathrm{d}\boldsymbol{\xi}^{(2)},$$

$$I_+=\int_{\substack{\partial S_{1+}^2\\ \text{all}\,\boldsymbol{\xi}^{(2)}}}(\boldsymbol{\xi}^{(2)}-\boldsymbol{\xi}^{(1)})\cdot\boldsymbol{n}_1^2\tilde{f}_1(\boldsymbol{X}^{(1)},\boldsymbol{\xi}^{(1)\prime},t)\tilde{f}_1(\boldsymbol{X}^{(1)},\boldsymbol{\xi}^{(2)\prime},t)\mathrm{d}^2\boldsymbol{X}^{(2)}\mathrm{d}\boldsymbol{\xi}^{(2)}.$$

The arguments $\boldsymbol{X}^{(1)}$ and t of $\tilde{f}_1$ being common, the integral over the hemisphere ∂S_{1-}^2 in I_- can be extended to the whole sphere ∂S_1^2 simply by replacing $(\boldsymbol{\xi}^{(2)}-\boldsymbol{\xi}^{(1)})\cdot\boldsymbol{n}_1^2$ by $-|(\boldsymbol{\xi}^{(2)}-\boldsymbol{\xi}^{(1)})\cdot\boldsymbol{n}_1^2|/2$, that is,

$$\begin{aligned}I_-&=-\frac{1}{2}\int_{\substack{\partial S_1^2\\ \text{all}\,\boldsymbol{\xi}^{(2)}}}|(\boldsymbol{\xi}^{(2)}-\boldsymbol{\xi}^{(1)})\cdot\boldsymbol{n}_1^2|\tilde{f}_1(\boldsymbol{X}^{(1)},\boldsymbol{\xi}^{(1)},t)\tilde{f}_1(\boldsymbol{X}^{(1)},\boldsymbol{\xi}^{(2)},t)\mathrm{d}^2\boldsymbol{X}^{(2)}\mathrm{d}\boldsymbol{\xi}^{(2)}\\ &=-\frac{d_m^2}{2}\int_{\substack{\text{all}\,\boldsymbol{e}\\ \text{all}\,\boldsymbol{\xi}^{(2)}}}|(\boldsymbol{\xi}^{(2)}-\boldsymbol{\xi}^{(1)})\cdot\boldsymbol{e}|\tilde{f}_1(\boldsymbol{X}^{(1)},\boldsymbol{\xi}^{(1)},t)\tilde{f}_1(\boldsymbol{X}^{(1)},\boldsymbol{\xi}^{(2)},t)\mathrm{d}\Omega(\boldsymbol{e})\mathrm{d}\boldsymbol{\xi}^{(2)},\end{aligned} \tag{A.9}$$

where $\boldsymbol{e}$ is a unit vector and $\mathrm{d}\Omega(\boldsymbol{e})$ is the solid angle element in the direction of $\boldsymbol{e}$. Similarly, the integral I_+ is transformed as

$$I_+=\frac{d_m^2}{2}\int_{\substack{\text{all}\,\boldsymbol{e}\\ \text{all}\,\boldsymbol{\xi}^{(2)}}}|(\boldsymbol{\xi}^{(2)}-\boldsymbol{\xi}^{(1)})\cdot\boldsymbol{e}|\tilde{f}_1(\boldsymbol{X}^{(1)},\boldsymbol{\xi}^{(1)\prime},t)\tilde{f}_1(\boldsymbol{X}^{(1)},\boldsymbol{\xi}^{(2)\prime},t)\mathrm{d}\Omega(\boldsymbol{e})\mathrm{d}\boldsymbol{\xi}^{(2)}, \tag{A.10}$$

where $\boldsymbol{e}=\boldsymbol{n}_1^2$ $[=(\boldsymbol{X}^{(2)}-\boldsymbol{X}^{(1)})/|\boldsymbol{X}^{(2)}-\boldsymbol{X}^{(1)}|]$. The parameter $\boldsymbol{\alpha}$ in the expressions $\boldsymbol{\xi}^{(1)\prime}$ and $\boldsymbol{\xi}^{(2)\prime}$ given by Eq. (A.6) is a function of $\boldsymbol{e}$ and $\boldsymbol{\xi}^{(2)}-\boldsymbol{\xi}^{(1)}$, that is,

$$\boldsymbol{\alpha}=\boldsymbol{\alpha}(\boldsymbol{e},\ \boldsymbol{\xi}^{(2)}-\boldsymbol{\xi}^{(1)}), \tag{A.11}$$

which is extended in such a way that $-\boldsymbol{e}$ gives the same $\boldsymbol{\alpha}$ or $-\boldsymbol{\alpha}$ as $\boldsymbol{e}$ in the extension of the range of integration. The $\boldsymbol{\alpha}$ is in the plane made by the two vectors $\boldsymbol{e}$ and $\boldsymbol{\xi}^{(2)}-\boldsymbol{\xi}^{(1)}$, and the functional form of $\boldsymbol{\alpha}(*,*)$ depends on the inter-particle potential; for example, $\boldsymbol{\alpha}=\boldsymbol{e}$ for a hard-sphere particle.

Then, with Eqs. (A.8)–(A.10) in Eq. (A.3a), the equation for $\tilde{f}_1(\boldsymbol{X}^{(1)}, \boldsymbol{\xi}^{(1)}, t)$ is given as

$$\frac{\partial \tilde{f}_1}{\partial t} + \xi^{(1)} \frac{\partial \tilde{f}_1}{\partial \boldsymbol{X}^{(1)}}$$
$$= \frac{N d_m^2}{2} \int_{\substack{\text{all } \boldsymbol{e} \\ \text{all } \boldsymbol{\xi}^{(2)}}} |(\boldsymbol{\xi}^{(2)} - \boldsymbol{\xi}^{(1)}) \cdot \boldsymbol{e}| [\tilde{f}_1(\boldsymbol{\xi}^{(1)\prime}) \tilde{f}_1(\boldsymbol{\xi}^{(2)\prime}) - \tilde{f}_1(\boldsymbol{\xi}^{(1)}) \tilde{f}_1(\boldsymbol{\xi}^{(2)})] \mathrm{d}\Omega(\boldsymbol{e}) \mathrm{d}\boldsymbol{\xi}^{(2)}, \tag{A.12}$$

where $(N-1)d_m^2$ is identified with Nd_m^2 for the Grad–Boltzmann limit, and the arguments for the space and time variables of the truncated density function $\tilde{f}_1$ are not shown, since they are commonly $\boldsymbol{X}^{(1)}$ and t.

Now consider the following marginal probability density functions $f_1, f_2, \ldots$:

$$f_1 = \int_{\substack{\text{all } \boldsymbol{X}^{(2)}, \ldots, \text{all } \boldsymbol{X}^{(N)} \\ \text{all } \boldsymbol{\xi}^{(2)}, \ldots, \text{all } \boldsymbol{\xi}^{(N)}}} f^N \mathrm{d}\boldsymbol{X}^{(2)} \cdots \mathrm{d}\boldsymbol{X}^{(N)} \mathrm{d}\boldsymbol{\xi}^{(2)} \cdots \mathrm{d}\boldsymbol{\xi}^{(N)}, \tag{A.13a}$$

$$f_2 = \int_{\substack{\text{all } \boldsymbol{X}^{(3)}, \ldots, \text{all } \boldsymbol{X}^{(N)} \\ \text{all } \boldsymbol{\xi}^{(3)}, \ldots, \text{all } \boldsymbol{\xi}^{(N)}}} f^N \mathrm{d}\boldsymbol{X}^{(3)} \cdots \mathrm{d}\boldsymbol{X}^{(N)} \mathrm{d}\boldsymbol{\xi}^{(3)} \cdots \mathrm{d}\boldsymbol{\xi}^{(N)}, \tag{A.13b}$$

$$\cdots\cdots\cdots.$$

The f_1 is the probability density of finding the particle 1 with velocity $\boldsymbol{\xi}^{(1)}$ at position $\boldsymbol{X}^{(1)}$ and at time t, irrespective of the positions and the velocities of the other particles (or the particles $2, 3, \ldots, N$) and so on.

The difference between f_1 and $\tilde{f}_1$ is only the domains of integration with respect to the space variables $\boldsymbol{X}^{(2)}, \ldots, \boldsymbol{X}^{(N)}$ in their definitions. Noting that the whole space of $\boldsymbol{X}^{(2)} \times \cdots \times \boldsymbol{X}^{(N)}$ is $(S_1^2 + D_1^2) \times \cdots \times (S_1^N + D_1^N)$ and that the function f^N is symmetric with respect to the N sets of the variables $(\boldsymbol{X}^{(r)}, \boldsymbol{\xi}^{(r)})$ $(r = 1, 2, \ldots, N)$, we have

$$\begin{aligned} f_1 &= \tilde{f}_1 + (N-1) \int_{S_1^2} \tilde{f}_2 \mathrm{d}\boldsymbol{X}^{(2)} \mathrm{d}\boldsymbol{\xi}^{(2)} + \cdots \\ &= \tilde{f}_1 + O((N-1)d_m^3) + \cdots. \end{aligned} \tag{A.14}$$

Thus, in the Grad–Boltzmann limit, we can identify $\tilde{f}_1$ with f_1. Put

$$f(\boldsymbol{X}, \boldsymbol{\xi}, t) = mN\tilde{f}_1(\boldsymbol{X}, \boldsymbol{\xi}, t). \tag{A.15}$$

Then $f(\boldsymbol{X}, \boldsymbol{\xi}, t)\mathrm{d}\boldsymbol{X}\mathrm{d}\boldsymbol{\xi}$ is the expected value of the mass of the particles in the position and velocity ranges $\mathrm{d}\boldsymbol{X}\mathrm{d}\boldsymbol{\xi}$ in the 6-dimensional space $(\boldsymbol{X}, \boldsymbol{\xi})$ at time t.[6]

[6]Let $\varphi(\boldsymbol{X}^{(1)}, \ldots, \boldsymbol{X}^{(N)}, \boldsymbol{\xi}^{(1)}, \ldots, \boldsymbol{\xi}^{(N)})$ be the function in the $6N$-dimensional space

From Eqs. (A.12) and (A.15), the equation for $f(\boldsymbol{X},\boldsymbol{\xi},t)$ is given in the form

$$\frac{\partial f}{\partial t}+\xi\frac{\partial f}{\partial \boldsymbol{X}}$$
$$=\frac{Nd_m^2}{2}\left(\frac{1}{mN}\right)\int_{\substack{\text{all } \boldsymbol{e}\\ \text{all } \boldsymbol{\xi}_*}}|(\boldsymbol{\xi}_*-\boldsymbol{\xi})\cdot\boldsymbol{e}|[f(\boldsymbol{\xi}')f(\boldsymbol{\xi}'_*)-f(\boldsymbol{\xi})f(\boldsymbol{\xi}_*)]\mathrm{d}\Omega(\boldsymbol{e})\mathbf{d}\boldsymbol{\xi}_*, \quad \text{(A.16)}$$

where

$$\boldsymbol{\xi}'=\boldsymbol{\xi}+[\boldsymbol{\alpha}\cdot(\boldsymbol{\xi}_*-\boldsymbol{\xi})]\boldsymbol{\alpha},\quad \boldsymbol{\xi}'_*=\boldsymbol{\xi}_*-[\boldsymbol{\alpha}\cdot(\boldsymbol{\xi}_*-\boldsymbol{\xi})]\boldsymbol{\alpha}, \quad \text{(A.17)}$$

and the relation between $\boldsymbol{e}$ and $\boldsymbol{\alpha}$ is given by Eq. (A.11). The following function $B(|\boldsymbol{\alpha}\cdot(\boldsymbol{\xi}_*-\boldsymbol{\xi})|/|\boldsymbol{\xi}_*-\boldsymbol{\xi}|,|\boldsymbol{\xi}_*-\boldsymbol{\xi}|)$, which is closely related to the differential cross-section (e.g., Reif [1965]), is often introduced to relate $\boldsymbol{e}$ to $\boldsymbol{\alpha}$:

$$|\boldsymbol{e}\cdot(\boldsymbol{\xi}_*-\boldsymbol{\xi})|\mathrm{d}\Omega(\boldsymbol{e})=\frac{2}{d_m^2}B(|\boldsymbol{\alpha}\cdot(\boldsymbol{\xi}_*-\boldsymbol{\xi})|/|\boldsymbol{\xi}_*-\boldsymbol{\xi}|,|\boldsymbol{\xi}_*-\boldsymbol{\xi}|)\mathrm{d}\Omega(\boldsymbol{\alpha}). \quad \text{(A.18)}$$

The choice of the arguments of B is allowed by the property of the inter-particle potential. For a hard-sphere particle, $B=d_m^2|\boldsymbol{\alpha}\cdot(\boldsymbol{\xi}_*-\boldsymbol{\xi})|/2$, since $\boldsymbol{\alpha}=\boldsymbol{e}$. Then,

$$\frac{\partial f}{\partial t}+\xi\frac{\partial f}{\partial \boldsymbol{X}}=\frac{1}{m}\int_{\substack{\text{all } \boldsymbol{\alpha}\\ \text{all } \boldsymbol{\xi}_*}}[f(\boldsymbol{\xi}')f(\boldsymbol{\xi}'_*)-f(\boldsymbol{\xi})f(\boldsymbol{\xi}_*)]B\mathrm{d}\Omega(\boldsymbol{\alpha})\mathbf{d}\boldsymbol{\xi}_*, \quad \text{(A.19)}$$

$$B=B(|\boldsymbol{\alpha}\cdot(\boldsymbol{\xi}_*-\boldsymbol{\xi})|/|\boldsymbol{\xi}_*-\boldsymbol{\xi}|,|\boldsymbol{\xi}_*-\boldsymbol{\xi}|).$$

Here, some comments may be in order about the relation between $\boldsymbol{e}$ and $\boldsymbol{\alpha}$. While $\boldsymbol{e}$ is moving over the unit sphere, $\boldsymbol{\alpha}$ also moves there and covers all the points there, but the correspondence between $\boldsymbol{e}$ and $\boldsymbol{\alpha}$ is not generally one-to one. Several $\boldsymbol{e}$ may correspond to a single $\boldsymbol{\alpha}$, though the opposite is unique. For such $\boldsymbol{\alpha}$, the sum of B defined by the differential relation at each $\boldsymbol{e}$ must be adopted as B in Eq. (A.19).

Equation (A.19) is the Boltzmann equation [Eq. (2.5) with (2.6), (2.7), and $F_i=0$] introduced in Section 2.2.

In the derivation of the Boltzmann equation, the molecular-chaos assumption is introduced for the two-particle truncated density function $\tilde{f}_2$ for the situation that two particles are going to interact (or in I_-). If this assumption is also introduced for the situation that two particles have just finished interaction (or in

---

$\boldsymbol{X}^{(1)},\ldots,\boldsymbol{X}^{(N)},\boldsymbol{\xi}^{(1)},\ldots,\boldsymbol{\xi}^{(N)}$ giving the number of the sets $(\boldsymbol{X}^{(i)},\boldsymbol{\xi}^{(i)})$ $(i=1,\ldots,N)$ that lie in a 6-dimensional domain I (for example, $\boldsymbol{X}_a\leq\boldsymbol{X}^{(r)}\leq\boldsymbol{X}_b$, $\boldsymbol{\xi}_a\leq\boldsymbol{\xi}^{(r)}\leq\boldsymbol{\xi}_b$). Then, the expected value $\bar{\varphi}$ of φ is given by

$$\bar{\varphi}=\int_{\text{all }\boldsymbol{X}^{(1)},\ldots,\text{all }\boldsymbol{X}^{(N)},\text{all }\boldsymbol{\xi}^{(1)},\ldots,\text{all }\boldsymbol{\xi}^{(N)}}\varphi f^N\mathbf{d}\boldsymbol{X}^{(1)}\cdots\mathbf{d}\boldsymbol{X}^{(N)}\mathbf{d}\boldsymbol{\xi}^{(1)}\cdots\mathbf{d}\boldsymbol{\xi}^{(N)}$$
$$=\int_I Nf_1\mathbf{d}\boldsymbol{X}^{(1)}\mathbf{d}\boldsymbol{\xi}^{(1)},$$

for a symmetric f^N in view of the decomposition $\varphi=\sum_{i=1}^N\varphi_I(\boldsymbol{X}^{(i)},\boldsymbol{\xi}^{(i)})$, where $\varphi_I(\boldsymbol{X}^{(i)},\boldsymbol{\xi}^{(i)})=1$ for $(\boldsymbol{X}^{(i)},\boldsymbol{\xi}^{(i)})$ in I and $=0$ outside I.

I_+),[7] then the interaction term $(I_- + I_+)$ vanishes[8] and the Boltzmann equation is reduced to the free molecular equation (or the equation without the collision term). The introduction of the molecular chaos inevitably requires that the initial density function f^N should be factorized. Further, it is implicitly assumed that the chance of collisions of the set of particles that experienced collision between them is at most of the order of d_m. The mathematical verification of the Boltzmann equation for a hard-sphere gas from the Liouville equation is given by Lanford [1975], Illner & Pulvirenti [1986] (see also Cercignani, Illner & Pulvirenti [1994]), and Ukai [2001], where the system of equations for $\tilde{f}_1$, $\tilde{f}_2$, ... (BBGKY hierarchy) is considered[9] but the piece-wise estimate of errors or the direct verification of the molecular chaos is wisely bypassed.

The extension of the preceding analysis of derivation of the Boltzmann equation to include a small effect Nd_m^3 (or the order of d_m in the Grad–Boltzmann limit) is, apparently, straightforward. The small difference of the order of Nd_m^3 (or d_m) between f_1 and $\tilde{f}_1$, which is expressed by $\tilde{f}_2$, should be taken into account for the evaluation of the expected value of the mass of the particles in $\mathbf{d}\boldsymbol{X}\mathbf{d}\boldsymbol{\xi}$ up to the order of d_m [see Eq. (A.14) and Footnote 6]; the analysis of the integral on the right-hand side of Eq. (A.3a) is required to be more precise so as for the quantities of the order of d_m to be estimated correctly, and thus $\tilde{f}_2$ there is kept without replacing it by the product of $\tilde{f}_1$. Therefore, Eq. (A.3b) for $\tilde{f}_2$ should be considered not only during the time of interaction of two particles but also for all the time. The three-particle truncated density function $\tilde{f}_3$ enters Eq. (A.3b) in the interaction integrals over ∂S_1^3 or $D_1^3 \cap S_2^3$. The situation ∂S_1^3 is the beginning or end of the interaction of the third particle $\boldsymbol{X}^{(3)}$ with the particle $\boldsymbol{X}^{(1)}$ for $|\boldsymbol{X}^{(2)} - \boldsymbol{X}^{(1)}| \geq d_m$ (two-particle interaction) or with the interacting particles $\boldsymbol{X}^{(1)}$ and $\boldsymbol{X}^{(2)}$ for $|\boldsymbol{X}^{(2)} - \boldsymbol{X}^{(1)}| < d_m$ (three-particle interaction); the integral over $D_1^3 \cap S_2^3$ is the contribution during interaction of the two particles $\boldsymbol{X}^{(2)}$ and $\boldsymbol{X}^{(3)}$. By a similar (but more careful) argument to that in the preceding derivation of the Boltzmann equation, $\tilde{f}_3$ is replaced by $\tilde{f}_1$ and $\tilde{f}_2$, that is, $\tilde{f}_3$ is, if necessary, approximated, with the aid of the equation for $\tilde{f}_3$ containing $\tilde{f}_4$ in the interaction term, by $\tilde{f}_3$ in the situation in a near-past time of the three particles $\boldsymbol{X}^{(1)}$, $\boldsymbol{X}^{(2)}$, and $\boldsymbol{X}^{(3)}$ system where the replacement of $\tilde{f}_3$ by $\tilde{f}_1$ and $\tilde{f}_2$ is reasonable. However, the more accurate estimate of the errors required and the explicit analysis of the three-particle interaction are not so simple.[10]

---

[7]Obviously from the form of the distribution function $\tilde{f}_2$ after interaction derived under the assumption of molecular chaos before the interaction, the assumption of molecular chaos after the interaction is not compatible with that before interaction.

[8]The slowly varying assumption on $\tilde{f}_1$ being used after the replacement of $\tilde{f}_2$ by the product of $\tilde{f}_1$ on the right-hand side of Eq. (A.3a) [or Eq. (A.8)], then the integral with respect to $\mathrm{d}^2\boldsymbol{X}^{(2)}$ over the sphere ∂S_1^2 is reduced to that of $(\boldsymbol{\xi}^{(2)} - \boldsymbol{\xi}^{(1)}) \cdot \boldsymbol{n}_1^2$ over the sphere, which is easily seen to vanish.

[9]The definitions of $\tilde{f}_1$, $\tilde{f}_2$, ... are slightly different from ours because hard-sphere molecules cannot be closer than d_m to each other. For example, the range Ω_1 of integration in Eq. (A.2a) should be modified so as to exclude all the spheres S_i^j for any pair (i, j) from $D_1^2 \times D_1^3 \times \cdots \times D_1^N$ in addition to S_1^j.

[10]To evaluate the interaction term on the right-hand side of Eq. (A.3a) up to the order of

Before closing this section, the derivation of the formula (A.4) is supplemented.

Lemma.

$$\frac{\mathrm{d}}{\mathrm{d}\alpha}\int_{\mathcal{D}(\alpha)} g(\boldsymbol{X},\alpha)\mathrm{d}\boldsymbol{X} = \int_{\mathcal{D}(\alpha)} \frac{\partial g(\boldsymbol{X},\alpha)}{\partial\alpha}\mathrm{d}\boldsymbol{X} + \int_{\partial\mathcal{D}(\alpha)} g(\boldsymbol{X},\alpha)\left(\frac{\partial \boldsymbol{X}_w}{\partial\alpha}\cdot\boldsymbol{n}_w\right)\mathrm{d}^2\boldsymbol{X},$$

where $\boldsymbol{X}_w$ is a point on the boundary $\partial\mathcal{D}(\alpha)$ of $\mathcal{D}(\alpha)$, $\boldsymbol{n}_w$ is the outward unit normal vector to the boundary, and $\mathrm{d}^2\boldsymbol{X}$ is the surface element on $\partial\mathcal{D}(\alpha)$; the quantity $(\boldsymbol{n}_w\cdot\partial\boldsymbol{X}_w/\partial\alpha)$ as a whole has a definite meaning for the deformable domain $\mathcal{D}(\alpha)$. This formula applies to the space $\boldsymbol{X}$ with arbitrary dimension (say N).

Proof. Let $\mathcal{D}(0)$ be a fixed domain and Y_j be a point in it, which has a one-to-one correspondence to the domain $\mathcal{D}(\alpha)$ as $X_i = X_i(Y_j,\alpha)$. Then, the integral over the variable domain $\mathcal{D}(\alpha)$ is expressed by that over the fixed domain $\mathcal{D}(0)$, for which the differentiation with respect to α is carried out:

$$\begin{aligned}
&\frac{\mathrm{d}}{\mathrm{d}\alpha}\int_{\mathcal{D}(\alpha)} g(\boldsymbol{X},\alpha)\mathrm{d}\boldsymbol{X}\\
&= \frac{\mathrm{d}}{\mathrm{d}\alpha}\int_{\mathcal{D}(0)} g(\boldsymbol{X}(\boldsymbol{Y},\alpha),\alpha)J(\boldsymbol{Y},\alpha)\mathrm{d}\boldsymbol{Y}\\
&= \int_{\mathcal{D}(0)}\left(\frac{\partial g(\boldsymbol{X},\alpha)}{\partial\alpha} + \frac{\partial g(\boldsymbol{X},\alpha)}{\partial X_i}\frac{\partial X_i}{\partial\alpha}\right)J(\boldsymbol{Y},\alpha)\mathrm{d}\boldsymbol{Y} + \int_{\mathcal{D}(0)} g(\boldsymbol{X},\alpha)\frac{\partial J}{\partial\alpha}\mathrm{d}\boldsymbol{Y}\\
&= \int_{\mathcal{D}(\alpha)}\left(\frac{\partial g(\boldsymbol{X},\alpha)}{\partial\alpha} + \frac{\partial g(\boldsymbol{X},\alpha)}{\partial X_i}\frac{\partial X_i}{\partial\alpha}\right)\mathrm{d}\boldsymbol{X} + \int_{\mathcal{D}(0)} g(\boldsymbol{X},\alpha)\frac{\partial J}{\partial\alpha}\mathrm{d}\boldsymbol{Y}, \qquad \text{(A.20)}
\end{aligned}$$

where J is the Jacobian of the transformation from Y_i to X_i. Let T be the transformation matrix whose (i,j) element T_{ij} is given by

$$T_{ij} = \frac{\partial X_i}{\partial Y_j}, \quad \text{that is,} \quad T = \langle T_{ij}\rangle = \left\langle\frac{\partial X_i}{\partial Y_j}\right\rangle.$$

---

d_m, we have to know $\bar{f}_2$ on ∂S^2_{1-} and ∂S^2_{1+} up to the order of d_m on the basis of Eq. (A.3b). The $\bar{f}_2$ on ∂S^2_{1-} contains the whole past history, but $\bar{f}_2$ on ∂S^2_{1+} is obtained by the analysis of a short interaction process from $\bar{f}_2$ on ∂S^2_{1-}. For $\bar{f}_2$ to be obtained correctly up to the order of d_m, the approximation of $\bar{f}_3$ is to be more careful. The contribution to the variation of $\bar{f}_2$ along the characteristic of Eq. (A.3b) from the region $|\boldsymbol{X}^{(2)} - \boldsymbol{X}^{(1)}| < O(d_m)$ is of the order of d_m. Thus, $\bar{f}_3$ has only to be estimated with the error of $O(d_m)$. On the other hand, when $|\boldsymbol{X}^{(2)} - \boldsymbol{X}^{(1)}| > d_m$, it should be estimated correctly up to the order of d_m. For a hard-sphere gas, the chance of three-particle collision is vanishingly small irrespective of the size of d_m, but the assumption of molecular chaos cannot be applied to the collisions between the particles that have collided with another common particle, which are practically treated as three-particle collision. This type of situation occurs likely for a non-hard-sphere gas. The situation of replacement of $\bar{f}_3$ by $\bar{f}_1$ and $\bar{f}_2$ should be chosen carefully. Furthermore, the replacement should be done accurately up to the order of d_m.

The Jacobian is the determinant derived from the matrix $\langle T_{ij}\rangle$,[11] i.e.,

$$J = \left\| \frac{\partial X_i}{\partial Y_j} \right\|, \text{ and thus, } \frac{\partial J}{\partial \alpha} = \frac{\partial}{\partial \alpha} \left\| \frac{\partial X_i}{\partial Y_j} \right\|.$$

Then its inverse T^{-1} (or $\langle T_{ij}\rangle^{-1}$) is given by

$$T^{-1} = \langle T_{ij}\rangle^{-1} = \left\langle \frac{\partial X_i}{\partial Y_j} \right\rangle^{-1} = \left\langle (-1)^{i+j} \frac{\Delta_{ji}}{J} \right\rangle,$$

where the element Δ_{ij} is the determinant of the submatrix of the matrix T with respect to its element T_{ij}.[12] The derivative $\partial J/\partial \alpha$ of the Jacobian is transformed into the following form:

$$\begin{aligned}
\frac{\partial J}{\partial \alpha} &= \frac{\partial}{\partial \alpha} \left\| \frac{\partial X_i}{\partial Y_j} \right\| \\
&= \sum_{j=1}^{N} \sum_{i=1}^{N} (-1)^{i+j} \Delta_{ij} \frac{\partial}{\partial \alpha} \left(\frac{\partial X_i}{\partial Y_j} \right) = \sum_{j=1}^{N} \sum_{i=1}^{N} (-1)^{i+j} \Delta_{ij} \frac{\partial}{\partial Y_j} \left(\frac{\partial X_i}{\partial \alpha} \right) \\
&= \sum_{j=1}^{N} \sum_{i=1}^{N} (-1)^{i+j} \Delta_{ij} \sum_{k=1}^{N} \frac{\partial X_k}{\partial Y_j} \frac{\partial}{\partial X_k} \left(\frac{\partial X_i}{\partial \alpha} \right) \\
&= \sum_{i=1}^{N} \sum_{k=1}^{N} \frac{\partial}{\partial X_k} \left(\frac{\partial X_i}{\partial \alpha} \right) \left(\sum_{j=1}^{N} (-1)^{i+j} \Delta_{ij} \frac{\partial X_k}{\partial Y_j} \right) = \sum_{i=1}^{N} \sum_{k=1}^{N} J \delta_{ik} \frac{\partial}{\partial X_k} \left(\frac{\partial X_i}{\partial \alpha} \right) \\
&= J \sum_{i=1}^{N} \frac{\partial}{\partial X_i} \left(\frac{\partial X_i}{\partial \alpha} \right),
\end{aligned} \tag{A.21}$$

where the following relation is used:

$$\sum_{j=1}^{N} (-1)^{i+j} \Delta_{ij} \frac{\partial X_k}{\partial Y_j} = J \delta_{ik},$$

and the summation sign $\sum$ is shown explicitly to make the order of summation explicit for clarity of deformation. With the aid of Eq. (A.21), the last integral in Eq. (A.20) is deformed into an integral over the domain $\mathcal{D}(\alpha)$:

$$\begin{aligned}
\int_{\mathcal{D}(0)} g(\boldsymbol{X}, \alpha) \frac{\partial J}{\partial \alpha} \mathbf{d}\boldsymbol{Y} &= \int_{\mathcal{D}(0)} g(\boldsymbol{X}, \alpha) \sum_{i=1}^{N} \frac{\partial}{\partial X_i} \left(\frac{\partial X_i}{\partial \alpha} \right) J \mathbf{d}\boldsymbol{Y} \\
&= \int_{\mathcal{D}(\alpha)} g(\boldsymbol{X}, \alpha) \sum_{i=1}^{N} \frac{\partial}{\partial X_i} \left(\frac{\partial X_i}{\partial \alpha} \right) \mathbf{d}\boldsymbol{X}.
\end{aligned}$$

---

[11] Consideration of the transformation T with a positive determinant is sufficient for the present purpose.

[12] The submatrix is defined as the matrix obtained by deleting the row and column containing the element T_{ij} from the matrix T. The $(-1)^{i+j}\Delta_{ij}$ is called the cofactor of the element T_{ij}.

With this relation in Eq. (A.20), we obtain

$$\frac{\mathrm{d}}{\mathrm{d}\alpha}\int_{\mathcal{D}(\alpha)} g(\boldsymbol{X},\alpha)\mathrm{d}\boldsymbol{X}$$
$$= \int_{\mathcal{D}(\alpha)} \left[\frac{\partial g(\boldsymbol{X},\alpha)}{\partial\alpha} + \sum_{i=1}^{N}\frac{\partial}{\partial X_i}\left(\frac{\partial X_i}{\partial\alpha}g(\boldsymbol{X},\alpha)\right)\right]\mathrm{d}\boldsymbol{X}$$
$$= \int_{\mathcal{D}(\alpha)} \frac{\partial g(\boldsymbol{X},\alpha)}{\partial\alpha}\mathrm{d}\boldsymbol{X} + \int_{\partial\mathcal{D}(\alpha)} g(\boldsymbol{X},\alpha)\left(\frac{\partial \boldsymbol{X}_w}{\partial\alpha}\cdot\boldsymbol{n}_w\right)\mathrm{d}^2\boldsymbol{X}.$$

The last equation is obtained with the aid of Gauss's divergence theorem. The transformation $X_i = X_i(Y_j,\alpha)$ is not unique for a given deformable $\mathcal{D}(\alpha)$, but the quantity $(\boldsymbol{n}_w\cdot\partial\boldsymbol{X}_w/\partial\alpha)$ in the last integral is uniquely determined by $\mathcal{D}(\alpha)$, irrespective of the choice of $X_i = X_i(Y_j,\alpha)$. ■

A.2 Solution of integral equation $\mathcal{L}(\phi) = Ih$

In this section, we discuss the structure of the solution of the integral equation related to the linearized collision integral $\mathcal{L}(\phi)$ defined by Eq. (2.58b) and its slight extension $\mathcal{L}_a(\phi)$:[13]

$$\mathcal{L}(\phi(\boldsymbol{\zeta})) = \int E(\zeta_*)(\phi' + \phi'_* - \phi - \phi_*)\widehat{B}\,\mathrm{d}\Omega(\boldsymbol{\alpha})\,\mathrm{d}\boldsymbol{\zeta}_*, \tag{A.22}$$
$$\widehat{B} = \widehat{B}(|\boldsymbol{\alpha}\cdot(\boldsymbol{\zeta}_* - \boldsymbol{\zeta})|/|\boldsymbol{\zeta}_* - \boldsymbol{\zeta}|, |\boldsymbol{\zeta}_* - \boldsymbol{\zeta}|),$$
$$\mathcal{L}_a(\phi(\boldsymbol{\zeta})) = \int E(\zeta_*)(\phi' + \phi'_* - \phi - \phi_*)\widehat{B}_a\mathrm{d}\Omega(\boldsymbol{\alpha})\,\mathrm{d}\boldsymbol{\zeta}_*, \tag{A.23}$$
$$\widehat{B}_a = \widehat{B}_a(|\boldsymbol{\alpha}\cdot(\boldsymbol{\zeta}_* - \boldsymbol{\zeta})|/|\boldsymbol{\zeta}_* - \boldsymbol{\zeta}|, |\boldsymbol{\zeta}_* - \boldsymbol{\zeta}|)$$
$$= \frac{\widehat{B}(|\boldsymbol{\alpha}\cdot(\boldsymbol{\zeta}_* - \boldsymbol{\zeta})|/|\boldsymbol{\zeta}_* - \boldsymbol{\zeta}|, |\boldsymbol{\zeta}_* - \boldsymbol{\zeta}|a^{1/2})}{a^{1/2}}, \tag{A.24}$$

where $a\,(>0)$ may depend on $\boldsymbol{x}$ but is independent of $\boldsymbol{\zeta}$, and

$$\phi = \phi(\boldsymbol{\zeta}),\quad \phi_* = \phi(\boldsymbol{\zeta}_*),\quad \phi' = \phi(\boldsymbol{\zeta}'),\quad \phi'_* = \phi(\boldsymbol{\zeta}'_*),$$
$$\boldsymbol{\zeta}' = \boldsymbol{\zeta} + [\boldsymbol{\alpha}\cdot(\boldsymbol{\zeta}_* - \boldsymbol{\zeta})]\boldsymbol{\alpha},\quad \boldsymbol{\zeta}'_* = \boldsymbol{\zeta}_* - [\boldsymbol{\alpha}\cdot(\boldsymbol{\zeta}_* - \boldsymbol{\zeta})]\boldsymbol{\alpha}.$$

The operator $\mathcal{L}(\phi)$ is the special case $a = 1$ of $\mathcal{L}_a(\phi)$, i.e.,

$$\mathcal{L}(\phi) = \mathcal{L}_1(\phi).$$

For a hard-sphere gas, $\widehat{B}_a = |\alpha_i(\zeta_{i*} - \zeta_i)|/4(2\pi)^{1/2}$. For the BKW equation, the operator $\mathcal{L}_a(*)$ is consistently defined as follows (see Appendix A.7):

$$\mathcal{L}_a(\phi) = \frac{1}{a^{1/2}}\left\{\int\left[1 + 2\zeta_i\zeta_{i*} + \frac{2}{3}\left(\zeta_i{}^2 - \frac{3}{2}\right)\left(\zeta_{j*}{}^2 - \frac{3}{2}\right)\right]\phi(\zeta_{k*})E(\zeta_*)\mathrm{d}\boldsymbol{\zeta}_* - \phi\right\}. \tag{A.25}$$

[13]The operator $\mathcal{L}_a(*)$ as well as $\mathcal{L}(*)$ generally depends on the reference temperature T_0 besides the parameter a, since $\widehat{B}_a$ as well as $\widehat{B}$ does (see Footnote 11 in Section 2.9). The $\mathcal{L}_a(*)$ itself for a hard-sphere gas and $a^{1/2}\mathcal{L}_a(*)$ for BKW are independent of T_0 and a.

Now take a group of integral equations

$$\mathcal{L}_a[\phi_{i_1,\dots,i_m}(\zeta_k)] = H_{i_1,\dots,i_m}(\zeta_k), \tag{A.26}$$

where the inhomogeneous term $H_{i_1,\dots,i_m}(\zeta_k)$, which satisfies the solvability conditions

$$\int (1, \zeta_i, \zeta_j^2) H_{i_1,\dots,i_m}(\zeta_k) E(\zeta) \mathbf{d}\boldsymbol{\zeta} = 0,$$

is assumed to be symmetric with respect to the subscripts $(i_1, \dots, i_m)$ and to satisfy the following property:

$$H_{i_1,\dots,i_m}(l_{kh}\zeta_h) = l_{i_1j_1} \cdots l_{i_mj_m} H_{j_1,\dots,j_m}(\zeta_k) \quad \text{with } l_{ik}l_{jk} = \delta_{ij}, \tag{A.27}$$

that is, the symmetric-tensor field $H_{i_1,\dots,i_m}(\zeta_k)$ is spherically symmetric. It is noted here that the linearized collision integral $\mathcal{L}_a$, as well as $\mathcal{L}$, also has this property. That is, let

$$F_{i_1,\dots,i_m}(\zeta_k) = \mathcal{L}_a[\zeta_{i_1} \cdots \zeta_{i_m} \phi(\zeta)], \tag{A.28}$$

and then

$$F_{i_1,\dots,i_m}(l_{kh}\zeta_h) = l_{i_1j_1} \cdots l_{i_mj_m} F_{j_1,\dots,j_m}(\zeta_k). \tag{A.29}$$

The function $F_{i_1,\dots,i_m}(\zeta_k)$ is, of course, symmetric with respect to the subscripts $(i_1, \dots, i_m)$. Incidentally, the tensor field $H_{i_1,\dots,i_m}(\zeta_i)$ is axially symmetric with respect to the axis a_i, when the relation (A.27) holds for the limited set of l_{ij} that satisfies the relation

$$a_i = l_{ij}a_j,$$

for a fixed a_i.

The solution $\phi_{i_1,\dots,i_m}(\zeta_k)$ of Eq. (A.26) that is orthogonal to 1, ζ_i, and ζ^2 (the solutions of the associated homogeneous equation),

$$\int (1,\ \zeta_i,\ \zeta^2) \phi_{i_1,\dots,i_m}(\zeta_k) E(\zeta) \mathbf{d}\boldsymbol{\zeta} = 0, \tag{A.30}$$

is easily shown to have the spherically symmetric property

$$\phi_{i_1,\dots,i_m}(l_{kh}\zeta_h) = l_{i_1j_1} \cdots l_{i_mj_m} \phi_{j_1,\dots,j_m}(\zeta_k), \tag{A.31}$$

and to be symmetric with respect to the subscripts $(i_1, \dots, i_m)$ (Sone & Aoki [1994]).

On the other hand, any function $\Phi_{i_1,\dots,i_m}(\zeta_k)$ that is symmetric with respect to the subscripts $(i_1, \dots, i_m)$ and that has the spherically symmetric property

$$\Phi_{i_1,\dots,i_m}(l_{kh}\zeta_h) = l_{i_1j_1} \cdots l_{i_mj_m} \Phi_{j_1,\dots,j_m}(\zeta_k), \tag{A.32}$$

is known to be expressed in the following form (see Appendix B for elementary derivation):

$$\Phi_{i_1,\dots,i_m}(\zeta_h) = \sum_{n=0}^{[m/2]} g_n(\zeta) \sum_{*(m,2n)} \underbrace{\zeta_{i_a} \cdots \zeta_{i_b}}_{m-2n} \underbrace{\delta_{i_ci_d} \cdots \delta_{i_ei_f}}_{n}, \tag{A.33}$$

where $g_n(\zeta)$ is an arbitrary function of ζ, the symbol $[m/2]$ is the largest integer that does not exceed $m/2$, and the summation $\sum_{*(m,2n)}$ is carried out over $m!/2^n n!(m-2n)!$ terms in the following way: Divide m subscripts $(i_1,\ldots,i_m)$ into two sets: one with $m-2n$ elements $i_a,\ldots,i_b$ and the other with $2n$ elements $i_c,\ldots,i_f$, for which there are $m!/(2n)!(m-2n)!$ ways of division. For each division, consider all possible ways to make n pairs $(i_c,i_d),\ldots,(i_e,i_f)$ from the $2n$ elements $i_c,\ldots,i_f$, which are $(2n)!/2^n n!$. Then there are $m!/(2n)!(m-2n)!\times(2n)!/2^n n!$ terms. The $\sum_{*(m,2n)}$ means to sum up all these terms. For some small values of m, $\Phi_{i_1,\ldots,i_m}(\zeta_h)$ is given as follows:

$$
\begin{aligned}
m=0:\ & \Phi(\zeta_h)=g_0(\zeta),\\
m=1:\ & \Phi_i(\zeta_h)=\zeta_i g_0(\zeta),\\
m=2:\ & \Phi_{i,j}(\zeta_h)=\zeta_i\zeta_j g_0(\zeta)+\delta_{ij}g_1(\zeta),\\
m=3:\ & \Phi_{i,j,k}(\zeta_h)=\zeta_i\zeta_j\zeta_k g_0(\zeta)+(\zeta_i\delta_{jk}+\zeta_j\delta_{ki}+\zeta_k\delta_{ij})g_1(\zeta),\\
m=4:\ & \Phi_{i,j,k,l}(\zeta_h)=\zeta_i\zeta_j\zeta_k\zeta_l g_0(\zeta)\\
& \quad+(\zeta_i\zeta_j\delta_{kl}+\zeta_i\zeta_k\delta_{jl}+\zeta_i\zeta_l\delta_{jk}+\zeta_j\zeta_k\delta_{il}+\zeta_j\zeta_l\delta_{ik}+\zeta_k\zeta_l\delta_{ij})g_1(\zeta)\\
& \quad+(\delta_{ij}\delta_{kl}+\delta_{ik}\delta_{jl}+\delta_{il}\delta_{jk})g_2(\zeta).
\end{aligned}
\tag{A.34}
$$

With this form of function as $\phi_{i_1,\ldots,i_m}(\zeta_k)$, solution of the integral equation (A.26) for a function of three independent variables ζ_1, ζ_2, and ζ_3 is reduced to that of an integral equation for a function or functions of one independent variable ζ. For example, when $H_{i_1,\ldots,i_m}(\zeta_k)$ in Eq. (A.26) is given by $\zeta_i(\zeta^2-5/2)$, then ϕ is expressed as $\zeta_i g(\zeta)$, where $g(\zeta)$ is the solution of the integral equation

$$\mathcal{L}_a[\zeta_1 g(\zeta)]=\zeta_1(\zeta^2-5/2),$$

with the orthogonal condition

$$\int_0^\infty \zeta^4 g(\zeta)E\mathrm{d}\zeta=0.$$

A.3 Derivation of the Stokes set of equations

In Section 3.2, we mentioned that the Stokes set of equations (3.18)–(3.19c) is derived from the condition (3.11) for solvability of Eq. (3.9) for any order of approximation, leaving its proof to this appendix.

For this purpose, we will prove the following statement about the form of the component function ϕ_{Gm} of the velocity distribution function of the Grad–Hilbert expansion (3.5) together with the Stokes set of equations (3.18)–(3.19c). The component function ϕ_{Gm}, which is the solution of the integral equation (3.8) or (3.9), is given by

$$\phi_{Gm}=P_{Gm}+2\zeta_i u_{iGm}+\left(\zeta^2-\frac{5}{2}\right)\tau_{Gm}+\hat{\phi}_{Gm}, \tag{A.35}$$

where $\hat{\phi}_{Gm}$, the particular solution of Eq. (3.9), is expressed in the form

$$\hat{\phi}_{G0} = 0, \tag{A.36a}$$

$$\begin{aligned}\hat{\phi}_{Gm} =& \sum_{s=1}^{m} \Phi_{i_1,\ldots,i_s}^{(m,1)}(\zeta)\frac{\partial^s P_{Gm-s}}{\partial x_{i_1}\cdots\partial x_{i_s}} \\ &+ 2\sum_{s=1}^{m} \zeta_{i_0}\zeta_{i_1}\cdots\zeta_{i_s} g_{(s,0)}^{(m,2)}(\zeta)\mathrm{sym}\left(\frac{\partial^s u_{i_0 Gm-s}}{\partial x_{i_1}\cdots\partial x_{i_s}}\right) \\ &+ \sum_{s=1}^{m} \Phi_{i_1,\ldots,i_s}^{(m,3)}(\zeta)\frac{\partial^s \tau_{Gm-s}}{\partial x_{i_1}\cdots\partial x_{i_s}},\end{aligned} \tag{A.36b}$$

and satisfies the orthogonal condition (3.14), i.e.,

$$\int (1, \zeta_i, \zeta^2)\hat{\phi}_{Gm} E \mathrm{d}\boldsymbol{\zeta} = 0. \tag{A.37}$$

Here, the symbol sym means the symmetrization with respect to the subscripts $(i_0, \ldots, i_s)$; $g_{(s,0)}^{(m,2)}$ is a function of ζ; and $\Phi_{i_1,\ldots,i_s}^{(m,r)}(\zeta)$ is a symmetric-tensor field with the spherically symmetric property (A.32), that is,

$$\Phi_{i_1,\ldots,i_s}^{(m,r)}(\zeta) = \sum_{p=0}^{[s/2]} g_{(s,p)}^{(m,r)}(\zeta) \sum_{*(s,2p)} \underbrace{\zeta_{i_a}\cdots\zeta_{i_b}}_{s-2p} \underbrace{\delta_{i_c i_d}\cdots\delta_{i_e i_f}}_{p}, \tag{A.38}$$

where $g_{(s,p)}^{(m,r)}(\zeta)$ is a function of ζ and the summation $\sum_{*(s,2p)}$ is carried out over $s!/2^p p!(s-2p)!$ terms as explained below Eq. (A.33). Especially, $g_{(1,0)}^{(m,2)}$ is given by the function $-B(\zeta)/2$ defined by Eq. (3.26):

$$g_{(1,0)}^{(m,2)} = -B(\zeta)/2. \tag{A.39}$$

The orthogonal condition (A.37) guarantees the relations (3.7a)–(3.7c).

The Stokes set of equations is written in the following form for the convenience of explanation:

$$\frac{\partial u_{iGm-1}}{\partial x_i} = 0, \tag{A.40}$$

$$\frac{\partial P_{Gm-1}}{\partial x_i} = \gamma_1 \frac{\partial^2 u_{iGm-2}}{\partial x_j^2}, \tag{A.41}$$

$$\frac{\partial^2 \tau_{Gm-2}}{\partial x_j^2} = 0, \tag{A.42}$$

$$(m = 0, 1, \ldots)$$

with the convention

$$u_{iG-2} = u_{iG-1} = 0, \quad \tau_{G-2} = \tau_{G-1} = 0, \quad P_{G-1} = 0,$$

where γ_1 is a constant defined by Eq. (3.33). From Eqs. (A.40) and (A.41),

$$\frac{\partial^2 P_{Gm-1}}{\partial x_j^2} = 0, \tag{A.43}$$

and further, from Eqs. (A.41) and (A.43),

$$\frac{\partial^2}{\partial x_k^2} \frac{\partial^2 u_{iGm-2}}{\partial x_j^2} = 0. \tag{A.44}$$

These relations will be used in the following analysis.

Equation (A.35) with Eq. (A.36a) or (A.36b) (or the functional form of ϕ_{Gm}) and the Stokes set of equations (A.40)–(A.42) will be proved by mathematical induction.

Before starting the proof, it should be mentioned here that we do not have to pay attention to the orthogonal condition (A.37) on $\hat{\phi}_{Gm}$ given by Eq. (A.36b) for the following reason. The symmetric functions $\Phi^{(m,1)}_{i_1,\dots,i_s}(\zeta)$ and $\Phi^{(m,3)}_{i_1,\dots,i_s}(\zeta)$ under the first and third summation signs of Eq. (A.36b) can be chosen to be orthogonal to 1, ζ_i, and ζ^2. That is, they are obviously orthogonal to 1 and ζ^2 for odd s and to ζ_i for even s. A special $\Phi^{(m,r)}_{i_1,\dots,i_s}(\zeta)$ with $g^{(m,r)}_{(s,s/2)} = a^{(m,r)}_s + b^{(m,r)}_s\zeta^2$ and $g^{(m,r)}_{(s,p)} = 0$ for the other p, say $(a^{(m,r)}_s + b^{(m,r)}_s\zeta^2)\Psi^0_{i_1,\dots,i_s}$, can be added to $\Phi^{(m,r)}_{i_1,\dots,i_s}(\zeta)$ for even s, and a special $\Phi^{(m,r)}_{i_1,\dots,i_s}(\zeta)$ with $g^{(m,r)}_{(s,(s-1)/2)} = e^{(m,r)}_s$ and $g^{(m,r)}_{(s,p)} = 0$ for the other p, say $e^{(m,r)}_s\Psi^1_{i_1,\dots,i_s}$, can be added to $\Phi^{(m,r)}_{i_1,\dots,i_s}(\zeta)$ for odd s, where $a^{(m,r)}_s$, $b^{(m,r)}_s$, and $e^{(m,r)}_s$ are arbitrary constants, because these special functions $\Phi^{(m,r)}_{i_1,\dots,i_s}(\zeta)$ are solutions of the homogeneous equation $\mathcal{L}(\phi) = 0$. Therefore, $\Phi^{(m,1)}_{i_1,\dots,i_s}(\zeta)$ and $\Phi^{(m,3)}_{i_1,\dots,i_s}(\zeta)$ in Eq. (A.36b) can be made orthogonal to 1, ζ_i, and ζ^2 by choosing the constants $a^{(m,r)}_s$, $b^{(m,r)}_s$, and $e^{(m,r)}_s$ properly. The function $\zeta_{i_0}\zeta_{i_1}\cdots\zeta_{i_s}g^{(m,2)}_{(s,0)}(\zeta)$ under the second summation sign in Eq. (A.36b) is obviously orthogonal to 1 and ζ^2 for even s and to ζ_i for odd s. For odd s,

$$\zeta_{i_0}\zeta_{i_1}\cdots\zeta_{i_s}g^{(m,2)}_{(s,0)}(\zeta)\mathrm{sym}\left(\frac{\partial^s u_{i_0Gm-s}}{\partial x_{i_1}\cdots\partial x_{i_s}}\right)$$

is also orthogonal to 1 and ζ^2 because the integration contains $\partial u_{i_tGm-s}/\partial x_{i_t}$. For even s (≥ 2), the orthogonal integration with ζ_i is proportional to

$$\frac{\partial^s u_{iGm-s}}{\partial x_{i_1}^2\cdots\partial x_{i_{s/2}}^2} \quad \text{or} \quad \frac{\partial^s u_{i_tGm-s}}{\partial x_{i_t}\partial x_i\cdots\partial x_{i_s}} \quad (=0).$$

The first one vanishes for $s \geq 4$ owing to Eq. (A.44), but is rewritten, with the aid of Eq. (A.41), as $\gamma_1^{-1}\partial P_{Gm-1}/\partial x_i$ for $s = 2$. This can be taken care of by the solution $\zeta_{i_1}\partial P_{Gm-1}/\partial x_{i_1}$, orthogonal to 1 and ζ^2, of the homogeneous equation, where ζ_{i_1} is a special $\Phi^{(m,r)}_{i_1,\dots,i_s}(\zeta)$ with $s = 1$ and $g^{(m,r)}_{(1,0)} = 1$. That is,

$$\zeta_{i_0}\zeta_{i_1}\zeta_{i_2}g^{(m,2)}_{(2,0)}(\zeta)\mathrm{sym}\left(\frac{\partial^2 u_{i_0Gm-2}}{\partial x_{i_1}\partial x_{i_2}}\right) + e^{(m,2)}_1\zeta_{i_1}\partial P_{Gm-1}/\partial x_{i_1}$$

can be made orthogonal to ζ_i for a properly chosen $e_1^{(m,2)}$. Thus, by adding $e_1^{(m,2)}\zeta_{i_1}\partial P_{Gm-1}/\partial x_{i_1}$ in the first summation sign in Eq. (A.36b), we can make the particular solution satisfy the orthogonal condition (A.37).

Equation (A.35) with (A.36a) or (A.36b) is shown to be true for $m = 0$ and 1 in Section 3.2. Let Eq. (A.35) with (A.36a) or (A.36b) and the Stokes set of equations (A.40)–(A.42) be true for $m = 0, 1, \ldots, n$. Then, ϕ_{Gn+1} is given as the solution of the integral equation

$$\mathcal{L}(\phi_{Gn+1}) = \zeta_i \frac{\partial \phi_{Gn}}{\partial x_i}, \tag{A.45}$$

where $\zeta_i \partial\phi_{Gn}/\partial x_i$ is[14]

$$\begin{aligned}
\zeta_i \frac{\partial \phi_{Gn}}{\partial x_i} &= \zeta_{i_1}\frac{\partial P_{Gn}}{\partial x_{i_1}} + 2\zeta_{i_0}\zeta_{i_1}\frac{\partial u_{i_0 Gn}}{\partial x_{i_1}} + \zeta_{i_1}\left(\zeta^2 - \frac{5}{2}\right)\frac{\partial \tau_{Gn}}{\partial x_{i_1}} \\
&\quad + \sum_{s=1}^{n} \zeta_{i_{s+1}} \Phi^{(n,1)}_{i_1,\ldots,i_s}(\zeta) \frac{\partial^{s+1} P_{Gn-s}}{\partial x_{i_1}\cdots\partial x_{i_{s+1}}} \\
&\quad + 2\sum_{s=1}^{n} \zeta_{i_0}\zeta_{i_1}\cdots\zeta_{i_{s+1}} g^{(n,2)}_{(s,0)}(\zeta)\frac{\partial}{\partial x_{i_{s+1}}}\mathrm{sym}\left(\frac{\partial^s u_{i_0 Gn-s}}{\partial x_{i_1}\cdots\partial x_{i_s}}\right) \\
&\quad + \sum_{s=1}^{n} \zeta_{i_{s+1}} \Phi^{(n,3)}_{i_1,\ldots,i_s}(\zeta) \frac{\partial^{s+1} \tau_{Gn-s}}{\partial x_{i_1}\cdots\partial x_{i_{s+1}}} \\
&= \zeta_{i_1}\frac{\partial P_{Gn}}{\partial x_{i_1}} + \zeta_{i_0}\zeta_{i_1}\left(\frac{\partial u_{i_0 Gn}}{\partial x_{i_1}} + \frac{\partial u_{i_1 Gn}}{\partial x_{i_0}}\right) + \zeta_{i_1}\left(\zeta^2 - \frac{5}{2}\right)\frac{\partial \tau_{Gn}}{\partial x_{i_1}} \\
&\quad + \sum_{s=1}^{n} \mathrm{sym}\left[\zeta_{i_{s+1}} \Phi^{(n,1)}_{i_1,\ldots,i_s}(\zeta)\right] \frac{\partial^{s+1} P_{Gn-s}}{\partial x_{i_1}\cdots\partial x_{i_{s+1}}} \\
&\quad + 2\sum_{s=1}^{n} \zeta_{i_0}\zeta_{i_1}\cdots\zeta_{i_{s+1}} g^{(n,2)}_{(s,0)}(\zeta)\mathrm{sym}\left(\frac{\partial^{s+1} u_{i_0 Gn-s}}{\partial x_{i_1}\cdots\partial x_{i_{s+1}}}\right) \\
&\quad + \sum_{s=1}^{n} \mathrm{sym}\left[\zeta_{i_{s+1}} \Phi^{(n,3)}_{i_1,\ldots,i_s}(\zeta)\right] \frac{\partial^{s+1} \tau_{Gn-s}}{\partial x_{i_1}\cdots\partial x_{i_{s+1}}}.
\end{aligned} \tag{A.46}$$

In the expression after the second equality sign, the symmetrization sym is with respect to the subscripts $(i_1, \ldots, i_{s+1})$ or $(i_0, \ldots, i_{s+1})$; and $\mathrm{sym}\left[\zeta_{i_{s+1}} \Phi^{(n,r)}_{i_1,\ldots,i_s}(\zeta)\right]$ obviously satisfies the spherically symmetric relation (A.32) and therefore is expressed in the form as Eq. (A.38), that is,

$$\mathrm{sym}\left[\zeta_{i_{s+1}} \Phi^{(n,r)}_{i_1,\ldots,i_s}(\zeta)\right] = \sum_{p=0}^{[(s+1)/2]} h^{(n,r)}_{(s+1,p)}(\zeta)\left(\sum_{*(s+1,2p)} \underbrace{\zeta_{i_a}\cdots\zeta_{i_b}}_{s+1-2p}\underbrace{\delta_{i_c i_d}\cdots\delta_{i_e i_f}}_{p}\right),$$

where $h^{(n,r)}_{(s+1,p)}(\zeta)$ is a function of ζ.

---

[14]For dummy indices, the letters can be chosen arbitrarily for our convenience.

The solvability condition of the integral equation (A.45) or (3.9) with $m = n+1$ is given by Eq. (3.11) with $m = n+1$, that is,

$$\int \begin{pmatrix} 1 \\ \zeta_i \\ \zeta^2 \end{pmatrix} \zeta_k \frac{\partial \phi_{Gn}}{\partial x_k} E(\zeta)\mathbf{d}\boldsymbol{\zeta} = 0. \tag{A.47}$$

The contribution of the first three terms of the right-hand side of Eq. (A.46) can be calculated as

$$\int \begin{pmatrix} 1 \\ \zeta_i \\ \zeta^2 \end{pmatrix} \left[\zeta_i \frac{\partial \phi_{Gn}}{\partial x_i}\right]_1 E(\zeta)\mathbf{d}\boldsymbol{\zeta} = \begin{pmatrix} \dfrac{\partial u_{iGn}}{\partial x_i} \\ \dfrac{1}{2}\dfrac{\partial P_{Gn}}{\partial x_i} \\ \dfrac{5}{2}\dfrac{\partial u_{iGn}}{\partial x_i} \end{pmatrix}, \tag{A.48}$$

where

$$\left[\zeta_i \frac{\partial \phi_{Gn}}{\partial x_i}\right]_1 = \zeta_{i_1}\frac{\partial P_{Gn}}{\partial x_{i_1}} + \zeta_{i_0}\zeta_{i_1}\left(\frac{\partial u_{i_0 Gn}}{\partial x_{i_1}} + \frac{\partial u_{i_1 Gn}}{\partial x_{i_0}}\right) + \zeta_{i_1}\left(\zeta^2 - \frac{5}{2}\right)\frac{\partial \tau_{Gn}}{\partial x_{i_1}}.$$

Before proceeding to the remaining terms of the solvability condition (A.47), some preparations are in order. Obviously,

$$\int \begin{pmatrix} 1 \\ \zeta^2 \end{pmatrix} \mathrm{sym}\left[\zeta_{i_{s+1}} \Phi^{(n,r)}_{i_1,\ldots,i_s}(\boldsymbol{\zeta})\right] E(\zeta)\mathbf{d}\boldsymbol{\zeta} = 0 \quad (s \text{ is even}), \tag{A.49a}$$

$$\int \zeta_i \mathrm{sym}\left[\zeta_{i_{s+1}} \Phi^{(n,r)}_{i_1,\ldots,i_s}(\boldsymbol{\zeta})\right] E(\zeta)\mathbf{d}\boldsymbol{\zeta} = 0 \quad (s \text{ is odd}). \tag{A.49b}$$

With the aid of the formula (B.37) with (B.38),

$$\begin{aligned}
&\int \begin{pmatrix} 1 \\ \zeta^2 \end{pmatrix} \mathrm{sym}\left[\zeta_{i_{s+1}} \Phi^{(n,r)}_{i_1,\ldots,i_s}(\boldsymbol{\zeta})\right] E(\zeta)\mathbf{d}\boldsymbol{\zeta} \\
&= \int \begin{pmatrix} 1 \\ \zeta^2 \end{pmatrix} \sum_{p=0}^{[(s+1)/2]} h^{(n,r)}_{(s+1,p)}(\zeta) \left(\sum_{*(s+1,2p)} \underbrace{\zeta_{i_a}\cdots\zeta_{i_b}}_{s+1-2p} \underbrace{\delta_{i_c i_d}\cdots\delta_{i_e i_f}}_{p} \right) E(\zeta)\mathbf{d}\boldsymbol{\zeta} \\
&= \begin{pmatrix} c^{(n,r)}_{(s,0)} \\ c^{(n,r)}_{(s,2)} \end{pmatrix} \sum_{k=1}^{Q_{(s+1)/2}} \prod_{\substack{\text{all the pairs} \\ \text{in } C_k((s+1)/2)}} \delta_{i_t i_u} \quad (s \text{ is odd}),
\end{aligned} \tag{A.50}$$

where $c^{(n,r)}_{(s,0)}$ and $c^{(n,r)}_{(s,2)}$ are some constants, $Q_N = (2N)!/2^N N!$, and $C_k((s+1)/2)$ is one of $Q_{(s+1)/2}$ sets of $(s+1)/2$ pairs $[(i_{a_1}, i_{a_2}), (i_{a_3}, i_{a_4}), \ldots, (i_{a_s}, i_{a_{s+1}})$, say$]$

of indices $(i_1, i_2, \ldots, i_{s+1})$;[15] and

$$\int \zeta_i \mathrm{sym}\left[\zeta_{i_{s+1}} \Phi^{(n,r)}_{i_1,\ldots,i_s}(\boldsymbol{\zeta})\right] E(\zeta)\mathbf{d}\boldsymbol{\zeta}$$

$$= \int \zeta_i \sum_{p=0}^{[(s+1)/2]} h^{(n,r)}_{(s+1,p)}(\zeta) \left(\sum_{*(s+1,2p)} \underbrace{\zeta_{i_a} \cdots \zeta_{i_b}}_{s+1-2p} \underbrace{\delta_{i_c i_d} \cdots \delta_{i_e i_f}}_{p} \right) E(\zeta)\mathbf{d}\boldsymbol{\zeta}$$

$$= c^{(n,r)}_{(s,1)} \sum_{k=1}^{Q_{(s+2)/2}} \prod_{\substack{\text{all the pairs} \\ \text{in } C_k((s+2)/2)}} \delta_{i_t i_u} \quad (s \text{ is even}), \tag{A.51}$$

where $c^{(n,r)}_{(s,1)}$ is a constant, and the pairs in $C_k((s+2)/2)$ are made among the $s+2$ indices $(i, i_1, \ldots, i_{s+1})$. For $s = 2$, for example,

$$\int \zeta_i \mathrm{sym}\left[\zeta_{i_{s+1}} \Phi^{(n,r)}_{i_1,\ldots,i_s}(\boldsymbol{\zeta})\right] E(\zeta)\mathbf{d}\boldsymbol{\zeta}$$

$$= c^{(n,r)}_{(2,1)} (\delta_{ii_1}\delta_{i_2 i_3} + \delta_{ii_2}\delta_{i_3 i_1} + \delta_{ii_3}\delta_{i_1 i_2}) \quad (s = 2).$$

Further, the following integrals, which also appear in the solvability condition (A.47), are special cases of the above integrals: Obviously,

$$\int \begin{pmatrix} 1 \\ \zeta^2 \end{pmatrix} \zeta_{i_0}\zeta_{i_1} \cdots \zeta_{i_{s+1}} g^{(n,2)}_{(s,0)} E(\zeta)\mathbf{d}\boldsymbol{\zeta} = \begin{pmatrix} 0 \\ 0 \end{pmatrix} \quad (s \text{ is odd}), \tag{A.52}$$

$$\int \zeta_i \zeta_{i_0}\zeta_{i_1} \cdots \zeta_{i_{s+1}} g^{(n,2)}_{(s,0)} E(\zeta)\mathbf{d}\boldsymbol{\zeta} = 0 \quad (s \text{ is even}), \tag{A.53}$$

and with the aid of the formula (B.37) with (B.38),

$$\int \begin{pmatrix} 1 \\ \zeta^2 \end{pmatrix} \zeta_{i_0}\zeta_{i_1} \cdots \zeta_{i_{s+1}} g^{(n,2)}_{(s,0)} E(\zeta)\mathbf{d}\boldsymbol{\zeta}$$

$$= \begin{pmatrix} d^{(n,2)}_{(s,0,0)} \\ d^{(n,2)}_{(s,0,2)} \end{pmatrix} \sum_{k=1}^{Q_{(s+2)/2}} \prod_{\substack{\text{all the pairs} \\ \text{in } C_k((s+2)/2)}} \delta_{i_t i_u} \quad (s \text{ is even}), \tag{A.54}$$

$$\int \zeta_i \zeta_{i_0}\zeta_{i_1} \cdots \zeta_{i_{s+1}} g^{(n,2)}_{(s,0)} E(\zeta)\mathbf{d}\boldsymbol{\zeta} = d^{(n,2)}_{(s,0,1)} \sum_{k=1}^{Q_{(s+3)/2}} \prod_{\substack{\text{all the pairs} \\ \text{in } C_k((s+3)/2)}} \delta_{i_t i_u} \quad (s \text{ is odd}), \tag{A.55}$$

where the pairs in $C_k((s+2)/2)$ in Eq. (A.54) and those in $C_k((s+3)/2)$ in Eq. (A.55) are made, respectively, among the $s+2$ indices $(i_0, i_1, \ldots, i_{s+1})$ and

[15] The notation $C_k((s+1)/2)$ will be used in the following discussion. The array of indices depends on cases, for example, $(i, i_1, \ldots, i_s)$. We will refer to the array of indices under concern in each place. The explanation of C_k is also given at the beginning of Appendix B.3.1.

among $s+3$ indices $(i, i_0, i_1, \dots, i_{s+1})$, and $d^{(n,2)}_{(s,0,l)}$ is a constant, especially, from Eqs. (A.39), (3.33), and (B.37) with (B.38), $d^{(n,2)}_{(1,0,1)}$ is related to γ_1:

$$d^{(n,2)}_{(1,0,1)} = \frac{-2}{15\sqrt{\pi}} \int_0^\infty \zeta^6 B(\zeta) \exp(-\zeta^2) \mathrm{d}\zeta = \frac{-1}{4}\gamma_1.$$

Most of the nondegenerate (nonvanishing) cases [Eqs. (A.50), (A.51), (A.54), and (A.55)] also vanish if we consider them with the combination of their factors in Eq. (A.46).

(i) With the aid of the relations (A.42) and (A.43) with $m \le n$, the case corresponding to Eq. (A.50) is reduced to

$$\left(\sum_{k=1}^{Q_{(s+1)/2}} \prod_{\substack{\text{all the pairs} \\ \text{in } C_k((s+1)/2)}} \delta_{i_t i_u} \right) \begin{pmatrix} \dfrac{\partial^{s+1} P_{Gn-s}}{\partial x_{i_1} \cdots \partial x_{i_{s+1}}} \\ \dfrac{\partial^{s+1} \tau_{Gn-s}}{\partial x_{i_1} \cdots \partial x_{i_{s+1}}} \end{pmatrix} = \begin{pmatrix} 0 \\ 0 \text{ if } s \ge 3, \quad \dfrac{\partial^2 \tau_{Gn-1}}{\partial x_{i_1}^2} \text{ if } s = 1 \end{pmatrix} \quad (s \text{ is odd}),$$

where the pairs in $C_k((s+1)/2)$ are made among the $s+1$ indices $(i_1, \dots, i_{s+1})$.

(ii) With the aid of the relations (A.42) and (A.43) with $m \le n$, the case corresponding to Eq. (A.51) is reduced to

$$\left(\prod_{\substack{\text{all the pairs} \\ \text{in } C_k((s+2)/2)}} \delta_{i_t i_u} \right) \begin{pmatrix} \dfrac{\partial^{s+1} P_{Gn-s}}{\partial x_{i_1} \cdots \partial x_{i_{s+1}}} \\ \dfrac{\partial^{s+1} \tau_{Gn-s}}{\partial x_{i_1} \cdots \partial x_{i_{s+1}}} \end{pmatrix} = \begin{pmatrix} 0 \\ 0 \end{pmatrix} \quad [s\ (\ge 2) \text{ is even}],$$

where the pairs in $C_k((s+2)/2)$ are made among the $s+2$ indices $(i, i_1, \dots, i_{s+1})$.

(iii) In view of Eq. (A.40) with $m \le n$, the case corresponding to Eq. (A.54) is reduced to

$$\left(\prod_{\substack{\text{all the pairs} \\ \text{in } C_k((s+2)/2)}} \delta_{i_t i_u} \right) \operatorname{sym} \left(\frac{\partial^{s+1} u_{i_0 Gn-s}}{\partial x_{i_1} \cdots \partial x_{i_{s+1}}} \right) = 0 \quad [s\ (\ge 2) \text{ is even}],$$

where the pairs in $C_k((s+2)/2)$ are made among the $s+2$ indices $(i_0, i_1, \dots, i_{s+1})$.

(iv) In view of Eqs. (A.40) and (A.44) with $m \le n$, the case corresponding to

Eq. (A.55) is reduced to

$$\left(\sum_{k=1}^{Q_{(s+3)/2}} \prod_{\substack{\text{all the pairs} \\ \text{in } C_k((s+3)/2)}} \delta_{i_t i_u}\right) \text{sym}\left(\frac{\partial^{s+1} u_{i_0 Gn-s}}{\partial x_{i_1}\cdots\partial x_{i_{s+1}}}\right) = 0 \quad [s\ (\geq 3) \text{ is odd}]$$

$$= \frac{\partial^2 u_{iGn-1}}{\partial x_{i_1}^2} + 2\frac{\partial^2 u_{i_1 Gn-1}}{\partial x_i \partial x_{i_1}}$$

$$= \frac{\partial^2 u_{iGn-1}}{\partial x_{i_1}^2} \quad (s=1),$$

where the pairs in $C_k((s+3)/2)$ are made among the $s+3$ indices $(i, i_0, \ldots, i_{s+1})$.

With these preparations, the solvability condition (A.47) is reduced as follows: The first relation of Eq. (A.47) is

$$\int \zeta_k \frac{\partial \phi_{Gn}}{\partial x_k} E(\zeta)\mathbf{d}\boldsymbol{\zeta} = \frac{\partial u_{iGn}}{\partial x_i} = 0,$$

where Eq.(3.7b) is used. [No literal manipulation using the formula (A.35) with (A.36b) for ϕ_{Gn} is required. This is the result of the second relation of the orthogonality condition (A.37) for $m = n$.] The third relation is

$$\int \zeta^2 \zeta_k \frac{\partial \phi_{Gn}}{\partial x_k} E(\zeta)\mathbf{d}\boldsymbol{\zeta} = \frac{5}{2}\frac{\partial u_{iGn}}{\partial x_i} + c_{(1,2)}^{(n,3)} \frac{\partial^2 \tau_{Gn-1}}{\partial x_{i_1}^2} = 0, \text{ or } \frac{\partial^2 \tau_{Gn-1}}{\partial x_{i_1}^2} = 0.$$

The second relation is

$$\int \zeta_i \zeta_k \frac{\partial \phi_{Gn}}{\partial x_k} E(\zeta)\mathbf{d}\boldsymbol{\zeta} = \frac{1}{2}\frac{\partial P_{Gn}}{\partial x_i} - \frac{\gamma_1}{2}\frac{\partial^2 u_{iGn-1}}{\partial x_{i_1}^2} = 0.$$

To summarize, from the solvability condition (A.47), we obtain the Stokes set with subscript n being increased by one:

$$\frac{\partial u_{iGn}}{\partial x_i} = 0, \tag{A.56}$$

$$\frac{\partial P_{Gn}}{\partial x_i} = \gamma_1 \frac{\partial^2 u_{iGn-1}}{\partial x_{i_1}^2}, \tag{A.57}$$

$$\frac{\partial^2 \tau_{Gn-1}}{\partial x_{i_1}^2} = 0. \tag{A.58}$$

Thus, Eqs. (A.40)–(A.44) hold up to $m = n+1$.

The solvability condition (A.47) or (A.56)–(A.58) being satisfied, according to the discussion in Appendix A.2, the solution of the integral equation (A.45) with the inhomogeneous term (A.46) is expressed in the following form:

$$\phi_{Gn+1} = P_{Gn+1} + 2\zeta_i u_{iGn+1} + \left(\zeta^2 - \frac{5}{2}\right)\tau_{Gn+1} + \hat{\phi}_{Gn+1}, \tag{A.59}$$

with the particular solution $\hat{\phi}_{Gn+1}$, satisfying $\int(1,\zeta_i,\zeta^2)\hat{\phi}_{Gn+1}E\mathbf{d}\boldsymbol{\zeta}=0$ (orthogonal relation), given by

$$\begin{aligned}\hat{\phi}_{Gn+1} &= \sum_{s=1}^{n+1}\Phi^{(n+1,1)}_{i_1,\dots,i_s}(\zeta)\frac{\partial^s P_{Gn+1-s}}{\partial x_{i_1}\cdots\partial x_{i_s}} \\ &+ 2\sum_{s=1}^{n+1}\Phi^{(n+1,2)}_{i_0,\dots,i_s}(\zeta)\mathrm{sym}\left(\frac{\partial^s u_{i_0 Gn+1-s}}{\partial x_{i_1}\cdots\partial x_{i_s}}\right) \\ &+ \sum_{s=1}^{n+1}\Phi^{(n+1,3)}_{i_1,\dots,i_s}(\zeta)\frac{\partial^s \tau_{Gn+1-s}}{\partial x_{i_1}\cdots\partial x_{i_s}},\end{aligned} \tag{A.60}$$

where $\Phi^{(n+1,1)}_{i_1,\dots,i_s}$ and $\Phi^{(n+1,3)}_{i_1,\dots,i_s}$ are symmetric with respect to the set of subscripts $(i_1,\dots,i_s)$, $\Phi^{(n+1,2)}_{i_0,\dots,i_s}$ is symmetric with respect to $(i_0,\dots,i_s)$, and they satisfy the spherically symmetric relation (A.32).

The second summation in the expression (A.60) is simplified in the following way. The symmetric function $\Phi^{(n+1,2)}_{i_0,\dots,i_s}$ is expressed as

$$\Phi^{(n+1,2)}_{i_0,\dots,i_s}(\zeta)=\sum_{p=0}^{[(s+1)/2]} g^{(n+1,2)}_{(s,p)}(\zeta)\sum_{*(s+1,2p)}\underbrace{\zeta_{i_a}\cdots\zeta_{i_b}}_{s+1-2p}\underbrace{\delta_{i_c i_d}\cdots\delta_{i_e i_f}}_{p},$$

where $g^{(n+1,2)}_{(1,0)}(\zeta)=-B(\zeta)/2$.[16] Consider the following expression in the summation sign:

$$\underbrace{\delta_{i_c i_d}\cdots\delta_{i_e i_f}}_{p}\,\mathrm{sym}\left(\frac{\partial^s u_{i_0 Gn+1-s}}{\partial x_{i_1}\cdots\partial x_{i_s}}\right). \tag{A.61}$$

When $s=1$ and $p=1$ (p can be 0 or 1 for $s=1$), the expression (A.61) is reduced to

$$\delta_{i_0 i_1}\left(\frac{\partial u_{i_0 Gn}}{\partial x_{i_1}}+\frac{\partial u_{i_1 Gn}}{\partial x_{i_0}}\right)=0\quad\text{[owing to Eq. (A.56)]}.$$

The expression (A.61) for $p\geq 1$ and $s\geq 2$ is reduced to the following two types:

$$\frac{\partial^s u_{i_t Gn+1-s}}{\partial x_{i_t}\underbrace{\partial^2 x_{i_u}\cdots\partial^2 x_{i_v}}_{p-1}\underbrace{\partial x_{i_r}\cdots\partial x_{i_q}}_{s-2p+1}}\quad\text{and}\quad\frac{\partial^s u_{i_t Gn+1-s}}{\underbrace{\partial x^2_{i_u}\cdots\partial^2 x_{i_v}}_{p}\underbrace{\partial x_{i_r}\cdots\partial x_{i_q}}_{s-2p}},$$

---

[16]In view of Eq. (A.56), the term proportional to $\mathrm{sym}(\partial u_{i_0 Gn+1}/\partial x_{i_1})$ in the particular solution (A.60) is reduced to $g^{(n+1,2)}_{(1,0)}(\zeta)\zeta_{i_0}\zeta_{i_1}(\partial u_{i_0 Gn}/\partial x_{i_1}+\partial u_{i_1 Gn}/\partial x_{i_0})$. Thus, this is the solution of Eq. (A.45) with the inhomogeneous term $\zeta_i\partial\phi_{Gn}/\partial x_i=\zeta_{i_0}\zeta_{i_1}(\partial u_{i_0 Gn}/\partial x_{i_1}+\partial u_{i_1 Gn}/\partial x_{i_0})$. By the comparison of this equation with Eq. (3.26), the relation $g^{(n+1,2)}_{(1,0)}(\zeta)=-B(\zeta)/2$ is derived.

the former of which vanishes because of Eq. (A.40), and the latter of which also vanishes for $p \geq 2$ because of Eq. (A.44)[17] and is reduced, for $p = 1$, to

$$\frac{1}{\gamma_1}\underbrace{\frac{\partial^{s-2}}{\partial x_{i_r}\cdots\partial x_{i_q}}}_{s-2}\frac{\partial P_{Gn+1-(s-1)}}{\partial x_{i_t}} = \frac{1}{\gamma_1}\frac{\partial^{s-1}P_{Gn+1-(s-1)}}{\underbrace{\partial x_{i_r}\cdots\partial x_{i_q}}_{s-1}},$$

because of Eq. (A.41). The terms arising from this type of term in the second summation sign of Eq. (A.60) are expressed as

$$\frac{s(s+1)}{2\gamma_1}\zeta_{i_1}\cdots\zeta_{i_{s-1}}g_{(s,1)}^{(n+1,2)}\frac{\partial^{s-1}P_{Gn+1-(s-1)}}{\partial x_{i_1}\cdots\partial x_{i_{s-1}}} \quad (s = 2,\dots,n+1), \tag{A.62}$$

and therefore can be incorporated in the first summation sign. The remaining terms in the second summation sign of Eq. (A.60) are those with $p = 0$, that is,

$$2\sum_{s=1}^{n+1}\zeta_{i_0}\zeta_{i_1}\cdots\zeta_{i_s}g_{(s,0)}^{(n+1,2)}(\zeta)\mathrm{sym}\left(\frac{\partial^s u_{i_0Gn+1-s}}{\partial x_{i_1}\cdots\partial x_{i_s}}\right).$$

With these deformations, the particular solution $\hat{\phi}_{Gn+1}$ is expressed in the form

$$\begin{aligned}\hat{\phi}_{Gn+1} = &\sum_{s=1}^{n+1}\Phi_{i_1,\dots,i_s}^{(n+1,1)}(\zeta)\frac{\partial^s P_{Gn+1-s}}{\partial x_{i_1}\cdots\partial x_{i_s}}\\ &+2\sum_{s=1}^{n+1}\zeta_{i_0}\zeta_{i_1}\cdots\zeta_{i_s}g_{(s,0)}^{(n+1,2)}(\zeta)\mathrm{sym}\left(\frac{\partial^s u_{i_0Gn+1-s}}{\partial x_{i_1}\cdots\partial x_{i_s}}\right)\\ &+\sum_{s=1}^{n+1}\Phi_{i_1,\dots,i_s}^{(n+1,3)}(\zeta)\frac{\partial^s \tau_{Gn+1-s}}{\partial x_{i_1}\cdots\partial x_{i_s}},\end{aligned} \tag{A.63}$$

where $g_{(1,0)}^{(n+1,2)}(\zeta) = -B(\zeta)/2$, and $\Phi_{i_1,\dots,i_s}^{(n+1,1)}$ modified by Eq. (A.62) is expressed with the same notation. This $\hat{\phi}_{Gn+1}$ can be made to satisfy the orthogonal relation (A.37) with $m = n+1$ as already explained. Then, the relations (3.7a)–(3.7c) for $m = n+1$ are guaranteed. That is, the choice of P_{Hn+1}, $2u_{iHn+1}$, and τ_{Hn+1} for the coefficients of 1, ζ_i, and $(\zeta^2 - 5/2)$ in the first three terms of Eq. (A.59) is legitimate.

A.4 Golse's theorem on a one-way flow

The proof of Golse's theorem introduced in Section 3.11.6 is given here.[18]

[17] Note that Eqs. (A.40)–(A.44) have been extended up to $m = n+1$.

[18] I am grateful to F. Golse for authorizing my use of his unpublished result in this monograph.

Consider a gas in a straight pipe whose temperature distribution is a periodic function with respect to the coordinate (x_1, say) along the pipe. Under a small temperature variation of the pipe, the time-independent behavior of the gas that is periodic with respect to the coordinate along the pipe is studied on the basis of the linearized Boltzmann equation (2.82) and boundary condition (2.90a) with (2.90b):

$$\zeta_i \frac{\partial \phi}{\partial x_i} = \frac{1}{k}\mathcal{L}(\phi), \tag{A.64}$$

$$\phi(x_i, \zeta_i) = (1-\alpha)\phi(x_i, \zeta_i - 2\zeta_j n_j n_i) + \alpha\left[\check{\sigma}_w + \left(\zeta_j^2 - \tfrac{3}{2}\right)\tau_w\right] \quad (\zeta_i n_i > 0), \tag{A.65a}$$

$$\check{\sigma}_w = -\frac{1}{2}\tau_w - 2\sqrt{\pi}\int_{\zeta_k n_k < 0} \zeta_j n_j \phi E \mathbf{d}\boldsymbol{\zeta}, \tag{A.65b}$$

where the perturbed temperature τ_w of the pipe is a periodic function of x_1 with the period s_0, the normal n_i to the pipe surface, pointed to the gas, is a function of x_2 and x_3, and the accommodation coefficient α is a constant $(0 < \alpha \leq 1)$. Without loss of generality, we can take

$$\int_0^{s_0} \tau_w \mathrm{d}x_1 = 0, \tag{A.66a}$$

$$\int \phi E \mathbf{d}\boldsymbol{\zeta} \mathrm{d}x_1 \mathrm{d}x_2 \mathrm{d}x_3 = 0. \tag{A.66b}$$

The integration in Eq. (A.66b) is carried out over the domain in the pipe from $x_1 = 0$ to $x_1 = s_0$ and the whole space of $\boldsymbol{\zeta}$.

Introduce a function

$$\Phi(x_2, x_3, \zeta_i) = \int_0^{s_0} \phi(x_i, \zeta_i)\mathrm{d}x_1, \tag{A.67}$$

where (x_2, x_3) is in the cross section S of the pipe. Integrating Eqs. (A.64), (A.65a), and (A.65b) with respect to x_1 from $x_1 = 0$ to $x_1 = s_0$ and noting the periodicity and linearity, we obtain the following equation and boundary condition for $\Phi(x_2, x_3, \zeta_i)$:

$$\zeta_2 \frac{\partial \Phi}{\partial x_2} + \zeta_3 \frac{\partial \Phi}{\partial x_3} = \frac{1}{k}\mathcal{L}(\Phi) \text{ in } S, \tag{A.68}$$

$$\Phi(x_2, x_3, \zeta_i) = (1-\alpha)\Phi(x_2, x_3, \zeta_i - 2\zeta_j n_j n_i) + \alpha\Gamma(x_2, x_3) \quad (\zeta_i n_i > 0), \tag{A.69a}$$

$$\Gamma(x_2, x_3) = -2\sqrt{\pi}\int_{\zeta_k n_k < 0} \zeta_j n_j \Phi(x_2, x_3, \zeta_i) E \mathbf{d}\boldsymbol{\zeta}. \tag{A.69b}$$

In Eqs. (A.69a) and (A.69b), the set (x_2, x_3) is on the boundary ∂S of the cross section S. Equations (A.68)–(A.69b) are a spatially two-dimensional boundary-value problem of the linearized Boltzmann equation for $\Phi(x_2, x_3, \zeta_i)$ in the domain S. The additional condition (A.66b) is reduced to the following condition on $\Phi(x_2, x_3, \zeta_i)$:

$$\int_{S\times R^3_\zeta} \Phi(x_2, x_3, \zeta_i) E \mathrm{d}\boldsymbol{\zeta} \mathrm{d}x_2 \mathrm{d}x_3 = 0, \tag{A.70}$$

where the integration is carried out over the cross section S and the whole space of $\boldsymbol{\zeta}$ (or $S \times R^3_\zeta$).

An obvious solution of this boundary-value problem (A.68)–(A.70) is

$$\Phi(x_2, x_3, \zeta_i) = 0. \tag{A.71}$$

This is shown to be the unique solution in the following lemma. Thus,

$$\int_{S\times R^3_\zeta} \zeta_i \Phi(x_2, x_3, \zeta_i) E \mathrm{d}\boldsymbol{\zeta} \mathrm{d}x_2 \mathrm{d}x_3 = 0. \tag{A.72}$$

On the other hand, from Eqs. (A.64)–(A.65b),

$$\frac{\mathrm{d}}{\mathrm{d}x_1} \int_{S\times R^3_\zeta} \zeta_1 \phi(x_i, \zeta_i) E \mathrm{d}\boldsymbol{\zeta} \mathrm{d}x_2 \mathrm{d}x_3 = 0. \tag{A.73}$$

From Eqs. (A.72) and (A.73),

$$\int_{S\times R^3_\zeta} \zeta_1 \phi(x_i, \zeta_i) E \mathrm{d}\boldsymbol{\zeta} \mathrm{d}x_2 \mathrm{d}x_3 = 0. \tag{A.74}$$

This relation shows that the mass flux through the pipe vanishes.

Lemma. *Assume $\alpha > 0$. Then, the only solution of Eqs. (A.68)–(A.70) is $\Phi = 0$.*[19]

Proof. Multiplying Eq. (A.68) by $2\Phi E$, integrating the result over the cross section S and the whole space of $\boldsymbol{\zeta}$ ($S \times R^3_\zeta$, say), and applying Gauss's divergence theorem, we obtain the relation

$$-\int_{\partial S\times R^3_\zeta} [\Phi(x_2, x_3, \boldsymbol{\zeta})]^2 \zeta_i n_i E(\zeta) \mathrm{d}s \mathrm{d}\boldsymbol{\zeta} = \frac{2}{k} \int_{S\times R^3_\zeta} \Phi \mathcal{L}(\Phi) E \mathrm{d}x_2 \mathrm{d}x_3 \mathrm{d}\boldsymbol{\zeta}, \tag{A.75}$$

where $\mathrm{d}s$ is the line element along the curve ∂S that bounds S. Thus, from the general inequality (2.64), the following inequality holds:

$$\int_{\partial S\times R^3_\zeta} [\Phi(x_2, x_3, \boldsymbol{\zeta})]^2 \zeta_i n_i E(\zeta) \mathrm{d}s \mathrm{d}\boldsymbol{\zeta} \geq 0. \tag{A.76}$$

[19]The lemma holds for the solution Φ in the class $\sqrt{E}\Phi$, $\sqrt{E}\mathcal{L}(\Phi)$, $\sqrt{E|\zeta_i n_i|}\Phi$ and $\sqrt{E}\zeta_i \partial\Phi/\partial x_i$ (the derivative in the sense of the distribution) in $L^2_{x_i,\zeta_i}$ (square integrable, e.g., $\int_{S\times R^3_\zeta} \Phi^2 E \mathrm{d}\boldsymbol{\zeta} \mathrm{d}x_2 \mathrm{d}x_3 < +\infty$). The following proof may be easily understood for the class of sectionally smooth solutions.

The integral on the left-hand side of the above inequality is first rewritten with the condition (A.69a) in the following form:

$$\begin{aligned}
&\int_{\partial S\times R^3_\zeta}[\Phi(x_2,x_3,\boldsymbol{\zeta})]^2\zeta_i n_i E(\zeta)\mathrm{d}s\mathrm{d}\boldsymbol{\zeta}\\
&=\int_{\partial S,\ \zeta_i n_i<0}\Phi^2\zeta_i n_i E\mathrm{d}s\mathrm{d}\boldsymbol{\zeta}+\int_{\partial S,\ \zeta_i n_i>0}\Phi^2\zeta_i n_i E\mathrm{d}s\mathrm{d}\boldsymbol{\zeta}\\
&=\int_{\partial S,\ \zeta_i n_i<0}\Phi^2\zeta_i n_i E\mathrm{d}s\mathrm{d}\boldsymbol{\zeta}\\
&\quad+(1-\alpha)^2\int_{\partial S,\ \zeta_i n_i>0}[\Phi(x_2,x_3,\zeta_i-2\zeta_j n_j n_i)]^2\zeta_k n_k E\mathrm{d}s\mathrm{d}\boldsymbol{\zeta}\\
&\quad+2\alpha(1-\alpha)\int_{\partial S,\ \zeta_i n_i>0}\Phi(x_2,x_3,\zeta_i-2\zeta_j n_j n_i)\Gamma(x_2,x_3)\zeta_k n_k E\mathrm{d}s\mathrm{d}\boldsymbol{\zeta}\\
&\quad+\alpha^2\int_{\partial S,\ \zeta_i n_i>0}[\Gamma(x_2,x_3)]^2\zeta_k n_k E\mathrm{d}s\mathrm{d}\boldsymbol{\zeta}.
\end{aligned}\tag{A.77}$$

With the change of variables $\zeta_i'=\zeta_i-2\zeta_j n_j n_i$, the integral with respect to $\boldsymbol{\zeta}$ in the second term of the last expression is reduced to

$$\begin{aligned}
&\int_{\zeta_i n_i>0}[\Phi(x_2,x_3,\zeta_i-2\zeta_j n_j n_i)]^2\zeta_k n_k E(\zeta)\mathrm{d}\boldsymbol{\zeta}\\
&\qquad=-\int_{\partial S,\ \zeta_i' n_i<0}[\Phi(x_2,x_3,\zeta_i')]^2\zeta_k' n_k E(\zeta')\mathrm{d}\boldsymbol{\zeta}',
\end{aligned}$$

and the integral with respect to $\boldsymbol{\zeta}$ in the third term is reduced to

$$\int_{\zeta_i n_i>0}\Phi(x_2,x_3,\zeta_i-2\zeta_j n_j n_i)\Gamma(x_2,x_3)\zeta_k n_k E(\zeta)\mathrm{d}\boldsymbol{\zeta}=\frac{1}{2\sqrt{\pi}}[\Gamma(x_2,x_3)]^2,$$

where the definition (A.69b) of Γ is used. With these two relations in Eq. (A.77), the integral of the left-hand side is expressed in the following form:

$$\begin{aligned}
&\int_{\partial S\times R^3_\zeta}[\Phi(x_2,x_3,\boldsymbol{\zeta})]^2\zeta_i n_i E(\zeta)\mathrm{d}s\mathrm{d}\boldsymbol{\zeta}\\
&=\alpha(2-\alpha)\int_{\partial S,\ \zeta_i n_i<0}\Phi^2\zeta_i n_i E\mathrm{d}s\mathrm{d}\boldsymbol{\zeta}+\frac{\alpha(1-\alpha)}{\sqrt{\pi}}\int_{\partial S}[\Gamma(x_2,x_3)]^2\mathrm{d}s\\
&\quad+\alpha^2\int_{\partial S}[\Gamma(x_2,x_3)]^2\left(\int_{\zeta_i n_i>0}\zeta_k n_k E\mathrm{d}\boldsymbol{\zeta}\right)\mathrm{d}s\\
&=-\frac{\alpha(2-\alpha)}{2\sqrt{\pi}}\int_{\partial S}\left[2\sqrt{\pi}\int_{\zeta_i n_i<0}\Phi^2|\zeta_i n_i|E\mathrm{d}\boldsymbol{\zeta}-\Gamma^2\right]\mathrm{d}s\\
&=-\alpha(2-\alpha)\int_{\partial S}\int_{\zeta_i n_i<0}(\Phi-\Gamma)^2|\zeta_i n_i|E\mathrm{d}\boldsymbol{\zeta}\mathrm{d}s\le 0,
\end{aligned}\tag{A.78}$$

where the expression before the last is derived with the aid of Eq. (A.69b).

On the other hand, the integral $\int_{\partial S\times R^3}\Phi^2\zeta_i n_i E\mathrm{d}s\mathrm{d}\boldsymbol{\zeta}$ is nonnegative from the inequality (A.76). Thus, the integral $\int_{\partial S}\int_{\zeta_i n_i<0}(\Phi-\Gamma)^2|\zeta_i n_i|E\mathrm{d}\boldsymbol{\zeta}\mathrm{d}s$ should be zero, for which the positive integrand $\int_{\zeta_i n_i<0}(\Phi-\Gamma)^2|\zeta_i n_i|E\mathrm{d}\boldsymbol{\zeta}$ should vanish almost everywhere (a.e., for short) on ∂S, i.e.,

$$\int_{\zeta_i n_i<0}(\Phi-\Gamma)^2|\zeta_i n_i|E\mathrm{d}\boldsymbol{\zeta}=0 \quad (\text{a.e. on } \partial S). \tag{A.79}$$

(The note a.e. is omitted in the following equations.) Thus,

$$\Phi(x_2,x_3,\boldsymbol{\zeta})=\Gamma(x_2,x_3) \quad \text{for } \zeta_i n_i<0 \text{ on } \partial S.$$

With this result in Eq. (A.69a), $\Phi(x_2,x_3,\boldsymbol{\zeta})$ for $\zeta_i n_i>0$ on ∂S is also reduced to be equal to $\Gamma(x_2,x_3)$. After all,

$$\Phi(x_2,x_3,\boldsymbol{\zeta})=\Gamma(x_2,x_3) \quad \text{on } \partial S,$$

that is, the function Φ is independent of $\boldsymbol{\zeta}$ on ∂S.

Then from Eq. (A.75),

$$\int_{S\times R^3_\zeta}\Phi\mathcal{L}(\Phi)E\mathrm{d}x_2\mathrm{d}x_3\mathrm{d}\boldsymbol{\zeta}=0. \tag{A.80}$$

In view of the general relation $\int_{R^3_\zeta}\Phi\mathcal{L}(\Phi)E\mathrm{d}\boldsymbol{\zeta}\leq 0$ [Eq. (2.64)],

$$\int_{R^3_\zeta}\Phi\mathcal{L}(\Phi)E\mathrm{d}\boldsymbol{\zeta}=0 \quad \text{in } S. \tag{A.81}$$

Equation (A.81) is transformed with the aid of the symmetry relation (2.61a) in the form

$$\int_{R^3_\zeta}\Phi\mathcal{L}(\Phi)E\mathrm{d}\boldsymbol{\zeta}=-\frac{1}{4}\int EE_*(\Phi'+\Phi'_*-\Phi-\Phi_*)^2\widehat{B}\mathrm{d}\Omega\mathrm{d}\boldsymbol{\zeta}_*\mathrm{d}\boldsymbol{\zeta}=0.$$

Therefore,

$$\Phi'+\Phi'_*-\Phi+\Phi_*,$$

that is, Φ is a summational invariant (Grad [1949]). Thus,

$$\mathcal{L}(\Phi)=0 \quad \text{in } S. \tag{A.82}$$

Then Eq. (A.68) is reduced to

$$\zeta_2\frac{\partial\Phi}{\partial x_2}+\zeta_3\frac{\partial\Phi}{\partial x_3}=0 \quad \text{in } S. \tag{A.83}$$

With the aid of Eq. (A.83), the boundary values of Φ or $\Gamma(x_2,x_3)$ at two points on ∂S that can be connected with a straight line without intersecting the boundary ∂S are equal. Thus, if the shape of the cross section S of the pipe is such a form that any two points on ∂S can be connected by a broken line inside S with broken points on ∂S,[20] the boundary value $\Gamma(x_2,x_3)$ of Φ is uniform (or independent of the position on ∂S). Thus, from Eq. (A.83), Φ is almost everywhere constant in both (x_2,x_3) and ζ on $S\times R^3_\zeta$. Because of the relation (A.70), this constant must be zero. ■

It may be noted that the condition $\alpha>0$ cannot be dispensed with in the uniqueness statement above. Indeed, the function $\Phi=\zeta^2-3/2$ satisfies Eq. (A.68) as well as Eq. (A.69a) with $\alpha=0$ and Eq. (A.70).

A.5 Functions $\mathcal{A}(\tilde{\zeta},\hat{T}_{SB0})$, $\mathcal{B}(\tilde{\zeta},\hat{T}_{V0})$, etc.

The functions $\mathcal{A}(\tilde{\zeta},\hat{T}_{SB0})$, $\mathcal{B}(\tilde{\zeta},\hat{T}_{V0})$, etc., encountered in Chapters 5 and 6, are expressed by linear combinations of solutions of the following integral equations related to the linearized collision operator $\mathcal{L}_a(*)$ defined by Eq. (A.23) with (A.24) in Appendix A.2:[21]

$$\mathcal{L}_a[\zeta_i\mathcal{A}(\zeta,a)]=-\zeta_i\left(\zeta^2-\frac{5}{2}\right), \tag{A.84}$$

with the subsidiary condition $\int_0^\infty \zeta^4\mathcal{A}(\zeta,a)E(\zeta)\mathrm{d}\zeta=0;$

$$\mathcal{L}_a\left[\left(\zeta_i\zeta_j-\frac{1}{3}\zeta^2\delta_{ij}\right)\mathcal{B}^{(m)}(\zeta,a)\right]=IB^{(m)}_{ij}; \tag{A.85}$$

$$\mathcal{L}_a\left[(\zeta_i\delta_{jk}+\zeta_j\delta_{ki}+\zeta_k\delta_{ij})\mathcal{T}^{(m)}_1(\zeta,a)+\zeta_i\zeta_j\zeta_k\mathcal{T}^{(m)}_2(\zeta,a)\right]=IT^{(m)}_{ijk}, \tag{A.86}$$

with the subsidiary condition $\int_0^\infty\left(5\zeta^4\mathcal{T}^{(m)}_1+\zeta^6\mathcal{T}^{(m)}_2\right)E(\zeta)\mathrm{d}\zeta=0;$

$$\mathcal{L}_a\left[\zeta_i\delta_{jk}\tilde{\mathcal{T}}^{(0)}_{11}(\zeta,a)+(\zeta_j\delta_{ki}+\zeta_k\delta_{ij})\tilde{\mathcal{T}}^{(0)}_{12}(\zeta,a)+\zeta_i\zeta_j\zeta_k\tilde{\mathcal{T}}^{(0)}_2(\zeta,a)\right]=I\tilde{T}^{(0)}_{i,jk}, \tag{A.87}$$

with the subsidiary conditions

$$\int_0^\infty\left(5\zeta^4\tilde{\mathcal{T}}^{(0)}_{12}+\zeta^6\tilde{\mathcal{T}}^{(0)}_2\right)E(\zeta)\mathrm{d}\zeta=0,\quad \int_0^\infty\left(5\zeta^4\tilde{\mathcal{T}}^{(0)}_{11}+\zeta^6\tilde{\mathcal{T}}^{(0)}_2\right)E(\zeta)\mathrm{d}\zeta=0;$$

---

[20] Obviously this assumption, which is called the bridge or ergodicity assumption (Sone [1984a, 1985], Aoki, Bardos, Golse, Kogan & Sone [1993]), does not hold only for very special cross sections.

[21] For simplicity, the notation ζ_i is used instead of $\tilde{\zeta}_i$ in Chapter 5 or 6. That is, this ζ_i is used as a variable and is not related to ξ_i by Eq. (2.36). As mentioned in Footnote 13, the operator $\mathcal{L}_a(*)$ generally depends on the reference temperature T_0 besides the parameter a. Thus, the solutions $\mathcal{A}(\zeta,a)$, $\mathcal{B}^{(m)}(\zeta,a)$, etc. and the nondimensional transport coefficients $\hat{\gamma}_1(a)$, $\hat{\gamma}_2(a)$, etc. defined with $\mathcal{A}$, $\mathcal{B}^{(m)}$, etc. also depend on the reference temperature T_0 besides the parameter a.

$$\mathcal{L}_a\Big[(\delta_{ij}\delta_{kl}+\delta_{ik}\delta_{jl}+\delta_{il}\delta_{jk})\,\mathcal{Q}_1^{(0)}(\zeta,a)$$
$$+(\zeta_i\zeta_j\delta_{kl}+\zeta_i\zeta_k\delta_{jl}+\zeta_i\zeta_l\delta_{jk}+\zeta_j\zeta_k\delta_{il}+\zeta_j\zeta_l\delta_{ik}+\zeta_k\zeta_l\delta_{ij})\mathcal{Q}_2^{(0)}(\zeta,a)$$
$$+\zeta_i\zeta_j\zeta_k\zeta_l\mathcal{Q}_3^{(0)}(\zeta,a)\Big]=IQ_{ijkl}^{(0)}, \tag{A.88}$$

with the subsidiary conditions

$$\int_0^\infty(1,\ \zeta^2)\left(15\zeta^2\mathcal{Q}_1^{(0)}+10\zeta^4\mathcal{Q}_2^{(0)}+\zeta^6\mathcal{Q}_3^{(0)}\right)E(\zeta)\mathrm{d}\zeta=0;$$

$$\mathcal{L}_a\Big[\delta_{ij}\delta_{kl}\tilde{\mathcal{Q}}_{11}^{(0)}(\zeta,a)+(\delta_{ik}\delta_{jl}+\delta_{il}\delta_{jk})\,\tilde{\mathcal{Q}}_{12}^{(0)}(\zeta,a)+(\zeta_i\zeta_j\delta_{kl}+\zeta_k\zeta_l\delta_{ij})\tilde{\mathcal{Q}}_{21}^{(0)}(\zeta,a)$$
$$+(\zeta_i\zeta_k\delta_{jl}+\zeta_i\zeta_l\delta_{jk}+\zeta_j\zeta_k\delta_{il}+\zeta_j\zeta_l\delta_{ik})\tilde{\mathcal{Q}}_{22}^{(0)}(\zeta,a)+\zeta_i\zeta_j\zeta_k\zeta_l\tilde{\mathcal{Q}}_3^{(0)}(\zeta,a)\Big]$$
$$=I\tilde{Q}_{ij,kl}^{(0)}, \tag{A.89}$$

with the subsidiary conditions

$$\int_0^\infty(1,\ \zeta^2)\left(15\zeta^2\tilde{\mathcal{Q}}_{11}^{(0)}+10\zeta^4\tilde{\mathcal{Q}}_{21}^{(0)}+\zeta^6\tilde{\mathcal{Q}}_3^{(0)}\right)E(\zeta)\mathrm{d}\zeta=0,$$
$$\int_0^\infty(1,\ \zeta^2)\left(15\zeta^2\tilde{\mathcal{Q}}_{12}^{(0)}+10\zeta^4\tilde{\mathcal{Q}}_{22}^{(0)}+\zeta^6\tilde{\mathcal{Q}}_3^{(0)}\right)E(\zeta)\mathrm{d}\zeta=0;$$

$$\mathcal{L}_a[\mathcal{N}^{(m)}(\zeta,a)]=IN^{(m)}, \tag{A.90}$$

with the subsidiary conditions $\int_0^\infty(1,\ \zeta^2)\zeta^2\mathcal{N}^{(m)}E(\zeta)\mathrm{d}\zeta=0.$

The inhomogeneous terms $IB_{ij}^{(m)}$, $IT_{ijk}^{(m)}$, etc. in Eqs. (A.85)–(A.90) are as follows:

$$IB_{ij}^{(0)}=-2\left(\zeta_i\zeta_j-\frac{\zeta^2}{3}\delta_{ij}\right), \tag{A.91a}$$

$$IB_{ij}^{(1)}=\left(\zeta_i\zeta_j-\frac{\zeta^2}{3}\delta_{ij}\right)\mathcal{A}(\zeta,a), \tag{A.91b}$$

$$IB_{ij}^{(2)}=\left(\zeta_i\zeta_j-\frac{\zeta^2}{3}\delta_{ij}\right)\left(2(\zeta^2-3)\mathcal{A}(\zeta,a)-\zeta\frac{\partial\mathcal{A}(\zeta,a)}{\partial\zeta}+2a\frac{\partial\mathcal{A}(\zeta,a)}{\partial a}\right), \tag{A.91c}$$

$$IB_{ij}^{(3)}=\mathcal{J}_a(\zeta_i\mathcal{A}(\zeta,a),\zeta_j\mathcal{A}(\zeta,a))-\frac{\delta_{ij}}{3}\sum_{k=1}^{3}\mathcal{J}_a(\zeta_k\mathcal{A}(\zeta,a),\zeta_k\mathcal{A}(\zeta,a)), \tag{A.91d}$$

$$IB_{ij}^{(4)}=\left(\zeta_i\zeta_j-\frac{\zeta^2}{3}\delta_{ij}\right)\mathcal{B}^{(0)}(\zeta,a); \tag{A.91e}$$

$$IT^{(0)}_{ijk} = -\zeta_i\zeta_j\zeta_k\mathcal{B}^{(0)}(\zeta, a) + \hat{\gamma}_1(a)\left(\zeta_i\delta_{jk} + \zeta_j\delta_{ki} + \zeta_k\delta_{ij}\right), \tag{A.92a}$$

$$IT^{(1)}_{ijk} = -\zeta_i\zeta_j\zeta_k\left(2\mathcal{A}(\zeta, a) - \frac{1}{\zeta}\frac{\partial \mathcal{A}(\zeta, a)}{\partial\zeta}\right), \tag{A.92b}$$

$$IT^{(2)}_{ijk} = -\zeta_i\zeta_j\zeta_k\left((\zeta^2 - 3)\mathcal{B}^{(0)}(\zeta, a) - \frac{\zeta}{2}\frac{\partial \mathcal{B}^{(0)}(\zeta, a)}{\partial\zeta}\right) + \frac{\hat{\gamma}_1(a)}{2}\left(\zeta_i\delta_{jk} + \zeta_j\delta_{ki} + \zeta_k\delta_{ij}\right), \tag{A.92c}$$

$$IT^{(3)}_{ijk} = -\zeta_i\zeta_j\zeta_k a\frac{\partial \mathcal{B}^{(0)}(\zeta, a)}{\partial a} + a\frac{\mathrm{d}\hat{\gamma}_1(a)}{\mathrm{d}a}\left(\zeta_i\delta_{jk} + \zeta_j\delta_{ki} + \zeta_k\delta_{ij}\right); \tag{A.92d}$$

$$I\tilde{T}^{(0)}_{i,jk} = \mathcal{J}_a(\zeta_i\mathcal{A}(\zeta, a),\, \zeta_j\zeta_k\mathcal{B}^{(0)}(\zeta, a)); \tag{A.93}$$

$$IQ^{(0)}_{ijkl} = \frac{1}{3}\left[\zeta^2\mathcal{B}^{(0)}(\zeta, a) + 2\hat{\gamma}_1(a)\left(\zeta^2 - \frac{3}{2}\right)\right]\left(\delta_{ij}\delta_{kl} + \delta_{ik}\delta_{jl} + \delta_{il}\delta_{jk}\right) - \zeta_i\zeta_j\zeta_k\zeta_l\left(2\mathcal{B}^{(0)}(\zeta, a) - \frac{1}{\zeta}\frac{\partial \mathcal{B}^{(0)}(\zeta, a)}{\partial\zeta}\right); \tag{A.94}$$

$$I\tilde{Q}^{(0)}_{ij,kl} = \mathcal{J}_a(\zeta_i\zeta_j\mathcal{B}^{(0)}(\zeta, a),\, \zeta_k\zeta_l\mathcal{B}^{(0)}(\zeta, a)); \tag{A.95}$$

$$IN^{(0)} = 2\zeta^2\left(\zeta^2 - \frac{7}{2}\right)\mathcal{A}(\zeta, a) - \zeta^3\frac{\partial \mathcal{A}(\zeta, a)}{\partial\zeta}, \tag{A.96a}$$

$$IN^{(1)} = 2a\frac{\partial \mathcal{A}(\zeta, a)}{\partial a}\zeta^2 - 5a\frac{\mathrm{d}\hat{\gamma}_2(a)}{\mathrm{d}a}\left(\zeta^2 - \frac{3}{2}\right), \tag{A.96b}$$

$$IN^{(2)} = \zeta^2\mathcal{A}(\zeta, a) - \frac{5}{2}\hat{\gamma}_2(a)\left(\zeta^2 - \frac{3}{2}\right), \tag{A.96c}$$

$$IN^{(3)} = \sum_{k=1}^{3}\mathcal{J}_a(\zeta_k\mathcal{A}(\zeta, a),\, \zeta_k\mathcal{A}(\zeta, a)). \tag{A.96d}$$

The summation sign $\sum$ is used in $IB^{(3)}_{ij}$ and $IN^{(3)}$, since the summation convention is a little difficult to distinguish there, and the comma inserted between the subscripts on the inhomogeneous terms $I\tilde{T}^{(0)}_{i,jk}$ and $I\tilde{Q}^{(0)}_{ij,kl}$ indicates the absence of symmetry with respect to the subscripts across it. The $\mathcal{J}_a(\phi, \psi)$, $\hat{\gamma}_1(a)$, and $\hat{\gamma}_2(a)$ in some of the above inhomogeneous terms are defined as follows:
(i) The $\mathcal{J}_a(\phi, \psi)$ is defined, as a slight extension of $\mathcal{J}(\phi, \psi)$, by

$$\mathcal{J}_a(\phi(\boldsymbol{\zeta}), \psi(\boldsymbol{\zeta})) = \frac{1}{2}\int E_*(\phi'_*\psi' + \phi'\psi'_* - \phi_*\psi - \phi\psi_*)\widehat{B}_a\,\mathrm{d}\Omega(\boldsymbol{\alpha})\,\mathrm{d}\boldsymbol{\zeta}_*, \tag{A.97}$$

$$\widehat{B}_a = \widehat{B}_a(|\boldsymbol{\alpha}\cdot(\boldsymbol{\zeta}_* - \boldsymbol{\zeta})|/|\boldsymbol{\zeta}_* - \boldsymbol{\zeta}|, |\boldsymbol{\zeta}_* - \boldsymbol{\zeta}|) = \frac{\widehat{B}(|\boldsymbol{\alpha}\cdot(\boldsymbol{\zeta}_* - \boldsymbol{\zeta})|/|\boldsymbol{\zeta}_* - \boldsymbol{\zeta}|, |\boldsymbol{\zeta}_* - \boldsymbol{\zeta}|a^{1/2})}{a^{1/2}}.$$

Thus,

$$\mathcal{J}(\phi,\psi)=\mathcal{J}_1(\phi,\psi),\quad \mathcal{L}_a(\phi)=2\mathcal{J}_a(1,\phi). \tag{A.98}$$

(ii) The $\hat{\gamma}_1(a)$ and $\hat{\gamma}_2(a)$ in Eqs. (A.92a), (A.96c), etc., as well as in the fluid-dynamic-type equations (5.42)–(5.44) or (6.68)–(6.71) and in the formulas (5.40a)-(5.41c) or (6.72)–(6.77b) of stress tensor and heat-flow vector, are defined by the following moments of $\mathcal{B}^{(0)}(\zeta,a)$ and $\mathcal{A}(\zeta,a)$:

$$\hat{\gamma}_1(a)=I_6(\mathcal{B}^{(0)}(\zeta,a)),\qquad \hat{\gamma}_2(a)=2I_6(\mathcal{A}(\zeta,a)), \tag{A.99}$$

where $I_n(Z)$ is introduced in Section 3.3, i.e.,

$$I_n(Z(\zeta))=\frac{8}{15\sqrt{\pi}}\int_0^\infty \zeta^n Z(\zeta)\exp(-\zeta^2)\mathrm{d}\zeta. \tag{A.100}$$

The inhomogeneous terms of the integral equations (A.84)–(A.90) satisfy the solvability conditions. The forms of the solutions are guaranteed by the discussion in Appendixes A.2 and B, with reservation for those of Eqs. (A.87) and (A.89). The inhomogeneous terms $I\tilde{T}^{(0)}_{i,jk}$ and $I\tilde{Q}^{(0)}_{ij,kl}$ are not symmetric with respect to some of their subscripts. Thus, the statement in Appendix A.2 is not directly applied, but the forms of the solutions corresponding to these inhomogeneous terms are verified by simple generalization of the discussion in Appendix B.[22] The subsidiary conditions accompanying Eqs. (A.84)–(A.90) except Eq. (A.85) are the conditions that the solutions are orthogonal to the solutions of the corresponding homogeneous equation [see Eq. (A.30)]. In Eq. (A.85), the orthogonal conditions are incorporated in the form of the solution in view of the fact that the traces, with respect to the subscript (i,j), of the inhomogeneous terms (A.91a)–(A.91e) vanish.

The functions $\mathcal{B}(\zeta,a)$, $\mathcal{B}_1(\zeta,a)$, $\mathcal{Q}_1(\zeta,a)$, etc. as well as $\mathcal{A}(\zeta,a)$ in the solutions $\hat{f}_{SBm}$ in Chapter 5 and $\hat{f}_{Vm}$ in Chapter 6 are expressed by linear combinations of solutions of the above integral equations as follows:[23]

$$\left.\begin{aligned}
&\mathcal{B}(\zeta,a)=\mathcal{B}^{(0)}(\zeta,a),\quad \mathcal{B}_1=-\mathcal{B}^{(1)},\quad \mathcal{B}_2=-\mathcal{B}^{(2)}-2\mathcal{B}^{(3)},\\
&\mathcal{N}^A=-\tfrac{1}{3}(\mathcal{N}^{(0)}+\mathcal{N}^{(1)}+\mathcal{N}^{(2)}+2\mathcal{N}^{(3)}),\\
&\mathcal{N}^B=-\tfrac{1}{3}\mathcal{N}^{(2)},\quad \mathcal{T}^A_N=\mathcal{T}^{(0)}_N\quad (N=1,2),\\
&\mathcal{T}^B_1=\mathcal{T}^{(1)}_1+\mathcal{T}^{(2)}_1+\mathcal{T}^{(3)}_1-2\tilde{\mathcal{T}}^{(0)}_{12},\\
&\mathcal{T}^B_2=\mathcal{T}^{(1)}_2+\mathcal{T}^{(2)}_2+\mathcal{T}^{(3)}_2-2\tilde{\mathcal{T}}^{(0)}_2,\\
&\mathcal{Q}_1=\mathcal{Q}^{(0)}_1-\tilde{\mathcal{Q}}^{(0)}_{12}-\tfrac{1}{3}\zeta^2\mathcal{B}^{(4)},\quad \mathcal{Q}_2=\mathcal{Q}^{(0)}_2-\tilde{\mathcal{Q}}^{(0)}_{22},\\
&\mathcal{Q}_3=\mathcal{Q}^{(0)}_2-\tilde{\mathcal{Q}}^{(0)}_{22}+\mathcal{B}^{(4)},\quad \mathcal{Q}_4=\mathcal{Q}^{(0)}_3-\tilde{\mathcal{Q}}^{(0)}_3,
\end{aligned}\right\} \tag{A.101}$$

---

[22]The present forms of the solutions are determined under the condition that the spherically symmetric relation (2.47) or (A.32) be satisfied for both the cases of the determinant $\|l_{ij}\|=1$ and -1.

[23]For the BKW equation, the following convention should be taken:

$$\mathcal{B}^{(3)}=\mathcal{N}^{(3)}=\tilde{\mathcal{T}}^{(0)}_{12}=\tilde{\mathcal{T}}^{(0)}_2=\tilde{\mathcal{Q}}^{(0)}_{12}=\tilde{\mathcal{Q}}^{(0)}_{22}=\tilde{\mathcal{Q}}^{(0)}_3=0.$$

This is due to the exceptional form of the BKW equation (see Footnote 1 in Section 4.2).

where the arguments omitted are common as shown in the first relation.

Similarly to $\hat{\gamma}_1$ and $\hat{\gamma}_2$, the $\hat{\gamma}_3$, $\hat{\gamma}_7$, $\hat{\gamma}_8$, $\hat{\gamma}_9$, and $\hat{\gamma}_{10}$ in the formulas of stress tensor and heat-flow vector [Eqs. (5.40a)–(5.41c) or Eqs. (6.72)–(6.77b)], as well as in the fluid-dynamic-type equations (5.42)–(5.44) or (6.68)–(6.71), are defined by the following moments of solutions of the integral equations:

$$\left.\begin{aligned}&\hat{\gamma}_3 = 2I_6(\mathcal{B}_1) = 5I_6(\mathcal{T}_1^A) + I_8(\mathcal{T}_2^A),\quad \hat{\gamma}_7 = I_6(\mathcal{B}_2),\quad \hat{\gamma}_8 = I_6(\mathcal{Q}_2) + \tfrac{1}{7}I_8(\mathcal{Q}_4),\\ &\hat{\gamma}_9 = -I_6(\mathcal{Q}_3 - \mathcal{Q}_2) = -I_6(\mathcal{B}^{(4)}),\quad \hat{\gamma}_{10} = \tfrac{5}{8}I_6(\mathcal{T}_1^B) + \tfrac{1}{8}I_8(\mathcal{T}_2^B).\end{aligned}\right\} \tag{A.102}$$

The second equality in the equation for $\hat{\gamma}_3$ is derived with the aid of the self-adjoint property of the linearized collision operator $\mathcal{L}_a(\phi)$.

For a hard-sphere gas, the solutions $\mathcal{A}(\zeta, a)$, $\mathcal{B}(\zeta, a)$, $\mathcal{B}^{(1)}(\zeta, a)$, etc. of Eqs. (A.84)–(A.90) and the $\hat{\gamma}_1(a)$, $\hat{\gamma}_2(a)$, $\hat{\gamma}_3(a)$, $\hat{\gamma}_7(a)$, $\hat{\gamma}_8(a)$, $\hat{\gamma}_9(a)$, and $\hat{\gamma}_{10}(a)$ are all independent of a and T_0. Some of $\mathcal{A}(\zeta, a)$, $\mathcal{B}(\zeta, a)$, etc. are related to the functions $A(\zeta)$, $B(\zeta)$, etc. defined by Eqs. (3.25)–(3.28) as

$$\left.\begin{aligned}&\mathcal{A}(\zeta,a) = A(\zeta),\qquad \mathcal{B}(\zeta,a) = B(\zeta),\qquad \mathcal{B}^{(1)}(\zeta,a) = F(\zeta),\\ &\mathcal{T}_1^{(0)}(\zeta,a) = D_1(\zeta),\quad \mathcal{T}_2^{(0)}(\zeta,a) = D_2(\zeta),\end{aligned}\right\} \tag{A.103}$$

and some of the data of $\hat{\gamma}_m$ are

$$\left.\begin{aligned}&\hat{\gamma}_1(a) = \gamma_1 = 1.270042427,\quad \hat{\gamma}_2(a) = \gamma_2 = 1.922284066,\\ &\hat{\gamma}_3(a) = \gamma_3 = 1.947906335,\quad \hat{\gamma}_7(a) = 0.189201.\end{aligned}\right\} \tag{A.104}$$

For the BKW model,

$$\left.\begin{aligned}&\mathcal{A}(\zeta,a)/a^{1/2} = \zeta^2 - \tfrac{5}{2},\quad \mathcal{B}(\zeta,a)/a^{1/2} = 2,\quad \mathcal{B}_1(\zeta,a)/a = \zeta^2 - \tfrac{5}{2},\\ &\mathcal{B}_2(\zeta,a)/a = 2\left(\zeta^4 - 6\zeta^2 + \tfrac{25}{4}\right),\quad \mathcal{N}^A(\zeta,a)/a = \tfrac{2}{3}\left(\zeta^6 - 6\zeta^4 + \tfrac{15}{4}\zeta^2 + \tfrac{15}{4}\right),\\ &\mathcal{N}^B(\zeta,a)/a = \tfrac{1}{3}\left(\zeta^4 - 5\zeta^2 + \tfrac{15}{4}\right),\quad \mathcal{T}_1^A(\zeta,a)/a = -1,\quad \mathcal{T}_2^A(\zeta,a)/a = 2,\\ &\mathcal{T}_1^B(\zeta,a)/a = -1,\quad \mathcal{T}_2^B(\zeta,a)/a = 4(\zeta^2-3),\quad \mathcal{Q}_1(\zeta,a)/a = -\tfrac{2}{3}\left(\zeta^2 - \tfrac{3}{2}\right),\\ &\mathcal{Q}_2(\zeta,a)/a = 0,\quad \mathcal{Q}_3(\zeta,a)/a = -2,\quad \mathcal{Q}_4(\zeta,a)/a = 4,\end{aligned}\right\} \tag{A.105}$$

and

$$\hat{\gamma}_1(a) = \hat{\gamma}_2(a) = a^{1/2},\ \hat{\gamma}_3(a) = \hat{\gamma}_7(a) = \hat{\gamma}_8(a) = \hat{\gamma}_9(a) = \hat{\gamma}_{10}(a) = a. \tag{A.106}$$

A.6 Viscosity and thermal conductivity

Transport coefficients, such as viscosity and thermal conductivity, are derived by rewriting the nondimensional stress tensor and heat-flow vector formulas in their dimensional forms, as we have done for the linear theory in Section 3.9. The transport coefficients should naturally be independent of the choice of reference variables. However, this is not so obvious from the formulas in nonlinear cases in Chapters 5 and 6, because the dependence of $\hat{\gamma}_m$ in the stress and heat flow formulas on the reference temperature T_0 and the nondimensional temperature, i.e., $\hat{T}_{SB0}$ in Chapter 5 and $\hat{T}_{V0}$ in Chapter 6, is not explicitly given for a general molecular model. Here we examine the independence of the transport coefficients on the choice of reference variables by taking the formulas (5.40a)–(5.41c) in Chapter 5 as an example.

From the formulas (5.40a)–(5.41c) and (2.36), the stress tensor p_{ij} and heat-flow vector q_i, correct up to the order k^2, are given as follows:

$$
\begin{aligned}
p_{ij} = p\delta_{ij} &- \frac{mRT_0^{1/2}\hat{\gamma}_1(\hat{T})}{B_0}T^{1/2}\left(\frac{\partial v_i}{\partial X_j}+\frac{\partial v_j}{\partial X_i}-\frac{2}{3}\frac{\partial v_k}{\partial X_k}\delta_{ij}\right)\\
&+\frac{(2RT_0)R^2m^2\hat{\gamma}_7(\hat{T})}{B_0^2}\frac{1}{p}\left[\frac{\partial T}{\partial X_i}\frac{\partial T}{\partial X_j}-\frac{1}{3}\left(\frac{\partial T}{\partial X_k}\right)^2\delta_{ij}\right]\\
&+\frac{(2RT_0)R^2m^2\hat{\gamma}_3(\hat{T})}{B_0^2}\frac{T}{p}\left(\frac{\partial^2 T}{\partial X_i\partial X_j}-\frac{1}{3}\frac{\partial^2 T}{\partial X_k^2}\delta_{ij}\right),\\
q_i = &-\frac{5R^2T_0^{1/2}m\hat{\gamma}_2(\hat{T})}{2B_0}T^{1/2}\frac{\partial T}{\partial X_i},
\end{aligned}
$$

where the relation (2.41b), i.e.,

$$k = (2RT_0)^{1/2}m/\rho_0 B_0 L,$$

is used. In these forms of p_{ij} and q_i, the reference temperature T_0 is contained in the quantities $T_0^{1/2}\hat{\gamma}_1(\hat{T})/B_0$, $T_0\hat{\gamma}_7(\hat{T})/B_0^2$, $T_0\hat{\gamma}_3(\hat{T})/B_0^2$, and $T_0^{1/2}\hat{\gamma}_2(\hat{T})/B_0$. We will confirm that these quantities are independent of T_0.

Consider the integral equation

$$\mathcal{L}_{\hat{T}}(\phi(\zeta)) = g(\zeta), \tag{A.107}$$

where the operator $\mathcal{L}_{\hat{T}}(\phi(\zeta))$ is defined by Eq. (A.23) with (A.24) in Appendix A.2 in the following way:

$$
\begin{aligned}
\mathcal{L}_{\hat{T}}(\phi(\zeta)) &= \int E(\zeta_*)(\phi'+\phi'_*-\phi-\phi_*)\widehat{B}_{\hat{T}}\mathrm{d}\Omega(\alpha)\,\mathrm{d}\zeta_*,\\
\widehat{B}_{\hat{T}} &= \widehat{B}_{\hat{T}}(|\alpha\cdot(\zeta_*-\zeta)|/|\zeta_*-\zeta|,|\zeta_*-\zeta|)\\
&= \widehat{B}(|\alpha\cdot(\zeta_*-\zeta)|/|\zeta_*-\zeta|,|\zeta_*-\zeta|\hat{T}^{1/2})/\hat{T}^{1/2}.
\end{aligned}
$$

This integral equation represents Eqs. (A.84)–(A.90), where $\hat{T}$ for the subscript of $\mathcal{L}_{\hat{T}}$ is chosen in accord with the argument of $\hat{\gamma}_m(\hat{T})$ in the above formulas of stress and heat flow. The $\hat{\gamma}_m(\hat{T})$ is obtained from an integral of the solution of this form of equation [see Eqs. (A.99), (A.100), and (A.102)]. From the definitions of B, $\widehat{B}$, and $\widehat{B}_{\hat{T}}$ [see Eqs. (2.6), (2.41c), and (A.24)],[24]

$$\widehat{B}_{\hat{T}}(|\alpha\cdot(\zeta_*-\zeta)|/|\zeta_*-\zeta|,|\zeta_*-\zeta|) = B(|\alpha\cdot(\xi_*-\xi)|/|\xi_*-\xi|,|\xi_*-\xi|)/\hat{T}^{1/2}B_0,$$

[24]When we try to find the relation between the present nondimensional $\widehat{B}_{\hat{T}}(|\alpha$

where

$$\zeta = \xi/\sqrt{2RT}, \qquad \hat{T} = T/T_0.$$

It is noted that the function $B(|\boldsymbol{\alpha}\cdot(\boldsymbol{\xi}_* - \boldsymbol{\xi})|/|\boldsymbol{\xi}_* - \boldsymbol{\xi}|, |\boldsymbol{\xi}_* - \boldsymbol{\xi}|)$ is independent of the reference quantities by definition. The integral equation (A.107) is rewritten in terms of $B(|\boldsymbol{\alpha}\cdot(\boldsymbol{\xi}_* - \boldsymbol{\xi})|/|\boldsymbol{\xi}_* - \boldsymbol{\xi}|, |\boldsymbol{\xi}_* - \boldsymbol{\xi}|)$ instead of $\widehat{B}_{\hat{T}}$ as follows:

$$\int E(\xi_*/\sqrt{2RT})\{\phi\}B(|\boldsymbol{\alpha}\cdot(\boldsymbol{\xi}_* - \boldsymbol{\xi})|/|\boldsymbol{\xi}_* - \boldsymbol{\xi}|, |\boldsymbol{\xi}_* - \boldsymbol{\xi}|)\mathrm{d}\Omega(\boldsymbol{\alpha})\,\mathrm{d}\boldsymbol{\xi}_*$$
$$= \frac{B_0}{T_0^{1/2}} g(\boldsymbol{\xi}/\sqrt{2RT})T^{1/2}(2RT)^{3/2},$$

where

$$\{\phi\} = \phi' + \phi'_* - \phi - \phi_*.$$

When $g(\boldsymbol{\zeta})$ is free from the reference quantities, the solution ϕ is expressed in the form

$$\phi = \frac{B_0}{(RT_0)^{1/2}}\Phi(\boldsymbol{\xi}, 2RT).$$

Thus, the functions $\mathcal{A}(\zeta, T/T_0)$ and $\mathcal{B}^{(0)}(\zeta, T/T_0)$, the solutions of Eqs. (A.84) and (A.85) for $m = 0$, are in the forms

$$\mathcal{A}(\zeta, T/T_0) = \frac{B_0}{(RT_0)^{1/2}}\Phi_A(\boldsymbol{\xi}, 2RT), \quad \mathcal{B}^{(0)}(\zeta, T/T_0) = \frac{B_0}{(RT_0)^{1/2}}\Phi_B(\boldsymbol{\xi}, 2RT),$$

and in view of these forms, the functions $\mathcal{B}_1(\zeta, T/T_0)$ and $\mathcal{B}_2(\zeta, T/T_0)$, expressed by linear combinations [see Eq. (A.101)] of the solutions of Eq. (A.85) for $m = 1$, 2 and 3 with $\mathcal{A}(\zeta, T/T_0)$ in the inhomogeneous term, are in the forms[25]

$$\mathcal{B}_1(\zeta, T/T_0) = \frac{B_0^2}{RT_0}\Phi_{B1}(\boldsymbol{\xi}, 2RT), \quad \mathcal{B}_2(\zeta, T/T_0) = \frac{B_0^2}{RT_0}\Phi_{B2}(\boldsymbol{\xi}, 2RT).$$

With these forms of $\mathcal{A}(\zeta, T/T_0)$, $\mathcal{B}^{(0)}(\zeta, T/T_0)$, $\mathcal{B}_1(\zeta, T/T_0)$, and $\mathcal{B}_2(\zeta, T/T_0)$ in the definitions of $\hat{\gamma}_1(T/T_0)$, $\hat{\gamma}_2(T/T_0)$, $\hat{\gamma}_3(T/T_0)$, and $\hat{\gamma}_7(T/T_0)$, we find that $T_0^{1/2}\hat{\gamma}_1(\hat{T})/B_0$, $T_0^{1/2}\hat{\gamma}_2(\hat{T})/B_0$, $T_0\hat{\gamma}_3(\hat{T})/B_0^2$, and $T_0\hat{\gamma}_7(\hat{T})/B_0^2$ [see Eqs. (A.99), (A.100), and (A.102)] are independent of T_0, since the integral (A.100) with $Z = \Phi_A(\boldsymbol{\xi}, 2RT)$ etc. is independent of T_0.

Similar discussion can be made about the slip coefficient and the Knudsen-layer correction.

$(\boldsymbol{\zeta}_* - \boldsymbol{\zeta})|/|\boldsymbol{\zeta}_* - \boldsymbol{\zeta}|, |\boldsymbol{\zeta}_* - \boldsymbol{\zeta}|)$ and the dimensional $B(|\boldsymbol{\alpha}\cdot(\boldsymbol{\xi}_* - \boldsymbol{\xi})|/|\boldsymbol{\xi}_* - \boldsymbol{\xi}|, |\boldsymbol{\xi}_* - \boldsymbol{\xi}|)$, some care is required for the relation between $\boldsymbol{\zeta}$ and $\boldsymbol{\xi}$; it is not given by Eq. (2.36), since $\boldsymbol{\zeta}$ used here is $\tilde{\boldsymbol{\zeta}}$ in the main text. That is,

$$\widehat{B}_{\hat{T}}(|\boldsymbol{\alpha}\cdot(\boldsymbol{\zeta}_* - \boldsymbol{\zeta})|/|\boldsymbol{\zeta}_* - \boldsymbol{\zeta}|, |\boldsymbol{\zeta}_* - \boldsymbol{\zeta}|) = \widehat{B}(|\boldsymbol{\alpha}\cdot(\boldsymbol{\zeta}_* - \boldsymbol{\zeta})|/|\boldsymbol{\zeta}_* - \boldsymbol{\zeta}|, |\boldsymbol{\zeta}_* - \boldsymbol{\zeta}|\hat{T}^{1/2})/\hat{T}^{1/2}.$$

Putting $\boldsymbol{\zeta}\hat{T}^{1/2} = \boldsymbol{\xi}/(2RT_0)^{1/2}$ [or $\boldsymbol{\zeta} = \boldsymbol{\xi}/(2RT)^{1/2}$], we have, from the definition of $\widehat{B}$ given in Eq. (2.41c),

$$\widehat{B}_{\hat{T}}(|\boldsymbol{\alpha}\cdot(\boldsymbol{\zeta}_* - \boldsymbol{\zeta})|/|\boldsymbol{\zeta}_* - \boldsymbol{\zeta}|, |\boldsymbol{\zeta}_* - \boldsymbol{\zeta}|) = B(|\boldsymbol{\alpha}\cdot(\boldsymbol{\xi}_* - \boldsymbol{\xi})|/|\boldsymbol{\xi}_* - \boldsymbol{\xi}|, |\boldsymbol{\xi}_* - \boldsymbol{\xi}|)/\hat{T}^{1/2}B_0.$$

[25] Some arrangements of the inhomogeneous terms (A.91c) and (A.91d) are required to derive the second relation.

A.7 Linear integral equations $\hat{J}(\hat{f}_{h0}, \hat{f}_{hm}) = Ih_{hm}$, $\hat{J}(\hat{f}_{SB0}, \hat{f}_{SBm}) = Ih_{SBm}$, etc.

The linear integral equations encountered in Chapters 5–7:[26]

$$2\hat{J}(\hat{f}_{SB0}, \hat{f}_{SBm}) = Ih_{SBm} \quad (m \geq 1), \tag{A.108a}$$

$$2\hat{J}(\hat{f}_{h0}, \hat{f}_{hm}) = Ih_{hm} \quad (m \geq 1), \tag{A.108b}$$

$$2\hat{J}(\hat{f}_{V0}, \hat{f}_{Vm}) = Ih_{Vm} \quad (m \geq 1), \tag{A.108c}$$

$$2\hat{J}(\hat{f}_{H0}, \hat{f}_{Hm}) = Ih_{Hm} \quad (m \geq 1), \tag{A.108d}$$

can be reduced to a common type of equation by introduction of the following variable $\tilde{\zeta}_i$ instead of the molecular velocity ζ_i:

$$\tilde{\zeta}_i = \zeta_i/\hat{T}_{SB0}^{1/2} \qquad \text{for } \hat{J}(\hat{f}_{SB0}, *), \tag{A.109a}$$

$$\tilde{\zeta}_i = (\zeta_i - \hat{v}_{ih0})/\hat{T}_{h0}^{1/2} \quad \text{for } \hat{J}(\hat{f}_{h0}, *), \tag{A.109b}$$

$$\tilde{\zeta}_i = (\zeta_i - \hat{v}_{iV0})/\hat{T}_{V0}^{1/2} \quad \text{for } \hat{J}(\hat{f}_{V0}, *), \tag{A.109c}$$

$$\tilde{\zeta}_i = (\zeta_i - \hat{v}_{iH0})/\hat{T}_{H0}^{1/2} \quad \text{for } \hat{J}(\hat{f}_{H0}, *). \tag{A.109d}$$

That is, for example,

$$\begin{aligned}
2\hat{J}(\hat{f}_{h0}, \hat{f}_{h0}\Phi) &= \int \hat{f}_{h0}\hat{f}_{h0*}(\Phi' + \Phi'_* - \Phi - \Phi_*) \\
&\qquad \times \widehat{B}(|\alpha_i(\zeta_{i*} - \zeta_i)|/|\zeta_{k*} - \zeta_k|, |\zeta_{i*} - \zeta_i|)\mathrm{d}\Omega(\alpha)\,\mathrm{d}\boldsymbol{\zeta}_* \\
&= \frac{\hat{\rho}_{h0}^2}{\hat{T}_{h0}^{3/2}} E(\tilde{\zeta}) \int E(\tilde{\zeta}_*)(\phi' + \phi'_* - \phi - \phi_*) \\
&\qquad \times \widehat{B}(|\alpha_i(\tilde{\zeta}_{i*} - \tilde{\zeta}_i)|/|\tilde{\zeta}_{k*} - \tilde{\zeta}_k|, |\tilde{\zeta}_{i*} - \tilde{\zeta}_i|\hat{T}_{h0}^{1/2})\mathrm{d}\Omega(\alpha)\mathrm{d}\tilde{\boldsymbol{\zeta}}_* \\
&= \frac{\hat{\rho}_{h0}^2}{\hat{T}_{h0}} E(\tilde{\zeta})\mathcal{L}_{\hat{T}_{h0}}(\phi(\tilde{\zeta}_i)),
\end{aligned} \tag{A.110}$$

where $\phi(\tilde{\zeta}_i) = \Phi(\zeta_i)$, i.e., the function derived from $\Phi(\zeta_i)$ by the change of variable (A.109b), and the operator $\mathcal{L}_{\hat{T}_{h0}}(\phi(\tilde{\zeta}_i))$ is defined by Eq. (A.23) with

[26] The integrals on the left-hand sides of Eqs. (A.108a)–(A.108d) are taken as the collision integrals linearized, respectively, around the Maxwellian $\hat{f}_{SB0}$, $\hat{f}_{h0}$, $\hat{f}_{V0}$, and $\hat{f}_{H0}$. That is, the linearized integral is given, for example, by $\hat{J}(\hat{f}_{SB0}, \hat{f}_{SB0}\Phi)$ for the collision integral $\hat{J}(\hat{f}_{SB0}(1+\Phi), \hat{f}_{SB0}(1+\Phi))$. Then, these integrals have their meanings for the BKW equation, and they are expressed by the same formulas as Eqs. (A.110), (A.113a)–(A.113c) with $\mathcal{L}_a$ defined by Eq. (A.25). See also Footnote 12 in Section 2.9.

Eq. (A.24) in Appendix A.2 in the following way:

$$\mathcal{L}_{\hat{T}_{h0}}(\phi(\tilde{\zeta}_i)) = \int E(\tilde{\zeta}_*)(\phi' + \phi'_* - \phi - \phi_*)\widehat{B}_{\hat{T}_{h0}} d\Omega(\boldsymbol{\alpha}) d\tilde{\boldsymbol{\zeta}}_*, \tag{A.111a}$$

$$\begin{aligned}\widehat{B}_{\hat{T}_{h0}} &= \widehat{B}_{\hat{T}_{h0}}(|\alpha_i(\tilde{\zeta}_{i*} - \tilde{\zeta}_i)|/|\tilde{\zeta}_{k*} - \tilde{\zeta}_k|, |\tilde{\zeta}_{i*} - \tilde{\zeta}_i|) \\ &= \widehat{B}(|\alpha_i(\tilde{\zeta}_{i*} - \tilde{\zeta}_i)|/|\tilde{\zeta}_{k*} - \tilde{\zeta}_k|, |\tilde{\zeta}_{i*} - \tilde{\zeta}_i|\hat{T}_{h0}^{1/2})/\hat{T}_{h0}^{1/2},\end{aligned} \tag{A.111b}$$

$$\phi = \phi(\tilde{\zeta}_i),\ \phi_* = \phi(\tilde{\zeta}_{i*}),\ \phi' = \phi(\tilde{\zeta}'_i),\ \phi'_* = \phi(\tilde{\zeta}'_{i*}), \tag{A.111c}$$

$$\tilde{\zeta}'_i = \tilde{\zeta}_i + \alpha_k(\tilde{\zeta}_{k*} - \tilde{\zeta}_k)\alpha_i, \quad \tilde{\zeta}'_{i*} = \tilde{\zeta}_{i*} - \alpha_k(\tilde{\zeta}_{k*} - \tilde{\zeta}_k)\alpha_i, \tag{A.111d}$$

and this operator is related to the linearized collision operator $\mathcal{L}(\phi(\tilde{\zeta}_i))$ defined by Eq. (2.58b) or (A.22) as

$$\mathcal{L}_1(\phi(\tilde{\zeta}_i)) = \mathcal{L}(\phi(\tilde{\zeta}_i)), \tag{A.112}$$

or the operator $\mathcal{L}_{\hat{T}_{h0}}(\phi(\tilde{\zeta}_i))$ is obtained by replacing the function $\widehat{B}(|\alpha_i(\tilde{\zeta}_{i*} - \tilde{\zeta}_i)|/|\tilde{\zeta}_{k*} - \tilde{\zeta}_k|, |\tilde{\zeta}_{i*} - \tilde{\zeta}_i|)$ in the operator $\mathcal{L}(\phi(\tilde{\zeta}_i))$ by the function $\hat{T}_{h0}^{-1/2}\widehat{B}(|\alpha_i(\tilde{\zeta}_{i*} - \tilde{\zeta}_i)|/|\tilde{\zeta}_{k*} - \tilde{\zeta}_k|, |\tilde{\zeta}_{i*} - \tilde{\zeta}_i|\hat{T}_{h0}^{1/2})$. Similarly,

$$2\hat{J}(\hat{f}_{SB0}, \hat{f}_{SB0}\Phi) = \frac{\hat{\rho}_{SB0}^2}{\hat{T}_{SB0}} E(\tilde{\zeta})\mathcal{L}_{\hat{T}_{SB0}}(\phi(\tilde{\zeta}_i)), \tag{A.113a}$$

$$2\hat{J}(\hat{f}_{V0}, \hat{f}_{V0}\Phi) = \frac{\hat{\rho}_{V0}^2}{\hat{T}_{V0}} E(\tilde{\zeta})\mathcal{L}_{\hat{T}_{V0}}(\phi(\tilde{\zeta}_i)), \tag{A.113b}$$

$$2\hat{J}(\hat{f}_{H0}, \hat{f}_{H0}\Phi) = \frac{\hat{\rho}_{H0}^2}{\hat{T}_{H0}} E(\tilde{\zeta})\mathcal{L}_{\hat{T}_{H0}}(\phi(\tilde{\zeta}_i)), \tag{A.113c}$$

where $\phi(\tilde{\zeta}_i) = \Phi(\zeta_i)$. Thus the integral equations (A.108a)–(A.108d) are reduced to the common form

$$\mathcal{L}_a(\phi(\tilde{\zeta}_i)) = IHM,$$

where IHM represents the inhomogeneous term.

The integrals $\hat{J}(\hat{f}_{SB0}\Phi, \hat{f}_{SB0}\Psi)$, $\hat{J}(\hat{f}_{h0}\Phi, \hat{f}_{h0}\Psi)$, etc., which appear in the inhomogeneous terms in the integral equations for $\hat{f}_{SBm}$, $\hat{f}_{hm}$, etc., are reduced to the following forms by the transformations (A.109a)–(A.109d):

$$\hat{J}(\hat{f}_{SB0}\Phi, \hat{f}_{SB0}\Psi) = \frac{\hat{\rho}_{SB0}^2}{\hat{T}_{SB0}} \hat{J}_{\hat{T}_{SB0}}(E(\tilde{\zeta})\phi(\tilde{\zeta}_i), E(\tilde{\zeta})\psi(\tilde{\zeta}_i)), \tag{A.114a}$$

$$\hat{J}(\hat{f}_{h0}\Phi, \hat{f}_{h0}\Psi) = \frac{\hat{\rho}_{h0}^2}{\hat{T}_{h0}} \hat{J}_{\hat{T}_{h0}}(E(\tilde{\zeta})\phi(\tilde{\zeta}_i), E(\tilde{\zeta})\psi(\tilde{\zeta}_i)), \tag{A.114b}$$

$$\hat{J}(\hat{f}_{V0}\Phi, \hat{f}_{V0}\Psi) = \frac{\hat{\rho}_{V0}^2}{\hat{T}_{V0}} \hat{J}_{\hat{T}_{V0}}(E(\tilde{\zeta})\phi(\tilde{\zeta}_i), E(\tilde{\zeta})\psi(\tilde{\zeta}_i)), \tag{A.114c}$$

$$\hat{J}(\hat{f}_{H0}\Phi, \hat{f}_{H0}\Psi) = \frac{\hat{\rho}_{H0}^2}{\hat{T}_{H0}} \hat{J}_{\hat{T}_{H0}}(E(\tilde{\zeta})\phi(\tilde{\zeta}_i), E(\tilde{\zeta})\psi(\tilde{\zeta}_i)), \tag{A.114d}$$

where $\phi(\tilde{\zeta}_i) = \Phi(\zeta_i)$, $\psi(\tilde{\zeta}_i) = \Psi(\zeta_i)$ and the integral $\hat{J}_a(\hat{f}(\zeta_i), \hat{g}(\zeta_i))$ is defined by[27]

$$\hat{J}_a(\hat{f}(\zeta_i), \hat{g}(\zeta_i)) = \frac{1}{2}\int (\hat{f}'_*\hat{g}' + \hat{f}'\hat{g}'_* - \hat{f}_*\hat{g} - \hat{f}\hat{g}_*)\widehat{B}_a \, d\Omega(\alpha)\, \mathbf{d}\boldsymbol{\zeta}_*, \tag{A.115}$$

$$\begin{aligned}\widehat{B}_a &= \widehat{B}_a(|\alpha_i(\zeta_{i*} - \zeta_i)|/|\zeta_{k*} - \zeta_k|, |\zeta_{i*} - \zeta_i|) \\ &= \widehat{B}(|\alpha_i(\zeta_{i*} - \zeta_i)|/|\zeta_{k*} - \zeta_k|, |\zeta_{i*} - \zeta_i|a^{1/2})\,/a^{1/2}.\end{aligned}$$

Thus,

$$\hat{J}(\hat{f}(\zeta_i), \hat{g}(\zeta_i)) = \hat{J}_1(\hat{f}(\zeta_i), \hat{g}(\zeta_i)),$$

which is a similar relation to that between $\mathcal{L}_a(\phi)$ and $\mathcal{L}(\phi)$ or $\mathcal{J}_a(\phi, \psi)$ and $\mathcal{J}(\phi, \psi)$. The integral $\hat{J}_a$ is related to $\mathcal{L}_a$ [Eq. (A.23)] and $\mathcal{J}_a$ [Eq. (A.97)] as

$$2\hat{J}_a(E, E\phi) = E(\zeta)\mathcal{L}_a(\phi), \tag{A.116}$$

$$\hat{J}_a(E\phi, E\psi) = E\mathcal{J}_a(\phi, \psi). \tag{A.117}$$

With the aid of Eq. (A.117), the right-hand sides of Eqs. (A.114a)–(A.114d) are rewritten in terms of $\mathcal{J}_a$, e.g.,

$$\hat{J}(\hat{f}_{V0}\Phi, \hat{f}_{V0}\Psi) = \frac{\hat{\rho}_{V0}^2 E(\tilde{\zeta})}{\hat{T}_{V0}}\mathcal{J}_{\hat{T}_{V0}}(\phi(\tilde{\zeta}_i), \psi(\tilde{\zeta}_i)). \tag{A.118}$$

As an example, consider the integral equation (5.14) for $m = 1$, i.e.,

$$2\hat{J}(\hat{f}_{SB0}, \hat{f}_{SB1}) = \frac{\zeta_i}{\hat{T}_{SB0}}\frac{\partial \hat{T}_{SB0}}{\partial x_i}\left(\frac{\zeta_j^2}{\hat{T}_{SB0}} - \frac{5}{2}\right)\hat{f}_{SB0}. \tag{A.119}$$

By the transformation (A.109a), the equation (A.119) is reduced to

$$\mathcal{L}_{\hat{T}_{SB0}}(\phi(\tilde{\zeta}_i)) = \frac{1}{\hat{p}_{SB0}}\frac{\partial \hat{T}_{SB0}}{\partial x_i}\tilde{\zeta}_i\left(\tilde{\zeta}^2 - \frac{5}{2}\right), \tag{A.120}$$

$$\text{with} \quad \hat{f}_{SB1} = \hat{f}_{SB0}\Phi, \quad \Phi(\zeta_i) = \phi(\tilde{\zeta}_i).$$

Thus, we find from Eq. (A.84) that the solution $\phi(\tilde{\zeta}_i)$ of Eq. (A.120) satisfying the orthogonal relation

$$\int \tilde{g}E(\tilde{\zeta})\phi(\tilde{\zeta}_i)\mathbf{d}\tilde{\boldsymbol{\zeta}} = 0, \quad \tilde{g} = 1,\ \tilde{\zeta}_i,\ \tilde{\zeta}_i^2, \tag{A.121}$$

which as a whole is equivalent to the orthogonal relation (5.21), is given by

$$\phi(\tilde{\zeta}_i) = -\frac{1}{\hat{p}_{SB0}}\frac{\partial \hat{T}_{SB0}}{\partial x_i}\tilde{\zeta}_i\mathcal{A}(\tilde{\zeta}, \hat{T}_{SB0}). \tag{A.122}$$

---

[27]The difference between the argument $\tilde{\zeta}_i$ of $\hat{J}_a(*,*)$ in Eqs. (A.114a)–(A.114d) and the argument ζ_i of $\hat{J}_a(*,*)$ in its definition (A.115) is just a matter of symbols and independent of the transformations (A.109a)–(A.109d). The simpler symbol ζ_i is used in the definition of $\hat{J}_a$.

Therefore, the particular solution $\hat{\psi}_{SB1}$ in Eq. (5.20) for $m = 1$ satisfying the orthogonal relation (5.21) is given by

$$\hat{\psi}_{SB1} = -\frac{1}{\hat{p}_{SB0}} \frac{\partial \hat{T}_{SB0}}{\partial x_i} \frac{\zeta_i}{\hat{T}_{SB0}^{1/2}} \mathcal{A}(\zeta/\hat{T}_{SB0}^{1/2}, \hat{T}_{SB0}) \hat{f}_{SB0}. \tag{A.123}$$

A.8 $\hat{J}_a(\zeta_i E, \zeta_j \mathcal{A}E)$, $\hat{J}_a(\zeta^2 E, (\zeta_i\zeta_j - \zeta^2\delta_{ij}/3)\mathcal{B}E)$, etc.

The collision integrals $\hat{J}_a(\zeta_i E, \zeta_j \mathcal{A}(\zeta, a)E)$, $\hat{J}_a(\zeta^2 E, (\zeta_i\zeta_j - \zeta^2\delta_{ij}/3)\mathcal{B}(\zeta, a)E)$, etc. are contained in the inhomogeneous term $\hat{J}(\hat{f}_{SB1}, \hat{f}_{SB1})$ or $\hat{J}(\hat{f}_{V1}, \hat{f}_{V1})$ of the integral equation (5.14) with $m = 2$ for $\hat{f}_{SB2}$ or (6.45) for $\hat{f}_{V2}$. These integrals can be transformed into simpler forms, for which the solution of the integral equation is more easily obtained.

For this purpose, consider the solution $\psi(\boldsymbol{\zeta})$ of the integral equation

$$2\hat{J}_a(E(\zeta), \psi(\boldsymbol{\zeta})E(\zeta)) = \hat{H}(\boldsymbol{\zeta}). \tag{A.124}$$

The equation is equivalent to the linear integral equation (A.26) [see Eq. (A.116)]. The equations for $\hat{f}_{SB2}$ and $\hat{f}_{V2}$ are reduced to this form (see Appendix A.7). The solution $\psi(\boldsymbol{\zeta})$ depends on a, which is explicitly shown as $\psi(\boldsymbol{\zeta}, a)$ if necessary. In some cases, it is convenient (or not confusing) to use the expression $\hat{J}(E(\zeta), \psi(\boldsymbol{\zeta})E(\zeta); \widehat{B}_a)$, instead of $\hat{J}_a(E(\zeta), \psi(\boldsymbol{\zeta})E(\zeta))$, in order to show more explicitly what type of $\widehat{B}$ is used in the collision integral. Thus,

$$2\hat{J}(E(\zeta), \psi(\boldsymbol{\zeta})E(\zeta); \widehat{B}_a) = \hat{H}(\boldsymbol{\zeta}). \tag{A.125}$$

Then, the following relations hold:

$$\hat{H}(\boldsymbol{\zeta} + \mathbf{c}) = 2\hat{J}([E(\zeta)]_{+\mathbf{c}}, [\psi(\boldsymbol{\zeta})E(\zeta)]_{+\mathbf{c}}; \widehat{B}_a), \tag{A.126}$$

$$\hat{H}(\sqrt{c}\boldsymbol{\zeta}) = 2c^2\hat{J}([E(\zeta)]_{\times c}, [\psi(\boldsymbol{\zeta})E(\zeta)]_{\times c}; \widehat{B}_{ca}), \tag{A.127}$$

$$2\hat{J}(E(\zeta), \psi(\boldsymbol{\zeta}, ca)E(\zeta); \widehat{B}_{ca}) = \hat{H}(\boldsymbol{\zeta}), \tag{A.128}$$

where c is a constant, $\mathbf{c}$ (also indicated by c_i below) is a constant vector and the symbols $[f(\boldsymbol{\zeta})]_{+\mathbf{c}}$ and $[f(\boldsymbol{\zeta})]_{\times c}$ mean that

$$[f(\boldsymbol{\zeta})]_{+\mathbf{c}} = f(\boldsymbol{\zeta} + \mathbf{c}), \quad [f(\boldsymbol{\zeta})]_{\times c} = f(\sqrt{c}\boldsymbol{\zeta}). \tag{A.129}$$

By differentiation of Eq. (A.128), with $\hat{H}(\boldsymbol{\zeta})$ independent of a (or c), with respect to c,

$$\hat{J}\left(E(\zeta), \frac{\partial\psi(\boldsymbol{\zeta}, ca)E(\zeta)}{\partial c}; \widehat{B}_{ca}\right) + \hat{J}\left(E(\zeta), \psi(\boldsymbol{\zeta}, ca)E(\zeta); \frac{\partial \widehat{B}_{ca}}{\partial c}\right) = 0.$$

Thus,

$$\left(\hat{J}\left(E(\zeta), \psi(\boldsymbol{\zeta}, ca)E(\zeta); \frac{\partial \widehat{B}_{ca}}{\partial c}\right)\right)_{c=1} = -\hat{J}\left(E(\zeta), a\frac{\partial\psi(\boldsymbol{\zeta}, a)E(\zeta)}{\partial a}; \widehat{B}_a\right). \tag{A.130}$$

Then it is easily seen from Eqs. (A.126), (A.127), and (A.130) that

$$\left(\frac{\partial \hat{H}(\zeta+\mathrm{c})}{\partial c_k}\right)_{c=0}$$
$$= -4\hat{J}(\zeta_k E(\zeta), \psi(\zeta)E(\zeta); \widehat{B}_a) + 2\hat{J}\left(E(\zeta), \frac{\partial \psi(\zeta)E(\zeta)}{\partial \zeta_k}; \widehat{B}_a\right), \tag{A.131}$$

$$\begin{aligned}\left(\frac{\partial}{\partial c}\frac{\hat{H}(\sqrt{c}\zeta)}{c^2}\right)_{c=1} &= -2\hat{J}(\zeta^2 E(\zeta), \psi(\zeta,a)E(\zeta); \widehat{B}_a) \\ &\quad + 2\hat{J}\left(E(\zeta), \frac{\zeta_k}{2}\frac{\partial \psi(\zeta,a)E(\zeta)}{\partial \zeta_k}; \widehat{B}_a\right) \\ &\quad + 2\hat{J}\left(E(\zeta), \psi(\zeta,a)E(\zeta); \left(\frac{\partial \widehat{B}_{ca}}{\partial c}\right)_{c=1}\right) \\ &= -2\hat{J}(\zeta^2 E(\zeta), \psi(\zeta,a)E(\zeta); \widehat{B}_a) + 2\hat{J}(E(\zeta), \Psi_0; \widehat{B}_a),\end{aligned} \tag{A.132}$$

where

$$\Psi_0 = \frac{\zeta_k}{2}\frac{\partial \psi(\zeta,a)E(\zeta)}{\partial \zeta_k} - a\frac{\partial \psi(\zeta,a)E(\zeta)}{\partial a}.$$

Let $\psi(\zeta)$ be $\zeta_i \mathcal{A}(\zeta,a)$ or $(\zeta_i\zeta_j - \zeta^2\delta_{ij}/3)\mathcal{B}(\zeta,a)$, then by the definitions of $\mathcal{A}(\zeta,a)$ and $\mathcal{B}(\zeta,a)$ [see Eqs. (A.84), (A.85), (A.91a), (A.101), and (A.116)],

$$\hat{H}(\zeta) = -\zeta_i\left(\zeta^2 - \frac{5}{2}\right)E, \quad \text{or} \quad \hat{H}(\zeta) = -2\left(\zeta_i\zeta_j - \frac{1}{3}\zeta^2\delta_{ij}\right)E. \tag{A.133}$$

With the aid of the relations (A.131) and (A.133), $\hat{J}_a(\zeta_i E(\zeta), \zeta_j \mathcal{A}(\zeta,a)E(\zeta))$ and $\hat{J}_a(\zeta_i E(\zeta), (\zeta_j\zeta_k - \zeta^2\delta_{jk}/3)\mathcal{B}(\zeta,a)E(\zeta))$ are transformed into

$$2\hat{J}_a(\zeta_i E(\zeta), \zeta_j \mathcal{A}(\zeta,a)E(\zeta))$$
$$= \hat{J}_a\left(E(\zeta), \frac{\partial \zeta_j \mathcal{A}(\zeta,a)E(\zeta)}{\partial \zeta_i}\right) + \frac{1}{2}\delta_{ij}\left(\zeta^2 - \frac{5}{2}\right)E(\zeta) - \zeta_i\zeta_j\left(\zeta^2 - \frac{7}{2}\right)E(\zeta), \tag{A.134}$$

$$2\hat{J}_a\left(\zeta_i E(\zeta), \left(\zeta_j\zeta_k - \frac{1}{3}\zeta^2\delta_{jk}\right)\mathcal{B}(\zeta,a)E(\zeta)\right)$$
$$= \hat{J}_a\left(E(\zeta), \frac{\partial}{\partial \zeta_i}\left[\left(\zeta_j\zeta_k - \frac{1}{3}\zeta^2\delta_{jk}\right)\mathcal{B}(\zeta,a)E(\zeta)\right]\right) + \frac{2}{3}\zeta_i\delta_{jk}(\zeta^2 - 1)E(\zeta)$$
$$+ (\zeta_j\delta_{ik} + \zeta_k\delta_{ij})E(\zeta) - 2\zeta_i\zeta_j\zeta_k E(\zeta). \tag{A.135}$$

With the aid of Eqs. (A.132) and (A.133), the integrals $\hat{J}_a(\zeta^2 E(\zeta), \zeta_i \mathcal{A}(\zeta,a)E(\zeta))$ and $\hat{J}_a(\zeta^2 E(\zeta), (\zeta_i\zeta_j - \zeta^2\delta_{ij}/3)\mathcal{B}(\zeta,a)E(\zeta))$ are transformed into

$$2\hat{J}_a(\zeta^2 E(\zeta), \zeta_i \mathcal{A}(\zeta,a)E(\zeta)) = 2\hat{J}_a(E(\zeta), \Psi_{1i}) - \zeta_i\left(\zeta^4 - 2\zeta^2 - \frac{15}{4}\right)E(\zeta), \tag{A.136}$$

where

$$\Psi_{1i} = \frac{\zeta_k}{2}\frac{\partial \zeta_i \mathcal{A}(\zeta, a)E(\zeta)}{\partial \zeta_k} - a\frac{\partial \zeta_i \mathcal{A}(\zeta, a)E(\zeta)}{\partial a},$$

and

$$2\hat{J}_a\left(\zeta^2 E(\zeta), \left(\zeta_i\zeta_j - \frac{1}{3}\zeta^2\delta_{ij}\right)\mathcal{B}(\zeta, a)E(\zeta)\right)$$
$$= 2\hat{J}_a(E(\zeta), \Psi_{2ij}) - 2\left(\zeta_i\zeta_j - \frac{1}{3}\zeta^2\delta_{ij}\right)(\zeta^2 + 1)E(\zeta), \tag{A.137}$$

where

$$\Psi_{2ij} = \frac{\zeta_k}{2}\frac{\partial}{\partial \zeta_k}\left[\left(\zeta_i\zeta_j - \frac{1}{3}\zeta^2\delta_{ij}\right)\mathcal{B}(\zeta, a)E(\zeta)\right]$$
$$- a\frac{\partial}{\partial a}\left[\left(\zeta_i\zeta_j - \frac{1}{3}\zeta^2\delta_{ij}\right)\mathcal{B}(\zeta, a)E(\zeta)\right].$$

With these transformations, the solution ψ of Eq. (A.124) with the inhomogeneous term of the types $\hat{J}_a(\zeta_i E, \zeta_j \mathcal{A}(\zeta, a)E)$, $\hat{J}_a(\zeta^2 E, (\zeta_i\zeta_j - \zeta^2\delta_{ij}/3)\mathcal{B}(\zeta, a)E)$, etc. becomes simpler. The particular solutions corresponding to $\hat{J}_a$ parts on the right-hand sides in Eqs. (A.134)–(A.137) are obvious. The remaining terms are much simpler than the original collision integrals.

A.9 Equation for the Knudsen layer and Bardos's theorem

The equations for the component functions $\hat{f}_{Km}$ of the Knudsen-layer correction $\hat{f}_K$ in Chapters 5 and 6 are given in the forms[28]

$$\zeta_i n_i \frac{\partial \hat{f}_{Km}}{\partial \eta} = 2\hat{J}((\hat{f}_{SB0})_0, \hat{f}_{Km}) + Ih_m, \tag{A.138a}$$

$$\zeta_i n_i \frac{\partial \hat{f}_{Km}}{\partial \eta} = 2\hat{J}((\hat{f}_{V0})_0, \hat{f}_{Km}) + Ih_m, \tag{A.138b}$$

where Ih_m (especially, $Ih_1 = 0$) represents the inhomogeneous term that decays rapidly (faster than any inverse power of η). These two equations are reduced to the same type of equation by the following change of variables:

$$x = (\hat{\rho}_{SB0})_0\eta, \quad \tilde{\zeta}_i = \frac{\zeta_i}{(\hat{T}_{SB0})_0^{1/2}} \quad \text{for } \hat{J}((\hat{f}_{SB0})_0, *), \tag{A.139a}$$

$$x = (\hat{\rho}_{V0})_0\eta, \quad \tilde{\zeta}_i = \frac{\zeta_i - \hat{v}_{wi}}{(\hat{T}_{V0})^{1/2}} \quad \text{for } \hat{J}((\hat{f}_{V0})_0, *). \tag{A.139b}$$

[28]The first terms on the right-hand sides of Eqs. (A.138a) and (A.138b) being the collision integrals linearized, respectively, around the Maxwellians $(\hat{f}_{SB0})_0$ and $(\hat{f}_{V0})_0$, Eq. (A.140) with Eq. (A.25) holds for the BKW equation. See Footnote 26 in Appendix A.7.

Noting that $(\hat{f}_{SB0})_0$ and $(\hat{f}_{V0})_0$ are expressed with the new variables in the forms

$$(\hat{f}_{SB0})_0 = \frac{(\hat{\rho}_{SB0})_0}{\hat{T}_{w0}^{3/2}} E(\tilde{\zeta}), \quad (\hat{f}_{V0})_0 = \frac{(\hat{\rho}_{V0})_0}{\hat{T}_{w}^{3/2}} E(\tilde{\zeta}),$$

we put

$$\hat{f}_{Km} = \frac{(\hat{\rho}_{SB0})_0}{\hat{T}_{w0}^{3/2}} E(\tilde{\zeta})\phi(x,\tilde{\zeta}_i) \quad \text{for } \hat{J}((\hat{f}_{SB0})_0, *),$$

$$\hat{f}_{Km} = \frac{(\hat{\rho}_{V0})_0}{\hat{T}_{w}^{3/2}} E(\tilde{\zeta})\phi(x,\tilde{\zeta}_i) \quad \text{for } \hat{J}((\hat{f}_{V0})_0, *).$$

Then with the aid of Eqs. (A.113a) and (A.113b), the Knudsen-layer equations (A.138a) and (A.138b) are reduced to the common form

$$\tilde{\zeta}_1 \frac{\partial \phi(x,\tilde{\zeta}_i)}{\partial x} = \mathcal{L}_a(\phi(x,\tilde{\zeta}_i)) + IHM, \tag{A.140}$$

where $a = \hat{T}_{w0}$ for Eq. (A.138a) and $a = \hat{T}_w$ for Eq. (A.138b), the operator $\mathcal{L}_a(*)$ is defined by Eq. (A.23), and IHM is the appropriately transformed form of the inhomogeneous term Ih_m.

Bardos, Caflisch & Nicolaenko [1986] considered the following half-space problem of the linearized Boltzmann equation for hard-sphere molecules: A function $\phi(x,\tilde{\zeta}_i)$ satisfies the one-dimensional linearized Boltzmann equation (A.140) in the domain $x > 0$ and the following boundary conditions at $x = 0$ and $x \to \infty$:

$$\phi = c_0 + c_2\tilde{\zeta}_2 + c_3\tilde{\zeta}_3 + c_4\tilde{\zeta}_j^2 + g(\tilde{\zeta}_i) \quad (\tilde{\zeta}_1 > 0) \quad \text{at } x = 0, \tag{A.141a}$$

$$\phi \to 0 \quad \text{as } x \to \infty, \tag{A.141b}$$

where c_0, c_2, c_3, and c_4 are some constants and $g(\tilde{\zeta}_i)$ is a given function of $\tilde{\zeta}_i$. According to them, the solution of this boundary-value problem, where exponential decay of the inhomogeneous term as $x \to \infty$ is assumed, exists when and only when the four undetermined constants c_0, c_2, c_3, and c_4 take special values.[29] The solution is unique including c_0, c_2, c_3, and c_4. It is also shown that the speed of decay of the solution as $x \to \infty$ is exponential.

The theorem is extended to more general molecular models (Cercignani [1986], Golse & Poupaud [1989]) and to more general boundary conditions (Coron, Golse & Sulem [1988]).

A.10 The boundary condition for the linearized Euler set of equations

Here we will give a simple explanatory example of the difference of the appropriate number of boundary conditions for the linearized Euler set of equations

[29] The solution including the constants c_0, c_2, c_3, and c_4 generally depends on the reference temperature T_0 as well as $g(\tilde{\zeta}_i)$ and the parameter a (see Footnote 13 in Appendix A.2).

depending on situations. Consider a two-dimensional case of an incompressible gas (ρ is constant). The flow is assumed here to be very close to a uniform flow. The uniform flow, naturally, satisfies the Euler set of equations. Let (X, Y) be the space coordinates, and let $p_0 + \tilde{p}$ and $(U + u(X,Y)\ ,\ v(X,Y))$ (p_0 and U : positive constants), respectively, be the pressure and the flow velocity, where $\tilde{p}$, u, and v are small quantities.

The equations for the perturbed quantities u, v, and $\tilde{p}$ are the linearized Euler set of equations

$$\frac{\partial u}{\partial X} + \frac{\partial v}{\partial Y} = 0, \tag{A.142}$$

$$U\frac{\partial u}{\partial X} = -\frac{1}{\rho}\frac{\partial \tilde{p}}{\partial X}, \tag{A.143}$$

$$U\frac{\partial v}{\partial X} = -\frac{1}{\rho}\frac{\partial \tilde{p}}{\partial Y}. \tag{A.144}$$

Note that the coefficients of $\partial u/\partial Y$ and $\partial v/\partial Y$ in the momentum equations (A.143) and (A.144) vanish. From Eqs. (A.143) and (A.144), we have

$$\frac{\partial}{\partial X}\left(\frac{\partial u}{\partial Y} - \frac{\partial v}{\partial X}\right) = 0. \tag{A.145}$$

Introducing the stream function Ψ defined by

$$u = \frac{\partial \Psi}{\partial Y}, \quad v = -\frac{\partial \Psi}{\partial X}, \tag{A.146}$$

then we have

$$\frac{\partial}{\partial X}\Delta\Psi = 0. \tag{A.147}$$

It is noted here that $\Delta\Psi$ is the vorticity. The above equation shows the invariance of vorticity along the streamlines of the unperturbed (or uniform) flow.

(I) Consider a uniform flow past a thin body set parallel to the flow. The body is expressed by

$$Y = f^{\pm}(X) \quad (0 \leq X \leq L), \tag{A.148}$$

where f^+ or f^- expresses the upper or lower surface of the body and $|\ f^{\pm}|$ is small. Then the boundary condition that the flow is along the body surface (or $v_i n_i = 0$) can be replaced by the condition $v = U\mathrm{d}f^{\pm}/\mathrm{d}X$ at $Y = 0_{\pm}$ $(0 \leq X \leq L)$. The condition at upstream infinity $(X_1 \to -\infty)$ is $u = v = 0$.

Then, from the condition at upstream infinity, we obtain the Laplace equation over the whole field, i.e.,

$$\Delta\Psi = 0. \tag{A.149}$$

The boundary condition on the body is

$$\frac{\partial \Psi}{\partial X} = -v = -U\frac{\mathrm{d}f^{\pm}}{\mathrm{d}X} \quad \text{at } Y = 0_{\pm} \quad (0 \leq X \leq L). \tag{A.150}$$

The boundary condition at upsteam infinity ($X_1 \to -\infty$) is $\partial\Psi/\partial X = \partial\Psi/\partial Y = 0$.

This boundary-value problem of the Laplace equation has unique solution Ψ except an additive constant if the circulation around the body is specified. In other words, the boundary-value problem for the linearized Euler set of equations under the boundary conditions that v is given on $Y = 0_+$ and 0_- ($0 \le X \le L$) and that u, v, and $\tilde{p}$ approach zero as $X \to -\infty$ determines the flow field except the circulation around the body, which cannot be determined within the framework of the time-independent Euler set of equations.

The problem of appropriate boundary conditions for the linearized Euler set of equations (6.24)–(6.26) in Section 6.6 is an extension of this example. Let the solution $\hat{\rho}_{h0}$, $\hat{v}_{ih0}$, and $\hat{T}_{h0}$ (or $\hat{p}_{h0}$) of the Euler set of equations (6.21)–(6.23) (say, the unperturbed flow) be given. From the linearized Euler set of equations (6.24)–(6.26), the equations governing the variation of perturbed vorticity and entropy along the streamlines of the unperturbed flow are derived. The streamlines do not pass the body and they are along the body surface, since $(\hat{v}_{ih0})_0 n_i = 0$ [Eq. (6.116)]. The perturbed vorticity and entropy are determined by their conditions at the upstream of the unperturbed flow. This is an important point. To make this point clearer, consider the following problem.

(II) Consider an artificial problem. The difference from problem (I) is the position and the property of the body. An artificial body, where some condition on u and v is imposed, is set at $X = 0$ and between $0 \le Y \le L$. Then from the condition $u = v = 0$ at upstream infinity ($X \to -\infty$), we obtain the Laplace equation

$$\Delta\Psi = 0,$$

in the domain

$$-\infty < X < \infty, \quad Y < 0 \text{ and } Y > L,$$
$$-\infty < X < 0, \quad 0 \le Y \le L,$$

and the Poisson-type equation

$$\Delta\Psi = F(Y),$$

in the rest of the whole domain of the gas, i.e.,

$$0 < X < \infty, \quad 0 \le Y \le L,$$

where $F(Y)$ is an undetermined function of Y. Then we need a condition (or an upstream condition) to determine $F(Y)$ on $X = 0_+$ ($0 \le Y \le L$), where the normal component of the velocity of the uniform flow is positive, in addition to the general boundary condition for the Laplace equation. Further, it may be noted that no additional condition is required on $X = 0_-$ ($0 \le Y \le L$), into which the streamlines of the unperturbed flow go.

Appendix B

Spherically Symmetric Field of Symmetric Tensor

In Chapters 3–6, we encountered solutions of the integral equation with the linearized collision operator and functions derived by application of the operator. These solutions and functions are characterized by a spherically symmetric field of symmetric tensors. Their special forms resulting from this property are used in the course of analysis leaving its proof to this appendix.[1] From the content of this appendix, the notation used here is independent of that in the foregoing chapters.

B.1 Problem

Consider a symmetric tensor field $T_{i_1,\dots,i_m}(\zeta_k)$ of order m in a three-dimensional space ζ_i that satisfies the following relation:

$$T_{i_1,\dots,i_m}(l_{kh}\zeta_h) = l_{i_1 j_1}\cdots l_{i_m j_m}T_{j_1,\dots,j_m}(\zeta_k), \tag{B.1}$$

where

$$l_{ik}l_{jk} = \delta_{ij}. \tag{B.2}$$

The general form of $T_{i_1,\dots,i_m}(\zeta_k)$ [or $T_I(\zeta_k), T_I$ for short] is of our interest here.[2]

We will discuss the possible form of T_I from the relation (B.1). In the discussion, the possible form of T_I is derived only for the orthogonal transformation (l_{ij}) with its determinant $\det(l_{ij}) = 1$. The resulting form will be found to

[1] This kind of subject was actively studied around 1900 (see, e.g., Ricci & Levi-Civita [1900], Weyl [1946], and Grad [1963a]). Here, we present a simple explanation of the subject that is sufficient for our purpose.

[2] The following discussion is based on Sone & Aoki [1994]. Apart from the extension to the axially symmetric case explained in Appendix B.4.2, another simple derivation of the formulas for $m = 1$ and 2 is given in Golse [1995]. According to Golse, this can be extended to arbitrary m by mathematical induction.

satisfy the relation (B.1) for both classes of orthogonal transformations with $\det(l_{ij}) = 1$ (rotation) and $\det(l_{ij}) = -1$ (rotation and reflection).

Equation (B.1) shows that T_I at the point $l_{ih}\zeta_h$, the result of rotation l_{ij} of ζ_i, is given by the rotation of T_I at the point ζ_i (or the field T_I is spherically symmetric). Thus, if T_I is known on a half-line originating at the origin, then, T_I at an arbitrary point ζ_i is obtained by a rotation l_{ij} bringing the half-line to that through the point ζ_i. The rotation, however, is not unique (there is a freedom of choice of a rotation around the half-line). Thus, T_I on the half-line cannot be arbitrary. It should take such a value that T_I is invariant with respect to a rotation around the half-line. More explicitly, let the point ζ_i be expressed in the form

$$\zeta_i = \zeta l_{i1}, \tag{B.3}$$

where l_{i1} is a unit vector. Two unit vectors l_{i2} and l_{i3} being chosen in such a way that l_{i1}, l_{i2}, and l_{i3} are orthogonal to each other, the tensor T_I at the point ζ_i is expressed in the form

$$T_{i_1,\ldots,i_m}(\zeta_i) = l_{i_1 j_1} \cdots l_{i_m j_m} T_{j_1,\ldots,j_m}(\zeta, 0, 0). \tag{B.4}$$

First we find the condition on $T_I(\zeta_i)$ on the ζ_1 axis [or $T_I(\zeta,0,0)$] by the condition that $T_I(\zeta_i)$ is invariant with respect to the rotation around the ζ_1 axis [or the transformation l_{ij} with $l_{i1} = (1,0,0)$, for which $T_{i_1,\ldots,i_m}(\zeta,0,0) = l_{i_1 j_1} \cdots l_{i_m j_m} T_{j_1,\ldots,j_m}(\zeta,0,0)$] in the next section. Then we derive the explicit form of T_I at an arbitrary point ζ_i from Eq. (B.4) and confirm the relation (B.1) for the result in Appendix B.3.

B.2 $T_I(\zeta_i)$ on ζ_1 axis

Take the following rotation around the ζ_1 axis:

$$R_{ij} = \begin{pmatrix} R_{11} & R_{12} & R_{13} \\ R_{21} & R_{22} & R_{23} \\ R_{31} & R_{32} & R_{33} \end{pmatrix} = \begin{pmatrix} 1 & 0 & 0 \\ 0 & \cos\theta & \sin\theta \\ 0 & -\sin\theta & \cos\theta \end{pmatrix}. \tag{B.5}$$

A point on the ζ_1 axis is invariant for this rotation. Thus,

$$T_I(\zeta,0,0) = R_{i_1 j_1} \cdots R_{i_m j_m} T_J(\zeta,0,0). \tag{B.6}$$

We will discuss the possible form of $T_I(\zeta,0,0)$ on the basis of Eq. (B.6). The arguments of T_I that appear in this section are always $(\zeta,0,0)$, and thus they are not shown explicitly. The components with p subscripts of 1 can be represented by

$$T_{\underbrace{1,\ldots,1}_{p},\bar{I}},$$

where $\bar{I}$ is a set $(i_{p+1},\ldots,i_m)$ of $M\ (= m-p)$ subscripts consisting only of 2 and 3, because of the symmetry of T_I. For these components, Eq. (B.6) is reduced to

$$
\begin{aligned}
T_{1,\ldots,1,I} &= R_{1j_1}\cdots R_{1j_p}R_{i_{p+1}j_{p+1}}\cdots R_{i_mj_m}T_J \\
&= \sum_{(2,3)} R_{i_{p+1}j_{p+1}}\cdots R_{i_mj_m}T_{1,\ldots,1,\bar{J}},
\end{aligned} \tag{B.7}
$$

where $R_{1j} = \delta_{1j}$, $R_{21} = R_{31} = 0$ are used and $\bar{J}$ is a set of subscripts $(j_{p+1},\ldots,j_m)$ consisting only of 2 and 3.[3] Put $F_{\bar{I}} = T_{1,\ldots,1,I}$ by omitting the subscripts 1 for brevity. Then

$$F_{\bar{I}} = \sum_{(2,3)} R_{i_{p+1}j_{p+1}}\cdots R_{i_mj_m}F_{\bar{J}}.$$

Let the numbers of the subscripts 2 and 3 be, respectively, q and r $(q+r=M)$. Then, $F_{\bar{I}}$ is represented by

$$F_{\bar{I}} = F_{\underbrace{2,\ldots,2}_{q},\underbrace{3,\ldots,3}_{r}},$$

which will be denoted by $F^{q,r}$. Then, $F^{M,0}$ $(= F_{\underbrace{2,\ldots,2}_{M}})$ is given as follows:

$$
\begin{aligned}
F^{M,0} &= \sum_{(2,3)} R_{2j_{p+1}}\cdots R_{2j_m}F_{\bar{J}} \\
&= \sum_{r=0}^{M} \binom{M}{r} \cos^{M-r}\theta \sin^r\theta F^{M-r,r}.
\end{aligned} \tag{B.8}
$$

On the basis of this expression, we will discuss the form of $F^{M-r,r}$ $(r = 0,\ldots,M)$:

(i) M is odd:

The $M+2$ functions $\{1, \cos^{M-r}\theta\sin^r\theta\ (r=0,\ldots,M)\}$ are linearly independent.[4] Thus, the coefficients of these functions in Eq. (B.8) should be zero.

---

[3] In the Cartesian-tensor notation, a double indices (e.g., a_ib_i) means the summation without $\sum$ sign (the summation convention). When the values of the set of double indices taken in the summation are limited, the $\sum$ sign is shown, for example,

$$\sum_{(2,3)} a_ib_i = a_2b_2 + a_3b_3, \quad \sum_{(2,3)} a_{ij}b_{ij} = a_{22}b_{22} + a_{23}b_{23} + a_{32}b_{32} + a_{33}b_{33}.$$

[4] Consider the identity:

$$a_0 + a_1\cos^M\theta + a_2\cos^{M-1}\theta\sin\theta + \cdots + a_M\cos\theta\sin^{M-1}\theta + a_{M+1}\sin^M\theta \equiv 0.$$

Taking $\theta = 0$ and $\theta = \pi$, we have, respectively, $a_0 + a_1 = 0$ and $a_0 - a_1 = 0$, since M is odd. Thus, $a_0 = a_1 = 0$. Therefore,

$$a_2\cos^{M-1}\theta\sin\theta + a_3\cos^{M-2}\theta\sin^2\theta + \cdots + a_M\cos\theta\sin^{M-1}\theta + a_{M+1}\sin^M\theta \equiv 0.$$

Division of this expression by $\sin\theta$ gives

$$a_2\cos^{M-1}\theta + a_3\cos^{M-2}\theta\sin\theta + \cdots + a_M\cos\theta\sin^{M-2}\theta + a_{M+1}\sin^{M-1}\theta \equiv 0.$$

Taking $\theta = 0$, we have $a_2 = 0$. Repeating a similar process, we have $a_3 = a_4 = \cdots = a_{M+1} = 0$.

That is,

$$F^{M-r,r} = 0 \quad (r = 0, \ldots, M), \tag{B.9}$$

or

$$F_{\bar{I}} = 0.$$

(ii) M is even:
Let $M = 2\bar{M}$. Then,

$$F^{2\bar{M},0} = \sum_{r=0}^{2\bar{M}} \binom{2\bar{M}}{r} \cos^{2\bar{M}-r}\theta \sin^r \theta F^{2\bar{M}-r,r}.$$

On the other hand,

$$\begin{aligned} F^{2\bar{M},0} &= (\cos^2\theta + \sin^2\theta)^{\bar{M}} F^{2\bar{M},0} \\ &= \sum_{r=0}^{\bar{M}} \binom{\bar{M}}{r} \cos^{2\bar{M}-2r}\theta \sin^{2r}\theta F^{2\bar{M},0}. \end{aligned}$$

Comparing the above two expressions of the $F^{2\bar{M},0}$ and noting that the $2\bar{M}+1$ functions $\{\cos^{2\bar{M}-r}\theta \sin^r\theta \ (r = 0, \ldots, 2\bar{M})\}$ are linearly independent,[5] we find that $F^{2\bar{M},r}$ $(r = 1, \ldots, 2\bar{M})$ is determined by $F^{2\bar{M},0}$ as follows:

$$F^{2\bar{M}-2r+1,2r-1} = 0 \quad (r = 1, \ldots, \bar{M}), \tag{B.10a}$$

$$F^{2\bar{M}-2r,2r} = \frac{\binom{\bar{M}}{r}}{\binom{2\bar{M}}{2r}} F^{2\bar{M},0} \quad (r = 0, \ldots, \bar{M}). \tag{B.10b}$$

Now we find that $T_I(\zeta, 0, 0)$ is determined by $[m/2]+1$ functions $T_{1,\ldots,1,\underbrace{2,\ldots,2}_{k}}$ of ζ, where $[a]$, Gauss's symbol, is the maximum integer that does not exceed a and $k = 0, 2, 4, \ldots, 2 \times [m/2]$.

B.3 $T_I(\zeta_i)$ at an arbitrary point

B.3.1 Preparation

Consider a sequence $(i_1, \ldots, i_{2S})$ of $2S$ indices. Let us compose a set of S pairs of indices [say $C_k(S)$], for example,

$$C_1(S) : \{\underbrace{(i_1, i_2), (i_3, i_4), \ldots, (i_{2S-3}, i_{2S-2}), (i_{2S-1}, i_{2S})}_{S}\}.$$

[5]Consider the identity:

$$a_1 \cos^M\theta + a_2 \cos^{M-1}\theta \sin\theta + \cdots + a_M \cos\theta \sin^{M-1}\theta + a_{M+1}\sin^M\theta \equiv 0.$$

Taking $\theta = 0$, we have $a_1 = 0$. The following process is similar to that of the latter part of Footnote 4.

Here, no distinction is made between the pairs (i, j) and (j, i), and the order of the pairs in the bracket $\{\cdots\}$ is indifferent. Let IP be the ensemble consisting of $C_k(S)$, and Q_S be the number of the elements of the ensemble IP. That is,

$$IP = \{C_k(S) \quad (k = 1, 2, \ldots, Q_S)\}. \tag{B.11}$$

Then, Q_S is given by

$$\begin{aligned} Q_S &= \frac{\binom{2S}{2}\binom{2S-2}{2}\cdots\binom{4}{2}\binom{2}{2}}{S!} \\ &= \frac{(2S)!}{2^S S!}. \end{aligned} \tag{B.12}$$

Let the expression $(\delta_{i_t i_u} - l_{i_t 1} l_{i_u 1})$ correspond to each pair (i_t, i_u), and $p_k(S)$ be defined by the product of the S expressions corresponding to the S pairs in $C_k(S)$. That is,

$$\begin{aligned} p_k(S) &= \prod_{\text{all the pairs } (i_t, i_u) \text{ in } C_k(S)} (\delta_{i_t i_u} - l_{i_t 1} l_{i_u 1}) \\ &= \prod_{\text{all the pairs } (i_t, i_u) \text{ in } C_k(S)} (l_{i_t 2} l_{i_u 2} + l_{i_t 3} l_{i_u 3}). \end{aligned} \tag{B.13}$$

For example,

$$\begin{aligned} p_1 &= (\delta_{i_1 i_2} - l_{i_1 1} l_{i_2 1})(l_{i_3 i_4} - \delta_{i_3 1}\delta_{i_4 1}) \cdots (\delta_{i_{2S-1} i_{2S}} - l_{i_{2S-1} 1} l_{i_{2S} 1}) \\ &= (l_{i_1 2} l_{i_2 2} + l_{i_1 3} l_{i_2 3})(l_{i_3 2} l_{i_4 2} + l_{i_3 3} l_{i_4 3}) \cdots (l_{i_{2S-1} 2} l_{i_{2S} 2} + l_{i_{2S-1} 3} l_{i_{2S} 3}). \end{aligned}$$

We define $P(S)$ as the sum of $p_k(S)$ for $k = 1, 2, \ldots, Q_S$, that is,

$$\begin{aligned} P(S) &= \sum_{k=1}^{Q_S} p_k(S) \\ &= \sum_{k=1}^{Q_S} \prod_{\text{all the pairs } (i_t, i_u) \text{ in } C_k(S)} (\delta_{i_t i_u} - l_{i_t 1} l_{i_u 1}) \\ &= \sum_{k=1}^{Q_S} \prod_{\text{all the pairs } (i_t, i_u) \text{ in } C_k(S)} (l_{i_t 2} l_{i_u 2} + l_{i_t 3} l_{i_u 3}). \end{aligned} \tag{B.14}$$

Now we will examine the expanded form of the last expression of $P(S)$. Each term of the expansion of $p_k(S)$ has the form

$$l_{i_{c_1} 2} \cdots l_{i_{c_{2S-2N}} 2} l_{i_{d_1} 3} \cdots l_{i_{d_{2N}} 3} \quad (N = 0, \ldots, S), \tag{B.15}$$

where the indices $c_1, \ldots, c_{2S-2N}, d_1, \ldots, d_{2N}$ are all different. Obviously, not all the terms of the form (B.15) appear, but the coefficients of the terms that appear in the expansion are all unity. Thus, $P(S)$, the sum of $p_k(S)$, is expressed by the linear combination of $l_{i_{c_1} 2} \cdots l_{i_{c_{2S-2N}} 2} l_{i_{d_1} 3} \cdots l_{i_{d_{2N}} 3}$, where all the possible forms

of the expression (B.15) appear. Their coefficients are no longer unity because some of the expressions appear for several $p_k(S)$, for example, $l_{i_1 2} \cdots l_{i_{2S} 2}$ is contained in all $p_k(S)$. Their coefficients are found in the following way.
(i) The case $N = 0$ or S : For $N = 0$, $l_{i_1 2} \cdots l_{i_{2S} 2}$ is the only possible form. This term appears once in each $p_k(S)$. Thus, the coefficient of this term in $P(S)$ is Q_S. A similar discussion applies to the case $N = S$.
(ii) The case $1 \leq N \leq S - 1$: Let N be given. Then, there are $\binom{2S}{2N}$ kinds of the form $l_{i_{c_1} 2} \cdots l_{i_{c_{2S-2N}} 2} l_{i_{d_1} 3} \cdots l_{i_{d_{2N}} 3}$ in $P(S)$ (the number of cases where $2N$ indices are chosen among $2S$ indices). On the other hand, there are $\binom{S}{N}$ kinds of such terms in $p_k(S)$ (the number of cases where N pairs are chosen among S pairs). Thus, the total number of such terms from all $p_k(S)$ is $Q_S\binom{S}{N}$ unless the terms of the same kind are combined into one. These terms are combined into $\binom{2S}{2N}$ kinds of terms as explained above. In view of the symmetry with respect to the indices $i_1, \ldots, i_{2S}$, therefore, the coefficient of each of such terms for a given N is $Q_S\binom{S}{N}/\binom{2S}{2N}$. This formula also covers the cases $N = 0$ and $N = S$.

Thus, $P(S)$ is expressed in the form

$$P(S) = Q_S \sum_{N=0}^{S} \frac{\binom{S}{N}}{\binom{2S}{2N}} \sum_{*1(2S,2N)} l_{i_{c_1} 2} \cdots l_{i_{c_{2S-2N}} 2} l_{i_{d_1} 3} \cdots l_{i_{d_{2N}} 3}, \tag{B.16}$$

where $\sum_{*1(2S,2N)}$ is the sum for all the cases $[\binom{2S}{2N}$ cases$]$ to separate the $2S$ indices $i_1, \ldots, i_{2S}$ into $2S - 2N$ and $2N$.

B.3.2 Derivation of the expression of $T_I(\zeta_i)$

Now we are ready to return to Eq. (B.4), i.e.,

$$T_{i_1,\ldots,i_m}(\zeta_i) = l_{i_1 j_1} \cdots l_{i_m j_m} T_{j_1,\ldots,j_m}(\zeta, 0, 0), \tag{B.17}$$

to determine the form of $T_{i_1,\ldots,i_m}(\zeta_i)$.

The terms on the right-hand side are classified by the number of 1s in $(j_1, \ldots, j_m)$. Let $\{J_M\}$ be the ensemble whose element is a sequence of indices $(j_1, \ldots, j_m)$ that contains $m - M$ of index 1. For example, $(\underbrace{1, \ldots, 1}_{m-M}, \underbrace{2, \ldots, 2}_{M})$ is one of the elements of $\{J_M\}$. Then, $\{J_M\} \cap \{J_N\} = \emptyset$ $(M \neq N)$, and the ensemble $\{J\}$ of all the possible J is the union of $\{J_M\}$ $(M = 0, \ldots, m)$, i.e., $\{J\} = \cup_{M=0}^{m} \{J_M\}$. Thus, we have

$$T_{i_1,\ldots,i_m} = \sum_{M=0}^{m} V^M_{i_1,\ldots,i_m}, \tag{B.18}$$

where

$$V^M_{i_1,\ldots,i_m} = \sum_{\{J_M\}} l_{i_1 j_1} \cdots l_{i_m j_m} T_{j_1,\ldots,j_m}(\zeta, 0, 0). \tag{B.19}$$

When M is odd, $T_{j_1,\ldots,j_m}(\zeta,0,0)=0$ from Eq. (B.9), and therefore $V^M_{i_1,\ldots,i_m}=0$. Thus, Eq. (B.18) is reduced to the sum of even M, that is,[6]

$$T_{i_1,\ldots,i_m}(\zeta_i)=\sum_{S=0}^{[m/2]} V^{2S}_{i_1,\ldots,i_m}. \tag{B.20}$$

Then $V^{2S}_{i_1,\ldots,i_m}$ (or V^{2S}_I for short) is discussed:
(i) The case $S=0$:
The set $\underbrace{(1,\ldots,1)}_{m}$ is the only element of $\{J_0\}$. Thus, from Eq. (B.19),

$$V^0_{i_1,\ldots,i_m}=l_{i_1 1}\cdots l_{i_m 1}T_{\underbrace{1,\ldots,1}_{m}}(\zeta,0,0). \tag{B.21}$$

(ii) The case $S\neq 0$:
The ensemble $\{J_{2S}\}$ is divided into $\binom{m}{2S}$ disjoint sous-ensembles (subsets) depending on which $m-2S$ indices of $j_1,\ldots,j_m$ take the value 1. Let $j_{a_1},\ldots,j_{a_{m-2S}}$ be the indices that take the value 1 and $j_{b_1},\ldots,j_{b_{2S}}$ be the ones that take the value 2 or 3 in a subset. The contribution to $l_{i_1 j_1}\cdots l_{i_m j_m}T_{j_1,\ldots,j_m}(\zeta,0,0)$ from the subset is given by

$$\begin{aligned}&\sum_{(2,3)} l_{i_{a_1}1}\cdots l_{i_{a_{m-2S}}1}l_{i_{b_1}j_{b_1}}\cdots l_{i_{b_{2S}}j_{b_{2S}}}T_{\underbrace{1,\ldots,1}_{m-2S},j_{b_1},\ldots,j_{b_{2S}}}(\zeta,0,0)\\ &=l_{i_{a_1}1}\cdots l_{i_{a_{m-2S}}1}\sum_{(2,3)} l_{i_{b_1}j_{b_1}}\cdots l_{i_{b_{2S}}j_{b_{2S}}}T_{\underbrace{1,\ldots,1}_{m-2S},j_{b_1},\ldots,j_{b_{2S}}}(\zeta,0,0),\end{aligned} \tag{B.22}$$

where the symmetry of T_J with respect to indices J is used. The sum on the right-hand side of Eq. (B.19), which is the sum over $\{J_M\}$, is expressed as the

[6]In this decomposition, each term, or $V^{2S}_{i_1,\ldots,i_m}$ is also symmetric with respect to the indices $i_1,\ldots,i_m$. In fact,

$$\begin{aligned}V^{2S}_{i_1,\ldots,i_q,\ldots,i_p,\ldots,i_m}&=\sum_{\substack{\text{all elements}\\ \text{in }\{J_{2S}\}}} l_{i_1 j_1}\cdots l_{i_q j_p}\cdots l_{i_p j_q}\cdots l_{i_m j_m}T_{j_1,\ldots,j_p,\ldots,j_q,\ldots,j_m}\\ &=\sum_{\substack{\text{all elements}\\ \text{in }\{J_{2S}\}}} l_{i_1 j_1}\cdots l_{i_p j_q}\cdots l_{i_q j_p}\cdots l_{i_m j_m}T_{j_1,\ldots,j_q,\ldots,j_p,\ldots,j_m}\\ &=\sum_{\substack{\text{all elements}\\ \text{in }\{\tilde{J}_{2S}\}}} l_{i_1 j_1}\cdots l_{i_p j_q}\cdots l_{i_q j_p}\cdots l_{i_m j_m}T_{j_1,\ldots,j_q,\ldots,j_p,\ldots,j_m}\\ &=\sum_{\substack{\text{all elements}\\ \text{in }\{J_{2S}\}}} l_{i_1 j_1}\cdots l_{i_p j_p}\cdots l_{i_q j_q}\cdots l_{i_m j_m}T_{j_1,\ldots,j_p,\ldots,j_q,\ldots,j_m}\\ &=V^{2S}_{i_1,\ldots,i_p,\ldots,i_q,\ldots,i_m},\end{aligned}$$

where $\{\tilde{J}_{2S}\}$ is the ensemble consisting of the elements of $\{J_{2S}\}$ where the pth element and the qth are interchanged. Thus, $\{\tilde{J}_{2S}\}=\{J_{2S}\}$. This fact is used in the transition from the 2nd line to the 3rd. The pth element and the qth in the ensemble $\{\tilde{J}_{2S}\}$ being interchanged, the transition from the 3rd line to the 4th is obtained.

sum of the expression (B.22):

$$V_I^{2S} = \sum_{*2(m,2S)} l_{i_{a_1}1} \cdots l_{i_{a_{m-2S}}1} \sum_{(2,3)} l_{i_{b_1}j_{b_1}} \cdots l_{i_{b_{2S}}j_{b_{2S}}} T_{\underbrace{1,\ldots,1}_{m-2S},j_{b_1},\ldots,j_{b_{2S}}}(\zeta,0,0), \tag{B.23}$$

where $\sum\limits_{*2(m,2S)}$ is the summation over the $\binom{m}{2S}$ cases to choose $m-2S$ indices $j_{a_1},\ldots,j_{a_{m-2S}}$ taking 1.

The sum $\sum_{(2,3)} l_{i_{b_1}j_{b_1}} \cdots l_{i_{b_{2S}}j_{b_{2S}}} T_{\underbrace{1,\ldots,1}_{m-2S},j_{b_1},\ldots,j_{b_{2S}}}(\zeta,0,0)$ in Eq. (B.23) is transformed. For this purpose, we consider it in the simplified indices $(i_1,j_1),\ldots,(i_{2S},j_{2S})$ instead of $(i_{b_1},j_{b_1}),\ldots,(i_{b_{2S}},j_{b_{2S}})$ and use the notation $F_{\bar{J}}$ and $F^{q,r}$ introduced in Appendix B.2. That is,

$$\begin{aligned}
&\sum_{(2,3)} l_{i_1j_1} \cdots l_{i_{2S}j_{2S}} T_{\underbrace{1,\ldots,1}_{m-2S},j_1,\ldots,j_{2S}}(\zeta,0,0) \\
&= \sum_{(2,3)} l_{i_1j_1} \cdots l_{i_{2S}j_{2S}} F_{j_1,\ldots,j_{2S}} \\
&= \sum_{N=0}^{S} \sum_{*1(2S,2N)} l_{i_{c_1}2} \cdots l_{i_{c_{2S-2N}}2} l_{i_{d_1}3} \cdots l_{i_{d_{2N}}3} F^{2S-2N,2N} \\
&= F^{2S,0} \sum_{N=0}^{S} \frac{\binom{S}{N}}{\binom{2S}{2N}} \sum_{*1(2S,2N)} l_{i_{c_1}2} \cdots l_{i_{c_{2S-2N}}2} l_{i_{d_1}3} \cdots l_{i_{d_{2N}}3} \\
&= \frac{1}{Q_S} F^{2S,0} P(S) \\
&= \frac{1}{Q_S} F^{2S,0} \left(\sum_{k=1}^{Q_S} \prod_{\text{all the pairs } (i_t,i_u) \text{ in } C_k(S)} (\delta_{i_t i_u} - l_{i_t1} l_{i_u1}) \right).
\end{aligned} \tag{B.24}$$

In the above transformation, some notes are required: $F_{j_1,\ldots,j_{2S}}$ vanishes when the number of index 3 is odd; $\sum\limits_{*1(2S,2N)}$ is defined below Eq. (B.16); $F^{2S-2N,2N}$ is expressed by $F^{2S,0}$ [see Eq. (B.10b)]; the relations (B.16) and (B.14) are used.

Now return to the expression $\sum\limits_{(2,3)} l_{i_{b_1}j_{b_1}} \cdots l_{i_{b_{2S}}j_{b_{2S}}} T_{\underbrace{1,\ldots,1}_{m-2S},j_{b_1},\ldots,j_{b_{2S}}}(\zeta,0,0)$ of our original interest. In the discussion of the preceding paragraph, the sequence $1,\ldots,2S$ of the indices is used instead of the sequence $b_1,\ldots,b_{2S}$. Thus, the corresponding formula for $\sum\limits_{(2,3)} l_{i_{b_1}j_{b_1}} \cdots l_{i_{b_{2S}}j_{b_{2S}}} T_{\underbrace{1,\ldots,1}_{m-2S},j_{b_1},\ldots,j_{b_{2S}}}(\zeta,0,0)$ is obtained from Eq. (B.24) by replacing the sequence $1,\ldots,2S$ by $b_1,\ldots,b_{2S}$, that is, by replacing C_k by the corresponding set (say C_{bk}) of S pairs made from the sequence

$b_1, \ldots, b_{2S}$. Then, from Eq. (B.23),

$$V_I^{2S} = \frac{F^{2S,0}}{Q_S} \sum_{*2(m,2S)} l_{i_{a_1}1} \cdots l_{i_{a_{m-2S}}1} \left(\sum_{k=1}^{Q_S} \prod_{\substack{\text{all the pairs } (i_t, i_u) \\ \text{in } C_{bk}(S)}} (\delta_{i_t i_u} - l_{i_t 1} l_{i_u 1}) \right). \tag{B.25}$$

This expression can easily be expressed in terms of ζ_i. That is, introducing the following $[m/2]+1$ functions of ζ:

$$f_0(\zeta) = T_{\underbrace{1,\ldots,1}_{m}}(\zeta, 0, 0),$$

$$f_S(\zeta) = \frac{F^{2S,0}}{Q_S} \quad (S = 1, \ldots, [m/2]),$$

and noting the relation (B.3):

$$l_{i1} = \zeta_i / \zeta,$$

we find that Eqs. (B.21) and (B.25) are rewritten as follows:

$$V_I^0 = \zeta^{-m} \zeta_{i_1} \cdots \zeta_{i_m} f_0(\zeta), \tag{B.26a}$$

$$V_I^{2S} = f_S(\zeta) \sum_{*1(m,2S)} \zeta^{2S-m} \zeta_{i_{a_1}} \cdots \zeta_{i_{a_{m-2S}}} \left(\sum_{k=1}^{Q_S} \prod_{\substack{\text{all the} \\ \text{pairs } (i_t, i_u) \\ \text{in } C_{bk}(S)}} (\delta_{i_t i_u} - \zeta^{-2} \zeta_{i_t} \zeta_{i_u}) \right). \tag{B.26b}$$

The expression (B.26b) for V_I^{2S} ($S = 1, \ldots, [m/2]$) is, then, expanded to arrange in powers of ζ_i. Each term of the expansion is given, except for a constant factor, in the form

$$f_S(\zeta)(-1)^{S-N} \zeta^{-m+2N} \underbrace{\zeta_{i_a} \cdots \zeta_{i_b}}_{m-2N} \underbrace{\delta_{i_c i_d} \cdots \delta_{i_e i_f}}_{N} \quad (N = 0, \ldots, S), \tag{B.27}$$

where $a, \ldots, f$ are all different m numbers taking $1, \ldots, m$. Because of symmetry of V_I^{2S} with respect to I, V_I^{2S} is arranged in the following form:

$$V_I^{2S} = f_S(\zeta) \sum_{N=0}^{S} (-1)^{S-N} A_S^N \zeta^{-m+2N} \sum_{*(m,2N)} \underbrace{\zeta_{i_a} \cdots \zeta_{i_b}}_{m-2N} \underbrace{\delta_{i_c i_d} \cdots \delta_{i_e i_f}}_{N}, \tag{B.28}$$

where A_S^N is a constant[7] and $\sum_{*(m,2N)}$ indicates the sum over the $\binom{m}{2N}(2N)!/2^N N!$ cases that is obtained by the following two steps: the $\binom{m}{2N}$ cases to divide the

[7] The constant A_S^N is obtained as follows. The term (B.27) with N and $\delta_{i_c i_d} \cdots \delta_{i_e i_f}$ being

m indices $i_1, \ldots, i_m$ into two groups: $m - 2N$ indices $i_a, \ldots, i_b$ and $2N$ indices $i_c, \ldots, i_f$ and the $(2N)!/2^N N!$ cases to make N pairs $(i_c, i_d), \ldots, (i_e, i_f)$. With the following notation:

$$H_I^N = \sum_{*(m,2N)} \underbrace{\zeta_{i_a} \cdots \zeta_{i_b}}_{m-2N} \underbrace{\delta_{i_c i_d} \cdots \delta_{i_e i_f}}_{N} \quad (N = 0, \ldots, [m/2]), \tag{B.29a}$$

$$B_S^N = (-1)^{S-N} A_S^N \quad (S = 1, \ldots, [m/2]), \quad B_0^0 = 1, \tag{B.29b}$$

$$f_S = f_S(\zeta), \tag{B.29c}$$

the function V_I^{2S} given in Eqs. (B.26a) and (B.28) is expressed in the following form:

$$V_I^0 = B_0^0 \zeta^{-m} H_I^0 f_0,$$

$$\cdots\cdots\cdots,$$

$$V_I^{2S} = (B_S^0 \zeta^{-m} H_I^0 + \cdots + B_S^N \zeta^{-m+2N} H_I^N + \cdots + B_S^S \zeta^{-m+2S} H_I^S) f_S,$$

$$\cdots\cdots\cdots,$$

$$V_I^{2[m/2]} = (B_{[m/2]}^0 \zeta^{-m} H_I^0 + \cdots + B_{[m/2]}^N \zeta^{-m+2N} H_I^N + \cdots + B_{[m/2]}^{[m/2]} \zeta^{-m+2[m/2]} H_I^{[m/2]}) f_{[m/2]}.$$

With these expressions for V_I^{2S} in Eq. (B.20), we obtain the expression for $T_I(\zeta_i)$ in the form

$$T_I(\zeta_i) = \sum_{N=0}^{[m/2]} g_N(\zeta) H_I^N, \tag{B.30}$$

where

$$g_{[m/2]}(\zeta) = B_{[m/2]}^{[m/2]} f_{[m/2]} \zeta^{-m+2[m/2]},$$

$$\cdots\cdots\cdots,$$

$$g_N(\zeta) = (B_N^N f_N + B_{N+1}^N f_{N+1} + \cdots + B_{[m/2]}^N f_{[m/2]}) \zeta^{-m+2N},$$

$$\cdots\cdots\cdots,$$

$$g_0(\zeta) = (B_0^0 f_0 + B_1^0 f_1 + \cdots + B_{[m/2]}^0 f_{[m/2]}) \zeta^{-m}.$$

---

given comes once from and only from the term

$$f_S(\zeta)\zeta^{2S-m}\zeta_{i_{a_1}} \cdots \zeta_{i_{a_{m-2S}}} \prod_{\substack{\text{all the pairs } (i_t, i_u) \\ \text{in } C_{bk}}} (\delta_{i_t i_u} - \zeta^{-2}\zeta_{i_t}\zeta_{i_u})$$

in Eq. (B.26b) where C_{bk} contains the N pairs $(i_c, i_d), \ldots, (i_e, i_f)$ that appear as the indices of the specified δ_{ij} term. The number of C_{bk} that satisfies this condition is the number of cases to choose $2S - 2N$ indices from $m - 2N$ indices obtained by excluding $2N$ indices $i_c, \ldots, i_f$ from m indices $i_1, \ldots, i_m$ and then to make $S - N$ pairs, that is, $\binom{m-2N}{2S-2N}(2S-2N)!/2^{S-N}(S-N)!$. This is the coefficient A_S^N.

The $g_N(\zeta)$ $(N = 0, \ldots, [m/2])$, which is a linear combination of $f_k\zeta^{-m+2N}$, is an arbitrary function of ζ, since f_N is an arbitrary function of ζ and $B_N^N \neq 0$. From Eqs. (B.29a) and (B.30), $T_{i_1,\ldots,i_m}(\zeta_i)$ is expressed as

$$T_{i_1,\ldots,i_m}(\zeta_i) = \sum_{N=0}^{[m/2]} g_N(\zeta) \sum_{*(m,2N)} \underbrace{\zeta_{i_a} \cdots \zeta_{i_b}}_{m-2N} \underbrace{\delta_{i_c i_d} \cdots \delta_{i_e i_f}}_{N}, \tag{B.31}$$

where $g_N(\zeta)$ is an arbitrary function of ζ. It is noted that Eq. (B.31) is derived with the aid of the orthogonal transformation (l_{ij}) with det $(l_{ij}) = 1$. Inversely, $T_I(\zeta_i)$ given by Eq. (B.31) satisfies the relation (B.1). In fact, each term of $T_I(\zeta_i)$, for example,

$$D_I^a(\zeta_i) = \zeta_{i_1} \cdots \zeta_{i_a} \delta_{i_{a+1} i_{a+2}} \cdots \delta_{i_{m-1} i_m} G(\zeta)$$

is easily shown to satisfy the relation (B.1) as follows:

$$\begin{aligned} D_I^a(l_{kh}\zeta_h) &= l_{i_1 j_1}\zeta_{j_1} \cdots l_{i_a j_a}\zeta_{j_a} \delta_{i_{a+1} i_{a+2}} \cdots \delta_{i_{m-1} i_m} G(\zeta) \\ &= l_{i_1 j_1} \cdots l_{i_a j_a} \zeta_{j_1} \cdots \zeta_{j_a} \delta_{i_{a+1} i_{a+2}} \cdots \delta_{i_{m-1} i_m} G(\zeta) \\ &= l_{i_1 j_1} \cdots l_{i_a j_a} l_{i_{a+1} j_{a+1}} \cdots l_{i_m j_m} \zeta_{j_1} \cdots \zeta_{j_a} \delta_{j_{a+1} j_{a+2}} \cdots \delta_{j_{m-1} j_m} G(\zeta) \\ &= l_{i_1 j_1} \cdots l_{i_m j_m} D_J^a(\zeta_k), \end{aligned}$$

where the following relation is used in the above process from the second line to the third:

$$\delta_{i_c i_{c+1}} = l_{i_c k} l_{i_{c+1} k} = l_{i_c j_c} l_{i_{c+1} j_{c+1}} \delta_{j_c j_{c+1}}.$$

It is noted that the relation holds for both transformations with det $(l_{ij}) = 1$ and det $(l_{ij}) = -1$.

B.3.3 Summary

To summarize the foregoing analysis, the spherically symmetric field of symmetric tensor $T_{i_1,\ldots,i_m}(\zeta_i)$, i.e.,

$$\begin{gathered} T_{i_1,\ldots,i_m}(l_{kh}\zeta_h) = l_{i_1 j_1} \cdots l_{i_m j_m} T_{j_1,\ldots,j_m}(\zeta_k), \\ l_{ik} l_{jk} = \delta_{ij}, \end{gathered} \tag{B.32}$$

is expressed in the form

$$T_{i_1,\ldots,i_m}(\zeta_i) = \sum_{N=0}^{[m/2]} g_N(\zeta) \sum_{*(m,2N)} \underbrace{\zeta_{i_a} \cdots \zeta_{i_b}}_{m-2N} \underbrace{\delta_{i_c i_d} \cdots \delta_{i_e i_f}}_{N}, \tag{B.33}$$

where $g_N(\zeta)$ is an arbitrary function of ζ and $\sum\limits_{*(m,2N)}$ is defined just below Eq. (B.28). For some small values of m, T_I is expressed as follows:

$$
\begin{aligned}
m=0: &\quad T(\zeta_h) = g_0(\zeta),\\
m=1: &\quad T_i(\zeta_h) = \zeta_i g_0(\zeta),\\
m=2: &\quad T_{i,j}(\zeta_h) = \zeta_i\zeta_j g_0(\zeta) + \delta_{ij} g_1(\zeta),\\
m=3: &\quad T_{i,j,k}(\zeta_h) = \zeta_i\zeta_j\zeta_k g_0(\zeta) + (\zeta_i\delta_{jk} + \zeta_j\delta_{ki} + \zeta_k\delta_{ij}) g_1(\zeta),\\
m=4: &\quad T_{i,j,k,l}(\zeta_h) = \zeta_i\zeta_j\zeta_k\zeta_l g_0(\zeta)\\
&\qquad + (\zeta_i\zeta_j\delta_{kl} + \zeta_i\zeta_k\delta_{jl} + \zeta_i\zeta_l\delta_{jk} + \zeta_j\zeta_k\delta_{il} + \zeta_j\zeta_l\delta_{ik} + \zeta_k\zeta_l\delta_{ij}) g_1(\zeta)\\
&\qquad + (\delta_{ij}\delta_{kl} + \delta_{ik}\delta_{jl} + \delta_{il}\delta_{jk}) g_2(\zeta).
\end{aligned} \tag{B.34}
$$

B.4 Applications

B.4.1 Definite integral $\iiint\limits_{-\infty}^{\infty} \zeta_{i_1}\cdots\zeta_{i_{2S}} \exp(-\zeta^2)\mathrm{d}\zeta_1\mathrm{d}\zeta_2\mathrm{d}\zeta_3$

The definite integral $I_{i_1,\dots,i_{2S}} = \iiint\limits_{-\infty}^{\infty} \zeta_{i_1}\cdots\zeta_{i_{2S}} \exp(-\zeta^2)\mathrm{d}\zeta_1\mathrm{d}\zeta_2\mathrm{d}\zeta_3$ is often encountered in analyses of kinetic theory. This integral is a constant tensor that satisfies the relation (B.1). Thus, it is expressed in the form

$$
I_{i_1,\dots,i_{2S}} = c_0 \sum_{k=1}^{Q_S} \prod_{\text{all the pairs } (i_t,i_u) \text{ in } C_k(S)} \delta_{i_t i_u}, \tag{B.35}
$$

where c_0 is an undetermined constant and the definitions of $C_k(S)$ and Q_S are given at the beginning of Appendix B.3.1. Taking $i_1 = \cdots = i_{2S} = 1$, we have, by its direct integration,

$$
\underbrace{I_{1,\dots,1}}_{2S} = \frac{(2S-1)!!\pi^{3/2}}{2^S},
$$

where

$$
(2S-1)!! = 1\cdot 3\cdots(2S-1), \quad (-1)!! = 1.
$$

On the other hand, from Eq. (B.35),

$$
\underbrace{I_{1,\dots,1}}_{2S} = c_0 Q_S.
$$

Thus, we have

$$
c_0 = \frac{(2S-1)!!\pi^{3/2}}{2^S Q_S} = \frac{\pi^{3/2}}{2^S}.
$$

That is,

$$\iiint_{-\infty}^{\infty} \zeta_{i_1} \cdots \zeta_{i_{2S}} \exp(-\zeta^2) \mathrm{d}\zeta_1 \mathrm{d}\zeta_2 \mathrm{d}\zeta_3 = \frac{\pi^{3/2}}{2^S} \sum_{k=1}^{Q_S} \prod_{\text{all the pairs } (i_t, i_u) \text{ in } C_k(S)} \delta_{i_t i_u}. \tag{B.36}$$

When the exponential function $\exp(-\zeta^2)$ is replaced by a function of ζ [say $g(\zeta)$] in the definite integral $\iiint_{-\infty}^{\infty} \zeta_{i_1} \cdots \zeta_{i_{2S}} \exp(-\zeta^2) \mathrm{d}\zeta_1 \mathrm{d}\zeta_2 \mathrm{d}\zeta_3$, the difference of the result is only the constant c_0, which is calculated in a similar way to the exponential function case, that is,

$$\iiint_{-\infty}^{\infty} \zeta_{i_1} \cdots \zeta_{i_{2S}} g(\zeta) \mathrm{d}\zeta_1 \mathrm{d}\zeta_2 \mathrm{d}\zeta_3 = \tilde{c}_0 \sum_{k=1}^{Q_S} \prod_{\text{all the pairs } (i_t, i_u) \text{ in } C_k(S)} \delta_{i_t i_u}, \tag{B.37}$$

where

$$\tilde{c}_0 = \frac{2^{S+2} S! \pi}{(2S+1)!} \int_0^{\infty} \zeta^{2S+2} g(\zeta) \mathrm{d}\zeta. \tag{B.38}$$

B.4.2 Axially symmetric field

Consider a tensor field T_I that is axially symmetric, that is, a tensor field that satisfies the relation (B.1) where the transformation l_{ij} is limited to the one that keeps an oriented straight line through the origin invariant, or l_{ij} satisfies the relation $a_i = l_{ij} a_j$ for a fixed a_i in addition to Eq. (B.2). The general form of the tensor when $a_1 = a_2 = 0$ and $a_3 = 1$, or when the transformation l_{ij} is subject to the restriction

$$l_{33} = 1, \quad l_{13} = l_{23} = l_{31} = l_{32} = 0, \tag{B.39}$$

is discussed here.

For a given ζ_i, the remaining l_{ij} is chosen in such a way that

$$\zeta_1 = \zeta_\perp l_{11}, \quad \zeta_2 = \zeta_\perp l_{21}, \quad \zeta_\perp = (\zeta_1^2 + \zeta_2^2)^{1/2}.$$

Then, T_I at an arbitrary point ζ_i is determined by the data of T_I on the half-plane $(\zeta_1 \geq 0,\ \zeta_2 = 0)$ as follows:

$$T_{\underbrace{i_1,\ldots,i_h}_{h},\underbrace{3,\ldots,3}_{m-h}}(\zeta_i) = l_{i_1 j_1} \cdots l_{i_h j_h} T_{\underbrace{j_1,\ldots,j_h}_{h},\underbrace{3,\ldots,3}_{m-h}}(\zeta_\perp, 0, \zeta_3) \quad (h = 0, \ldots, m), \tag{B.40}$$

where any of the indices $i_1, \ldots, i_h$ does not take the value 3. Then, the contribution from the terms with some of the indices $j_1, \ldots, j_h$ on the right-hand side of Eq. (B.40) being 3 vanishes because of Eq. (B.39). For the transformation l_{ij} where $l_{11} = 1$ and $l_{22} = -1$ [$l_{12} = l_{21} = 0$ because of Eq. (B.2)],

$$T_{\underbrace{1,\ldots,1}_{h-r},\underbrace{2,\ldots,2}_{r},\underbrace{3,\ldots,3}_{m-h}}(\zeta_\perp, 0, \zeta_3) = (-1)^r T_{\underbrace{1,\ldots,1}_{h-r},\underbrace{2,\ldots,2}_{r},\underbrace{3,\ldots,3}_{m-h}}(\zeta_\perp, 0, \zeta_3).$$

Thus, $T_I(\zeta_\perp, 0, \zeta_3)$ with an odd number of index 2 vanishes.

With the above preparation, in the formula (B.40), giving T_I at an arbitrary point ζ_i, the factor $l_{i_1 j_1} \cdots l_{i_h j_h}$ can be arranged in such a way that the second indices with 2 of l_{ij} are paired (e.g., $l_{i_t 2} l_{i_u 2}$) and the expression is symmetrized. The paired expression $l_{i_t 2} l_{i_u 2}$ is rewritten as $l_{i_t 2} l_{i_u 2} = \delta_{i_t i_u} - l_{i_t 1} l_{i_u 1} = \delta_{i_t i_u} - \zeta_{i_t} \zeta_{i_u} \zeta_\perp^{-2}$. Thus, we find, by a similar discussion to the spherically symmetric case, that the tensor $T_{\underbrace{i_1, \ldots, i_h}_{h}, \underbrace{3, \ldots, 3}_{m-h}}(\zeta_i)$, where any of the indices $i_1, \ldots, i_h$ does not take the value 3, is given by

$$T_{\underbrace{i_1, \ldots, i_h}_{h}, \underbrace{3, \ldots, 3}_{m-h}}(\zeta_i) = \sum_{N=0}^{[h/2]} g_N(\zeta, \zeta_3) \sum_{*(h,2N)} \underbrace{\zeta_{i_a} \cdots \zeta_{i_b}}_{h-2N} \underbrace{\delta_{i_c i_d} \cdots \delta_{i_e i_f}}_{N}, \tag{B.41}$$

where $g_N(\zeta, \zeta_3)$ is an arbitrary function of ζ and ζ_3 (or $\zeta_\perp$ and ζ_3) and the definition of $\sum_{*(h,2N)}$ just below Eq. (B.28) is applied here with h for m there.

In the derivation of Eq. (B.41), the transformation l_{ij} with det $(l_{ij}) = -1$ is required, since that with $l_{11} = 1$, $l_{22} = -1$, and $l_{33} = 1$ is introduced.

Appendix C

Kinetic-Equation Approach to Fluid-Dynamic Equations

There is a trend to make use of kinetic-type equations in obtaining numerical solution of the Euler or Navier–Stokes set of equations or more generally of a set of partial differential equations of conservation form. In this appendix, a method constructing a kinetic equation replacing the fluid-dynamic-type equation exactly is explained. On the basis of this method, the validity of the numerical kinetic-equation system of the time-discrete free-molecular type is discussed. This approach can be applied more generally to giving a quick practical estimate of the validity of various heuristic kinetic systems. The discussion in this appendix will also help to avoid the confusion of the kinetic method for fluid-dynamic equations with the numerical computation of the Boltzmann system for small Knudsen numbers.

C.1 Introduction

There is a recent trend to make use of kinetic-type equations in obtaining numerical solutions of the Euler or Navier–Stokes set of equations (e.g., Pullin [1980], Deshpande [1986], Perthame [1990, 1992], Inamuro, Yoshino & Ogino [1997], Vasseur [1999], Xu [2000]; see Junk [1997] for detailed references). The way of computation is different among the authors, but they noticed the simple linear form of the differential term of the kinetic equation and tried to avoid some difficulties in numerical analysis of the Euler or Navier–Stokes set (e.g., instability problems in shock wave computations, the pressure term in the incompressible Navier–Stokes set, preference of a quick computation of complex problems to a detailed precise one). They made considerable success in mathematical theories and applications. Some of the approaches are given rigorous mathematical proof of their validity. Various solutions are successfully obtained by some heuristic schemes, but their theoretical backgrounds are not so obvious.

In this appendix, we explain a simple way to construct a system of a kinetic equation and its initial condition from a set of partial differential equations of a conservation form (e.g., Euler and Navier–Stokes sets for a compressible fluid) in such a way that some moments of the solution of the kinetic system satisfy the set of partial differential equations exactly. On the basis of this kinetic system, we discuss the numerical system of the time-discrete free-molecular type, which will be explained in Appendix C.3, as an example. This type of discussion has wide application to construction of efficient schemes and a quick evaluation of their validity in a practical sense. This appendix is based on Sone [2000c]. From the content of this appendix, the notation used here is independent of that in the foregoing chapters.

C.2 Exact kinetic-equation approach

Let x_i, ξ_i, η, and t $(i = 1, 2, \ldots, N)$ be independent variables, and $f(x_i, \xi_i, \eta, t)$ be a dependent variable (or the velocity distribution function). The (macroscopic) variables ρ_r $(r = 0, 1, \ldots, N+1)$ are defined as

$$\rho_r = \int \gamma_r f \mathrm{d}\xi_1 \mathrm{d}\xi_2 \cdots \mathrm{d}\xi_N \mathrm{d}\eta, \tag{C.1}$$

where γ_r $(r = 0, 1, \ldots, N+1)$ are functions of ξ_k or (ξ_j^2, η^2) chosen appropriately depending on the specific situation:

$$\gamma_0 = \gamma_0(\xi_k),\ \gamma_i = \gamma_i(\xi_k),\ \gamma_{N+1} = \gamma_{N+1}(\xi_j^2, \eta^2) \quad (i = 1, 2, \ldots, N), \tag{C.2}$$

and the flux H_i^r of the macroscopic variables is defined as

$$H_i^r = \int \xi_i \gamma_r f \mathrm{d}\xi_1 \mathrm{d}\xi_2 \cdots \mathrm{d}\xi_N \mathrm{d}\eta. \tag{C.3}$$

Take a velocity distribution function f_c of the form

$$f(x_i, \xi_i, \eta, t) = f_c(\rho_r, \nabla\rho_r, \xi_i, \eta), \tag{C.4}$$

where the variables t and x_i enter only through the macroscopic variables ρ_r or their derivatives $\nabla\rho_r$ with ∇ being the representative of a collection of the derivatives $\partial^s/\partial x_{i_1} \cdots \partial x_{i_s}$. This type of function will be called a function of Chapman–Enskog type.[1] Naturally,

$$\rho_r = \int \gamma_r f_c(\rho_r, \nabla\rho_r, \xi_i, \eta) \mathrm{d}\xi_1 \mathrm{d}\xi_2 \cdots \mathrm{d}\xi_N \mathrm{d}\eta.$$

[1] It should not be confused with the Chapman–Enskog solution of the Boltzmann equation.

The flux H_i^r corresponding to the velocity distribution function $f_c(\rho_r, \nabla\rho_r, \xi_i, \eta)$ of Chapman–Enskog type is denoted by $\hat{H}_i^r$, that is,

$$\hat{H}_i^r(\rho_r, \nabla\rho_r) = \int \xi_i \gamma_r f_c(\rho_r, \nabla\rho_r, \xi_j, \eta) \mathrm{d}\xi_1 \mathrm{d}\xi_2 \cdots \mathrm{d}\xi_N \mathrm{d}\eta. \tag{C.5}$$

Now, choosing a function $f_c(\rho_r, \nabla\rho_r, \xi_i, \eta)$ of Chapman–Enskog type, we discuss a solution of the following functional equation:[2]

$$\begin{aligned}\frac{\partial f}{\partial t} + \xi_j \frac{\partial f}{\partial x_j} = \xi_j & \left(\frac{\partial f_c(\rho_r, \nabla\rho_r, \xi_i, \eta)}{\partial \rho_s} \frac{\partial \rho_s}{\partial x_j} + \frac{\partial f_c(\rho_r, \nabla\rho_r, \xi_i, \eta)}{\partial \nabla\rho_s} \frac{\partial \nabla\rho_s}{\partial x_j} \right) \\ & - \frac{\partial f_c(\rho_r, \nabla\rho_r, \xi_i, \eta)}{\partial \rho_q} \left(\frac{\partial \hat{H}_j^q(\rho_r, \nabla\rho_r)}{\partial \rho_s} \frac{\partial \rho_s}{\partial x_j} + \frac{\partial \hat{H}_j^q(\rho_r, \nabla\rho_r)}{\partial \nabla\rho_s} \frac{\partial \nabla\rho_s}{\partial x_j} \right) \\ & - \frac{\partial f_c(\rho_r, \nabla\rho_r, \xi_i, \eta)}{\partial \nabla\rho_q} \nabla \left(\frac{\partial \hat{H}_j^q(\rho_r, \nabla\rho_r)}{\partial \rho_s} \frac{\partial \rho_s}{\partial x_j} + \frac{\partial \hat{H}_j^q(\rho_r, \nabla\rho_r)}{\partial \nabla\rho_s} \frac{\partial \nabla\rho_s}{\partial x_j} \right),\end{aligned} \tag{C.6}$$

with the initial condition

$$f(x_i, \xi_i, \eta, t) = f_c(\rho_r^0, \nabla\rho_r^0, \xi_i, \eta) \quad \text{at } t = t_0. \tag{C.7}$$

This equation has the same form of conservation equation as the equation without the right-hand side terms:

$$\frac{\partial \rho_r}{\partial t} + \frac{\partial H_j^r(\rho_r, \nabla\rho_r)}{\partial x_j} = 0. \tag{C.8}$$

This is derived from the following manipulation of the right-hand side:

$$\begin{aligned}& \int \gamma_r \xi_j \left(\frac{\partial f_c}{\partial \rho_s} \frac{\partial \rho_s}{\partial x_j} + \frac{\partial f_c}{\partial \nabla\rho_s} \frac{\partial \nabla\rho_s}{\partial x_j} \right) \mathrm{d}\xi_1 \mathrm{d}\xi_2 \cdots \mathrm{d}\xi_N \mathrm{d}\eta \\ & = \frac{\partial \hat{H}_j^r}{\partial \rho_s} \frac{\partial \rho_s}{\partial x_j} + \frac{\partial \hat{H}_j^r}{\partial \nabla\rho_s} \frac{\partial \nabla\rho_s}{\partial x_j},\end{aligned}$$

and

$$\begin{aligned}& \int \gamma_r \frac{\partial f_c}{\partial \rho_q} \left(\frac{\partial \hat{H}_j^q}{\partial \rho_s} \frac{\partial \rho_s}{\partial x_j} + \frac{\partial \hat{H}_j^q}{\partial \nabla\rho_s} \frac{\partial \nabla\rho_s}{\partial x_j} \right) \mathrm{d}\xi_1 \mathrm{d}\xi_2 \cdots \mathrm{d}\xi_N \mathrm{d}\eta \\ & = \left(\frac{\partial \hat{H}_j^q}{\partial \rho_s} \frac{\partial \rho_s}{\partial x_j} + \frac{\partial \hat{H}_j^q}{\partial \nabla\rho_s} \frac{\partial \nabla\rho_s}{\partial x_j} \right) \frac{\partial}{\partial \rho_q} \int \gamma_r f_c \mathrm{d}\xi_1 \mathrm{d}\xi_2 \cdots \mathrm{d}\xi_N \mathrm{d}\eta \\ & = \left(\frac{\partial \hat{H}_j^q}{\partial \rho_s} \frac{\partial \rho_s}{\partial x_j} + \frac{\partial \hat{H}_j^q}{\partial \nabla\rho_s} \frac{\partial \nabla\rho_s}{\partial x_j} \right) \frac{\partial \rho_r}{\partial \rho_q} \\ & = \frac{\partial \hat{H}_j^r}{\partial \rho_s} \frac{\partial \rho_s}{\partial x_j} + \frac{\partial \hat{H}_j^r}{\partial \nabla\rho_s} \frac{\partial \nabla\rho_s}{\partial x_j},\end{aligned}$$

[2]The summation convention is applied also to the superscript or subscripts q, s (q, s = $0, 1, \ldots, N+1$) of $\hat{H}_j^q$ and $\rho_{q \text{ or } s}$.

and

$$\int \gamma_r \frac{\partial f_c}{\partial \nabla \rho_q} \nabla \left(\frac{\partial \hat{H}_j^q}{\partial \rho_s} \frac{\partial \rho_s}{\partial x_j} + \frac{\partial \hat{H}_j^q}{\partial \nabla \rho_s} \frac{\partial \nabla \rho_s}{\partial x_j} \right) \mathrm{d}\xi_1 \mathrm{d}\xi_2 \cdots \mathrm{d}\xi_N \mathrm{d}\eta$$

$$= \nabla \left(\frac{\partial \hat{H}_j^q}{\partial \rho_s} \frac{\partial \rho_s}{\partial x_j} + \frac{\partial \hat{H}_j^q}{\partial \nabla \rho_s} \frac{\partial \nabla \rho_s}{\partial x_j} \right) \frac{\partial}{\partial \nabla \rho_q} \int \gamma_r f_c \mathrm{d}\xi_1 \mathrm{d}\xi_2 \cdots \mathrm{d}\xi_N \mathrm{d}\eta$$

$$= \nabla \left(\frac{\partial \hat{H}_j^q}{\partial \rho_s} \frac{\partial \rho_s}{\partial x_j} + \frac{\partial \hat{H}_j^q}{\partial \nabla \rho_s} \frac{\partial \nabla \rho_s}{\partial x_j} \right) \frac{\partial \rho_r}{\partial \nabla \rho_q}$$

$$= 0.$$

Proposition. *The solution of the initial-value problem (C.6) with (C.7) is assumed to be unique. Then the set of the macroscopic variables ρ_r calculated from the solution $f(x_i, \xi_i, \eta, t)$ is the solution of the initial-value problem of the partial differential equations*

$$\frac{\partial \rho_r}{\partial t} + \frac{\partial \hat{H}_j^r(\rho_r, \nabla \rho_r)}{\partial x_j} = 0, \tag{C.9}$$

with the initial condition

$$\rho_r = \rho_r^0 \quad \text{at } t = t_0. \tag{C.10}$$

Proof. Take a function

$$f(x_i, \xi_i, \eta, t) = f_c(\rho_r^c(x_i, t), \nabla \rho_r^c(x_i, t), \xi_i, \eta), \tag{C.11}$$

where $\rho_r^c(x_i, t)$ is the solution of the partial differential equations

$$\frac{\partial \rho_r^c}{\partial t} + \frac{\partial \hat{H}_j^r(\rho_r^c, \nabla \rho_r^c)}{\partial x_j} = 0, \tag{C.12}$$

with the initial condition

$$\rho_r^c = \rho_r^0 \quad \text{at } t = t_0. \tag{C.13}$$

In view of the above choice of f [or Eq. (C.11)],

$$\rho_r(x_i, t) = \rho_r^c(x_i, t),$$

$$\xi_j \frac{\partial f}{\partial x_j} = \xi_j \frac{\partial f_c(\rho_r^c, \nabla \rho_r^c, \xi_i, \eta)}{\partial \rho_s^c} \frac{\partial \rho_s^c}{\partial x_j} + \xi_j \frac{\partial f_c(\rho_r^c, \nabla \rho_r^c, \xi_i, \eta)}{\partial \nabla \rho_s^c} \frac{\partial \nabla \rho_s^c}{\partial x_j}$$

$$= \xi_j \left(\frac{\partial f_c(\rho_r, \nabla \rho_r, \xi_i, \eta)}{\partial \rho_s} \frac{\partial \rho_s}{\partial x_j} + \frac{\partial f_c(\rho_r, \nabla \rho_r, \xi_i, \eta)}{\partial \nabla \rho_s} \frac{\partial \nabla \rho_s}{\partial x_j} \right),$$

and

$$\begin{aligned}\frac{\partial f}{\partial t} &= \frac{\partial f_c(\rho_r^c, \nabla\rho_r^c, \xi_i, \eta)}{\partial \rho_q^c}\frac{\partial \rho_q^c}{\partial t} + \frac{\partial f_c(\rho_r^c, \nabla\rho_r^c, \xi_i, \eta)}{\partial \nabla\rho_q^c}\frac{\partial \nabla\rho_q^c}{\partial t} \\
&= -\frac{\partial f_c}{\partial \rho_q^c}\frac{\partial \hat{H}_j^q(\rho_r^c, \nabla\rho_r^c)}{\partial x_j} - \frac{\partial f_c}{\partial \nabla\rho_q^c}\nabla\left(\frac{\partial \hat{H}_j^q(\rho_r^c, \nabla\rho_r^c)}{\partial x_j}\right) \\
&= -\frac{\partial f_c}{\partial \rho_q^c}\left(\frac{\partial \hat{H}_j^q(\rho_r^c, \nabla\rho_r^c)}{\partial \rho_s^c}\frac{\partial \rho_s^c}{\partial x_j} + \frac{\partial \hat{H}_j^q(\rho_r^c, \nabla\rho_r^c)}{\partial \nabla\rho_s^c}\frac{\partial \nabla\rho_s^c}{\partial x_j}\right) \\
&\quad - \frac{\partial f_c}{\partial \nabla\rho_q^c}\nabla\left(\frac{\partial \hat{H}_j^q(\rho_r^c, \nabla\rho_r^c)}{\partial \rho_s^c}\frac{\partial \rho_s^c}{\partial x_j} + \frac{\partial \hat{H}_j^q(\rho_r^c, \nabla\rho_r^c)}{\partial \nabla\rho_s^c}\frac{\partial \nabla\rho_s^c}{\partial x_j}\right) \\
&= -\frac{\partial f_c(\rho_r, \nabla\rho_r, \xi_i, \eta)}{\partial \rho_q}\left(\frac{\partial \hat{H}_j^q(\rho_r, \nabla\rho_r)}{\partial \rho_s}\frac{\partial \rho_s}{\partial x_j} + \frac{\partial \hat{H}_j^q(\rho_r, \nabla\rho_r)}{\partial \nabla\rho_s}\frac{\partial \nabla\rho_s}{\partial x_j}\right) \\
&\quad - \frac{\partial f_c(\rho_r, \nabla\rho_r, \xi_i, \eta)}{\partial \nabla\rho_q}\nabla\left(\frac{\partial \hat{H}_j^q(\rho_r, \nabla\rho_r)}{\partial \rho_s}\frac{\partial \rho_s}{\partial x_j} + \frac{\partial \hat{H}_j^q(\rho_r, \nabla\rho_r)}{\partial \nabla\rho_s}\frac{\partial \nabla\rho_s}{\partial x_j}\right).\end{aligned}$$

Thus,

$$\begin{aligned}\frac{\partial f}{\partial t} + \xi_j\frac{\partial f}{\partial x_j} &= \xi_j\left(\frac{\partial f_c(\rho_r, \nabla\rho_r, \xi_i, \eta)}{\partial \rho_s}\frac{\partial \rho_s}{\partial x_j} + \frac{\partial f_c(\rho_r, \nabla\rho_r, \xi_i, \eta)}{\partial \nabla\rho_s}\frac{\partial \nabla\rho_s}{\partial x_j}\right) \\
&\quad - \frac{\partial f_c(\rho_r, \nabla\rho_r, \xi_i, \eta)}{\partial \rho_q}\left(\frac{\partial \hat{H}_j^q(\rho_r, \nabla\rho_r)}{\partial \rho_s}\frac{\partial \rho_s}{\partial x_j} + \frac{\partial \hat{H}_j^q(\rho_r, \nabla\rho_r)}{\partial \nabla\rho_s}\frac{\partial \nabla\rho_s}{\partial x_j}\right) \\
&\quad - \frac{\partial f_c(\rho_r, \nabla\rho_r, \xi_i, \eta)}{\partial \nabla\rho_q}\nabla\left(\frac{\partial \hat{H}_j^q(\rho_r, \nabla\rho_r)}{\partial \rho_s}\frac{\partial \rho_s}{\partial x_j} + \frac{\partial \hat{H}_j^q(\rho_r, \nabla\rho_r)}{\partial \nabla\rho_s}\frac{\partial \nabla\rho_s}{\partial x_j}\right).\end{aligned}$$

Thus the function f defined by Eq. (C.11) with the subsidiary conditions (C.12) and (C.13) is a solution of the initial-value problem (C.6) with (C.7). From the uniqueness, the set of the macroscopic variables ρ_r obtained from the solution f of the initial-value problem is the solution of the partial differential equations (C.9) with the initial condition (C.10). ■

We will give several examples.[3]

1) Let

$$\gamma_0 = 1,\ \gamma_i = \xi_i,\ \gamma_4 = \xi_j^2 \quad (i = 1, 2, 3), \tag{C.14}$$

and

$$f_c = \frac{\rho}{(2\pi RT)^{3/2}}\exp\left(-\frac{(\xi_j - u_j)^2}{2RT}\right), \tag{C.15}$$

[3] In some of the following examples where f_c is independent of η, the modification of the above procedure may be obvious. That is, it is to cross out the process concerning η.

as a velocity distribution function of Chapman–Enskog type, where the macroscopic variables ρ_r are related to the parameters ρ, u_i, and T in the Maxwellian as

$$\rho_0 = \rho, \quad \rho_i = \rho u_i, \quad \rho_4 = \rho(3RT + u_j^2), \tag{C.16}$$

and the flux $\hat{H}_i^r$ for the Maxwellian is given by

$$\hat{H}_i^0 = \rho u_i, \quad \hat{H}_i^j = \rho(u_i u_j + RT\delta_{ij}), \quad \hat{H}_i^4 = \rho u_i(5RT + u_j^2). \tag{C.17}$$

The macroscopic variables ρ_r (or ρ, u_i, and T) derived from the solution of this kinetic equation satisfy the Euler set of equations:

$$\frac{\partial \rho}{\partial t} + \frac{\partial \rho u_i}{\partial x_i} = 0, \tag{C.18a}$$

$$\frac{\partial \rho u_i}{\partial t} + \frac{\partial \rho u_i u_j}{\partial x_j} + \frac{\partial p}{\partial x_i} = 0, \tag{C.18b}$$

$$\frac{\partial \rho(3RT + u_j^2)}{\partial t} + \frac{\partial \rho u_i(3RT + u_j^2) + 2pu_i}{\partial x_i} = 0, \tag{C.18c}$$

where

$$p = R\rho T.$$

2) Let

$$\gamma_0 = 1, \ \gamma_i = \xi_i, \ \gamma_4 = \xi_j^2 \quad (i = 1, 2, 3), \tag{C.19}$$

and

$$f_c = \frac{\rho}{(2\pi RT)^{3/2}} \exp\left(-\tilde{\zeta}^2\right) \left[1 - \frac{a(T,\rho)}{(2RT)^{1/2}} \frac{\partial T}{\partial x_i} \tilde{\zeta}_i(\tilde{\zeta}^2 - \frac{5}{2}) - b(T,\rho)\left(\frac{\partial u_i}{\partial x_j} + \frac{\partial u_j}{\partial x_i}\right)\left(\tilde{\zeta}_i\tilde{\zeta}_j - \frac{\tilde{\zeta}^2}{3}\delta_{ij}\right)\right], \tag{C.20}$$

where

$$\tilde{\zeta}_i = \frac{\xi_i - u_i}{(2RT)^{1/2}}, \qquad \tilde{\zeta}^2 = \tilde{\zeta}_i^2, \tag{C.21}$$

and $a(T,\rho)$ and $b(T,\rho)$ are functions of T and ρ, which can be chosen according to our convenience. The macroscopic variables ρ_r are related to the parameters ρ, u_i, and T in the distribution function (C.20) as

$$\rho_0 = \rho, \quad \rho_i = \rho u_i, \quad \rho_4 = \rho(3RT + u_j^2), \tag{C.22}$$

and the flux $\hat{H}_i^r$ is given by

$$\hat{H}_i^0 = \rho u_i, \tag{C.23a}$$

$$\hat{H}_i^j = \rho(u_i u_j + RT\delta_{ij}) - b(T,\rho)R\rho T\left[\left(\frac{\partial u_i}{\partial x_j} + \frac{\partial u_j}{\partial x_i}\right) - \frac{2}{3}\frac{\partial u_k}{\partial x_k}\delta_{ij}\right], \tag{C.23b}$$

$$\hat{H}_i^4 = \rho u_i(5RT + u_j^2) - 2b(T,\rho)R\rho T u_j\left[\left(\frac{\partial u_i}{\partial x_j} + \frac{\partial u_j}{\partial x_i}\right) - \frac{2}{3}\frac{\partial u_k}{\partial x_k}\delta_{ij}\right]$$
$$- \frac{5a(T,\rho)R\rho T}{2}\frac{\partial T}{\partial x_i}. \tag{C.23c}$$

The macroscopic variables ρ_r (or ρ, u_i, and T) derived from the solution of this kinetic equation satisfy the following (compressible) Navier–Stokes set of equations with viscosity $b(T,\rho)R\rho T$ and thermal conductivity $5a(T,\rho)R\rho T/4$:

$$\frac{\partial \rho}{\partial t} + \frac{\partial \rho u_i}{\partial x_i} = 0, \tag{C.24a}$$

$$\frac{\partial \rho u_i}{\partial t} + \frac{\partial \rho u_i u_j}{\partial x_j} + \frac{\partial p}{\partial x_i} = \frac{\partial}{\partial x_j}\left\{b(T,\rho)R\rho T\left[\left(\frac{\partial u_i}{\partial x_j} + \frac{\partial u_j}{\partial x_i}\right) - \frac{2}{3}\frac{\partial u_k}{\partial x_k}\delta_{ij}\right]\right\}, \tag{C.24b}$$

$$\frac{\partial \rho(3RT + u_j^2)}{\partial t} + \frac{\partial \rho u_i(3RT + u_j^2) + 2pu_i}{\partial x_i}$$
$$= \frac{\partial}{\partial x_j}\left\{2b(T,\rho)R\rho T u_i\left[\left(\frac{\partial u_i}{\partial x_j} + \frac{\partial u_j}{\partial x_i}\right) - \frac{2}{3}\frac{\partial u_k}{\partial x_k}\delta_{ij}\right]\right\}$$
$$+ \frac{\partial}{\partial x_j}\left(\frac{5a(T,\rho)R\rho T}{2}\frac{\partial T}{\partial x_j}\right), \tag{C.24c}$$

where

$$p = R\rho T.$$

Choosing $a(T,\rho)$ and $b(T,\rho)$ properly, we obtain the Navier–Stokes set with arbitrary viscosity and thermal conductivity. It should be noted that f_c need not be the same distribution function of the Chapman–Enskog expansion that gives the Navier–Stokes set with the same viscosity and thermal conductivity.

3) Let

$$\gamma_0 = 1,\ \gamma_i = \xi_i,\ \gamma_4 = \xi_j^2 + (\alpha - 3)\eta^2 \quad (i = 1, 2, 3), \tag{C.25}$$

and

$$f_c = \frac{\rho}{(2\pi RT)^2}\exp\left(-\tilde{\zeta}^2 - \tilde{\eta}^2\right), \tag{C.26}$$

where

$$\tilde{\zeta}_i = \frac{\xi_i - u_i}{(2RT)^{1/2}},\quad \tilde{\eta} = \frac{\eta}{(2RT)^{1/2}},\quad \tilde{\zeta}^2 = \tilde{\zeta}_i^2,$$

and α is a constant, which can be chosen at our discretion. The macroscopic variables ρ_r are related to the parameters ρ, u_i, and T in the distribution function (C.26) as

$$\rho_0 = \rho, \quad \rho_i = \rho u_i, \quad \rho_4 = \rho(\alpha RT + u_j^2), \tag{C.27}$$

and the flux $\hat{H}_i^r$ is given by

$$\hat{H}_i^0 = \rho u_i, \tag{C.28a}$$

$$\hat{H}_i^j = \rho(u_i u_j + RT\delta_{ij}), \tag{C.28b}$$

$$\hat{H}_i^4 = \rho u_i[(2+\alpha)RT + u_j^2]. \tag{C.28c}$$

The macroscopic variables ρ_r (or ρ, u_i, and T) derived from the solution of this kinetic equation satisfy the following (compressible) Euler set of equations with specific internal energy $\alpha RT/2$:

$$\frac{\partial \rho}{\partial t} + \frac{\partial \rho u_i}{\partial x_i} = 0, \tag{C.29a}$$

$$\frac{\partial \rho u_i}{\partial t} + \frac{\partial \rho u_i u_j}{\partial x_j} + \frac{\partial p}{\partial x_i} = 0, \tag{C.29b}$$

$$\frac{\partial \rho(e + u_i^2/2)}{\partial t} + \frac{\partial \rho u_j(e + u_i^2/2) + p u_j}{\partial x_j} = 0, \tag{C.29c}$$

where

$$p = R\rho T, \quad e = \alpha RT/2.$$

4) Let

$$\gamma_0 = 1, \ \gamma_i = \xi_i, \ \gamma_4 = \xi_j^2 + (\alpha - 3)\eta^2 \quad (i = 1, 2, 3), \tag{C.30}$$

and

$$f_c = \frac{\rho}{(2\pi RT)^2} \exp\left(-\tilde{\zeta}^2 - \tilde{\eta}^2\right) \left\{ 1 - \frac{4}{5(2RT)^{1/2}} \frac{\lambda(T,\rho)}{R\rho T} \frac{\partial T}{\partial x_i} \tilde{\zeta}_i (\tilde{\zeta}^2 - \frac{5}{2}) \right.$$
$$- \frac{\mu(T,\rho)}{R\rho T} \left(\frac{\partial u_i}{\partial x_j} + \frac{\partial u_j}{\partial x_i} \right) \left(\tilde{\zeta}_i \tilde{\zeta}_j - \frac{\tilde{\zeta}^2}{3} \delta_{ij} \right)$$
$$\left. - \frac{\mu_B(T,\rho)}{R\rho T} \frac{\partial u_k}{\partial x_k} \left[\tilde{\zeta}^2 - \frac{3}{\alpha - 3} \tilde{\eta}^2 - \frac{3}{2} \left(\frac{\alpha - 4}{\alpha - 3} \right) \right] \right\}, \tag{C.31}$$

where

$$\tilde{\zeta}_i = \frac{\xi_i - u_i}{(2RT)^{1/2}}, \quad \tilde{\eta} = \frac{\eta}{(2RT)^{1/2}}, \quad \tilde{\zeta}^2 = \tilde{\zeta}_i^2,$$

and α is a constant, and $\mu(T,\rho)$, $\mu_B(T,\rho)$, and $\lambda(T,\rho)$ are functions of T and ρ. They can be chosen at our discretion. The macroscopic variables ρ_r are related to the parameters ρ, u_i, and T in the distribution function (C.31) as

$$\rho_0 = \rho, \quad \rho_i = \rho u_i, \quad \rho_4 = \rho(\alpha RT + u_j^2), \tag{C.32}$$

and the flux $\hat{H}_i^r$ is given by

$$\hat{H}_i^0 = \rho u_i, \tag{C.33a}$$

$$\hat{H}_i^j = \rho(u_i u_j + RT\delta_{ij}) - \mu(T,\rho)\left[\left(\frac{\partial u_i}{\partial x_j} + \frac{\partial u_j}{\partial x_i}\right) - \frac{2}{3}\frac{\partial u_k}{\partial x_k}\delta_{ij}\right]$$
$$- \mu_B(T,\rho)\frac{\partial u_k}{\partial x_k}\delta_{ij}, \tag{C.33b}$$

$$\hat{H}_i^4 = \rho u_i[(2+\alpha)RT + u_j^2] - 2\mu(T,\rho)u_j\left[\left(\frac{\partial u_i}{\partial x_j} + \frac{\partial u_j}{\partial x_i}\right) - \frac{2}{3}\frac{\partial u_k}{\partial x_k}\delta_{ij}\right]$$
$$- 2\mu_B(T,\rho)u_i\frac{\partial u_k}{\partial x_k} - 2\lambda(T,\rho)\frac{\partial T}{\partial x_i}. \tag{C.33c}$$

The macroscopic variables ρ_r (or ρ, u_i, and T) derived from the solution of this kinetic equation satisfy the following (compressible) Navier–Stokes set of equations with specific internal energy $\alpha RT/2$, viscosity $\mu(T,\rho)$, bulk viscosity $\mu_B(T,\rho)$, and thermal conductivity $\lambda(T,\rho)$:

$$\frac{\partial \rho}{\partial t} + \frac{\partial \rho u_i}{\partial x_i} = 0, \tag{C.34a}$$

$$\frac{\partial \rho u_i}{\partial t} + \frac{\partial \rho u_i u_j}{\partial x_j} + \frac{\partial p}{\partial x_i} = \frac{\partial}{\partial x_j}\left[\mu(T,\rho)\left(\frac{\partial u_i}{\partial x_j} + \frac{\partial u_j}{\partial x_i} - \frac{2}{3}\frac{\partial u_k}{\partial x_k}\delta_{ij}\right)\right]$$
$$+ \frac{\partial}{\partial x_i}\left(\mu_B(T,\rho)\frac{\partial u_k}{\partial x_k}\right), \tag{C.34b}$$

$$\frac{\partial \rho(e + u_i^2/2)}{\partial t} + \frac{\partial \rho u_j(e + u_i^2/2) + p u_j}{\partial x_j}$$
$$= \frac{\partial}{\partial x_j}\left[\mu(T,\rho)u_i\left(\frac{\partial u_i}{\partial x_j} + \frac{\partial u_j}{\partial x_i} - \frac{2}{3}\frac{\partial u_k}{\partial x_k}\delta_{ij}\right)\right]$$
$$+ \frac{\partial}{\partial x_i}\left(\mu_B(T,\rho)u_i\frac{\partial u_k}{\partial x_k}\right) + \frac{\partial}{\partial x_j}\left(\lambda(T,\rho)\frac{\partial T}{\partial x_j}\right), \tag{C.34c}$$

where

$$p = R\rho T, \quad e = \alpha RT/2.$$

In the examples, the function f_c somewhat resembles the Chapman–Enskog expansion of the Boltzmann equation, but it is a much simpler function. We can choose freely according to the request that the function gives the required set of equations. Furthermore, the variables ξ_i and η need not be continuous variables, as in various kinetic equation approaches.

C.3 Discussion on numerical systems

In the preceding section, a kinetic-equation system that describes fluid-dynamic equations exactly is explained. We will discuss the validity of numerical kinetic-equation systems on the basis of the kinetic equation, taking the system of the time-discrete free-molecular type as an example. The kinetic-equation approach of the time-discrete free-molecular type, which is of interest here, is the following one or its variant: The solution of an initial-value problem of the Euler set of equations is obtained with the kinetic equation of the free-molecular type by the following procedure. Consider the solution of the equation

$$\frac{\partial f}{\partial t} + \xi_i \frac{\partial f}{\partial x_i} = 0, \tag{C.35}$$

in a continuous sequence of time intervals $(t_0, t_1], (t_1, t_2], (t_2, t_3], \ldots, \ldots$ under the following initial condition for each interval $(t_m, t_{m+1}]$:

$$f = \frac{\rho_{(m)}}{(2\pi RT_{(m)})^{3/2}} \exp\left(-\frac{(\xi_i - u_{i(m)})^2}{2RT_{(m)}}\right) \quad \text{at } t = t_m, \tag{C.36}$$

where $\rho_{(0)}$, $u_{i(0)}$, and $T_{(0)}$ are the real initial values at $t = t_0$ for the Euler set, and $\rho_{(m)}$, $u_{i(m)}$, and $T_{(m)}$ $(m \neq 0)$ are ρ, u_i, and T calculated from the solution f at $t = t_m$ of Eq. (C.35) in the preceding interval $(t_{m-1}, t_m]$. The set ρ, u_i, and T obtained from the solution f at an arbitrary time is taken as the solution of the corresponding initial-value problem of the Euler set.

The above system is compared with the exact system (C.6) and (C.7) with Eq. (C.15) for f_c on their right-hand sides. The difference between the two kinetic equations (C.6) and (C.35) lies on their right-hand sides, and their difference is of the order of unity. With the same initial condition, the error of the velocity distribution function f is of the order of $t_{m+1} - t_m$ in the interval $(t_m, t_{m+1}]$. That is, $|f - f_c| = O(t_{m+1} - t_m)$, where f_c is the solution of the exact kinetic system. Our interest is the error of the variables ρ, u_i, and T. These variables or their equivalents ρ_r of the free-molecular system satisfy the conservation equations

$$\frac{\partial \rho_r}{\partial t} + \frac{\partial H_i^r}{\partial x_i} = 0, \tag{C.37}$$

and the variables ρ^c, u_i^c, and T^c (or ρ_r^c) of the exact system satisfy

$$\frac{\partial \rho_r^c}{\partial t} + \frac{\partial \hat{H}_i^r(\rho_r^c, \nabla \rho_r^c)}{\partial x_i} = 0. \tag{C.38}$$

The difference between fluxes H_i^r and $\hat{H}_i^r(\rho_r^c, \nabla\rho_r^c)$ is bounded by the difference between f and f_c by definition [see Eqs. (C.3) and (C.5)], that is,

$$|H_i^r - \hat{H}_i^r| = O(|f - f_c|) = O(t_{m+1} - t_m). \tag{C.39}$$

Thus from Eqs. (C.37) and (C.38),

$$|\rho_r - \rho_r^c| = O((t_{m+1} - t_m)^2). \tag{C.40}$$

Then the initial distribution function $f_{(m+1)}$ at $t = t_{(m+1)}$, i.e.,

$$f_{(m+1)} = \frac{\rho_{(m+1)}}{(2\pi RT_{(m+1)})^{3/2}} \exp\left(-\frac{(\xi_i - u_{i(m+1)})^2}{2RT_{(m+1)}}\right), \tag{C.41}$$

differs from f_c at $t = t_{m+1}$ only by $O((t_{m+1} - t_m)^2)$. Continuing the discussion to the next interval, we obtain the same error estimate with the subscripts $m+1$ and m shifted by unity. Thus, the errors of the distribution function f and the macroscopic variables ρ_r at a finite time t from f_c and ρ_r^c are of the order of $\max|t_{m+1} - t_m|$. That is, by modifying the distribution function f to the corresponding Maxwellian $f_{(m)}$ at every end t_m of the time intervals, we can limit the error within the order of the maximum length of divided time intervals. Thus, the kinetic system of the time-discrete free-molecular type described at the beginning of this section is an appropriate one to obtain the solution of the Euler set of equations (the first-order system). By choosing an appropriate term on the right-hand side of Eq. (C.35), we can construct a higher-order system.

By a similar discussion, we find that the solution of the Navier–Stokes set of equations (C.24a)–(C.24c) is obtained by the kinetic system of the time-discrete free-molecular type with modification of the initial condition at the beginning of every interval to the function f^c given by Eq. (C.20). For the first-order system, the kinetic system of the time-discrete free-molecular type is sufficient. It often happens that the viscous and thermal conduction terms [b and a in Eq. (C.20)] are small corrections [$O(\varepsilon)$, say] and that we would like to obtain the solution correct up to that order. Then, for a reasonable computation avoiding too small a time step, we should modify Eq. (C.35) by adding the right-hand side of Eq. (C.6) with f_c given by Eq. (C.15) to the right-hand side of Eq. (C.35).[4] Further, the solution of the Euler set of equations (C.29a)–(C.29c) with a modified specific internal energy or the Navier–Stokes set of equations (C.34a)–(C.34c) with a modified specific internal energy and a bulk viscosity can be obtained by the free-molecular-type system by taking f_c given by Eq. (C.26) or (C.31) as the initial condition for every interval, where a new variable η enters parametrically. In this way the exact kinetic equation introduced in the preceding section can make a quick and practical evaluation of heuristic systems possible.

The direct-simulation Monte-Carlo (DSMC) method developed by Bird [1963] (see also Bird [1994], Sone & Aoki [1994]) is widely used in engineering problems of a rarefied gas flow. Hybrid methods that combine the DSMC and fluid-dynamic (Euler or Navier–Stokes) schemes were developed by several authors (e.g., Tiwari & Klar [1998], Garcia, Bell, Crutchfield & Alder [1999], and references there) to save the computing time of the DSMC computation in the fluid-dynamic region, where the effective Knudsen number (the local mean free path divided by the local length scale of variation of the variables) is small. The main difference among various works is the technique of transition from one to the other between the two quite different schemes. The kinetic-equation approach of

[4] By this introduction to the right-hand side of Eq. (C.35), the error $|f - f_c|$ in the time interval $(t_m, t_{m+1}]$ is reduced to $O(\varepsilon(t_{m+1} - t_m))$.

the time-discrete free-molecular type reminds us of the DSMC method: (i) the process of solution of the free-molecular equation (C.35) in the kinetic-equation approach corresponds to the process of replacement of the particles by their free motion in DSMC, and (ii) the modification of the distribution function to the local Maxwellian at the beginning of every time interval in the former corresponds to the renewal of the velocities of some pairs of the particles by statistical collision simulation in the latter. If the second process in the DSMC computation to renew particle velocities by the statistical process is replaced by the procedure of modifying the distribution of the particles to the local Maxwellian (C.36) or the distribution given, e.g., by Eq. (C.20), the DSMC computation can be carried out more efficiently in the fluid-dynamic region. By combining the original DSMC method and the DSMC version of the kinetic system of the time-discrete free-molecular type, the transition is done naturally, since the two methods are of the same character. The combined system can be another candidate for a time-saving technique in addition to the existing domain splitting methods.

Finally, it may be added that it is not efficient to solve the Boltzmann equation (or the BKW equation) faithfully for small Knudsen numbers in order to obtain the solution of the fluid-dynamic-type equations (Euler set of equations etc.), which are derived from the Boltzmann equation in the limit that the Knudsen number tends to zero. In this case, a fine lattice system is required not only in the Knudsen layer but also in the fluid-dynamic region. In the fluid-dynamic region, the variation of the velocity distribution function is moderate and is close to the local Maxwellian. Therefore, in the collision integral, which determines the time and spatial variations of the velocity distribution function, many digits are cancelled out in the subtraction of its gain and loss terms, and the resulting small difference is magnified by the inverse of the Knudsen number. This collision term has fewer significant figures than the velocity distribution function, that is, its last significant figure is at a much higher position than that of the velocity distribution function (e.g., $0.abc\cdots$ for the former and $0.ABCDEF\cdots$ for the latter; c and F: their last significant figures). Thus, in order to obtain the distribution function at the next time or space step with this collision term without losing the original significant figures, the time or space step must be taken so small as to make the last significant figure of the product of the collision term and the time or space step not to exceed the position of the last significant figure of the original distribution function [e.g., $0.abc\cdots \times$ (time step) $= 0.000efg\cdots$]. Thus, the step size should be taken smaller for smaller Knudsen numbers.

Bibliography

[1] Abramowitz, M. and I. Stegun (1972), *Handbook of Mathematical Functions with Formulas, Graphs, and Mathematical Tables* (Dover, New York), 1001–1003.

[2] Andries, P., P. Le Tellec., J. P. Perlat, and B. Perthame (2000), The Gaussian-BGK model of Boltzmann equation with small Prandtl number, *Eur. J. Mech. B/Fluids* **19**, 813–830.

[3] Aoki, K. and C. Cercignani (1983), Evaporation and condensation on two parallel plates at finite Reynolds numbers, *Phys. Fluids* **26**, 1163–1164.

[4] Aoki, K., C. Bardos, F. Golse, M. N. Kogan, and Y. Sone (1993), Steady flows of a rarefied gas around arbitrary obstacle distributions, *Eur. J. Mech. B/Fluids* **12**, 565–577.

[5] Aoki, K., T. Inamuro, and Y. Onishi (1979), Slightly rarefied gas flow over a body with small accommodation coefficient, *J. Phys. Soc. Jpn* **47**, 663–671.

[6] Aoki, K. and N. Masukawa (1994), Gas flows caused by evaporation and condensation on two parallel condensed phases and the negative temperature gradient: Numerical analysis by using a nonlinear kinetic equation, *Phys. Fluids* **6**, 1379–1395.

[7] Aoki, K., K. Nishino, Y. Sone, and H. Sugimoto (1991), Numerical analysis of steady flows of a gas condensing on or evaporating from its plane condensed phase on the basis of kinetic theory: Effect of gas motion along the condensed phase, *Phys. Fluids A* **3**, 2260–2275.

[8] Aoki, K. and Y. Sone (1991), Gas flows around the condensed phase with strong evaporation or condensation–Fluid dynamic equation and its boundary condition on the interface and their application–, in: R. Gatignol and Soubbaramayer, eds., *Advances in Kinetic Theory and Continuum Mechanics* (Springer-Verlag, Berlin), 43–54.

[9] Aoki, K., Y. Sone, and N. Masukawa (1995), A rarefied gas flow induced by a temperature field, in: J. Harvey and G. Lord, eds., *Rarefied Gas Dynamics* (Oxford Univ. Press, Oxford), 35–41.

[10] Aoki, K., Y. Sone, K. Nishino, and H. Sugimoto (1991), Numerical analysis of unsteady motion of a rarefied gas caused by sudden changes of wall temperature with special interest in the propagation of a discontinuity in the velocity distribution function, in: A. E. Beylich, ed., *Rarefied Gas Dynamics* (VCH Verlagsgesellschaft, Weinheim), 222–231.

[11] Aoki, K., Y. Sone, S. Takata, K. Takahashi, and G. A. Bird (2001), One-way flow of a rarefied gas induced in a circular pipe with a periodic temperature distribution, in: T. J. Bartel and M. A. Gallis, eds., *Rarefied Gas Dynamics* (AIP, New York), 940–947.

[12] Aoki, K., Y. Sone, and T. Yamada (1990), Numerical analysis of gas flows condensing on its plane condensed phase on the basis of kinetic theory, *Phys. Fluids A* **2**, 1867–1878. See Sone, Aoki & Doi [1992] for the effect of a noncondensable gas.

[13] Aoki, K., Y. Sone, and T. Yano (1989), Numerical analysis of a flow induced in a rarefied gas between noncoaxial circular cylinders with different temperatures for the entire range of the Knudsen number, *Phys. Fluids A* **1**, 409–419.

[14] Arkeryd, L. (1972), On the Boltzmann equation Part II: The full initial value problem, *Arch. Rational Mech. Anal.* **45**, 17–34.

[15] Arkeryd, L. and A. Nouri (2000), On the Milne problem and the hydrodynamic limit for a steady Boltzmann equation model, *J. Stat. Phys.* **99**, 993–1019.

[16] Bakanov, S. P. (1991), Thermophoresis in gases at small Knudsen numbers, *Aeros. Sci. Tech.* **15**, 77–92.

[17] Bakanov, S. P., V. V. Vysotskij, B. V. Derjaguin, and V. I. Roldughin (1983), Thermal polarization of bodies in the rarefied gas flow, *J. NonEquilib. Thermodyn.* **8**, 75–83.

[18] Bardos, C., R. E. Caflisch, and B. Nicolaenko (1986), The Milne and Kramers problems for the Boltzmann equation of a hard sphere gas, *Commun. Pure Appl. Math.* **39**, 323–352.

[19] Bardos, C., F. Golse, and D. Levermore (1993), Fluid dynamic limits of kinetic equations II: Convergence proofs for the Boltzmann equation, *Commun. Pure Appl. Math.* **46**, 667–753.

[20] Bardos, C., F. Golse, and D. Levermore (1998), Acoustic and Stokes limits for the Boltzmann equation, *C. R. Acad. Sci.* **327**, Série I, 323–328.

[21] Bardos, C., F. Golse, and D. Levermore (2000), Acoustic limit for the Boltzmann equation, *Arch. Rational Mech. Anal.* **153**, 177–204.

[22] Bellomo, L. (1995), *Lecture Notes on the Mathematical Theory of the Boltzmann Equation* (World Scientific, Singapore). The article by M. Lachowics is the mathematical discussion of the continuum limit.

[23] Bhatnagar, P. L., E. P. Gross, and M. Krook (1954), A model for collision processes in gases. I. Small amplitude processes in charged and neutral one-component systems, *Phys. Rev.* **94**, 511–525.

[24] Bird, G. A. (1963), Approach to translational equilibrium in a rigid sphere gas, *Phys. Fluids* **6**, 1518–1519.

[25] Bird, G. A. (1994), *Molecular Gas Dynamics and the Direct Simulation of Gas Flows* (Oxford University Press, Oxford).

[26] Bobylev, A. V., R. Grzhibovskis, and A. Heinz (2001), Entropy inequalities for evaporation/condensation problem in rarefied gas, *J. Stat. Phys.* **102**, 1151–1176. (announced at the Oberwolfach workshop, Dec. 1999)

[27] Boltzmann, L. (1896–98), *Vorlesungen über Gastheorie*, 2 Vols. (Barth, Leipzig) [English translation by S. G. Bush: (1964), *Lectures on Gas Theory* (University of California Press, California)].

[28] Bouchut, F., F. Golse, and M. Pulvirenti (2000), *Kinetic Equation and Asymptotic Theory* (Gauthier-Villars, Paris). The article by F. Golse is the mathematical discussion of the continuum limit.

[29] Brillouin, L. (1938), *Les Tenseurs en Mécanique et en Élasticité* (Masson, Paris); Réimpression (Éditions Jacques Gabay, 1987).

[30] Caflisch, R. E. and B. Nicolaenko (1982), Shock profiles solutions of the Boltzmann equation, *Commun. Math. Phys.*, **86**, 161–194.

[31] Carslaw, H. S. and J. C. Jaeger (1959), *Conduction of Heat in Solids, 2nd ed.* (Oxford University Press, London).

[32] Cartan, É. (1946), *Géometrie des espaces de Riemann* (Gauthier-Villars, Paris); Réimpression (Édition Jacques Gabay, Paris, 1988).

[33] Cartan, H. (1977), *Cours de Calcul Différentiel* (Hermann, Paris), 329.

[34] Cercignani, C. (1962), Elementary solutions of the linearized gas dynamics Boltzmann equation and their application to the slip flow problem, *Annals of Physics* **20**, 219–233.

[35] Cercignani, C. (1968), Stokes paradox in kinetic theory, *Phys. Fluids* **11**, 303–308.

[36] Cercignani, C. (1986), Half-space problems in the kinetic thecry of gases, in: E. Kröner and K. Kirchgässner, eds., *Trends in Applications of Pure Mathematics to Mechanics* (Springer-Verlag, Berlin), 35–50.

[37] Cercignani, C. (1988), *The Boltzmann Equation and Its Applications* (Springer-Verlag, Berlin).

[38] Cercignani, C. (2000), *Rarefied Gas Dynamics: from basic concepts to actual calculations* (Cambridge University Press, Cambridge).

[39] Cercignani, C., R. Illner, and M. Pulvirenti (1994), *The Mathematical Theory of Dilute Gases* (Springer-Verlag, Berlin).

[40] Cercignani, C. and C. Lancellotti (1991), Velocity slip for curved shock waves, in: A. E. Beylich, ed., *Rarefied Gas Dynamics* (VCH Verlagsgesellschaft, Weinheim), 153–160.

[41] Chandrasekhar, S. (1961), *Hydrodynamic and Hydromagnetic Stability* (Oxford University Press, London).

[42] Chapman, S. (1916), On the law of distribution of molecular velocities, and on the theory of viscosity and thermal conduction, in a non-uniform simple monoatomic gas, *Philos. Trans. R. Soc. London* **216**, 279–348.

[43] Chapman, S. and T. G. Cowling (1952), *The Mathematical Theory of Non-uniform Gases* (Cambridge University Press, Cambridge).

[44] Coron, F., F. Golse, and C. Sulem (1988), A classification of well-posed kinetic layer problems, *Commun. Pure Appl. Math.* **41**, 409–435.

[45] Darrozes, J. S. (1969), Approximate solutions of the Boltzmann equation for flows past bodies of moderate curvature, in: L. Trilling and H. Y. Wachman, eds., *Rarefied Gas Dynamics* (Academic Press, New York), Vol. I, 111–120.

[46] Darrozes, J. S. and J.-P. Guiraud (1966), Généralisation formelle du théorème H en présence de parois. Applications, *C. R. Acad. Sci., Paris A* **262**, 1368–1371.

[47] Deshpande, S. M. (1986), Kinetic theory based new upwind methods for inviscid compressible flows, *AIAA Paper 86-0275* (AIAA, New York).

[48] Diu, B., C. Guthmann, D. Lederer, and B. Roulet (1989), *Éléments de Physique Statistique* (Hermann, Paris).

[49] Edwards, R. H. and H. K. Cheng (1966), Steady expansion of a gas into a vacuum, *AIAA J.* **4**, 558–561.

[50] Enskog, D. (1917), *Kinetische Theorie der Vorgänge in mässig verdünnten Gasen*, dissertation, Uppsala.

[51] Freeman, N. C. (1967), Solution of the Boltzmann equation for expanding flows, *AIAA J.* **5**, 1696–1698.

[52] Garcia, A., J. Bell, Wm. Y. Crutchfield, and B. Alder (1999), Adaptive mesh and algorithm refinement using direct simulation Monte Carlo, *J. Comp. Phys.* **154**, 134–155.

[53] Goldstein, H. (1950), *Classical Mechanics* (Addison-Wesley, Massachusetts).

[54] Golse, F. (1995), Incompressible hydrodynamic limits of the Boltzmann equation: a survey of mathematical results. R.C.P. 25, Vol. 47 (Strasbourg, 1993–1995), 1–31, Prépubl. Inst. Rech. Math. Av., 1995/24, Univ. Louis Pasteur, Strasbourg.

[55] Golse, F. and F. Poupaud (1989), Stationary solutions of the linearized Boltzmann equation in a half-space, *Math. Methods Appl. Sci.* **11**, 483–502.

[56] Grad, H. (1949), On the kinetic theory of rarefied gases, *Commun. Pure Appl. Math.* **2**, 331–407.

[57] Grad, H. (1958), Principles of the kinetic theory of gases, in: S. Flügge, ed., *Handbuch der Physik* (Springer-Verlag, Berlin), Band XII, 205–294.

[58] Grad, H. (1960), Theory of rarefied gases, in: F. M. Devienne, ed., *Rarefied Gas Dynamics* (Pergamon, London), 100–138.

[59] Grad, H. (1963a), Asymptotic theory of the Boltzmann equation, *Phys. Fluids* **6**, 147–181.

[60] Grad, H. (1963b), Asymptotic theory of the Boltzmann equation II, in: J. A. Laurmann, ed., *Rarefied Gas Dynamics* (Academic Press, New York), 26–59.

[61] Grad, H. (1969), Singular and nonuniform limits of solutions of the Boltzmann equation, in: R. Bellman, G. Birkhoff, and I. Abu-Shumays, eds., *Transport Theory* (American Mathematical Society, Providence), 269–308. See also Grad [1960]

[62] Grundy, R. E. (1969), Axially symmetric expansion of a monatomic gas from an orifice into a vacuum, *Phys. Fluids* **12**, 2011–2018.

[63] Guiraud, J.-P. (1973), Gas dynamics from the point of view of kinetic theory, in: E. Becker and G. K. Mikhailov, eds., *Theoretical and Applied Mechanics* (Springer-Verlag, Heidelberg), 104–123.

[64] Hamel, B. B. and D. R. Willis (1966), Kinetic theory of source flow expansion with application to free jet, *Phys. Fluids* **9**, 829–841.

[65] Hasegawa, M. and Y. Sone (1988), Poiseuille and thermal transpiration flows of a rarefied gas for various pipes, *J. Vac. Soc. Jpn* **31**, 416–419. (in Japanese)

[66] Hermans, L. J. F. and J. J. M. Beenakker (1986), The temperature paradox in the kinetic theory of evaporation, *Phys. Fluids* **29**, 4231–4232.

[67] Hilbert, D. (1900), Mathematische Probleme (*Vortrag, gehalten auf dem internationalen Mathematiker-Congress zu Paris 1900*), Gött. Nachr. (Vadenhöck & Ruprecht, Göttingen), 253–297; English translation by M. Winston: (2000), Bulletin (New Series) of the American Mathematical Society **37**, 407–436.

[68] Hilbert, D. (1912), *Grundzüge einer Allgemeinen Theorie der Linearen Integralgleichungen* (Teubner, Leipzig), Chap. 22.

[69] Hill, T. L. (1987), *Statistical Mechanics, Principles and Selected Applications* (Dover, New York).

[70] Huber, C. A. (1995), Nanowire array composite, *Science* **263**, 800–802.

[71] Hudson, M. L. and T. J. Bartel (1999), DSMC simulation of thermal transpiration and accommodation pumps, in: R. Brun, R. Campargue, R. Gatignol, and J.-C. Lengrand, eds., *Rarefied Gas Dynamics* (CÉPADUÈS, Toulouse), Vol. I, 719–726.

[72] Illner, R. and M. Pulvirenti (1986), Global validity of the Boltzmann equation for a two and three-dimensional rare gas into vacuum, *Commun. Math. Phys.* **105**, 189–203; (1989), Erratum and improved result, *Commun. Math. Phys.* **121**, 143–146.

[73] Inamuro, T., M. Yoshino, and F. Ogino (1997), Accuracy of the lattice Boltzmann method for small Knudsen number with finite Reynolds number, *Phys. Fluids* **9**, 3535–3542.

[74] Jeffreys, H. (1965), *Cartesian Tensors* (Cambridge University Press, Cambridge, U.K.).

[75] Junk, M. (1997), *Kinetic Schemes: A New Approach and Application* (Shaker Verlag, Postfach).

[76] Kennard, E. H. (1938), *Kinetic Theory of Gases* (McGraw-Hill, New York).

[77] Knudsen, M. (1910a), Eine Revision der Gleichgewichtsbedingung der Gase. Thermische Molekularströmung, *Ann. Phys.* (Leipzig) **31**, 205–229.

[78] Knudsen, M. (1910b), Thermischer Molekulardruck der Gase in Röhren, *Ann. Phys.* (Leipzig) **33**, 1435–1448.

[79] Kogan, M. N. (1958), On the equations of motion of a rarefied gas, *Appl. Math. Mech.* **22**, 597–607.

[80] Kogan, M. N. (1969), *Rarefied Gas Dynamics* (Plenum, New York).

[81] Kogan, M. N. (1972), Molecular gas dynamics, in: ***Annual Review of Fluid Mechanics* 5** (Annual Reviews, Palo Alto), 383–404. The related subject is discussed in Guiraud [1973].

[82] Kogan, M. N. (1985), Transport processes in Knudsen layers, in: O. M. Belotserkovskii, M. N. Kogan, S. S. Kutateladze, and A. K. Rebrov, eds., Rarefied Gas Dynamics (Plenum, New York), 545–560.

[83] Kogan, M. N. (1992), Kinetic theory in aerothermodynamics, *Prog. Aerosp. Sci.* **29**, 271–354. The reviews by Kogan [1972, 1985, 1992, 1995] are related to each other.

[84] Kogan, M. N. (1995), Evaporation/condensation kinetics, in: J. Harvey and G. Lord, eds., *Rarefied Gas Dynamics* (Oxford University Press, Oxford), 253–262.

[85] Kogan, M. N. and A. A. Abramov (1991), Direct simulation solution of the strong evaporation and condensation problem, in: A. E. Beylich, ed., *Rarefied Gas Dynamics* (VCH Verlagsgesellschaft, Weinheim), 1251–1257.

[86] Kogan, M. N., V. S. Galkin, and O. G. Fridlender (1976), Stresses produced in gases by temperature and concentration inhomogeneities. New type of free convection, *Sov. Phys. Usp.* **19**, 420–438. (The last author spells his last name Friedlander when he writes in English by himself.) The related materials are found also in the above reviews by Kogan.

[87] Kogan, M. N. and N. K. Makashev (1971), Role of the Knudsen layer in the theory of heterogeneous reactions and in flows with surface reactions, *Fluid Dynamics* **6**, 913–920. The related materials are found in Kogan [1995] and Trilling [1975].

[88] Kryukov, A. P. (1991), Strong subsonic and supersonic condensation on a plane surface, in: A. E. Beylich, ed., *Rarefied Gas Dynamics* (VCH Verlagsgesellschaft, Weinheim), 1278–1284.

[89] Landau, L. D. and E. M. Lifshitz (1963), *Course of Theoretical Physics Vol. 5: Statistical Physics* (Pergamon Press, London), Section 82.

[90] Landau, L. D. and E. M. Lifshitz (1987), *Course of Theoretical Physics Vol. 6: Fluid Mechanics, 2nd edition* (Pergamon Press, London), Section 83.

[91] Landau, L. D. and E. M. Lifshitz (1960), *Course of Theoretical Physics Vol. 1: Mechanics* (Pergamon Press, London).

[92] Lanford III, O. (1975), The evolution of large classical system, in: J. Moser, ed., Lecture Notes in Physics (Springer, Heidelberg), 1–111.

[93] Liepmann, H. W. and A. Roshko (1957), *Elements of Gasdynamics* (Wiley, New York).

[94] Loeb, L. (1961), *Kinetic Theory of Gases* (Dover, New York), Sections 83 and 84.

[95] Loyalka, S. K. (1983), Mechanics of aerosols in nuclear reactor safety: A review, *Prog. Nucl. Energ.* **12**, 1–56.

[96] Loyalka, S. K. (1989), Temperature jump and thermal creep slip: Rigid sphere gas, *Phys. Fluids A* **1**, 403–408.

[97] Loyalka, S. K. (1991), Kinetic theory of planar condensation and evaporation, *Transp. Theory Stat. Phys.* **20**, 237–249.

[98] Loyalka, S. K. and K. A. Hickey (1989), Velocity slip and defect: Hard sphere gas, *Phys. Fluids A* **1**, 612–614.

[99] Mager, R., G. Adomeit, and G. Wortberg (1989), Theoretical and experimental investigation of the strong evaporation of solids, in: E. P. Muntz, D. P. Weaver, and D. H. Campell, eds., *Rarefied Gas Dynamics* (AIAA, Washington, DC), 460–469.

[100] Maslova, N. (1993), *Nonlinear Evolution Equations, Kinetic Approach* (World Scientific, Singapore).

[101] Maxwell, J. C. (1879), On stresses in rarefied gases arising from inequalities of temperature, *Philos. Trans. R. Soc.* **170**, 231–256.

[102] Murakami, M. and K. Oshima (1974), Kinetic approach to the transient evaporation and condensation problem, in: M. Becker and M. Fiebig, eds., *Rarefied Gas Dynamics* (DFVLR-Press, Porz-Wahn), F. 6.

[103] Nishida, T. (1978), Fluid dynamical limit of the nonlinear Boltzmann equation to the level of the compressible Euler equation, *Comm. Math. Phys.* **61**, 119–148.

[104] Ohwada, T. and Y. Sone (1992), Analysis of thermal stress slip flow and negative thermophoresis using the Boltzmann equation for hard-sphere molecules, *Eur. J. Mech. B/Fluids* **11**, 389–414.

[105] Ohwada, T., Y. Sone, and K. Aoki (1989a), Numerical analysis of the shear and thermal creep flows of a rarefied gas over a plane wall on the basis of the linearized Boltzmann equation for hard-sphere molecules, *Phys. Fluids A* **1**, 1588–1599.

[106] Ohwada, T., Y. Sone, and K. Aoki (1989b), Numerical analysis of the Poiseuille and thermal transpiration flows between two parallel plates on the basis of the Boltzmann equation for hard-sphere molecules, *Phys. Fluids A* **1**, 2042–2049; (1990), Erratum, *Phys. Fluids A* **2**, 639.

[107] Onishi, Y. and Y. Sone (1979), Kinetic theory of slightly strong evaporation and condensation – Hydrodynamic equation and slip boundary condition for finite Reynolds number , *J. Phys. Soc. Jpn* **47**, 1676–1685.

[108] Onishi, Y. and Y. Sone (1983), Kinetic theory of evaporation and condensation for a cylindrical condensed phase, *Phys. Fluids* **26**, 659–664.

[109] Oswatitsch, K. (1956), *Gasdynamics* (Academic Press, New York).

[110] Pao, Y. P. (1971), Application of kinetic theory to the problem of evaporation and condensation, *Phys. Fluids* **14**, 306–312.

[111] Pekeris, C. L. and Z. Alterman (1957), Solution of the Boltzmann-Hilbert integral equation II. the coefficients of viscosity and heat conduction, *Proceedings of the National Academy of Sciences of the U. S. A.* **43**, 998–1007.

[112] Perthame, B. (1990), Boltzmann type scheme for gas dynamics and the entropy property, *SIAM J. Numer. Anal.* **27**, 1405–1421.

[113] Perthame, B. (1992), Second-order Boltzmann schemes for compressible Euler equations in one and two space dimensions, *SIAM J. Numer. Anal.* **29**, 1–19.

[114] Pham-Van-Diep, G., P. Keeley, E. P. Muntz, and D. P. Weaver (1995), A micromechanical Knudsen compressor, in: J. Harvey and G. Lord, eds., *Rarefied Gas Dynamics* (Oxford University Press, London), Vol. I, 715–721.

[115] Pullin, D. I. (1980), Direct simulation method for compressible inviscid ideal-gas flow, *J. Comput. Phys.* **34**, 231–244.

[116] Reif, F. (1965), *Fundamentals of Statistical and Thermal Physics* (McGraw-Hill, New York), Section 8.5.

[117] Reinhardt, F. and H. Soeder (1997), *Atlas des Mathématiques* (Librairie Générale Française, Paris), 410.

[118] Ricci, M. M. G. and T. Levi-Civita (1900), Méthodes de calcul différentiel absolu et leurs applications, *Mathematische Annalen* **54**, 125–201.

[119] Schlichting, H. (1979), *Boundary-Layer Theory* (McGraw-Hill, New York), Chapters VII e, XIV b.

[120] Shankar, P. N. and M. D. Deshpande (1990), On the temperature distribution in liquid-vapor phase change between plane liquid surfaces, *Phys. Fluids A* **2**, 1030–1038.

[121] Sone, Y. (1964), Kinetic theory analysis of linearized Rayleigh problem, *J. Phys. Soc. Jpn* **19**, 1463–1473.

[122] Sone, Y. (1965), Effect of sudden change of wall temperature in a rarefied gas, *J. Phys. Soc. Jpn* **20**, 222–229.

[123] Sone, Y. (1966a), Some remarks on Knudsen layer, *J. Phys. Soc. Jpn* **21**, 1620–1621.

[124] Sone, Y. (1966b), Thermal creep in rarefied gas, *J. Phys. Soc. Jpn* **21**, 1836–1837.

[125] Sone, Y. (1968), Asymptotic behavior of diffusion of tangential velocity discontinuity in rarefied gas, *Phys. Fluids* **11**, 1935–1937.

[126] Sone, Y. (1969), Asymptotic theory of flow of rarefied gas over a smooth boundary I, in: L. Trilling and H. Y. Wachman, eds., *Rarefied Gas Dynamics* (Academic Press, New York), Vol. I, 243–253.

[127] Sone, Y. (1971), Asymptotic theory of flow of rarefied gas over a smooth boundary II, in: D. Dini, ed., *Rarefied Gas Dynamics* (Editrice Tecnico Scientifica, Pisa), Vol. II, 737–749.

[128] Sone, Y. (1972), Flow induced by thermal stress in rarefied gas, *Phys. Fluids* **15**, 1418–1423.

[129] Sone, Y. (1973), New kind of boundary layer over a convex solid boundary in a rarefied gas, *Phys. Fluids* **16**, 1422–1424. See also Trilling [1975].

[130] Sone, Y. (1974), Rarefied gas flow induced between non-parallel plane walls with different temperatures, in: M. Becker and M. Fiebig, eds., *Rarefied Gas Dynamics* (DFVLR Press, Porz-Wahn), Vol. II, D. 23.

[131] Sone, Y. (1978), Kinetic theory of evaporation and condensation–Linear and nonlinear problems–, *J. Phys. Soc. Jpn* **45**, 315–320.

[132] Sone, Y. (1984a), Highly rarefied gas around a group of bodies with various temperature distributions I. small temperature variation, *J. de Mécanique Théorique et Appliquée* **3**, 315–328.

[133] Sone, Y. (1984b), Analytical studies in rarefied gas dynamics, in: H. Oguchi, ed., *Rarefied Gas Dynamics* (University of Tokyo Press, Tokyo), Vol. I, 71–87.

[134] Sone, Y. (1984c), Force, its moment, and energy transfer on a closed body in a slightly rarefied gas, in: H. Oguchi, ed., *Rarefied Gas Dynamics* (University of Tokyo Press, Tokyo), Vol. I, 117–126.

[135] Sone, Y. (1985), Highly rarefied gas around a group of bodies with various temperature distributions II. arbitrary temperature variation, *J. de Mécanique Théorique et Appliquée* **4**, 1–14.

[136] Sone, Y. (1991a), Asymptotic theory of a steady flow of a rarefied gas past bodies for small Knudsen numbers, in: R. Gatignol and Soubbaramayer, eds., *Advances in Kinetic Theory and Continuum Mechanics* (Springer-Verlag, Berlin), 19–31. See also Sone [1991c].

[137] Sone, Y. (1991b), A simple demonstration of a rarefied gas flow induced over a plane wall with a temperature gradient, *Phys. Fluids A* **3**, 997–998.

[138] Sone, Y. (1991c), Analytical and numerical studies of rarefied gas flows on the basis of the Boltzmann equation for hard-sphere molecules, in: A. E. Beylich, ed., *Rarefied Gas Dynamics* (VCH, Weinheim), 489–504.

[139] Sone, Y. (1997), Continuum gas dynamics in the light of kinetic theory and new features of rarefied gas flows, in: Ching Shen, ed., *Rarefied Gas Dynamics* (Peking University Press, Beijing), 3–24.

[140] Sone, Y. (1998), *Theoretical and Numerical Analyses of the Boltzmann Equation–Theory and Analysis of Rarefied Gas Flows–Part I*, Lecture Notes, Department of Aeronautics and Astronautics, Graduate School of Engineering, Kyoto University, Kyoto (http://www.users.kudpc.kyoto-u.ac.jp/~a50077).

[141] Sone, Y. (2000a), Flows induced by temperature fields in a rarefied gas and their ghost effect on the behavior of a gas in the continuum limit, in: *Annual Review of Fluid Mechanics* **32** (Annual Reviews, Palo Alto), 779–811.

[142] Sone, Y. (2000b), Kinetic theoretical studies of the half-space problem of evaporation and condensation, *Transp. Theory Stat. Phys.* **29**, 227–260.

[143] Sone, Y. (2000c), Notes on kinetic-equation approach of fluid-dynamic equations, *Mekanik-KTH 2000:09* Royal Institute of Technology, Department of Mechanics, Stockholm. (can be downloaded at http://www2.mech.kth.se/~lhs/kinetic2 00.pdf)

[144] Sone, Y. and K. Aoki (1977a), Slightly rarefied gas flow over a specularly reflecting body, *Phys. Fluids* **20**, 571–576.

[145] Sone, Y. and K. Aoki (1977b), Forces on a spherical particle in a slightly rarefied gas, in: J. L. Potter, ed., *Rarefied Gas Dynamics* (AIAA, New York), 417–433.

[146] Sone, Y. and K. Aoki (1979), Thermal force and drag on a volatile particle in a slightly rarefied gas, in: R. Campargue, ed., *Rarefied Gas Dynamics* (Commissariat a l'energie atomique, Paris), 1207–1218.

[147] Sone, Y. and K. Aoki (1981), Negative thermophoresis: Thermal stress slip flow around a spherical particle in a rarefied gas, in: S. S. Fisher, ed., *Rarefied Gas Dynamics* (AIAA, New York), 489–503.

[148] Sone, Y. and K. Aoki (1983), A similarity solution of the linearized Boltzmann equation with application to thermophoresis of a spherical particle, *J. de Mécanique Théorique et Appliquée* **2**, 3–12.

[149] Sone, Y. and K. Aoki (1987), Steady gas flows past bodies at small Knudsen numbers–Boltzmann and hydrodynamic systems–, *Transp. Theory Stat. Phys.* **16**, 189–199; Asymptotic theory of slightly rarefied gas flow and force on a closed body, *Memoirs of the Faculty of Engineering, Kyoto University* **49**, 237–248.

[150] Sone, Y. and K. Aoki (1994), *Molecular Gas Dynamics* (Asakura, Tokyo). (in Japanese)

[151] Sone, Y., K. Aoki, and T. Doi (1992), Kinetic theory analysis of gas flows condensing on a plane condensed phase: Case of a mixture of a vapor and a noncondensable gas, *Transp. Theory Stat. Phys.* **21**, 297–328.

[152] Sone, Y., K. Aoki, and T. Ohwada (1986), Forces on heated bodies in a highly rarefied gas, *Jpn J. Aero. Astro.* **34**, 386–395. (in Japanese)

[153] Sone, Y., K. Aoki, and Y. Onishi (1977), Imperfect accommodation effects on the flow induced by thermal stresses in a rarefied gas, *Heat Transfer Japanese Research* **6**, 1–12.

[154] Sone, Y., K. Aoki, H. Sugimoto, and T. Yamada (1988), Steady evaporation and condensation on a plane condensed phase, *Theoretical and Applied Mechanics* (Bulgarian Academy of Sciences), Year XIX, No. 3, 89–93.

[155] Sone, Y., K. Aoki, S. Takata, H. Sugimoto, and A. V. Bobylev (1996), Inappropriateness of the heat-conduction equation for description of a temperature field of a stationary gas in the continuum limit: Examination by asymptotic analysis and numerical computation of the Boltzmann equation, *Phys. Fluids* **8**, 628–638; (1996), Erratum, *Phys. Fluids* **8**, 841.

[156] Sone, Y., K. Aoki, and I. Yamashita (1986), A study of unsteady strong condensation on a plane condensed phase with special interest in formation of steady profile, in: V. Boffi and C. Cercignani, eds., *Rarefied Gas Dynamics* (Teubner, Stuttgart), Vol. II, 323–333. See Sone, Aoki & Doi [1992] for the effect of a noncondensable gas.

[157] Sone, Y., C. Bardos, F. Golse, and H. Sugimoto (2000), Asymptotic theory of the Boltzmann system, for a steady flow of a slightly rarefied gas with a finite Mach number: General theory, *Eur. J. Mech. B/Fluids* **19**, 325–360.

[158] Sone, Y. and T. Doi (2000), Analytical study of bifurcation of a flow of a gas between coaxial circular cylinders with evaporation and condensation, *Phys. Fluids* **12**, 2639–2660.

[159] Sone, Y., T. Fukuda, T. Hokazono, and H. Sugimoto (2001), Experiment on a one-way flow of a rarefied gas through a straight circular pipe without average temperature and pressure gradients, in: T. J. Bartel and M. A. Gallis, eds., *Rarefied Gas Dynamics* (AIP, New York), 948–955.

[160] Sone, Y., F. Golse, T. Ohwada, and T. Doi (1998), Analytical study of transonic flows of a gas condensing onto its plane condensed phase on the basis of kinetic theory, *Eur. J. Mech. B/Fluids* **17**, 277–306.

[161] Sone, Y., M. Handa, and H. Sugimoto (2002), Bifurcation studies of flows of a gas between coaxial circular cylinders with evaporation and condensation by the Boltzmann system, *Transp. Theory Stat. Phys.* **31**, (to be published).

[162] Sone, Y. and M. Hasegawa (1987), Poiseuille and thermal transpiration flows of a rarefied gas through a rectangular pipe, *J. Vac. Soc. Jpn* **30**, 425–428. (in Japanese)

[163] Sone, Y., E. Itakura, and M. Handa (1998), Analysis of Poiseuille and thermal transpiration flows for arbitrary Knudsen numbers by a modified Knudsen number expansion method and their database. http://www.users.kudpc.kyoto-u.ac.jp/~a50077.

[164] Sone, Y., T. Kataoka, T. Ohwada, H. Sugimoto, and K. Aoki (1994), Numerical examination of applicability of the linearized Boltzmann equation, *Eur. J. Mech. B/Fluids* **13**, 573–589.

[165] Sone, Y., T. Ohwada, and K. Aoki (1989a), Temperature jump and Knudsen layer in a rarefied gas over a plane wall: Numerical analysis of the linearized Boltzmann equation for hard-sphere molecules, *Phys. Fluids A* **1**, 363–370.

[166] Sone, Y., T. Ohwada, and K. Aoki (1989b), Evaporation and condensation on a plane condensed phase: Numerical analysis of the linearized Boltzmann equation for hard-sphere molecules, *Phys. Fluids A* **1**, 1398–1405.

[167] Sone, Y., T. Ohwada, and K. Aoki (1991), Evaporation and condensation of a rarefied gas between its two parallel plane condensed phases with different temperatures and negative temperature-gradient phenomenon – Numerical analysis of the Boltzmann equation for hard-sphere molecules–, in: G. Toscani, V. Boffi, and S. Rionero, eds., *Mathematical Aspects of Fluid and Plasma Dynamics, Lecture Notes in Mathematics Vol. 1460* (Springer-Verlag, Berlin), 186–202.

[168] Sone, Y. and Y. Onishi (1973), Kinetic theory of evaporation and condensation, *J. Phys. Soc. Jpn* **35**, 1773–1776.

[169] Sone, Y. and Y. Onishi (1977), Heat transfer through a rarefied gas between coaxial circular cylinders with large temperature difference, in: J. L. Potter ed., *Rarefied Gas Dynamics* (AIAA, New York), Part 1, 295–310.

[170] Sone, Y. and Y. Onishi (1978), Kinetic theory of evaporation and condensation – Hydrodynamic equation and slip boundary condition –, *J. Phys. Soc. Jpn* **44**, 1981–1994.

[171] Sone, Y. and K. Sato (2000), Demonstration of a one-way flow of a rarefied gas induced through a pipe without average pressure temperature gradients, *Phys. Fluids* **12**, 1864–1868.

[172] Sone, Y., K. Sawada, and H. Hirano (1994), A simple experiment on the strength of thermal creep flow of a rarefied gas over a flat wall, *Eur. J. Mech. B/Fluids* **13**, 299–303.

[173] Sone, Y. and H. Sugimoto (1990), Strong evaporation from a plane condensed phase, in: G. E. A. Meier and P. A. Thompson, eds., *Adiabatic Waves in Liquid-Vapor System* (Springer, Berlin), 293–304.

[174] Sone, Y. and H. Sugimoto (1993), Kinetic theory analysis of steady evaporating flows from a spherical condensed phase into a vacuum, *Phys. Fluids A* **5**, 1491–1511; (1995), Erratum, *Phys. Fluids* **7**, 2096. See also Y. Sone and H. Sugimoto (1993), Steady evaporating flows from a spherical condensed phase into vacuum, in: R. Brun and A. A. Chkhaoui, eds, *Proceedings of the IUTAM Symposium: Aerothermochemistry of Spacecraft and Associated Hypersonic Flows* (Le Palais du Pharo, September 1–4, 1992), 67–72.

[175] Sone, Y. and H. Sugimoto (1995), Evaporation of a rarefied gas from a cylindrical condensed phase into a vacuum, *Phys. Fluids* **7**, 2072–2085.

[176] Sone, Y. and H. Sugimoto (2001), (a): Knudsen compressor I, *Proceedings of the 2001 Annual Meeting of Japan Society of Fluid Mechanics* (Japan Society of Fluid Mechanics, Tokyo), 629–630; (b): (2002) Knudsen compressor, *J. Vac. Soc. Jpn* **45**, 138–141.

[177] Sone, Y., H. Sugimoto, and K. Aoki (1999), Cylindrical Couette flows of a rarefied gas with evaporation and condensation: Reversal and bifurcation of flows, *Phys. Fluids* **11**, 476–490.

[178] Sone, Y. and S. Takata (1992), Discontinuity of the velocity distribution function in a rarefied gas around a convex body and the S layer at the bottom of the Knudsen layer, *Transp. Theory Stat. Phys.* **21**, 501–530.

[179] Sone, Y., S. Takata, and F. Golse (2001), Notes on the boundary conditions for fluid-dynamic equations on the interface of a gas and its condensed phase, *Phys. Fluids* **13**, 324–334.

[180] Sone, Y., S. Takata, and H. Sugimoto (1996), The behavior of a gas in the continuum limit in the light of kinetic theory: The case of cylindrical Couette flows with evaporation and condensation, *Phys. Fluids* **8**, 3403–3413; (1998), Erratum, *Phys. Fluids* **10**, 1239.

[181] Sone, Y., S. Takata, and M. Wakabayashi (1994), Numerical analysis of a rarefied gas flow past a volatile particle using the Boltzmann equation for hard-sphere molecules, *Phys. Fluids* **6**, 1914–1928.

[182] Sone, Y. and S. Tanaka (1980), Thermal stress slip flow induced in rarefied gas between noncoaxial circular cylinders, in: F. P. J. Rimrott and B. Tabarrok, eds., *Theoretical and Applied Mechanics* (North-Holland, Amsterdam), 405–416.

[183] Sone, Y. and M. Wakabayashi (1988), Flow induced by nonlinear thermal stress in a rarefied gas, *Preprint of Symposium on Flight Mechanics and*

Astrodynamics (Institute of Space and Aeronautical Science, Tokyo) (in Japanese). The content is available in a more easily accessible textbook by Sone & Aoki [1994].

[184] Sone, Y., Y. Waniguchi, and K. Aoki (1996), One-way flow of a rarefied gas induced in a channel with a periodic temperature distribution, *Phys. Fluids* **8**, 2227–2235.

[185] Sone, Y. and K. Yamamoto (1968), Flow of rarefied gas through a circular pipe, *Phys. Fluids* **11**, 1672–1678; (1970), Erratum, *Phys. Fluids* **13**, 1651.

[186] Sone, Y. and K. Yamamoto (1970), Flow of rarefied gas over plane wall, *J. Phys. Soc. Jpn* **29**, 495–508; Y. Sone and Y. Onishi (1979), Flow of rarefied gas over plane wall, *J. Phys. Soc. Jpn* **47**, 672.

[187] Sone, Y. and M. Yoshimoto (1997), Demonstration of a rarefied gas flow induced near the edge of a uniformly heated plate, *Phys. Fluids* **9**, 3530–3534.

[188] Sugimoto, H. and Y. Sone (1992), Numerical analysis of steady flows of a gas evaporating from its cylindrical condensed phase on the basis of kinetic theory, *Phys. Fluids A* **4**, 419–440.

[189] Takata, S. and K. Aoki (1999), Two-surface problems of a multicomponent mixture of vapors and noncondensable gases in the continuum limit in the light of kinetic theory, *Phys. Fluids* **11**, 2743–2756.

[190] Takata, S. and K. Aoki (2001), The ghost effect in the continuum limit for a vapor-gas mixture around condensed phases: Asymptotic analysis of the Boltzmann equation, *Transp. Theory Stat. Phys.* **30**, 205–237.

[191] Takata, S., K. Aoki, and T. Muraki (1999), Behavior of a vapor-gas mixture between two parallel plane condensed phases in the continuum limit, in: R. Brun, R. Campargue, R. Gatignol, and J.-C. Lengrand, eds., *Rarefied Gas Dynamics* (CÉPADUÈS, Toulouse), Vol. I, 479–486.

[192] Takata, S., K. Aoki, and Y. Sone (1994), Thermophoresis of a sphere with a uniform temperature: Numerical analysis of the Boltzmann equation for hard-sphere molecules, in: B. D. Shizgal and D. P. Weaver, eds., *Rarefied Gas Dynamics: Theory and Simulations* (AIAA, Washington, DC), 626–639.

[193] Takata, S. and Y. Sone (1995), Flow induced around a sphere with a non-uniform temperature in a rarefied gas, with application to the drag and thermal force problems of a spherical particle with an arbitrary thermal conductivity, *Eur. J. Mech. B/Fluids* **14**, 487–518.

[194] Takata, S., Y. Sone, and K. Aoki (1993), Numerical analysis of a uniform flow of a rarefied gas past a sphere on the basis of the Boltzmann equation for hard-sphere molecules, *Phys. Fluids A* **5**, 716–737.

[195] Takens, W. B., W. Mischke, J. Korving, and J. J. M. Beenakker (1984), A spectroscopic study of free evaporation of sodium, in: H. Oguchi, ed., *Rarefied Gas Dynamics* (University of Tokyo Press, Tokyo), Vol. II, 967–974.

[196] Taylor, G. I. (1923), Stability of a viscous liquid contained between two rotating cylinders, *Philos. Trans. R. Soc. London, Ser. A* **223**, 289–343.

[197] Tiwari, T. and A. Klar (1998), An adaptive domain decomposition procedure for Boltzmann and Euler equation, *J. Comp. Appl. Math.* **90**, 223-237.

[198] Tolman, R. C. (1979), *Principles of Statistical Mechanics* (Dover, New York).

[199] Trilling, L. (1975), The Knudsen boundary layer in low density gas flow past a crystal, *SIAM J. Appl. Math.* **28**, 757–777.

[200] Ukai, S. (2001), The Boltzmann–Grad limit and Cauchy–Kovalevskaya theorem, *Japan J. Industrial Appl. Math.* **18**, 383–392.

[201] Ukai, S. and K. Asano (1983), The Euler limit and initial layer of the nonlinear Boltzmann equation, *Hokkaido Math. J.* **12**, 311–332.

[202] Vargo, S. E. and E. P. Muntz (1997), An evaluation of a multiple-stage micromechanical Knudsen compressor and vacuum pump, in: Ching Shen, ed., *Rarefied Gas Dynamics* (Peking University Press, Beijing), 995–1000.

[203] Vasseur, A. (1999), Convergence of a semi-discrete kinetic scheme for the system of isentropic gas dynamics with $\gamma = 3$, *Indiana University Mathematics J.* **48**, 347–363.

[204] Vincenti, W. G. and C. H. Kruger Jr. (1965), *Introduction to Physical Gas Dynamics* (John Wiley and Sons, New York).

[205] von Karman, T. (1963), *From Low-speed Aerodynamics to Astronautics* (Pergamon Press, Oxford).

[206] Wakabayashi, M., T. Ohwada, and F. Golse (1996), Numerical analysis of the shear and thermal creep flows of a rarefied gas over the plane wall of a Maxwell-type boundary condition on the basis of the linearized Boltzmann equation for hard-sphere molecules, *Eur. J. Mech. B/Fluids* **15**, 175–201.

[207] Welander, P. (1954), On the temperature jump in a rarefied gas, *Ark. Fys.* **7**, 507–553.

[208] Weyl, H. (1946), *The Classical Groups* (Princeton University Press, Princeton, New Jersey).

[209] Xu, K. (2000), A gas-kinetic BGK schemes for the compressible Navier–Stokes equations, *NASA/CR-2000-21054 ICSE Report No. 2000-38.*

[210] Ytrehus, T. (1977), Theory and experiments on gas kinetics in evaporation, in: J. L. Potter, ed., *Rarefied Gas Dynamics* (AIAA, New York), 1197–1212.

[211] Ytrehus, T. (1997), Molecular-flow effects in evaporation and condensation at interfaces, *Multiphase Science and Technology* **9**, 205–327.

Index